Springer Tracts in Advanced Robotics

Volume 17

Springer Tracts in Advanced Robotics

Edited by B. Siciliano, O. Khatib, and F. Groen

Vol. 16: Cuesta, F.; Ollero, A.
Intelligent Mobile Robot Navigation
224 p. 2005 [3-540-23956-1]

Vol. 15: Dario, P.; Chatila R. (Eds.)
Robotics Research – The Eleventh International Symposium
595 p. 2005 [3-540-23214-1]

Vol. 14: Prassler, E.; Lawitzky, G.; Stopp, A.; Grunwald, G.; Hägele, M.; Dillmann, R.; Iossifidis. I. (Eds.)
Advances in Human-Robot Interaction
414 p. 2005 [3-540-23211-7]

Vol. 13: Chung, W.
Nonholonomic Manipulators
115 p. 2004 [3-540-22108-5]

Vol. 12: Iagnemma K.; Dubowsky, S.
Mobile Robots in Rough Terrain – Estimation, Motion Planning, and Control with Application to Planetary Rovers
123 p. 2004 [3-540-21968-4]

Vol. 11: Kim, J.-H.; Kim, D.-H.; Kim, Y.-J.; Seow, K.-T.
Soccer Robotics
353 p. 2004 [3-540-21859-9]

Vol. 10: Siciliano, B.; De Luca, A.; Melchiorri, C.; Casalino, G. (Eds.)
Advances in Control of Articulated and Mobile Robots
259 p. 2004 [3-540-20783-X]

Vol. 9: Yamane, K.
Simulating and Generating Motions of Human Figures
176 p. 2004 [3-540-20317-6]

Vol. 8: Baeten, J.; De Schutter, J.
Integrated Visual Servoing and Force Control
198 p. 2004 [3-540-40475-9]

Vol. 7: Boissonnat, J.-D.; Burdick, J.; Goldberg, K.; Hutchinson, S. (Eds.)
Algorithmic Foundations of Robotics V
577 p. 2004 [3-540-40476-7]

Vol. 6: Jarvis, R.A.; Zelinsky, A. (Eds.)
Robotics Research – The Tenth International Symposium
580 p. 2003 [3-540-00550-1]

Vol. 5: Siciliano, B.; Dario, P. (Eds.)
Experimental Robotics VIII
685 p. 2003 [3-540-00305-3]

Vol. 4: Bicchi, A.; Christensen, H.I.; Prattichizzo, D. (Eds.)
Control Problems in Robotics
296 p. 2003 [3-540-00251-0]

Vol. 3: Natale, C.
Interaction Control of Robot Manipulators – Six-degrees-of-freedom Tasks
120 p. 2003 [3-540-00159-X]

Vol. 2: Antonelli, G.
Underwater Robots – Motion and Force Control of Vehicle-Manipulator Systems
209 p. 2003 [3-540-00054-2]

Vol. 1: Caccavale, F.; Villani, L. (Eds.)
Fault Diagnosis and Fault Tolerance for Mechatronic Systems – Recent Advances
191 p. 2002 [3-540-44159-X]

M. Erdmann · D. Hsu · M. Overmars · F. van der Stappen (Eds.)

Algorithmic Foundations of Robotics VI

With 200 Figures and 15 Tables

Springer

Professor Bruno Siciliano, Dipartimento di Informatica e Sistemistica, Università degli Studi di Napoli Federico II, Via Claudio 21, 80125 Napoli, Italy, email: siciliano@unina.it

Professor Oussama Khatib, Robotics Laboratory, Department of Computer Science, Stanford University, Stanford, CA 94305-9010, USA, email: khatib@cs.stanford.edu

Professor Frans Groen, Department of Computer Science, Universiteit van Amsterdam, Kruislaan 403, 1098 SJ Amsterdam, The Netherlands, email: groen@science.uva.nl

STAR (Springer Tracts in Advanced Robotics) has been promoted under the auspices of EURON (European Robotics Research Network)

Editors

Prof. Michael Erdmann
Computer Science Department
& Robotics Institute
Carnegie Mellon University
5000 Forbes Avenue
Pittsburgh, PA 15213, USA

Prof. Dr. Mark Overmars
Utrecht University
Department of Information
and Computing Sciences
P.O.Box 80.089
3508 TB Utrecht, The Netherlands

Prof. David Hsu
National University of Singapore
Department Computer Science
3 Science Drive 2
Singapore 117543, Singapore

Dr. A. Frank van der Stappen
Utrecht University
Department of Information
and Computing Sciences
P.O.Box 80.089
3508 TB Utrecht, The Netherlands

ISSN print edition: 1610-7438
ISSN electronic edition: 1610-742X

ISBN-10 3-540-25728-4 **Springer Berlin Heidelberg New York**
ISBN-13 978-3-540-25728-8 **Springer Berlin Heidelberg New York**

Library of Congress Control Number: 2005926497

Springer is a part of Springer Science+Business Media
springeronline.com

Printed in Germany

Typesetting: Digital data supplied by editors.
Data-conversion and production: PTP-Berlin Protago-TEX-Production GmbH, Germany
Cover-Design: design & production GmbH, Heidelberg
Printed on acid-free paper 89/3141/Yu - 5 4 3 2 1 0

Foreword

At the dawn of the new millennium, robotics is undergoing a major transformation in scope and dimension. From a largely dominant industrial focus, robotics is rapidly expanding into the challenges of unstructured environments. Interacting with, assisting, serving, and exploring with humans, the emerging robots will increasingly touch people and their lives.

The goal of the new series of Springer Tracts in Advanced Robotics (STAR) is to bring, in a timely fashion, the latest advances and developments in robotics on the basis of their significance and quality. It is our hope that the wider dissemination of research developments will stimulate more exchanges and collaborations among the research community and contribute to further advancement of this rapidly growing field.

The biennial Workshop on Algorithmic Foundations of Robotics (WAFR) has traditionally published archival volumes of high reference value. Since its latest edition, WAFR has found a more suitable home within STAR, together with other thematic symposia devoted to excellence in robotics research.

The Sixth edition of Algorithmic Foundations of Robotics edited by Michael Erdmann, David Hsu, Mark Overmars, and Frank van der Stappen offers a collection of a broad range of topics in advanced robotics. The contents of these contributions represent a cross-section of the current state of robotics research from one particular aspect: algorithms, and how they reflect on the theoretical basis of subsequent developments. Validation of algorithms, design concepts, or techniques is the common thread running through this focused collection.

Rich by topics and authoritative contributors, WAFR culminates with this unique reference on the current developments and new directions in the field of algorithmic foundations. A fine addition to the series!

Naples, Italy
March 2005

Bruno Siciliano
STAR Editor

Preface

Algorithms play a crucial role in describing the computational processes that plan, control, or reason about motion, manipulation, and perception, both in the physical world and in simulated virtual environments. Such algorithms form a foundation for robotics, but also for processes in other application domains, ranging from structural molecular biology to computer games.

The biannual Workshop on the Algorithmic Foundations of Robotics focuses on such algorithmic issues. The Sixth incarnation of the workshop was held from July 11 till July 13 2004 in Zeist, the Netherlands, organized by Utrecht University. Approximately 50 international researchers participated. The program contained three invited talks, by Bruce Donald (Dartmouth College), Pieter Jonker (Delft University of Technology) and Ken Goldberg (University of California at Berkeley). There were twenty-seven additional papers presented, carefully selected by an international program committee. The papers study a large range of algorithmic issues, including significant new results in sampling-based path planning, online searching, location, tracking, coverage and grasping, to name just a few. Important applications in for example computing protein structures, finding camera motions and robot soccer were discussed. This volume collects all these contributions.

In addition to the editors of this volume, the program committee of WAFR 2004 included: Karl Böhringer, Greg Chirikjian, Yan-Bin Jia, Oussama Khatib, Rolf Klein, Steven LaValle, Ming Lin, Kevin Lynch, Mark Moll, Nancy Pollard, Elon Rimon, Thierry Siméon, Jack Snoeyink, and Michael Yu Wang.

As the co-chairs of WAFR 2004 we would like to thank the Dutch Organization for Scientific Research (N.W.O.) and Utrecht University for their financial support. Also we would like to thank Thomas Ditzinger of Springer Verlag for his assistance in publishing this volume. Finally, our thanks go to all the participants and authors for their enthusiasm in making WAFR 2004 such a wonderful research meeting.

Utrecht, The Netherlands
July 2004

Michael Erdmann
David Hsu
Mark Overmars
Frank van der Stappen

Contents

Invited Papers

Contributed Papers

Algorithmic Challenges in Structural Molecular Biology and Proteomics

Bruce Randall Donald [1–4]

[1] Dartmouth Computer Science Department
[2] Dartmouth Chemistry Department
[3] Dartmouth Biological Sciences Department
[4] Dartmouth Center for Structural Biology and Computational Chemistry

This paper reviews our research in computational biology and chemistry. Some of the most challenging and influential opportunities for Physical Geometric Algorithms (PGA) arise in developing and applying information technology to understand the molecular machinery of the cell. Our recent work (e.g., [1-20]) shows that many PGA techniques may be fruitfully applied to the challenges of computational molecular biology. PGA research may lead to computer systems and algorithms that are useful in structural molecular biology, proteomics, and rational drug design.

Concomitantly, a wealth of interesting computational problems arise in proposed methods for discovering new pharmaceuticals. I'll briefly discuss some recent results from my lab, including new algorithms for interpreting X-ray crystallography [14, 17, 16] and NMR (nuclear magnetic resonance) data [3, 9, 6, 19, 10, 5, 7, 18, 4], disease classification using mass spectrometry of human serum [12], and protein redesign [13]. Our algorithms have recently been used, respectively, to reveal the enzymatic architecture of organisms high on the CDC bioterrorism watch-list [17, 16], for probabilistic cancer classification from human peripheral blood [12], and to redesign an antibiotic-producing enzyme to bind a novel substrate [13]. I'll overview these projects, and highlight some of the algorithmic and computational challenges.

1 Background and Significance

In the post-genomic era, key problems in molecular biology center on the determination and exploitation of three-dimensional protein structure and function. For example, modern drug design techniques use protein structure to understand how a drug can bind to an enzyme and inhibit its function.

Author's address: 6211 Sudikoff Laboratory, Dartmouth Computer Science Department, Hanover, NH 03755, USA. Email: brd@cs.dartmouth.edu.

M. Erdmann et al. (Eds.): Algorithmic Foundations of Robotics VI, STAR 17, pp. 1–10, 2005.

Structural proteomics will require high-throughput experimental techniques, coupled with sophisticated computer algorithms for data analysis and experiment planning.

My laboratory develops novel computational methods to enable high-throughput structural and functional studies of proteins. A key focus is structural genomics, whose goal is (in the broadest terms) to determine the three-dimensional structures of all proteins in nature, through a combination of direct experiments and theoretical analysis. Proteins are the worker molecules in every living thing. By determining the structures of proteins, we are better able to understand how each protein functions normally and how faulty protein structures can cause disease. Scientists can use the structures of disease-related proteins to help develop new medicines and diagnostic techniques.

PGA research in computational structural biology and proteomics can assist in our long-range goal of understanding biopolymer interactions in systems of significant biochemical as well as pharmacological interest. At the molecular level, many genes provide the blueprint for proteins, and it remains very expensive and time-consuming to determine what these proteins do, and how they do it. This paper reviews some novel algorithms to build three-dimensional models of proteins to better understand protein mechanism and function.

Our work spans a number of projects in computational structural biology, including algorithms for automated assignment and structure determination in NMR structural biology [3, 9, 6, 19, 10, 5, 7, 18, 4], protein redesign [13], computer-aided drug design [1, 15, 13], computational molecular replacement in X-ray crystallography [14, 17, 16], and other related projects [2, 11, 8]. To pursue research in the field of structural genomics is to study the geometric structures of proteins. Structural genomics is a field born of the the marriage between computer science and biology. The goal is to develop new technology – specifically, computer algorithms – to enable the determination of 10,000 new protein structures in 10 years. This would have an enormous impact on our understanding of disease mechanisms, and our ability to target drugs to specific protein targets. If successful, the impact would be comparable to that of the human genome project.

Modern automated techniques are revolutionizing many aspects of biology, for example, supporting extremely fast gene sequencing and massively-parallel gene expression testing. Protein structure determination, however, remains a long, hard, and expensive task. High-throughput, automated, algorithmic methods are required in order to apply modern techniques such as computer-aided drug design on a much larger scale. For example, to analyze non-crystallographic symmetry in X-ray diffraction data of biopolymers, one must "recognize" a finite subgroup of $SO(3)$ (the Lie group of 3D rotations) out of a large set of molecular orientations. The problem may be reduced to clustering in $SO(3)$ modulo a finite group, and solved efficiently by "factoring" into a clustering on the unit circle followed by clustering on the 2-sphere S^2, plus some group-theoretic calculations [14]. This yields a polynomial-time algorithm that is efficient in practice, and which enabled us to collaborate with biological crystallographers to reveal the architecture of a parasite's enzyme, which could help researchers reduce the threat of certain diseases among those with weak immune systems [17, 16]. As another example, we recently employed geometric techniques from statistical estimation and machine learning to develop an algorithm for cancer proteomics, in which we use data from a mass spectrometer to distinguish between healthy and diseased blood in humans [12]. Geometry pervades our work: [12] can be viewed as an investigation of the geometry of the oncoproteome (the space of cancer proteins) as projected onto the mass-to-charge ratios of their proteolytic digest. Nuclear vector replacement for automated NMR resonance assignments [6, 10] is essentially a matching problem on a quotient space of orientations, induced by a quadratic form ξ on S^2. ξ is parameterized by $SO(3)$, and 3D structural homology detection from *unassigned* NMR data (enabling rapid fold determination) can be performed by combinatorial optimization, searching over $SO(3)$ to minimize a functional that compares distributions generated by ξ's image of the bond vectors from putative database protein models [9, 10, 5, 7].

In each of our research projects, computational techniques are central, and the applications present intriguing problems to computer scientists who design algorithms and implement systems. For example, the techniques we introduced for automated NMR resonance assignment [6, 10, 3, 9, 7] and protein structure determination [19, 18, 5, 7] are an instance a general approach to combinatorial problem solving in which constraints on the solution are enforced in an order determined by the strength of the evidence for them. This approach, which has analogies to the Celera whole-genome shotgun sequencing algorithm, also presents a flock of fascinating questions from the point of view of theoretical computer science (*cf.* Richard Karp's Keynote address, Computational Systems Bioinformatics Conference, 2003).

2 Results in NMR Structural Genomics

In this section (2), I attempt to illustrate the general themes introduced above, by way of specific examples and results. The NMR biophysics requires some technical description; some readers may prefer to skip to the higher-level discussion on *Protein Fold Determination* (p. 4) or even to Section 3 (p. 5).

Nuclear Vector Replacement for Automated NMR Resonance Assignment and Structure Determination. In X-ray crystallography, the molecular replacement technique (used in [14, 17, 16]) allows solution of the crystallographic phase problem when a "close" or homologous structural model is known *a priori*, thereby facilitating rapid structure determination. In contrast, a key bottleneck in NMR structural biology is the resonance assignment problem. One would hope that knowing a structural model ahead of time could expedite assignments. We recently reported an automated procedure for high-throughput NMR resonance assignment for a protein of known structure, or of an homologous structure [6]. Our algorithm performs *Nuclear Vector Replacement* (NVR) by *Expectation/Maximization* (EM) to compute assignments. NVR correlates experimentally-measured NH residual dipolar couplings (RDCs) and chemical shifts to a given *a priori* whole-protein 3D structural model. The algorithm requires only uniform ^{15}N-labeling of the protein, and processes unassigned H^N-^{15}N HSQC spectra, H^N-^{15}N RDCs, and sparse H^N-H^N NOE's (d_{NN}s), all of which can be acquired in a fraction of the time needed to record the traditional suite of experiments used to perform resonance assignments. NVR runs in minutes and efficiently assigns the (H^N,^{15}N) backbone resonances as well as the sparse d_{NN}s from the 3D ^{15}N-NOESY spectrum, in $O(n^3)$ time. We tested NVR on NMR data from 3 proteins using 20 different alternative structures, all determined either by X-ray crystallography or by *different* NMR experiments (without RDCs). When NVR was run on NMR data from the 76-residue protein, human ubiquitin (matched to four structures, including one mutant/homolog), we achieved 100% assignment accuracy. Similarly good results were obtained in experiments with the 56-residue streptococcal protein G (99%) and the 129-residue hen lysozyme (100%) when they were matched by NVR to 16 3D structural models. Our success in assigning 1UD7, a mutant of ubiquitin, suggests that NVR could be applied more broadly to assign spectra based on homologous structures. Thus, NVR could play a role in structural genomics.

Protein Fold Determination via *Unassigned* Residual Dipolar Couplings. We extended NVR to a second application, 3D structural homology detection, and demonstrated that NVR is able to identify structural homologies between remote amino acid sequences from a database of structural models [9, 10, 5, 7]. One goal of the structural genomics initiative is the identification of new protein folds. Sequence-based structural homology prediction methods are an important means for prioritizing unknown proteins for structure determination. However, an important challenge remains: two highly dissimilar sequences can have similar folds — how can we detect this rapidly, in the context of structural genomics? High-throughput NMR experiments, coupled with novel algorithms for data analysis, can address this challenge. We reported an automated procedure for detecting 3D structural homologies from sparse, *unassigned* protein NMR data. Our method identifies the 3D structural models in a protein structural database whose geometries best fit the unassigned experimental NMR data. It does not use sequence information and is thus not limited by sequence homology. The method can also be used to confirm or refute structural predictions made by other techniques such as protein threading or sequence homology. The method requires only uniform ^{15}N-labeling of the protein and processes unassigned H^N-^{15}N residual dipolar couplings, which can be acquired in a couple of hours. Our experiments on NMR data from 5 different proteins demonstrate that the method identifies closely related protein folds, despite low-sequence homology between the target protein and the computed model.

Novel NMR Structure Determination Algorithms. Three recent papers [19, 20, 18] make contributions to the method of determining protein structures by solution NMR spectroscopy using residual dipolar couplings (RDCs) as the main restraints. These contributions, I believe, may be valuable not only to the NMR community in particular and structural genomics in general, but also to structural biologists more broadly. This is because in both experimental and computational structural biology, exact computational methods have been, for the most part, elusive to date. Second, rigorous comparisons of structures derived from NMR vs. X-ray crystallography are made possible by our techniques, and these comparisons should be of general interest.

In contrast to (e.g.) simulated annealing approaches, our algorithm is combinatorially-precise [18], and is built upon the exact solutions for computing backbone (ϕ, ψ) angles from RDC data and systematic search. [19] is the first NMR structure determination algorithm that simultaneously uses exact solutions, systematic search and only 2 RDCs per residue. (A systematic search is a search over all possible conformations (solutions) that employs a provable pruning strategy that guarantees pruned conformations need not be considered further). Our first contribution is the derivation of low-degree polynomial equations for computing, *exactly* and *in constant time*, dihedral (ϕ, ψ) angles from RDCs measured on a single internuclear vector $\mathbf{v}$ in two different aligning media. The easily computable exact solutions eliminate the need for one-dimensional grid-search previously employed to compute the directions of $\mathbf{v}$ or two-dimensional grid-search to compute (ϕ, ψ) angles. Furthermore, these equations are very general and can easily be extended to compute both the backbone and sidechain dihedral angles from RDC data measured on any single vector in two aligning media. And, our method can be applied *mutatis mutandis* to derive similar equations for computing dihedral angles from RDCs in nucleic acids. Compared with other algorithms for computing backbone structures using RDCs, our algorithm achieves similar accuracies but requires less data, relies less on statistics from the PDB and does not depend on molecular dynamics. Since RDCs can be acquired and assigned much more quickly than NOEs in general, our results show it is possible to compute structures very rapidly and inexpensively using mainly RDC restraints.

3 Analytic versus Synthetic

Most of the work described above concentrates on algorithms and system for *analyzing* biological data and biological problems. However, the techniques we develop can also be applied to *synthetic* problems such as a protein engineering. For example, in collaboration with Prof. Amy Anderson, we recently developed a novel ensemble-based scoring and search algorithm for protein redesign, and applied it to modify the substrate specificity of an antibiotic-producing enzyme in the *non-ribosomal peptide synthetase (NRPS)* pathway [13]. Realization of novel molecular function requires the ability to alter molecular complex formation. Enzymatic function can be altered by changing enzyme-substrate interactions via modification of an enzyme's active site. A redesigned enzyme may either perform a novel reaction on its native substrates or its native reaction on novel substrates. A number of computational approaches have been developed to address the combinatorial nature of the protein redesign problem. These approaches typically search for the global minimum energy conformation among an exponential number of protein conformations. We developed a novel algorithm for protein redesign, which combines a statistical mechanics-derived ensemble-based approach to computing the binding constant with the speed and completeness of a branch-and-bound pruning algorithm. In addition, we developed an efficient deterministic approximation algorithm, capable of approximating our scoring function to arbitrary precision. Our algorithm is the first provable ϵ-approximation algorithm for estimating partition functions for protein flexibility.

In practice, our approximation algorithm decreases the execution time of the mutation search by a factor of ten. To test our method, we examined the Phe-specific adenylation domain of the NRPS gramicidin synthetase A (GrsA-PheA). We used ensemble scoring, via a rotameric approximation to the partition functions of the bound and unbound states for GrsA-PheA, to predict binding of the wildtype protein and a previously described mutant (selective for leucine), and second, to switch the enzyme specificity toward leucine, using two novel active site sequences computationally predicted by searching through the space of possible active site mutations. The top scoring *in silico* mutants were created in the wetlab and dissociation/binding constants were determined by fluorescence quenching. These tested mutations exhibit the desired change in specificity from Phe to Leu. Our ensemble-based algorithm which flexibly models both protein and ligand using rotamer-based partition functions, has application in enzyme redesign, the prediction of protein-ligand binding, and computer-aided drug design.

This result represents a computational approach to reprogramming enzyme specificity, with the ultimate goal of combinatorial biosynthesis for small-molecule diversity. We are studying a family of enzymes responsible for the biosynthesis of hundreds of pharmaceutically-active peptide-like products. Understanding these enzyme functions will elucidate how natural biological products (e.g., antibiotics) are synthesized *in vivo*. To modify enzyme function, we are developing computational techniques to plan structurally-based site-directed mutations. By reëngineering the active site(s) to operate on different substrates, we hope to modify these different enzymatic steps with an eye to potential reprogramming of those steps for combinatorial biosynthesis. This opens the door to the possibility of using our redesigned enzymes for *in vivo* combinatorial chemistry, to create candidate drug leads for new antibiotics and other drugs. We are applying our algorithms to NRPS modules, whose products include natural antibiotics, antifungals, antivirals, immunosuppressants, and siderophores. NRPS have multiple domains with individual functions acting in an assembly-line fashion. We are modifying the active sites to switch the specificity of the amino acid-accepting domains from their natural substrates, to different amino acids. The modifications are planned and analyzed *in silico*, by developing new algorithms based on techniques from geometric algorithms, robotics, machine vision, and scientific computation. Our "enzyme reprogramming" could allow the modified NRPS to synthesize different modified peptides. Exploration of the combinatorial space of new NRPS "programs" will generate a large number of new compounds, which could then be screened for pharmaceutical activity.

4 Future Work

There is much to be done. A primary focus will be to explore novel computational methods in structural biology, specifically, new algorithms for NMR resonance assignments and protein structure determination, with applications to structural genomics. I am particularly interested in new algorithms for structural biology using only a minimal number of inexpensive, fast experiments. I'm also interested in collaborating with structural biologists to develop novel algorithms and apply them to biological systems of significant biochemical and pharmacological importance. A model for this kind of work is our collaboration on the structure of dihydrofolate reductase-thymidylate synthase (DHFR-TS) from *Cryptosporidium hominis* [14, 17, 16]. *Cryptosporidium* is an organism high on the bioterrorism list of the Center for Disease Control (CDC), a Category B bioterrorist threat. Agents/diseases that fall under this category are given the second-highest priority, because they are moderately easy to disseminate; result in moderate morbidity rates and low mortality rates; and require specific enhancements of CDC's diagnostic and treatment capacities. There is currently no drug therapy for cryptosporidosis. The enzyme DHFR-TS is in the sole *de novo* biosynthetic pathway for the pyrimidine deoxyribonucleotide dTMP, and therefore an attractive drug target. Solving the structure of DHFR-TS from *C. hominis* opens the door to species-specific drug design, exploiting both structural and biophysical differences between the human enzyme and *Cryptosporidium* DHFR-TS.

We collaborated with Dr. Amy Anderson's lab to determine the structure of DHFR-TS from *C. hominis*, revealing a unique linker domain containing an 11-residue alpha helix that has extensive interactions with the opposite DHFR-TS monomer of the homodimeric enzyme [17, 16]. Analysis of the structure of DHFR-TS from *C. hominis* and of previously solved structures of DHFR-TS from *Plasmodium falciparum* (a.k.a. malaria) and *Leishmania major* reveals that the linker domain primarily controls the relative orientation of the DHFR and TS domains. Using the tertiary structure of the linker domains, we have been able to place a number of protozoa in two distinct and dissimilar structural families corresponding to two evolutionary families and provide the first structural evidence validating the use of DHFR-TS as a tool of phylogenetic classification.

I am also interested more broadly in proteomics and functional genomics, both in developing novel algorithms for proteomic problems and bringing to bear structural, geometric, and biophysical insights (and algorithms) for proteomics. For example, in collaboration with the Norris-Cotton Cancer Center at Dartmouth, we are exploring oncoproteomic target selection using mass spectrometry [12]. We developed an algorithm called Q5 for probabilistic classification of healthy vs. disease whole serum samples using mass spectrometry. Q5 is the first closed-form, exact solution to the problem of classification of complete mass spectra of a complex protein mixture. It employs a discriminant back-projection algorithm to compute clues as to the molecular identities of differentially-expressed proteins and peptides. Q5 analyzes whole spectrum Surface-Enhanced Laser Desorption/Ionization Time of Flight (SELDI-TOF) Mass Spectrometry (MS) data, and was demonstrated on four real datasets from complete, complex SELDI spectra of human blood serum. We achieved sensitivity, specificity, and positive predictive values above 97% on three ovarian cancer datasets and one prostate cancer dataset. The Q5 method outperforms previous full-spectrum complex sample spectral classification techniques, and represents the first attempt to compute the molecular identities of the differentially-expressed proteins in two important MS data sets for ovarian and prostate cancer. Further investigation of our lead proteins and peptide fragments may enhance our understanding of the molecular basis of oncogenisis and could potentially lead to new therapeutic targets.

As discussed above (Sec. 3), we are designing and implementing planning algorithms to propose site-directed mutations for protein redesign. We are developing a general planner that can reprogram the specificity of many NRPS domains, from many biological systems. The results of our algorithms are being compared to known crystal structures, to biochemical activity assays, and to crystal structures of the modified domains bound to the proposed substrates. We are developing a predictive model for when and how well our planner will work, by characterizing the complexity, correctness, and completeness of our algorithms. We believe these algorithms may be generally useful to the structural biology community, for studies of protein-ligand binding and protein redesign. Our provable approximation algorithm represents a new technique for computer-assisted drug design, and a novel approach for docking flexible ligands to flexible active sites [13]. In the future, we will extend and apply our algorithms to the redesign of other enzymes, including polyketide synthase (PKS) systems, which synthesize polyketide products such as erythromycin, rapamycin and tetracycline. I believe there are broad potential applications of our techniques for modeling protein flexibility, redesigning enzymes, and evaluating the biophysical processes of binding and catalysis in protein biochemistry, and that these goals can (hopefully) be realized after a lot more hard work in developing geometric algorithms, provably-good approximation algorithms, statistical methods, and an array of algorithmic techniques for handling noise and uncertainty in combinatorial geometry and computational biophysics.

Acknowledgments. I would like to thank all members of my laboratory, past and present, for their contributions to the research reviewed here, and for many helpful discussions. Particular thanks go to: Chris Bailey-Kellogg, Chris Langmead, Ryan Lilien, Ram Mettu, Lincong Wang, Elisheva Werner-Reiss, and Anthony Yan. I would like to thank Amy Anderson, Hany Farid, Rob McClung, Brian Stevens, Robert O'Neil, Veljko Popov, and Siwei Lyu for years of fruitful collaboration and discussions about science. Finally, thanks to Jeff Hoch, Tomás Lozano-Pérez, Gerhard Wagner, Bruce Tidor, and Brian Hare for many helpful discussions and suggestions on our work.

This work is supported by grants from the National Institutes of Health (R01 GM-65982 and R01 GM-67542), and the National Science Foundation (IIS-9906790, EIA-0102710, EIA-0102712, EIA-9818299, and EIA-9802068, EIA-0305444).

References

1. C. Bailey-Kellogg, J. J. Kelley III, R. Lilien, and B. R. Donald. Physical geometric algorithms for structural molecular biology. In *the Special Session on Computational Biology & Chemistry, Proceedings IEEE Int'l Conf. on Robotics and Automation (ICRA-2001)*, pages 940–947, May 2001. .
2. C. Bailey-Kellogg, J. J. Kelley III, C. Stein, and B. R. Donald. Reducing mass degeneracy in SAR by MS by stable isotopic labeling. *Jour. Comp. Biol.*, 8(1):19–36, 2001.
3. C. Bailey-Kellogg, A. Widge, J. J. Kelley III, M. J. Berardi, J. H. Bushweller, and B. R. Donald. The NOESY Jigsaw: Automated protein secondary structure and main-chain assignment from sparse, unassigned NMR data. *Jour. Comp. Biol.*, 3-4(7):537–558, 2000.
4. C. Langmead and B. R. Donald. Extracting structural information using time-frequency analysis of protein NMR data. In *Proceedings of The Fifth Annual International Conference on Research in Computational Molecular Biology (RECOMB)*, pages 164–175. ACM Press, April 2001.
5. C. Langmead and B. R. Donald. 3D structural homology detection via unassigned residual couplings. In *Proceedings of the IEEE Computer Society Bioinformatics Conference (CSB)*, pages 209–217, Stanford, August 2003. .
6. C. Langmead and B. R. Donald. An expectation/maximization nuclear vector replacement algorithm for automated NMR resonance assignments. *Jour. Biomolecular NMR*, 29(2):111–138, 2004.
7. C. Langmead and B. R. Donald. High-throughput 3D structural homology detection via NMR resonance assignment. In *Proceedings of the IEEE Computational Systems Bioinformatics Conference (CSB)*, Stanford, CA, August 2004. In Press.
8. C. Langmead, C. R. McClung, and B. R. Donald. A maximum entropy algorithm for rhythmic analysis of genome-wide expression patterns. In *Proceedings of the IEEE Computer Society Bioinformatics Conference (IEEE CSB)*, pages 237–245, August 2002.

9. C. Langmead, A. Yan, R. Lilien, L. Wang, and B. R. Donald. A polynomial-time nuclear vector replacement algorithm for automated NMR resonance assignments. In *Proceedings of The Seventh Annual International Conference on Research in Computational Molecular Biology (RECOMB)*, pages 176–187, Berlin, Germany, April 2003. ACM Press. .
10. C. Langmead, A. Yan, R. Lilien, L. Wang, and B. R. Donald. A polynomial-time nuclear vector replacement algorithm for automated NMR resonance assignments. *Jour. Comp. Biol.*, 11(2-3):277–298, 2004.
11. C. Langmead, A. Yan, C. R. McClung, and B. R. Donald. Phase-independent rhythmic analysis of genome-wide expression patterns. *Journal of Computational Biology*, 10(3-4):521–536, 2003.
12. R. Lilien, H. Farid, and B. R. Donald. Probabilistic disease classification of expression-dependent proteomic data from mass spectrometry of human serum. *Journal of Computational Biology*, 10(6):925–946, 2003.
13. R. Lilien, B. Stevens, A. Anderson, and B. R. Donald. A novel ensemble-based scoring and search algorithm for protein redesign, and its application to modify the substrate specificity of the gramicidin synthetase A phenylalanine adenylation enzyme. In *Proceedings of the Eighth Annual International Conference on Research in Computational Molecular Biology (RECOMB)*, pages 46–57, San Diego, March 2004. .
14. Ryan H. Lilien, Chris Bailey-Kellogg, Amy C. Anderson, and Bruce R. Donald. A subgroup algorithm to identify cross-rotation peaks consistent with non-crystallographic symmetry. *Acta Crystallographica Section D*, 60(6):1057–1067, Jun 2004.
15. Ryan H. Lilien, Mohini Sridharan, and Bruce R. Donald. Identification of Novel Small Molecule Inhibitors of Core-Binding Factor Dimerization by Computational Screening against NMR Molecular Ensembles. Technical Report TR2004-492, Dartmouth College, Computer Science, Hanover, NH, March 2004.
16. R. O'Neil, R. Lilien, B. R. Donald, R. Stroud, and A. Anderson. The crystal structure of dihydrofolate reductase-thymidylate synthase from *Cryptosporidium hominis* reveals a novel architecture for the bifunctional enzyme. *Jour. Eukaryotic Microbiology*, 50(6):555–556, 2003.
17. R. O'Neil, R. Lilien, B. R. Donald, R. Stroud, and A. Anderson. Phylogenetic classification of protozoa based on the structure of the linker domain in the bifunctional enzyme, dihydrofolate reductase-thymidylate synthase. *Jour. Biol. Chem.*, 278(52):52980–52987, 2003.
18. L. Wang and B. R. Donald. Analysis of a systematic search-based algorithm for determining protein backbone structure from a minimal number of residual dipolar couplings. In *Proceedings of the IEEE Computational Systems Bioinformatics Conference (CSB)*, Stanford, CA, August 2004. In Press.
19. L. Wang and B. R. Donald. Exact solutions for internuclear vectors and backbone dihedral angles from NH residual dipolar couplings in two media, and their application in a systematic search algorithm for determining protein backbone structure. *Jour. Biomolecular NMR*, 29(3):223–242, 2004.
20. L. Wang, R. Mettu, R. Lilien, and B. R. Donald. An exact algorithm for determining protein backbone structure from NH residual dipolar couplings. In *Proceedings of the IEEE Computer Society Bioinformatics Conference (CSB)*, pages 611–612, Stanford, August 2003.

Networked Robots: Ten Years of Experiments

Ken Goldberg[1]

IEOR and EECS Departments, UC Berkeley, Berkeley, CA, USA

Abstract. As a feasibility study in 1994, we built a system that allows a robot manipulator to be teleoperated by the public via the Internet. Access control became a primary issue: in subsequent networked robot systems we explore techniques ranging from user queues, multitasking, vector averaging, and geometric optimization. These systems are also designed to suggest questions about human perception and expectations about technology.

1 Introduction

Although the field of teleoperation dates back over 50 years, the hypertext transfer protocol developed at CERN in 1992 [4], provides a low-cost and publicly available network interface. In the Spring of 1994, we conjectured that we could use it to offer public access to a teleoperated robot via the Internet.

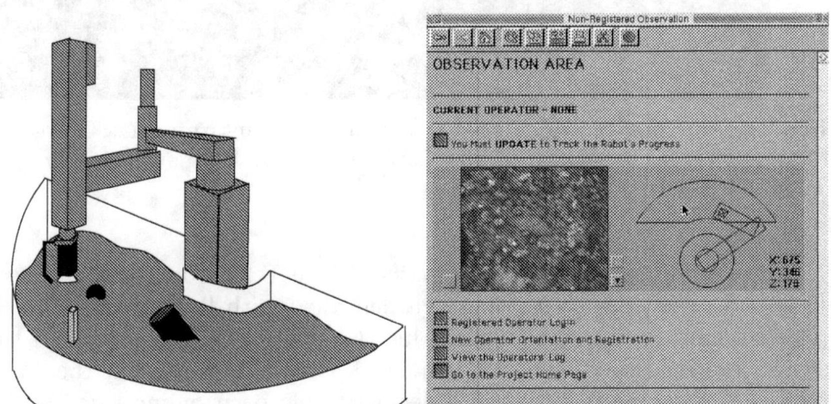

Fig. 1. Mercury Project (1994-1995). Left: Robot, camera and air nozzle above buried artifacts. Right: Browser Interface using vanilla HTTP 1.0.

As illustrated in Figure 1, we set up an IBM SCARA robot arm over a semi-annular workspace containing sand and buried artifacts. We attached a CCD camera to the end of the arm along with a nozzle to direct air bursts into the sand. We then developed a browser interface to the hardware. In addition

M. Erdmann et al. (Eds.): Algorithmic Foundations of Robotics VI, STAR 17, pp. 11–16, 2005.

to the challenges associated with dynamic time delay, the issue of access control became a primary issue. For the Mercury Project we implemented a User Queue; each user was allocated five minutes to operate the robot. The Mercury Project was online for seven months and was operated by over 10,000 people.

Fig. 2. The Tele-Garden (1995-). (with Joseph Santarromana, George Bekey, Steven Gentner, Rosemary Morris Carl Sutter, Jeff Wiegley, Erich Berger, and Thomas Steindl).

Our second networked robot system is the Telegarden. It allows users to view and interact with a remote garden filled with living plants. We incorporated a much faster Adept-1 industrial robot arm, and modified the the system so that the robot is multi-tasked. Currently hosted at the Ars Electronica Center in Austria, the Telegarden has been online since in August 1995 and has been viewed by close to 100,000 people.

2 Related Work

Since Nikola Tesla demonstrated the first radio-controlled boat in 1898 and Goertz demonstrated a bilateral manipulator in 1954 [7], remotely operated machines have been widely desired for use in inhospitable environments such as radiation sites, undersea [1] and space exploration [3,24,32]. Today, teleoperation is being developed for medical diagnosis [2], manufacturing [6] and micromanipulation [27]. See Sheridan [28] for an excellent review of the extensive literature on teleoperation and telerobotics. Most of these systems require fairly complex hardware at the human interface: exoskeleton master linkages are operated by highly trained specialists. In contrast, the Internet can provide public access by using only the interface available in standard browsers.

To our knowledge, the Mercury Project was the first Internet robot [11,10]; it went online in August 1994. Working independently, a team led by K. Taylor and J. Trevelyan at the University of Western Australia demonstrated a remotely controlled six-axis telerobot in September 1994 [5,17]. There are now dozens of Internet robots online, a book from MIT Press [12], and an IEEE Technical Committee on Networked Robots that has over 100 members. See [18,26,20,19,21,23,15,25,22] examples of recent projects.

3 The Tele-Actor and ShareCam

In 1999 we began exploring other models of access control, where user inputs are combined rather than sequenced. In In [8,9], we describe an Internet-based Multiple Operator Single Robot system that use vector averaging to combine multiple mouse inputs to simultaneously control a single industrial robot arm. In [13,14], we describe a Java-based "Spatial Dynamic Voting" (SDV) interface that collects, displays, and analyzes a sequence of spatial votes from multiple online operators at their Internet browsers. The votes can drive the motion of a single mobile robot or, for increased mobility and agility, a human "Tele-Actor".

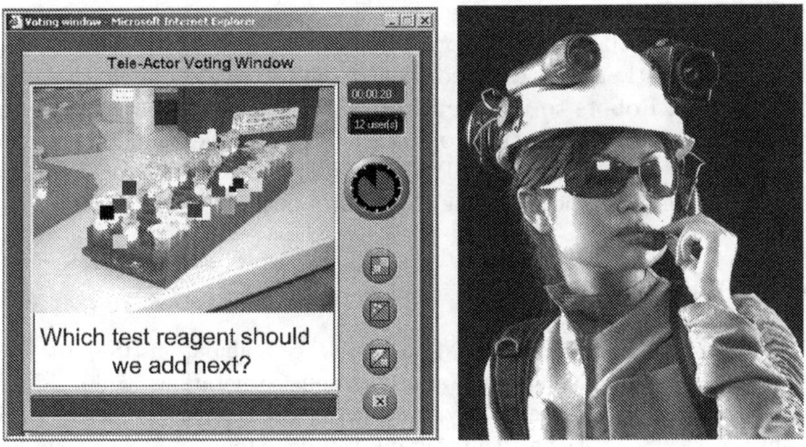

Fig. 3. Spatial Dynamic Voting Interface and the Tele-Actor www.tele-actor.net

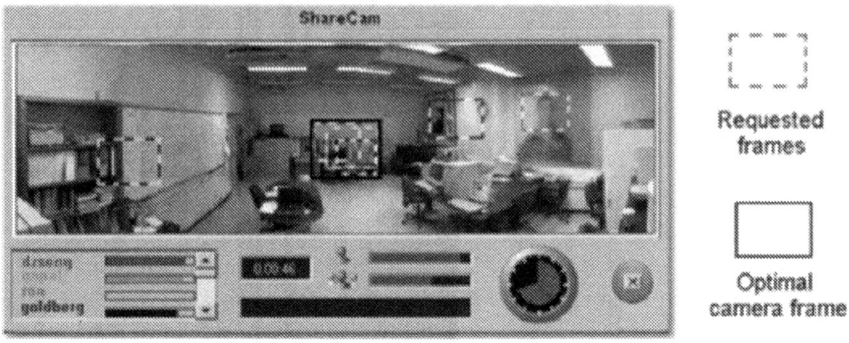

Fig. 4. ShareCam. (with Dez Song): www.co-opticon.net

In 2002, we developed Sharecam, a robotic pan, tilt, zoom camera controllable by many simultaneous users over the Internet. We proposed the ShareCam Problem: choosing the optimal camera frame based on simultaneous frame requests from n online users [31]. We propose a new piecewise linear metric, Intersection Over Maximum (IOM), for the degree of satisfaction for each user. To maximize overall satisfaction, we developed a sequence of several algorithms. For a discrete set of m distinct zoom levels, we developed an exact algorithm that runs in $O(mn^2)$ time.

Har-Peled et al. [16] proposed a near-linear $\epsilon-$approximation algorithm with fixed resolution based on a stair-shape approximation of the objective functions. In [30], we proposed a variant for access control of satellite-based cameras and presented an $O(n^3)$ algorithm. In [29], we report $\epsilon-$approximation algorithms for the ShareCam Problem with continuous pan, tilt and zoom that runs in $O(n/\epsilon^3)$ time.

These and other projects are accessible online at:
http://alpha.ieor.berkeley.edu/projects.html

In June 2004, the IEEE Technical Committee expanded the definition of Networked Robots to include a range of devices connected to a range of communications networks that could be wired or wireless, and based on any of of a variety of protocols such as TCP, UDP, or 802.11. Many new applications are now being developed ranging from automation to exploration.

Acknowledgments

This work was funded in part by National Science Foundation IIS-0113147, by Intel, Adept, Panasonic, Qualcomm, and Microsoft Corporations, and UC Berkeley's Center for Information Technology Research in the Interest of Society (CITRIS). Thanks to Dez Song, Dana Plautz, Frank van der Stappen, Steven Gentner, Carl Sutter, Jeff Wiegley, Michael Mascha, and Nick Rothenberg. Joseph Santarromana, George Bekey, Steven Gentner, Rosemary Morris Carl Sutter, Jeff Wiegley, Erich Berger, Thomas Steindl, Howard Moraff, John Canny, Eric Paulos, Judith Donath,

Bob Farzin, Eric Mankin, Peter Lunenfeld, Roland Siegwart, Ruzena Bajcsy, Paul Wright, Anthony Levandowski, and everyone who participated by operating the robots.

References

1. R. D. Ballard. A last long look at titanic. *National Geographic*, 170(6), December 1986.
2. A. Bejczy, G. Bekey, R. Taylor, and S. Rovetta. A research methodology for tele-surgery with time delays. In *First International Symposium on Medical Robotics and Computer Assisted Surgery*, Sept 1994.
3. A. K. Bejczy. Sensors, controls, and man-machine interface for advanced teleoperation. *Science*, 208(4450), 1980.
4. Tim Berners-Lee, Robert Cailliau, Jean-Francios Groff, and Bernd Pollerman. World-wide web: The information universe. *Electronic Networking:Research, Applications and Policy*, 1(2), Westport CT, Spring 1992.
5. Barney Dalton and Ken Taylor. A framework for internet robotics. In *IEEE International Conference On Intelligent Robots and Systems (IROS): Workshop on Web Robots*, Victoria, Canada, 1998.
6. Matthew Gertz, David Stewart, and Pradeep Khosla. A human-machine interface for distributed virtual laboratories. *IEEE Robotics and Automation Magazine*, December 1994.
7. Raymond Goertz and R. Thompson. Electronically controlled manipulator. *Nucleonics*, 1954.
8. K. Goldberg, B. Chen, R. Solomon, S. Bui, B. Farzin, J. Heitler, D. Poon, and G. Smith. Collaborative teleoperation via the internet. In *IEEE International Conference on Robotics and Automation (ICRA)*, volume 2, pages 2019–2024, April 2000.
9. K. Goldberg and Billy Chen. Collaborative control of robot motion: Robustness to error. In *International Conference on Intelligent Robots and Systems (IROS)*, 2001.
10. K. Goldberg, M. Mascha, S. Gentner, N. Rothenberg, C. Sutter, and Jeff Wiegley. Beyond the web: Manipulating the physical world via the www. *Computer Networks and ISDN Systems Journal*, 28(1), December 1995. Archives can be viewed at http://www.usc.edu/dept/raiders/.
11. K. Goldberg, M. Mascha, S. Gentner, N. Rothenberg, C. Sutter, and Jeff Wiegley. Robot teleoperation via www. In *International Conference on Robotics and Automation*. IEEE, May 1995.
12. K. Goldberg and Roland Siegwart, editors. *Beyond Webcams: An Introduction to Online Robots*. MIT Press, 2002.
13. K. Goldberg, D. Song, Y. Khor, D. Pescovitz, A. Levandowski, J. Himmelstein, J. Shih, A. Ho, E. Paulos, and J. Donath. Collaborative online teleoperation with spatial dynamic voting and a human "tele-actor". In *IEEE International Conference on Robotics and Automation (ICRA)*, volume 2, pages 1179–84, May 2002.
14. K. Goldberg, D. Song, and A. Levandowski. Collaborative teleoperation using networked spatial dynamic voting. *The Proceedings of The IEEE*, 91(3):430–439, March 2003.

15. K. Han, Y. Kim, J. Kim, and S.Hsia. Internet control of personal robot between kaist and uc davis. In *IEEE International Conference on Robotics and Automation (ICRA)*, 2002.
16. Sariel Har-Peled, Vladlen Koltun, Dezhen Song, and Ken Goldberg. Efficient algorithms for shared camera control. In *19th ACM Symposium on Computational Geometry, San Diego, CA*, June 2003.
17. **http://telerobot.mech.uwa.edu.au/**.
18. H. Hu, L. Yu, P. W. Tsui, and Q. Zhou. Internet-based robotic systems for teleoperation. *Assemby Automation*, 21(2):143–151, May 2001.
19. S. Jia, Y. Hada, G. Ye, and K. Takase. Distributed telecare robotic systems using corba as a communication architecture. In *IEEE International Conference on Robotics and Automation (ICRA)*, 2002.
20. S. Jia and K. Takase. A corba-based internet robotic system. *Advanced Robotics*, 15(6):663–673, Oct 2001.
21. J. Kim, B. Choi, S. Park, K.Kim, and S. Ko. Remote control system using real-time mpeg-4 streaming technology for mobile robot. In *IEEE International Conference on Consumer Electronics*, 2002.
22. R.C. Luo and T. M. Chen. Development of a multibehavior-based mobile robot for remote supervisory control through the internet. *IEEE/ASME Transactions on Mechatronics*, 5(4):376–385, 2000.
23. T. Mirfakhrai and S. Payandeh. A delay prediction approach for teleoperation over the internet. In *IEEE International Conference on Robotics and Automation (ICRA)*, 2002.
24. R. S. Mosher. Industrial manipulators. *Scientific American*, 211(4), 1964.
25. L. Ngai, W.S. Newman, and V. Liberatore. An experiment in internet-based, human-assisted robotics. In *IEEE International Conference on Robotics and Automation (ICRA)*, 2002.
26. R. Safaric, M. Debevc, R. Parkin, and S. Uran. Telerobotics experiments via internet. *IEEE Transactions on Industrial Electronics*, 48(2):424–31, April 2001.
27. T. Sato, J. Ichikawa, M. Mitsuishi, and Y. Hatamura. A new microteleoperation system employing a hand-held force feedback pencil. In *International Conference on Robotics and Automation*. IEEE, May 1994.
28. Thomas B. Sheridan. *Telerobotics, Automation, and Human Supervisory Control*. MIT Press, 1992.
29. D. Song, A. Pashkevich, and K. Goldberg. Sharecam part II: Approximate and distributed algorithms for a collaboratively controlled robotic webcam. In *IEEE/RSJ International Conference on Intelligent Robots (IROS)*, Nov. 2003.
30. D. Song, A. F. van der Stappen, and K. Goldberg. An exact algorithm optimizing coverage-resolution for automated satellite frame selection. In *IEEE International Conference on Robotics and Automation (ICRA), New Orleans, LA*, May 2004.
31. D. Song, F. van der Stappen, and K. Goldberg. Exact and distributed algorithms for collaborative camera control. In *International Workshop on Algorithmic Foundations of Robotics*, 2002.
32. R. Tomovic. On man-machine control. *Automatica*, 5, 1969.

Algorithmic Foundation of the Clockwork Orange Robot Soccer Team

Pieter Jonker[1] Bram van Driel[1] Jev Kuznetsov[1] Bas Terwijn[2]

[1] Quantitative Imaging Group, Faculty of Applied Sciences, Delft University of Technology {pieter,bram,jev}@qi.tnw.tudelft.nl

[2] Intelligent Autonomous Systems Group, Science Faculty, University of Amsterdam bterwijn@science.uva.nl

Summary. The Dutch robot soccer team Clockwork Orange is a research project of the Delft University of Technology and the University of Amsterdam. In this project we study multi-robot collaboration, robot vision and robot control. Our research focus, overviewed in this paper, is on automatic calibration for robust colour segmentation, behaviour based robotics and reinforcement trained dribbling with the ball.

1 Introduction

Robot Soccer is an environment to research, develop and test algorithms for co-operating autonomous robots. The robot teams operate in a simple physical world with simple rules. A world wide competition (the RoboCup) [1] takes care of evolutionary steps in the development of the robots. Concepts and algorithms that are not good enough do not survive. The RoboCup has several leagues, such as the simulation, small size, middle size, four legged (AIBO) and humanoid robot leagues. The middle size league includes development of robots, sensors, actuators, hard and software and is as such the most challenging league. World wide about 32 teams take part in the competition in this league. In a robot soccer game 4 robots autonomously compete with 4 other robots during 2 x 10 minutes. Each year the rules of the game are updated to be more and more in accordance with the FIFA soccer rules and the world of the robots is made more alike the real physical world. The end-goal is to beat in 2050 with a humanoid robot team a human soccer team, i.e. the 2050 world champion. This goal is in analogy with the chess computer Deep Blue that has "proven to be a better chess player" than humans. The Dutch robot soccer team Clockwork Orange [2] is a research project of the Delft University of Technology and the University of Amsterdam. In this project we study multi-robot collaboration, robot vision and robot control. For our team we use a single type of robot, formerly known as Nomad Scouts, but with

M. Erdmann et al. (Eds.): Algorithmic Foundations of Robotics VI, STAR 17, pp. 17–26, 2005.

re-engineered robot hardware so that only the housing, the power-electronics and motor driving electronics remained original. The system has an embedded processing board with a 1 GHz Pentium, a PCI-bus frame-grabber and several PC104 boards for motion control, sensing control, etc. Each robot has a colour camera, simple sonar sensors and collision bumpers, and uses 2.4GHz WLAN for communication.

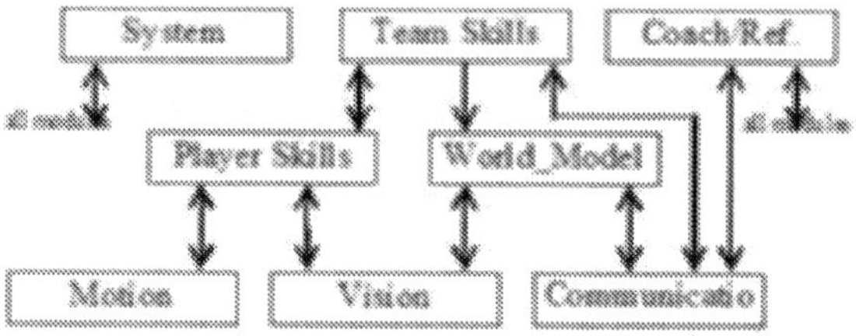

Fig. 1. The software architectur

Our software architecture, see Fig. 1 is a layered and modular system in which higher layers represent higher abstraction processes [3]. The software of all modules in all layers runs in parallel, and within each module multiple threads can exist. The software consists of three layers: the intelligent devices (motion, vision, communication), the skills (player skills, world observation) and the top layer (team play, system health monitoring and automatic error recovery, and coach/referee interaction). The coach and referee-box are laptops at the side of the field. The coach sends commands to the robot indicating the direction of play, the initial roles of the players,... The referee sends commands such as: kickoff-team-A, start, pause, continue, free-kick-A,... As the aim of the software is an annually moving target, we describe below our latest developments.

2 Robust, lighting independent colour vision system

One of the rule changes that is about to come in the middle size league is the changing of lighting conditions, the allowance of daylight spots and large shadows on the field, as our ultimate goal in 2050 is to be able to play in open air. Since object detection is done based on colour segmentation, the vision of the robot is severely hindered by a chang in colour of the illumination [4]. Moreover, a colour classifier built and calibrated before each game and each different field is not a sensible thing for a fully autonomous agent. Common auto-iris and auto-whitebalancing systems use the global statistics of the entire image. Fig. 2 (left original, right white-balanced system) shows the result.

Also a system in which the camera permanently sees a grey or color wig, attached to the camera housing, at the bottom of its image, did not gave satisfying results. Acutally, one is interested in true colours of objects, independent

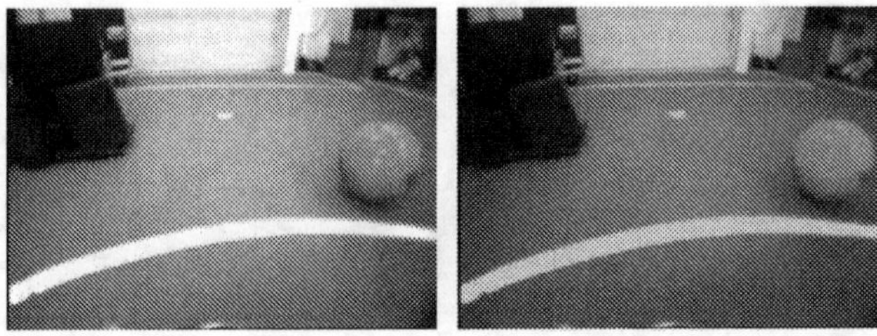

Fig. 2. Adapting the camera to changing lighting conditions

of the colour of the light, such as the ball on the field, typically a few meters in front of the robot and not just in front of the lens. Consequently, we made a self calibrating vision system using a strategy of emergent behaviour [5]. Emergence can basically be described as using information from various orthogonal but overlapping sensors, constantly observing the world. Currently, we use the robot's Polaroid sonars to estimate the free space around the robot. Mapping this estimation on the camera data (images) we can make an estimate of which pixels and colours belong to the field. Fig. 3 shows the sonar data mapped onto the image, indicating the green open space in front of the robot. This space is used to perform the auto iris and auto white (or, better, green) balancing on: by adjusting the R,G and B amplifiers in the camera through its serial interface, and further refined in software.

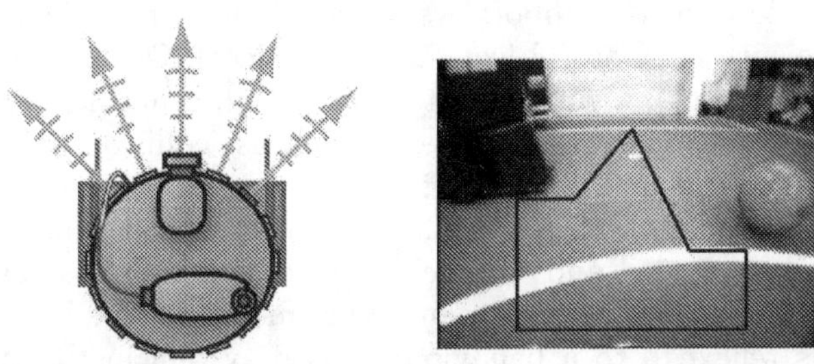

Fig. 3. Using the sonar sensors to detect the free green floor in the image

With this information we can continuously recalibrate the colour segmentation by growing, shrinking, resizing or moving the colour classifier in the appropriate colour space. Since the classifier is a 3D look up table, this method works just as well for YUV as for RGB colour spaces. The colour segmentation only uses this classifier which is in fact a (temporal) generalization of the data from the sensors. There is no separate learning and testing phase, so if lighting conditions change, the robot starts forgetting what it knows and learns new colour labels. This way it adapts itself in a few seconds to the new situation. If it does not see sufficiently enough green floor, it does not recalibrate.

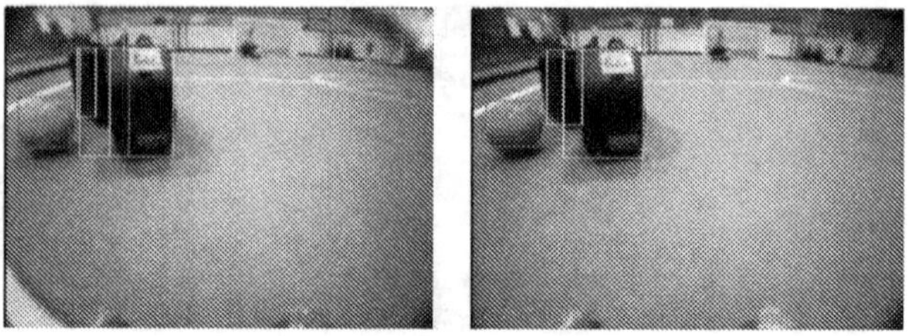

Fig. 4. Camera adapting itself to brown sunglasses

Fig. 4 shows left a camera on which we have put brown sunglasses. It can be clearly seen that the classification of objects indicated by the found bounding boxes, such as ball, goal and robots, goes wrong. Right, it shows the adaptation of the camera due to adjusting the parameters on the camera as well as by further fine tuning the illumination level and the white balance in software. Similar results are obtained by holding blue, pink and grey sunglasses in front of the camera. Note that the same notion of free floor space can be obtained and used to perform the auto-calibration on, by using a stereo pair of camera's instead of one camera and depth sensors.

3 Behaviours based soccer skills

The player skills of our robots was formed by a module that could execute actions such as: Goto(object), Search(object), Dribble-to(object), Turn-and-Shoot-to(object),... with object = ball, goal, centerfield, (x,y,phi),.... All actions were conditioned with reflexes such as: collision avoidance, at-line-behaviour, shoot-to-goal, etc. Within the player skills, many (> 50) parameters were used, much with a non-physical meaning [6]. This module was too complex and entangled and we decided to change the system from a single action that is selected by the team skills module (possibly interrupted by autonomous reflexes), to a behaviours based system in the tradition of the Subsumption Architecture. See e.g. [5]. Using a behaviour based approach has advantages such as reduced complexity due to the separation into simple behaviours and relative ease to add and test new behaviours. The current action behaviours are: goto(object), get-ball, dribble, side-tick-ball, avoid-collision, retract-from-collision, stay-on-field. The current sense behaviours are: displacement, freespace, ball, goal, static-objects. Behaviours are scheduled based on the belief state of the individual robot. For the scheduling we have chosen a finite state machine with fixed activation values, in contrast to dynamic activation functions, to allow for easier control and fine-tuning. In order to fuse the output of the behaviours we use a voting mechanism where each behaviour votes on a large set of possible movements by assigning a value between 0 and 1. The votes of the behaviours are multiplied by the activation value of the respective behaviour and for each movement all the votes are added. After addition, the movement with the highest sum of votes is executed. The module has shown good results in dealing with complex situations where there are multiple obstacles and the position of the ball to consider. Especially good results come from the specially designed 'rotate with ball' behaviour which enables the robot to turn with a relatively short radius while keeping control of the ball. The 'goal sensor' and 'ball sensor' gives the robot a good measure of the chance of scoring at any time to allow for long-distance shooting when appropriate.

4 Training a robot to dribble with the ball

Present software heavily relies on hand coded algorithms along with many empirical constants. This approach produces fairly good results in the short term, but as the software becomes increasingly complex, human comprehension of what is going on becomes more difficult. At a certain point so many empirical factors have been introduced [5], that it is almost impossible to find their optimal values, as many of them are correlated. Consequently, we implemented and tested Q-learning [7],[8],[9],[10] in our dribble behaviour. The requirements for this behaviour are quite complex due to the rules that only one third of the ball may be covered by the robot and while trying to score, holding the ball is not allowed. At the same time, the robot should avoid obstacles and orient itself with its vision system on its absolute position on the field and its relative position with respect to the other players. This need to look around independently generates motions. A typical lay-out of a ball handler's geometry and the involved variables is presented in Fig. 5.

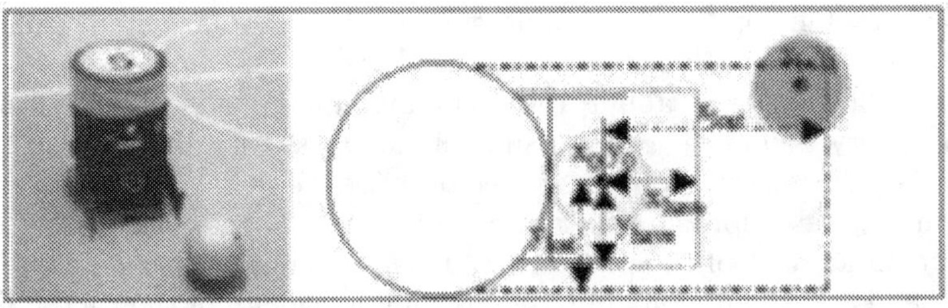

Fig. 5. a) Clockwork Orange robot b) Layout of the ball handling

Using reinforcement learning, the robot can try to optimize its behaviour by learning what to do in each situation on the field. Moreover, it can be made to adapt itself to new situations, e.g., for changes in its ball handler geometry, or for the friction of the ball to the field. In general, a learning algorithm for a system with a large number of parameters needs several hours to converge to its optimal performance. Each training episode starts with positioning the robot and the ball on arbitrary positions of the field, which is impractical for a real physical robot and a human trainer. With an adequate simulator this training can be done considerably faster than with a real robot. Pre-trained with the simulation and downloaded onto a real robot, we should only need a limited number of extra training runs to correct for the factors that were not present in the simulation. This probably also holds for small adaptations to the ball handler geometry and to the field-ball friction on the various soccer fields.

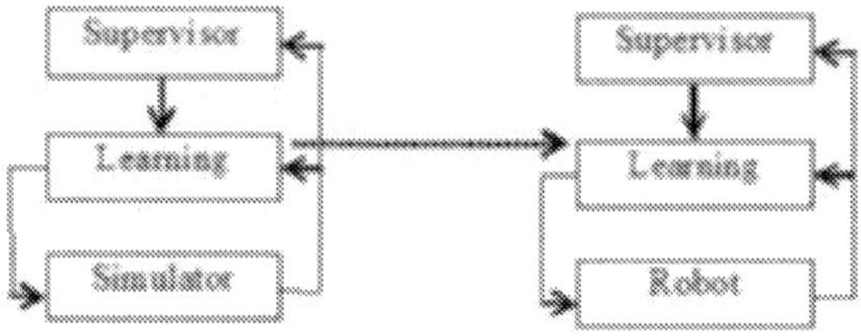

Fig. 6. The learning procedure; a) pre-training b) post-training

So, in order to develop a learning algorithm for a robot, a dynamic system model was made describing the kinematics of the robot and its interaction with the ball. When robot and ball are colliding elastically, both momentum and kinetic energy are conserved. We derived equations for the (angular) velocity of the robot and the (angular) velocity of the ball resolved in components parallel and perpendicular to the ball handler, both before and after the collision. We have set-up a set of equations for a partially elastic collision, to simulate the collisions between robot with piecewise linear kicking device and ball. We proved that the simulator and real robot controller showed the same behaviour. On top of the dribble simulator we implemented reinforcement learning. The algorithm is equal to the Watkin's Q() algorithm, as described in [11], [8], [9]. After about 1500 training runs, our simulated robot could dribble very well. The achieved simulated success rate on a field of 15 x 25m. was about 80%, with a performance of 60%. We defined the performance as the time it takes for the agent to complete a task, relative to the theoretical minimum amount of time necessary. This minimum was calculated by dividing the distance to the goal by the maximum velocity of the robot. This gives the shortest time it can take the robot to complete its task. Training on a 10x20 m field resulted in a decrease of success rate from 80% to 70% and on a 4x8 field to only 40%. We then tested the post-training on the real robot, and the robot could dribble using the pre-learned behaviour. Using an algorithm of no more than 30 logical statements and almost no expert data, a complex behaviour was achieved, resulting in a robot chasing the ball and pushing it into goal.

5 Training a robot to dribble while avoiding obstacles

In our behaviours module we fuse simple behaviours, up until now all hand coded algorithms. Dribble to goal and collision avoidance are two disjunct simple behaviours. In section 4 we showed that we could replace the dribble to goal behaviour with a version based on Q-learning. We want to augment this now with collision avoidance. The main problem with reinforcement learning algorithms, however, is scalability. Every extra input variable adds an extra dimension to the state space. During learning, the agent should visit all the states repeatedly to find the optimal policy. Increasing the number of inputs/outputs adds extra dimensions to the state space, which increases the learning times exponentially. Still, long convergence times to an optimal policy are not even the biggest problem. In high dimensional spaces the agent is very likely to land in local minima. Different policies tend to interfere with each other. With Q-learning in particular, extra inputs which are un- or weakly correlated to the rewards and therefore irrelevant, make the system perform less. This is inherent to the algorithm, as only the positive part of the 'noise' is propagated forward, biasing the system. High dimensionality along with the inability to reject irrelevant data and poor generalization, results in poor performances.

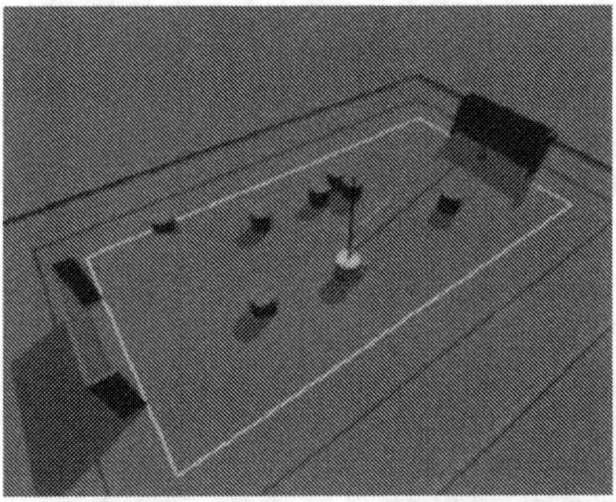

Fig. 7. Dribbling with collision avoidance requires 6 inputs: the vector to the ball, the goal, and to the nearest obstacle, and 2 outputs: the velocities of the wheels

For a real life robot soccer game, fast reaction is required, meaning that time steps (the rate at which the decisions are taken) should be kept as small as possible. Ideally, a continuous decision making is desirable. Unfortunately reinforcement learning algorithms do not support continuous time. In a robot soccer game, robots move at rates of about 1 m/s, so when using a 1s sampling time, a decision to stop just in front of an obstacle can come too late, the robot has smashed itself into its opponent. Using smaller time steps, however, makes learning more problematic, as table 1 shows.

Table 1. Success rate and performance in relation with the time step for the dribble to goal task

Time step size [sec]	Success Rate mean	Performance mean
0.25	0.42	0.20
0.50	0.64	0.26
1.00	0.69	0.38

This is caused by the fact that the agent can 'feel' the rewards only during a certain number of time steps from reinforcement. This number of steps is dependant on the length of the eligibility trace [11]. Making traces longer enables the robot to feel the rewards further from a goal. But long traces also pose a problem, as an uncertainty is introduced at each step. Choosing a trace length of about 5 time steps appeared to be reasonable, provided that the time steps are small enough to enable the agent to keep up with the changes in its environment. If we want to make a combined behaviour based on Q-learning that involves dribbling while performing collision avoidance, this means in terms of sampling rates, that reaction times of less than 0.2 s., or sampling frequencies of more than 5 Hz, are needed. An input and an output (action) is a continuous one dimensional vector. In Q-learning the actions to perform are just considered an extra input. Our actions, setting the speed of the two wheels, are sampled using the CMACS tile coding. Originally we used the vector (angle, distance) to the ball and to the goal as inputs. Now for each obstacle we want to avoid, another vector (2 inputs) is needed. The many attempts we made, revealed the inability of the Q learning algorithm to cope with more than 8 inputs, given the sampling of the space due to the 5 Hz sampling rate we need. But this is not enough to keep track of the ball and multiple obstacles simultaneously. At least two obstacles must be avoided simultaneously, to overcome closing in by opponents or corners. This problem of the large input/output space can be reduced by using scaling functions, filtering, clipping, etc. But this imposes a complexity, which the reinforcement system should solve in addition. The great advantage of a learning system is that the designer only tells it what to do, and not how. From a simple reward structure the agent should be able to figure out the policy its designer had in mind. Simple rewards soon prove to be inefficient. Take as an example a simple 'go to ball' behaviour. The simplest reward structure is a reward for getting to the ball and a penalty for every time step. This reward structure soon proved to be inefficient as an effect of a reward at the end of the entire training episode is propagated back in time only for a certain amount of steps according to the eligibility trace, meaning that the agent cannot 'feel' the goal when he is more than the eligibility trace length (in time) away from it. Increasing the sampling rate makes such simple rewards even less efficient. To solve this, a designer decides to give reinforcements at every time step, positive for getting closer to the ball, and negative for going away from it. This results in rewards resembling a potential field, pushing the robot towards the goal. This forms an important caveat: when expecting complex behaviours and the agent fails to find the policy that satisfies its designer, the designer tends to start telling the agent how to accomplish its task, implying complex potential reward structures and completely robbing the agent of its freedom to explore. So the solution is the same as we have experienced before: decouple behaviours into simple behaviours and fuse them to obtain transitions in the behaviours. This should also be done when Q-learning behaviours; use a sort of "prioritized learning": Our desired behaviour is to approach the ball at a predefined angle while avoiding obstacles. While far away from the ball, keeping the heading is important. However, only close by, an approach at a right angle should be achieved. So our distributed learning system learns one policy on top of an already learned behaviour. See Figure 8.

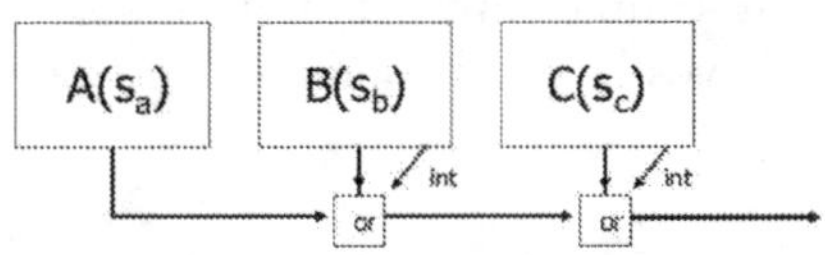

Fig. 8. Cascading Behaviors

Let $A(s_a)$ be a first and 'base' behaviour like 'go to ball', learned only with its relevant states s_a. Let $B(s_b)$ be the obstacle avoidance behaviour, etc. Let S be the total state and v be an input variable. First, behaviour $A(s_a)$ chooses an appropriate action based on the data it receives. The second behaviour $B(s_b)$ uses only inputs which are relevant for it (i.e. obstacle avoidance behaviour does not need to know where the ball is). Along with its action to drive the wheels, behaviour $B(s_b)$ has a special 'intervention' action. This is a binary action, determining whether the actions of all previous behaviours (in this case $A(s_a)$ will be overridden. By making this intervention very costly, the agent learns to use it only in the cases where the collision is most probably unavoidable. The total state S, optimal action value $\tilde{a}$ and relevant states s_i become:

$$S = v_1 + v_2 + \ldots + v_n \tag{1}$$

$$\tilde{a} = A(s_a) \vee B(s_b) \vee C(s_c) \ldots \tag{2}$$

$$s_i = \Sigma v_{relevant} \tag{3}$$

We need $B(s_b)$ (and all subsequent behaviours) to be active only in the relevant situations, with back propagation times of only a couple of steps, making a simple reward structure possible. One great advantage for using such compensators is that their preferred action when learning starts is always 0, meaning one can fill all Q values accordingly before starting the actual learning.

6 Conclusions

Setting up and maintaining a team of autonomous co-operating robots, albeit functioning in a simple world with simple rules, involves many different algorithms, all with their own parameters settings. As each robot's algorithmic "install base" currently contains about 200,000 lines of C/C++ code and an estimated state space of over 3000 states, the calibration of all involved parameters is a tedious job. Apart from a proper layered and modular architecture, auto-calibration, learning and emergent behaviour must be applied in order to cope with this complexity. In this paper we showed some techniques we used: 1) Overlapping sensors of different kind that measure the same item; i.e. measuring the colour of the free floor space, to perform auto-gain and white balancing for colour segmentation under outdoor lighting conditions. A first step towards emergent behaviour. 2) Q-learning in a cascaded way to learn combined simple behaviours, without stepping into the pitfall of large state spaces with high sample rates. It performs at least equivalent to hand coded algorithms, involves less code and lacks magic numbers.

References

1. : Robocup foundation (2005) `www.robocup.org`.
2. : Clockwork orange team (2004) `www.ph.tn.tudelft.nl/~phrobosoc`.
3. Jonker, P.P., Terwijn, B.: Clockwork orange team description paper. In: Proceedings of the RoboCup International Symposium, July 4-5, Lissabon. Lecture Notes in Computer Science, Springer, Berlin (2004)
4. Jonker, P.P., Caarls, J., Bokhove, W.: Fast and accurate robot vision for vision based motion. In Stone, P., Balch, T., Kraetzschmar, G., eds.: RoboCup 2000: Robot Soccer World Cup IV. Volume 2019 of Lecture Notes in Artificial Intelligence., Springer, Berlin (2001) 149–159
5. Pfeiffer, R., Scheierl, C.: Understanding Intelligence. The MIT Press, Cambridge, Massachusetts (1999) ISBN 0-262-16181-8.
6. Altewischer, W.: Implementation of robot soccer player skills using vision based motion. MSc. thesis, Delft University of Technology (2001) `ftp://ftp.ph.tn.tudelft.nl/pub/ROBOCUP/afstudeerders/verslagen`.
7. Santamaria, J.C., Sutton, R.S., Ram, A.: Experiments with reinforcement learning in problems with continuous state and action spaces. Technical report, Lederle Graduate Research Center, University of Massachussetts, Amherst (1996)
8. Watkins, C., Dayan, P.: Technical note: Q-learning. In: Machine Learning. Volume 8(3-4). (1992) 279–292
9. Watkins, C.J.C.H.: Learning from delayed rewards. PhD thesis, Cambridge University (1989)
10. Kok, J.R., Vlassis, N.: Sparse tabular multiagent q-learning. In Steenhaut, K., Now, A., Lenaerts, T., eds.: Benelearn 2004: Proceedings of the Annual Machine Learning Conference of Belgium and The Netherlands, Brussels, Belgium. (2004) 65–71
11. Sutton, R.S., Barto, A.G.: Reinforcement Learning. The MIT Press, Cambridge, Massachusetts (1998) ISBN 0-262-19398-18.

Uniform Coverage of Simple Surfaces Embedded in $\mathbb{R}^3$ for Auto-Body Painting

Prasad N. Atkar, David C. Conner, Aaron Greenfield, Howie Choset, and Alfred A. Rizzi

Carnegie Mellon University, Pittsburgh PA 15213, USA

Abstract. We develop a procedure for automated trajectory generation for robotic spray painting applications. Painting requires that the spray gun deposit paint at each point on the target surface such that the variation of the resultant paint deposition is within acceptable limits; we term this the uniform coverage problem. To understand the key issues in the uniform coverage problem, we consider surface patches that are geodesically convex and topologically simple to represent realistic automotive surfaces. Our goal is to understand the relationship between the spray gun trajectory and the following output characteristics: deposition uniformity, cycle time, and paint waste. Our planning approach decomposes the coverage trajectory generation problem into three subproblems: selection of the start curve, selection of the speed profiles along each pass, and selection of the spacing between the passes. Using concepts such as area magnification, the Gauss-Bonnet theorem from differential geometry, and standard optimization procedures, we present procedures to solve each subproblem independently of the others. Finally, we demonstrate our trajectory planning procedures in simulation as well as experimentally on simple surfaces that approximate real automobile surfaces.

1 Introduction

Today, robots are widely used for spray-painting in the automotive industry. Among the different kinds of spray painting mechanisms available, electrostatic rotating bell (ESRB) atomizers are one of the most popular. However, the distribution of paint generated by an ESRB atomizer is relatively complex. This complexity is further compounded as the spray gun moves over non-planar automotive surfaces, making the task of planning paths for such atomizers challenging. Paint specialists typically produce the coverage paths based on their experience, often requiring 3 to 5 months to completely plan trajectories on a new automobile model. This programming time is a critical bottleneck in the "concept-to-consumer" timeline for bringing a new automobile to the market. Automating the process of path planning will help the paint specialists reduce this programming time significantly by offering them reasonable guidelines for effective paths.

In this work, we develop procedures for automated generation of spray gun trajectories to optimize the output characteristics over simple surfaces. By "simple surfaces," we mean surfaces that are geodesically convex and have no holes. A surface is termed geodesically convex if the shortest curve joining any two points on the surface is a geodesic (see Fig. 1(a) and 1(b)). We use such surfaces to closely approximate the auto-body surface parts.

M. Erdmann et al. (Eds.): Algorithmic Foundations of Robotics VI, STAR 17, pp. 27–42, 2005.

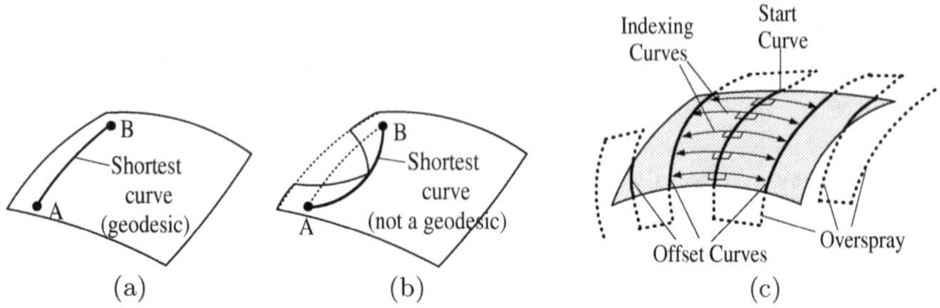

Fig. 1. (a) A geodesically convex bent sheet; here the shortest curve joining any two points is a geodesic. (b) the surface is not geodesically convex because the shortest curve in the surface joining two points A and B is not a geodesic. (c) We generate a coverage path on arbitrary simple surfaces by selecting a start curve and offsetting it sideways within the surface to generate new passes.

To generate a spray gun trajectory termed a *coverage trajectory* that uniformly covers (i.e., deposits paint on) the target surface, the path planning algorithm must determine the orientation of the passes in the path, the spacing between them and the speed along the passes. The dimensionality of the variables required to specify these three entities is huge. Therefore, global optimization procedures that attempt to determine all coverage variables simultaneously are computationally expensive and often not practical.

We seek to make the optimization of coverage variables tractable by decomposing the coverage problem into three relatively independent subproblems: i) selection of a "seed" pass termed the *start curve* on the surface (see Fig. 1(c)), ii) selection of the speed profile along a given pass, and iii) selection of the optimum spacing between a given pass and its adjacent pass. Based on these three subproblems, we generate the coverage trajectory using the procedure described in Algorithm 1.

Data : Target surface CAD model, parameters for deposition model
Result : Spray gun trajectory
Select a pass termed the start curve on the surface;
Optimize end-effector speed along the start curve;
repeat
  Offset the most recently generated pass within the surface to its "right" side to obtain a new offset pass;
  Optimize end-effector speed along the new offset pass;
until *Offset pass lies completely outside the surface*;
Designate the start curve as the most recently generated pass;
repeat
  Offset the most recently generated pass within the surface to its "left" side to obtain a new offset pass;
  Optimize end-effector speed along the new offset pass;
until *Offset pass lies completely outside the surface*;

Algorithm 1: Path Planning algorithm

The rationale behind decomposing the coverage problem into the three subproblems is that each subproblem can be solved in a reasonable amount of time independently from the others. The cycle time for the coverage trajectory is primarily dependent on the orientation of the passes in the path. The selection of a particular start curve essentially determines the orientation of the rest of the passes, and consequently the cycle time required. The structure of the paint profiles measured in the direction orthogonal to a pass is relatively consistent along the direction of the pass (see Fig. 2(a)). Therefore, the uniformity of paint deposition can be seen as having two components: 1) uniformity along the direction of the passes, and 2) uniformity in the direction orthogonal to the passes. Speed optimization attempts to produce similar paint profiles along a pass, whereas the optimization of the spacing between passes, termed the *index width*, attempts to overlap the paint profiles of two adjacent passes appropriately to produce uniform paint deposition in the direction orthogonal to the passes.

In this work, we continue our prior work on deposition modelling and start curve selection, briefly discussed in Sections 2.2 and 2.3 respectively, by providing procedures for speed optimization and index width optimization. We formulate the speed optimization problem in Section 3. We address the index optimization problem in Section 4 for surfaces with increasing geometric complexity: planar surfaces, extruded sheets and surfaces with non-zero Gaussian curvature. Finally, we demonstrate our coverage procedures for a variety of surfaces in simulation as well as experimentally in Section 5.

2 Prior Work

2.1 Related Work

Most prior researchers typically focus on a particular subproblem for the path generation procedure: 1) start curve selection, 2) speed optimization, or 3) index width optimization. However, only a few researchers study all the three problems together.

For start curve selection, most prior researchers [1–4] select the pass orientation that aligns with one of the faces of a bounding box that fits the surface, while choosing the relative position of the start curve arbitrarily. Such an approach of start curve selection implicitly tries to minimize cycle time, but does not consider the effects of the relative position of the start curve on paint uniformity and can lead to poor uniformity results. For a planar surface, Huang [5] gives a sophisticated approach for minimizing the cycle time for coverage by minimizing number of passes in the path. Kim and Sarma [6] use vector fields to choose the pass orientation that minimizes the cycle time with an implicit constraint on the paint deposition uniformity. The effect of the vector field orientation on the paint uniformity is not explicitly considered. In an attempt to maximize uniformity, Smith *et al.* [7] select the orientation of the passes by determining the section plane that is maximally orthogonal to the surface. Their approach does not consider minimizing the cycle time or the paint waste and is sensitive to small curvature changes in the target surface.

One of the most general frameworks for efficiently optimizing the speed profile of a coverage path is given by Antonio and Ramabhadran [8]. Their work assumes that the coverage path is already known and the deposition

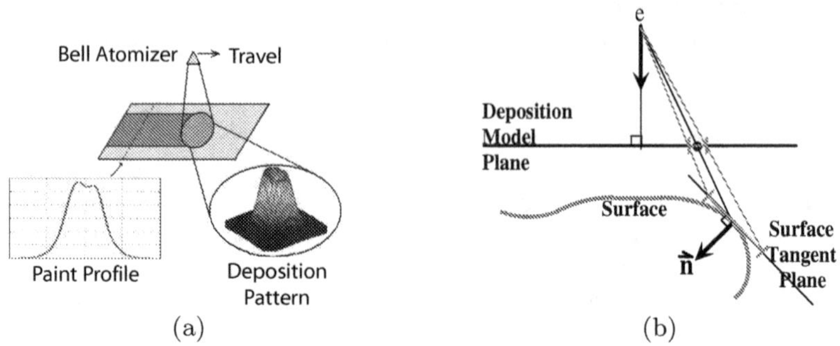

Fig. 2. (a) A typical deposition model and the paint profile across the paint swath on a flat planar sheet. (b) Our current model determines the deposition on any given point on an arbitrary surface using concept of area magnification under the linear projection model.

model is either a bivariate Cauchy or a Gaussian distribution applied to a flat panel. Kim and Sarma [6] also present a speed optimization model for a coverage path generation framework based on vector fields with a particular focus on the process cycle time.

Most prior path planning approaches consider simplistic deposition models such as circular [1], parabolic [4,9,10], or beta distributions [2]. Likewise, most of these approaches make first order approximations to the surface geometry, thus limiting their use for realistic auto surfaces. In these cases, the selection of index widths between adjacent passes is easier [1–4,6,7], but fails to capture the realistic scenario.

Finally, commercially available path planning systems such as RobCAD™ [1] project a user-specified planar path on to the target surface and yield the output characteristics of the resultant coverage path. The limited scope for the specification of the deposition model and the requirement for manual specification of the coverage path limit the utility of such software tools.

2.2 Our Prior Work: Deposition Modelling

In automated path planning systems, it is necessary to have a paint deposition model that can predict paint deposition on an arbitrary surface with a reasonable accuracy to effectively determine the suitability of a given coverage trajectory. In [11], we develop a simple model that provides a significant improvement in paint prediction over the earlier models while retaining sufficient tractability for use in our planning tools. The deposition model captures the shape of the paint distribution from a spray gun in an analytical representation. The deposition model is composed of one bivariate and two revolved Gaussians [11,12]. We extract the parameters for the deposition model by applying data fitting techniques to the experimental data obtained by painting flat panels (see Fig. 2(a)). We then determine the deposition at any point on a given arbitrary curved surface by using the area magnification concept from differential geometry (see Fig. 2(b)).

[1] A product of Tecnomatix Technologies Ltd.

For simplicity, our current model assumes that the paint particles flow in a straight line after leaving the spray gun nozzle. While this is an incorrect assumption, our experience shows us that this assumption is reasonable for surfaces with low curvature. In the future, we plan to extend our techniques to include more realistic projection models. We would like to emphasize that although deposition models are required by the planning procedure, the coverage planning procedures we develop in this work are independent of any particular deposition model.

2.3 Our Prior Work: Start Curve Selection

The choice of start curve impacts the entire coverage path and its cycle time. Unfortunately, the search-space for the start curve selection problem is huge. In [13], we select a start curve such that the resultant coverage path attempts to minimize the cycle time and yields the desired paint deposition uniformity. Uniformity of paint deposition suffers severely in the worst possible case where the passes in the coverage path self-intersect. The more the start curve bends "sideways" in the surface, or equivalently, the higher the geodesic curvature of the start curve, the greater the risk that the subsequent offset curves will exhibit a self-intersection [13,14]. Therefore, to minimize the possibility of self-intersections on the offset curves, we seek to select start curves that have low geodesic curvature, or ideally, are themselves geodesic. This restriction on the choice of start curve to the family of geodesics reduces the start curve selection problem to determining two variables: 1) the orientation of the (geodesic) start curve, and 2) the relative position of start curve with respect to the surface boundary.

Our path construction procedure (see Algorithm 1), first offsets the start curve and then continues to offset newly generated passes until the surface is covered completely. To ensure that the subsequent offset passes are also free from self-intersections, we must ensure that the geodesic curvature on any newly generated pass is also minimal. Thus, ideally we want to select a start curve such that all the resultant passes are geodesics. However, on a surface with non-zero Gaussian curvature, the offset of a geodesic curve is, in general, not a geodesic. In such a case, the position of the start curve relative to the boundary dictates the resultant geodesic curvature on the offset curves.

To examine the effect of surface Gaussian curvature K on the geodesic curvature κ_g of the offset curves, we apply the Gauss-Bonnet theorem to the region B bounded between the start curve, C_{st}, and its offset curve, C_{of}, (see Fig. 3(a)) and arrive at

$$\int_{C_{\mathrm{of}}} \kappa_g = \int_B K + \int_{C_{\mathrm{st}}} \kappa_g. \tag{1}$$

Equation 1 tells us that the more the surface bounded between the offset curve and the geodesic start curve bends, the more the geodesic curvature of the offset curve increases. On a surface where the sign of the Gaussian curvature stays the same, the further away we place the offset curve from the start curve, the higher the geodesic curvature of the offset curve is. Therefore, the possibility of self-intersection is maximal on the bounding passes

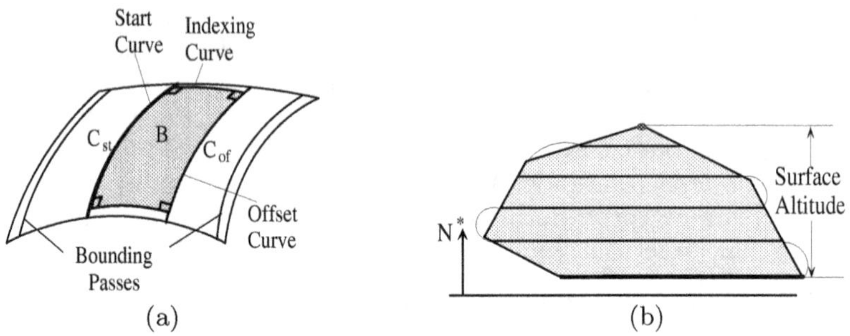

Fig. 3. (a) Application of Gauss-Bonnet theorem to the region bounded between the start curve and the offset curve. (b) We select the orientation of the start curve that minimizes the surface altitude in the indexing direction, and consequently the cycle time.

(see Fig. 3(a)). To minimize the geodesic curvature on the bounding passes, Equation 1 tells us that the start curve should divide the surface into two parts such that the integral of the Gaussian curvature is equal over each part. This approach determines the relative position of the start curve on the surface as defined by the *geodesic Gaussian curvature divider* curve.

To determine the orientation of the start curve, we approximate the geodesic start curve by a curve of planar intersection that is also a Gaussian curvature divider. To ensure that such an approximation is sufficiently close to being a geodesic, we require that the normal to the section plane chosen for intersection is orthogonal to the average target surface normal. We generalize the concept of surface "altitude" from planar surfaces [5] to non-planar surfaces, and determine the orientation of the section plane that minimizes the surface altitude in the direction normal to the section plane (see Fig. 3(b)). The resultant orientation of the start curve yields a coverage path that minimizes the number of turns in the coverage path (for the given constraint on section plane normal) and equivalently, the cycle time and the paint waste.

3 Speed Optimization

In the automobile industry, constant speed trajectories are typically used. Because of the relatively large size of the ESRB deposition pattern with respect to a typical automotive surface, constant speed profiles typically require long oversprays to ensure that the boundary effects, which produce non-uniform paint deposition near surface boundaries, are minimized. If there is a restriction on the maximum amount of paint waste, shorter oversprays could be used, but not without a compromise in the paint deposition uniformity. Additionally, on non-planar surfaces, the changes in the surface curvature along the pass result in non-uniform paint deposition along the pass. Speed optimization attempts to compensate for these curvature related results and improves the uniformity of paint deposition in the direction of the passes.

Unlike prior approaches [6,8] that use speed optimization techniques over the entire path, our approach uses a semi-global method that optimizes speed profiles on a pass-by-pass basis. The underlying assumption in our approach

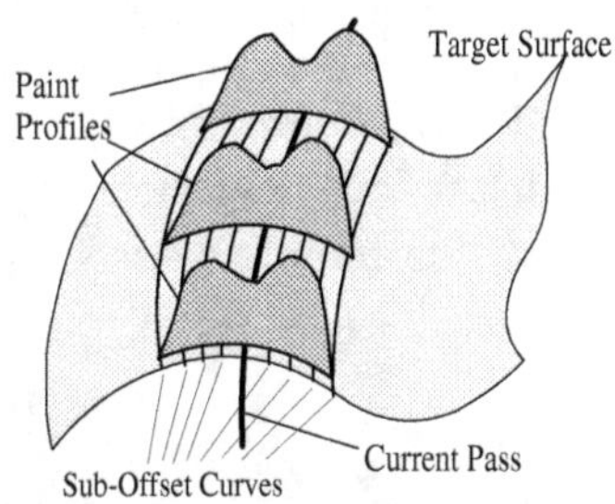

Fig. 4. Speed optimization attempts to minimize the variation of paint profiles in the direction of the pass.

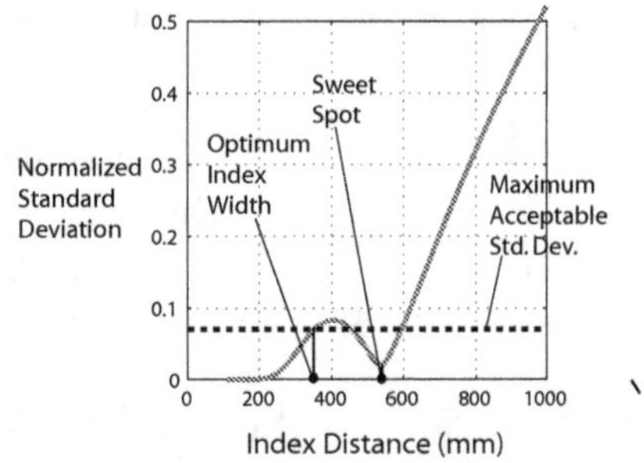

Fig. 5. Graph of normalized standard deviation of paint deposition vs. the index width for a typical ESRB deposition pattern.

is that the speed optimization improves the uniformity in the direction of the passes and a local change in the speed profile affects the paint deposition uniformity on only a subset of the surface. Such an approach tries to combine the best of global and local optimization techniques.

Our implementation first samples a "band" of the surface around a given pass by constructing several "sub-offset" curves which are the similar to the offset curves of the current pass, but with a much smaller spacing between them (see Fig. 4). Our objective here is to minimize the weighted sum of standard deviations of paint deposition along all sub-offset curves, thus minimizing the variations in the paint profile along the pass.

Our implementation first discretizes the given spray gun pass into n linear segments of length s_i. We assume that the speed of the spray gun tool center point (TCP) does not vary in a given segment i and is given by v_i. The speed profile along the pass is represented by the n-tuple $\mathbf{V} = \{v_i : i \in [1, n]\}$. The time t_i spent by the spray gun in each segment i of the pass is equal to $\frac{s_i}{v_i}$. Instead of optimizing the speed profile $\mathbf{V}$ explicitly, we optimize the time profile $\mathbf{T} = \{t_i : i \in [1, n]\}$, and then calculate $\mathbf{V}$ accordingly.

To determine the standard deviation of paint deposition along each sub-offset curve j, we discretize the jth sub-offset curve into m_j linear segments of length s_{m_j}. Let $\mathbf{D_j}$ represent the matrix $[d_{ef}]$ which gives the deposition flux on segment e of the jth sub-offset curve when the spray gun tool center point lies at the center of segment f of the current pass.

For a given time profile $\mathbf{T}$, the resultant paint deposition on the jth sub-offset curve is given by $\mathbf{D_j T}$. Note that $\mathbf{D_j T}$ is a m_j-tuple whose ith component gives the deposition on ith segment of jth sub-offset curve. Let S_j be the sum of all elements of vector $\mathbf{D_j T}$. The average paint deposition k_j along the jth sub-offset curve is then equal to $k_j = \frac{S_j}{m_j}$. The normalized standard deviation of paint deposition along jth sub-offset curve is then computed as $||\frac{\mathbf{D_j T} - \mathbf{K_j}}{k_j}||$, where $\mathbf{K_j}$ is a m_j-tuple with elements of constant values equal to k_j.

The constraints on the maximum and minimum speed are specified at each segment along the spray gun pass by requiring $\frac{s_i}{v_{\max}} \le t_i \le \frac{s_i}{v_{\min}} \quad \forall i$.

The acceleration constraint has a non-linear form in each segment which we linearize by using a conservative linear constraint that always satisfies the intended acceleration constraint. Here, the bounds for the maximum acceleration and deceleration are assumed to be a_{max} and $-a_{max}$ respectively.

Let $w_j = \frac{k_j}{\sum_i k_i}$ be the weight of the normalized standard deviation along jth sub-offset curve in the objective function. Then, the complete speed optimization problem is formulated as:

$$\min_{\mathbf{T}} \sum_j \frac{w_j}{k_j} ||\mathbf{D_j T} - \mathbf{K_j}|| \tag{2}$$

$$\text{such that } \mathbf{D_0 T} = \mathbf{K_0}$$

$$\frac{s_i}{v_{\max}} \leq t_i \leq \frac{s_i}{v_{\min}}, \text{ and } -a_{\max}\frac{s_{i+1}{s_i}^2}{{v_{\max}}^3} \leq s_{i+1}t_i - s_i t_{i+1} \leq a_{\max}\frac{s_{i+1}s_i^2}{{v_{\max}}^3}.$$

At the beginning of the optimization procedure, we specify an initial time trajectory $\mathbf{T^{beg}}$ that corresponds to a constant speed profile with speed v_{nom}, where v_{nom} is chosen according to the desired average paint thickness. Each vector $\mathbf{K_j}$ is then calculated using $\mathbf{T^{beg}}$, that is, $\mathbf{K_j} = \mathbf{D_j T^{beg}}$. We then have all the initial conditions required for the optimization, and execute the optimization algorithm. Without loss of generality, we assume that the 0th sub-offset curve is along the spray gun pass. The equality constraint $\mathbf{D_0 T} = \mathbf{K_0}$ enforces that all speed profiles yield approximately the same average paint deposition on the surface.

The objective function in our speed optimization formulation is a quadratic form in the variable $\mathbf{T}$ with all constraints (i.e., equality and non-equality) that are linear in $\mathbf{T}$. This quadratic constrained optimization can be solved using standard optimization techniques with superlinear convergence. Note that a fine path resolution $\{s_i\}$ is necessary for accurate uniformity estimation, but results in long running times due to large number of inequality constraints. We address this issue by using a fine path resolution for evaluation of the objective function, but optimizing the speed profile over only a sub-sample of non-contiguous segments. The speed at the rest of segments in the path is determined by assuming a linear variation in speed between the segments for which we optimize the speed. This approach solves the optimization problem in a relatively short amount of time, and yet is acceptable as the industrial robots are typically "taught" with only a few points.

4 Index Width Optimization

The goal of speed optimization is to produce acceptable paint uniformity along the direction of a pass. Given speed optimized passes, the remaining problem is to decide how to place passes next to one another on the surface – that is, select *index widths* between the passes. The objective is to select index widths such that the paint profiles of adjacent passes overlap appropriately and produce acceptable uniformity orthogonal to the direction of the passes. At the same time, the index widths should be as wide as possible in order to reduce the number of passes in the coverage path, thereby reducing the cycle time and paint waste. In this section, we present procedures for index width optimization on surfaces with increasing geometric complexity: planar surfaces, extruded surfaces, and surfaces with non-zero Gaussian curvature.

4.1 Determining Index Widths on Planar Surfaces

On a planar surface, geodesics are simply straight lines. Therefore the start curve, chosen as described in Section 2.3, will be a straight line. The offsets of the start curve are parallel straight lines and accordingly the resultant coverage path consists of a family of parallel lines.

To study the effect of index width on paint deposition uniformity, we plot the graph of the normalized standard deviation of paint deposition versus index width between passes (see Fig. 5) [12]. To evaluate the paint uniformity, we consider the interactions between the deposition profile curves of a sufficiently large number of passes spaced at constant index width.

From the graph, we observe that there is a "sweet spot" of index width that corresponds to a local minimum of the standard deviation (around 525 mm index width for the atomizer whose deposition pattern was considered in Fig. 5). Painting the target surface at this higher index width is desirable because higher index widths reduce the process cycle time. Unfortunately, the sensitivity of the standard deviation of paint deposition to the index width is high at the sweet spot; in other words, small changes in index width at the sweet spot produce high variations in paint deposition uniformity. In order to ensure that the paint deposition uniformity over the surface is not sensitive to small changes in index width, we typically do not use the sweet spot spacing between the passes.

Our implementation defines the index width search range, $\mathcal{SR}$, as a finite collection of potential index widths taken from the closed interval $[w_{min}, w_{max}]$ sampled at an appropriate resolution w_{res}. The minimum index width, w_{min}, is chosen according to the cycle time consideration, while the maximum index width, w_{max}, is chosen according to the standard deviation sensitivity. To reduce the computational costs, we choose w_{res} as large as possible, yet sufficiently small to meaningfully capture the variation of uniformity as a function of index width. Next, from the index width search range, we determine the set of feasible index widths that yield the normalized standard deviation of paint deposition below the user specified limit.

We then establish a cost function over the set of feasible index widths by assigning costs inversely proportional to the index width, thus penalizing the process cycle time. The feasible index width that minimizes the cost function is chosen as the optimum index width. By the design of the cost function, the optimum index width is the largest index width that yields an acceptable uniformity. Since the planar surface locally appears the same everywhere, the same optimum index width is chosen for all passes, assuming that there is sufficient overspray.

4.2 Determining Index Widths on Extruded Sheets

To lift the index width selection framework from planar surfaces to non-planar surfaces, we first consider a special class of target surfaces – extruded sheets. Extruded surfaces have zero Gaussian curvature, and are non-planar in general. Many automobile surfaces, such as doors, are designed based on extruded surfaces, making this a useful class of surfaces.

Although different choices of start curve orientation are available on the extruded surface as described in Section 2.3, in this section we assume that the passes are along the zero curvature direction (see Fig. 6). Thus, the surface curvature orthogonal to each pass varies from pass to pass. In general, the paint profiles along any two passes on an extruded surface are different due to the variation in surface curvature. As a result, the optimal spacing between passes varies as surface curvature changes.

Our path planning approach generates new passes by offsetting known passes as described in Algorithm 1. At a given instance of time, we term the known pass we are offsetting the *current pass.* We then number all known passes relative to the current pass along the indexing curve that is orthogonal to the current pass. Let the known passes be denoted by $\{\mathcal{C}_{-m}, \ldots, \mathcal{C}_0\}$, where $\mathcal{C}_0$ is the current pass and $\mathcal{C}_{-m}$ is the mth known pass from the current pass. Similarly, we denote n number of *future passes*, whose locations are unknown, by $\{\mathcal{C}_1, \ldots, \mathcal{C}_n\}$. We term the first future pass, $\mathcal{C}_1$ as the *candidate pass.*

Let the index width between pass $\mathcal{C}_i$ and pass $\mathcal{C}_{i+1}$, measured along the indexing curve, be denoted by w_i. For m known passes, we define the ordered collection of known index widths as $\mathcal{K} = \{w_{-m+1}, w_{-m+2}, ..., w_{-1}\}$. Similarly, the ordered collection of potential index widths for the n unknown future passes is defined as $\mathcal{U} = \{w_0, w_1, ..., w_{n-1}\}$.

We are interested in generating the candidate pass, or equivalently determine the index width w_0 between the current pass and the candidate pass, such that the resultant paint deposition on the optimization profile is acceptably uniform. For the size of a typical deposition pattern of an ESRB atomizer, the paint deposition uniformity on the optimization profile is a function of the index widths of subsequently generated future passes, in addition to index widths of known passes.

We require that each potential index width belongs to index width search range $\mathcal{SR}$ as defined in Section 4.1. The ordered collection $\mathcal{W} \equiv \{\mathcal{K}, \mathcal{U}\}$ denotes index widths between all passes, known and future (unknown). For

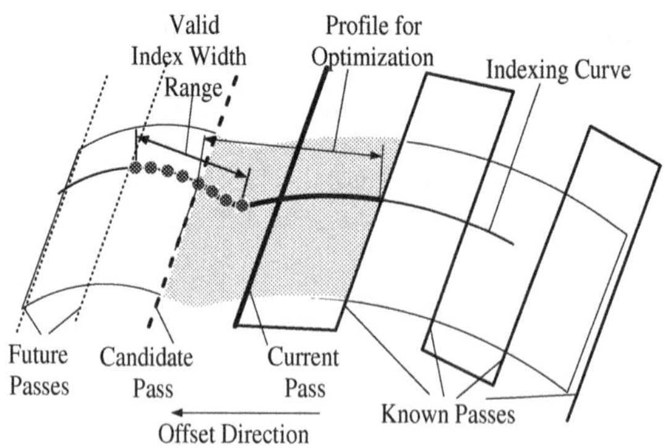

Fig. 6. Index width optimization on extruded sheets: the paint deposition uniformity is calculated along the indexing curve bounded between the previous pass and the candidate pass.

a given collection $\mathcal{W}$ of index widths between all passes, we evaluate the normalized standard deviation, $u(\mathcal{W})$, of the paint deposition on the segment of the indexing curve bounded between pass $\mathcal{C}_{-1}$ and the candidate pass $\mathcal{C}_1$. We term this segment of the indexing curve bounded between passes $\mathcal{C}_1$ and $\mathcal{C}_{-1}$ the *optimization profile.*

Our brute force approach varies the positions of the candidate and subsequently generated passes by using various collections of elements taken from $\mathcal{SR}$ to compute $\mathcal{U}$. For each resultant collection of index widths $\mathcal{W}$, we evaluate the uniformity $u(\mathcal{W})$ over the optimization profile. We define the set of feasible candidate index widths, $\mathcal{F}$, as

$$\mathcal{F} = \{w_0^{'} : \exists \mathcal{W} = \{w_{-m+1}, ..., w_0, ..., w_{n-1}\}, w_0^{'} = w_0,\ u(\mathcal{W}) \leq u_{max}\}, \quad (3)$$

where u_{max} is the maximum allowed normalized standard deviation. An index width in the feasible index set yields acceptable uniformity of paint deposition on the optimization profile for some set of future passes. In our implementation, for each value $w_0 \in \mathcal{SR}$, we vary the positions of future passes to find a set that yields acceptable uniformity on optimization profile. If such a set is found, we mark the index width w_0 as feasible. On the other hand, if the paint deposition uniformity over the optimization profile is not acceptable for any possible sets of future passes, the index width w_0 is marked as infeasible.

After the feasible index width set $\mathcal{F}$ is determined, we establish a cost function for each index width in the feasible set, as in the planar case. Here, the cost function not only penalizes the smaller index widths, but also the difference between the current index width and the previous index width (i.e., $w_0 - w_{-1}$). This additional cost component ensures an appropriate balance between the deposition on the region between the current pass and the previous pass, and the deposition on the region between the candidate pass and the candidate pass. Then, the feasible index width that yields the minimal cost is chosen as the optimum index width, and accordingly the new candidate offset pass is generated.

4.3 Determining Index Widths on Surfaces with Non-zero Curvature

On surfaces with non-zero Gaussian curvature, the curvature of the surface, in general, changes not only as we move along an indexing curve (as in extruded surface case), but also as we move along a given pass. As such, the geometry of the indexing curve changes as we move along a given pass. Here, in order to determine the offset of the current pass, we sample the current pass at a finite number of "marker" points spaced at intervals based on the total curvature of the pass. At each marker point, we then determine the *indexing curve* as the intersection curve of a plane that is orthogonal to both the surface and the tangent to the pass at that marker point. We determine the optimum index width along each indexing curve and obtain the corresponding offset marker point by tracing along the indexing curve a distance equal to the optimum index width from the current marker point. The offset curve is then determined by interpolating between the collection of offset marker points.

To determine the optimum index width at each marker point, we first approximate the surface locally with a surface of extrusion generated by

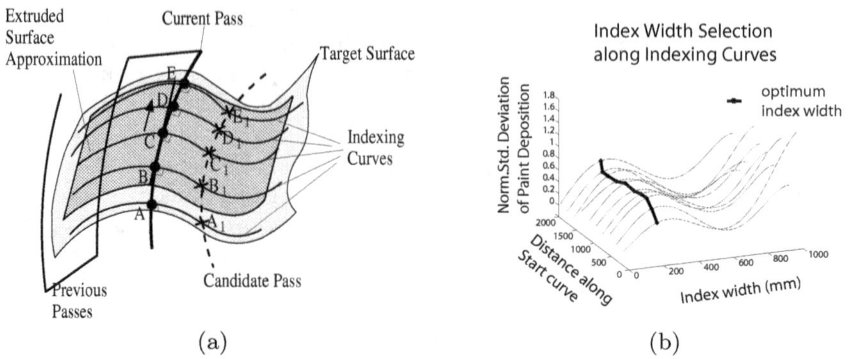

Fig. 7. Index width optimization on arbitrary surfaces: (a) at each of marker point A, B, C, D, and E, we approximate the surface with an extruded surface using the indexing curve, and determine the uniformity graph along the indexing curve. (b) The optimization procedure determines the optimum index width at each marker point. A_1, B_1, C_1, D_1 and E_1 are the corresponding offsets of the marker points.

extruding the indexing curve along the direction tangent to the pass at the marker point (see Fig. 7(a)). Employing the same approach used for the extruded sheets, we then determine the set of feasible index widths $\mathcal{F}^i$ at each marker point i. For l marker points, we construct *index sets* formed by a combination operation by picking a single element from each $\mathcal{F}^i$ at a time. That is, an index set is represented as $\mathcal{I} = \{w^1, w^2, ..., w^l\}$, where $w^i \in \mathcal{F}^i$. Note that each index set, formed by choosing a different combination of elements from each $\mathcal{F}^i$, represents a different offset curve.

For each index set, we assign a cost that penalizes smaller index widths in the set, change in the index width from previous passes to current passes, as well as the variation of index width along the current pass. The last component of the cost function, in some sense, attempts to minimize the geodesic curvature of the offset curve.

Selecting the weights for the cost function components that produce satisfactory results for a general class of surfaces depends on the relationships between the costs for non-uniformity, cycle time and computational time; and sometimes this choice may not be readily apparent. In our implementation, we were principally motivated to reduce the occurrence of the zig-zag structure on the offset passes that not only renders the motion of the real-world end-effector impractical but may also lead to self-intersections of subsequent offset curves. An obvious way to achieve this goal is to assign heavy penalties for variation in the index width along the current pass. However, for simplicity in our implementation, we used a hard constraint that the index widths at all marker points should be equal. In other words, we consider only those offset passes which maintain a constant spacing from the current pass at all marker points.

This restriction of optimizing over only those index sets whose elements have equal values helps us to: 1) select the cost function with relative ease, 2) comply with the "constant index" offset assumption in selecting the start curve as discussed in Section 2.3, and 3) reduce the computational cost for index width selection for l marker points and index resolution w_{res} from

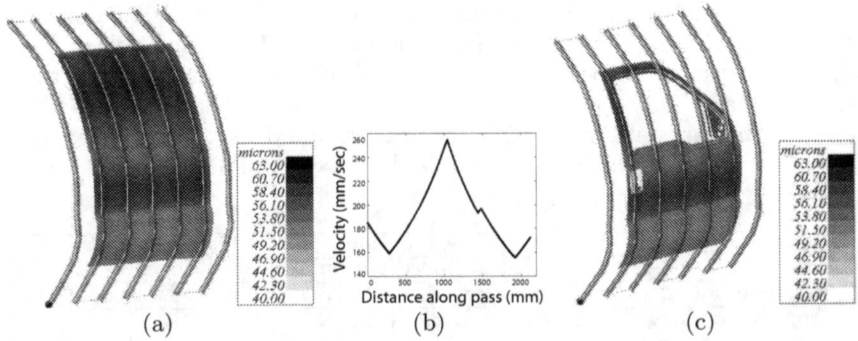

Fig. 8. (a) Simple surface approximation to the Ford Excursion door and the resultant paint deposition on the surface. The colormap shown is specifically designed to bring out the paint deposition variation. (b) A typical optimized speed profile along a spray gun pass. (c) Once the path on the approximation surface is generated, we simulate the paint deposition on the CAD model of a Ford Excursion door.

$O\left(\left(\frac{1}{w_{res}}\right)^{nl}\right)$ to $O\left(\left(l\,\frac{1}{w_{res}}\right)^{n}\right)$, where n (the number of future passes considered) is typically 4. We then select the optimum index set that minimizes the other two components of cost function from all possible constant-index sets (see Fig. 7(b)) and accordingly generate the candidate pass.

Note that although this "constant index" set solution that may not be globally optimal, it enables us to solve the index width selection problem satisfactorily within a practical amount of time. Moreover, the constant index set solution can serve as the starting point for a gradient-descent based method that considers all index sets for further improvement in uniformity.

5 Simulation and Experimental Results

To validate the utility of our coverage procedures, we generate spray gun trajectories using our planning tools on a variety of simple surfaces and study the effects on the resultant paint deposition uniformity. To evaluate paint deposition uniformity yielded by our generated trajectories, we simulate the paint deposition process on the corresponding surfaces. We also experimentally determine the paint uniformity on some real automobile surfaces. The surfaces we consider have varied geometric complexity including planar sheets, cylindrical surfaces, a door panel from a Ford Excursion and a fender from a Ford Crown Victoria. We model each of the two automobile surfaces by a slightly simplified single C^2-continuous NURBS surface, by removing holes and merging multiple NURBS patches together from the corresponding CAD data. We then generate paths on the corresponding approximation surfaces, and use the same paths for experimentally painting the surfaces and for simulating paint deposition on the surface CAD models (see Fig. 8).

5.1 Speed Optimization Results

To examine the improvement in the resultant paint deposition uniformity using speed optimized trajectories over constant speed trajectories, we evaluate the resultant paint uniformity in each case over a few surfaces and list the results in Table 1. We measure the uniformity in terms of normalized standard deviation of resultant paint deposition. The results show that in simulation, the speed optimization substantially improves the uniformity of resultant paint deposition. Also, the planar surface example shows that to produce similar uniformity, the overspray required by the speed optimized trajectories is shorter than that required by a constant speed trajectory.

Table 1. Speed optimization results

Surface	Simulation		Experiment	
	Const.Speed	Optimized	Const.Speed	Optimized
Excursion door	21.82 %	15.73 %	17.80 %	15.88 %
Approx. to Crown Victoria fender	16.62 %	11.92 %	-	-
Plane 100mm overspray	8.60 %	3.88 %	-	-
Plane 235mm overspray	3.88 %	-	-	-

The experimental data on the Ford Excursion door, however, shows that the speed optimized trajectory yields only a small improvement in the paint deposition uniformity over constant speed trajectory. Note that the speed optimization algorithm inherently relies on simulation of paint deposition to evaluate the objective function for a given speed profile. If there is a significant difference between simulated deposition and the experimental deposition, just as we observed, the effectiveness of speed optimization diminishes for the real-world application. More realistic deposition models will enable us to effectively use the speed optimization procedures in real systems.

5.2 Index Width Optimization Results

To study the effect of index width optimization on paint deposition uniformity, we consider a variety of target surfaces, generate passes for each of them and evaluate the paint deposition uniformity by simulating the paint deposition process (see Table 2). In each case, we set the maximum allowed normalized standard deviation at 4% and assume sufficient overspray to minimize the boundary effects. We observe that the index optimization yields the desired uniformity in cases where the surface curvature remains the same along the indexing curve. On the other hand, if the surface curvature changes along the indexing curve, the resultant paint deposition uniformity does not match the desired levels of uniformity. This is the case for Ford Excursion (horizontal passes) and the Ford Crown Victoria fender. The higher standard deviations observed on these surfaces are not unexpected because the index optimization process is, at best, semi-global in nature, and cannot globally guarantee the desired level of uniformity over the entire surface.

We also studied the effect of index optimization on paint uniformity experimentally by painting the middle and lower portion of a Ford Excursion door. During the experiments, our planning tools generated the index optimized path with a sideways overspray pass near the door bottom. In simulation, this path yielded a paint deposition uniformity of 7.16% on the door. However, because of robot workspace constraints, we had to manually remove the bottom overspray pass from the coverage path. For the coverage path without the overspray pass, the uniformity of resultant paint deposition was 13.13% in simulation, whereas experimentally it was measured to be 16.5%. If we had optimized the index width with the constraint that there would be no bottom overspray pass, the resultant paint deposition uniformity could be improved to 9.01% in simulation. Thus, our planning tools perform with limited success when there is no overspray pass. Nonetheless, our procedures yield desired uniformity results (when sufficient overspray is available) in a reasonable amount of time, and can offer guidelines to the paint specialists to substantially reduce the robot programming time.

6 Conclusion and Future Work

In this work, our approach examines the relationship between the variables and the output characteristics of the uniform coverage problem on simple yet realistic surfaces, and identifies the underlying key issues. We then decompose the coverage problem into three relatively independent sub-problems, thus significantly reducing the huge dimensionality of the search space for the coverage problem. This dimensionality reduction enables us to solve the coverage problem in a reasonable amount of time without over-trivializing assumptions about the deposition pattern or the surface geometry. Although we consider the deposition models only for ESRB atomizers, our coverage procedure is independent of the deposition model and can be used for any other type of deposition patterns.

The simulation results show that in presence of a sufficient overspray, the sub-problems of speed optimization along the passes and the index width optimization stay relatively independent. In the absence of sufficient overspray, deposition constraints in speed optimization problem may affect the index optimization results adversely, thus warranting a reformulation of the speed optimization and index width optimization problems. Our future work will address this reformulation of speed and index optimization subproblems, and will focus on developing path planning algorithms for broader class of surfaces, including surfaces with holes.

Table 2. Index Width Optimization Results

Target Surface	Norm. std. dev. (%)
Planar Sheet	2.64%
100cm Radius Cylinder passes parallel to axis	2.21%
100cm Radius Cylinder passes around cylinder	4.19%
Approximation to Ford Excursion Door Passes along zero curvature direction	6.81%
Approximation to Ford Excursion Door Passes along max. curvature direction	4.48%
Approx. to Crown Victoria fender	11.92%

Acknowledgements

This work was supported by the National Science Foundation and the Ford Motor Company through grant IIS-9987972. The authors gratefully acknowledge the input and support provided by Dr. Jake Braslaw and Mr. Jeff Petty, our collaborators at the Ford Motor Company. We also sincerely thank Mr. Tie Wang from the Ford Motor Company and Mr. Greg Marsh, Mr. Noel Gauci, Mr. Michael Bichel, Mr. Ron Mata and Mr. Dale Smith from the ABB for the help provided in carrying out the experiments at the ABB facility in Auburn Hills, Michigan, USA.

References

1. Suk-Hwan Suh, In-Kee Woo, and Sung-Kee Noh. Development of An Automated Trajectory Planning System (ATPS) for Spray Painting Robots. In *IEEE Int'l. Conf. on Robotics and Automation*, pages 1948–55, Sacremento, California, USA, April 1991.
2. M. A. Sahir and Tuna Balkan. Process Modeling, Simulation, and Paint Thickness Measurement for Robotic Spray Painting. *Journal of Robotic Systems*, Vol. 17(9):479–94, 2000.
3. Naoki Asakawa and Yoshimi Takeuchi. Teachless Spray-Painting of Sculptured Surface by an Industrial Robot. In *IEEE Int'l. Conf. on Robotics and Automation*, volume 3, pages 1875–79, Albuquerque, New Mexico, USA, April 1997.
4. Weihua Sheng, Ning Xi, Mumin Song, Yifan Chen, and Perry MacNeille. Automated CAD-Guided Robot Path Planning for Spray Painting of Compound Surfaces. In *IEEE/RSJ Int'l. Conf. on Intelligent Robots and Systems*, volume 3, pages 1918–23, 2000.
5. Wesley H. Huang. Optimal Line-sweep-based Decompositions for Coverage Algorithms. In *IEEE Int'l. Conf. on Robotics and Automation*, volume 1, pages 27–32, Seoul, Korea, May 2001.
6. Taejung Kim and S.E.Sarma;. Optimal Sweeping Paths on a 2-Manifold: A New Class of Optimization Problems Defined by Path Structures. *IEEE Transactions on Robotics and Automation*, 19(4):613–636, August 2003.
7. Tait S. Smith, Rida T. Farouki, Mohammad al Kandari, and Helmut Pottman. Optimal Slicing of free-form surfaces. *Computer Aided Geometric Design*, 19:43–64, 2002.
8. Ramanujam Ramabhadran and John K. Antonio. Fast Solution Techniques for a Class of Optimal Trajectory Planning Problems with Applications to Automated Spray Coating. *IEEE Transactions on Robotics and Automation*, Vol. 13(4):519–30, August 1997.
9. Eckhard Freund, Dirk Rokossa, and Jürgen Roßmann. Process-Oriented Approach to an Efficient Off-line Programming of Industrial Robots. In *IECON '98: Proceedings of the 24th Annual Conference of the IEEE Industrial Electronics Society*, volume 1, pages 208–13, 1998.
10. Heping Chen, Weihua Sheng, Ning Xi, Mumin Song, and Yifan Chen. Automated Robot Trajectory Planning for Spray Painting of Free-Form Surfaces in Automotive Manufacturing. In *IEEE Int'l. Conf. on Robotics and Automation*, volume 1, pages 450–55, Washington, D.C., USA, May 2002.
11. David C. Conner, Prasad N. Atkar, Alfred A. Rizzi, and Howie Choset. Development of Deposition Models for Paint Application on Surfaces Embedded in $\mathbb{R}^3$ for Use in Automated Path Planning. In *IEEE/RSJ Int'l. Conf. on Intelligent Robots and Systems*, volume 2, pages 1844–49, Lausanne, Switzerland, 2002.
12. David C. Conner, Prasad N. Atkar, Alfred A. Rizzi, and Howie Choset. Deposition Modeling for Paint Application on Surfaces Embedded in $\mathbb{R}^3$ for use in Automated Trajectory Planning. Technical Report CMU-RI-TR-02-08, Carnegie Mellon, Robotics Institute, Pittsburgh, Pennsylvania, USA, June 2002.
13. Prasad N. Atkar, Howie Choset, and Alfred A. Rizzi. Towards Optimal Coverage of 2-Dimensional Surfaces Embedded $\mathbb{R}^3$: Choice of Start Curve. In *IEEE/RSJ Int'l. Conf. on Intelligent Robots and Systems*, volume 3, pages 3581–87, Las Vegas,USA, 2003.
14. Thomas Rausch, Franz-Erich Wolter, and Oliver Sniehotta. Computation of Medial Curves on Surfaces. *The Mathematics of Surfaces VII*, Information Geometers, 1997.

Stealth Tracking of an Unpredictable Target among Obstacles

Tirthankar Bandyopadhyay[1], Yuanping Li[1], Marcelo H. Ang Jr.[1], and David Hsu[1,2]

[1] National University of Singapore, Singapore
[2] Singapore-MIT Alliance, Singapore

Abstract. This papers introduces the *stealth tracking* problem, in which a robot, equipped with visual sensors, tries to track a moving target among obstacles and, at the same time, remain hidden from the target. Obstacles impede both the tracker's motion and visibility, but on the other hand, provide hiding places for the tracker. Our proposed tracking algorithm uses only local information from the tracker's visual sensors and assumes no prior knowledge of target motion or a global map of the environment. It applies a local greedy strategy, and chooses the best move at each time step by minimizing the target's escape risk, subject to the stealth constraint. The algorithm is efficient, taking $O(n)$ time at each step, where n is the complexity of the tracker's visibility region. Simulation experiments show that the algorithm performs well in difficult environments.

1 Introduction

Target tracking is an important task for mobile robots. In this work, we investigate motion strategies for an autonomous mobile robot to track a moving target among obstacles and, at same time, remain hidden from the target. We call this problem *stealth tracking*. It has many applications. For example, in surveillance, a tracker follows a target to observe the target's action and may fail to acquire useful information or endanger itself if exposed. Other examples include observing wildlife and gaming. Stealth tracking is a common tracking behavior. In all these applications, a tracker must not only follow the target, but also avoid detection by the target.

More specifically, both the tracker and the target are equipped with visual sensors. The tracker has two objectives: *tracking*—keeping the target inside the field of view—and *stealth*—staying outside the target's field of view. Obstacles in the environment complicate tracking, as they introduce two types of constraints. Motion constraints prevent the tracker and the target from crossing the obstacles. Visibility constraints prevent the tracker and the target from seeing each other.

We assume that the tracker has no prior knowledge of target motion or a global map of the environment. This assumption is valid in many real-life situations. The lack of prior information limits the use of off-line pre-computation to plan the tracker's motion. The tracker must rely on its visual sensors to acquire information on the target and the local environment, and decide the necessary motion on-line.

Ideally the tracker always moves to keep the target in the middle of its visibility region so that the target cannot escape easily. At the same time, the tracker stays

M. Erdmann et al. (Eds.): Algorithmic Foundations of Robotics VI, STAR 17, pp. 43–58, 2005.

outside the target's visibility region to remain hidden. However, this is impossible to achieve, if the target has the same sensing capability as the tracker, *i.e.*, the visibility relationship is symmetric: if the tracker sees the target, the target sees the tracker. This apparent dilemma can be resolved by allowing some asymmetry. Although the tracker and the target have the same visual sensors, the visibility relationship may not be exactly symmetric. The target is initially unaware of the tracker and may not detect it even if the line of visibility between the target and the tracker is clear. So we assume that the tracker can operate safely near the boundary of the target's visibility region and quickly run outside if it detects risk of exposure. This provides the tracker the "slack" needed to achieve the dual objectives of tracking and stealth.

The stealth tracking problem introduced here is related to the more common tracking problem without the stealth requirement [8]. However, the stealth requirement makes the problem more difficult. We have already mentioned the conflict between the dual objectives of tracking and stealth. In addition, good tracking strategies may not work for stealth tracking at all. For example, in a star-shaped environment, without the stealth requirement, the tracker can simply stay in the middle of the environment, but this does not work for stealth tracking. Fortunately, as we will see, a suitable formulation of the problem translates the stealth requirement as a constraint on the motion of the tracker's motion. The resulting algorithm is almost as efficient as that for tracking without the stealth requirement.

In the following, Section 2 reviews previous work. Section 3 formulates the stealth tracking problem. Section 4 describes our tracking algorithm. Section 5 gives an efficient algorithm for computing the target's visibility region. Section 6 presents experimental results. Section 7 discusses limitations of our approach and possible solutions. Section 8 summarizes main results.

2 Previous Work

The objective of tracking is to find the motion for the tracker to keep the target within the tracker's field of view. Different aspects and variants of the problem have been studied in control theory [5], computer vision [10], and robotics [9]. In comparison, our goal here is to keep the target inside the field of view by moving the tracker actively to strategic locations under motion and visibility constraints.

The stealth tracking problem is related the work in [2,6,8,13,15,16], which studies tracking without the stealth requirement. The earlier work is based on two main approaches. The off-line approach assumes that the target's trajectory or behavior is known in advance. This allows one to derive stronger theoretical guarantees of the tracking algorithm [2,6,13]. The on-line approach assumes very little knowledge about the target's behavior. This is more realistic in many applications. Our work takes the second approach.

3 Problem Formulation

We assume that the robot tracker and the target operate in a planar environment $\mathcal{W}$ cluttered with obstacles (Figure 1*a*). The free space $\mathcal{F}$ is the subset of $\mathcal{W}$ not occu-

(a)

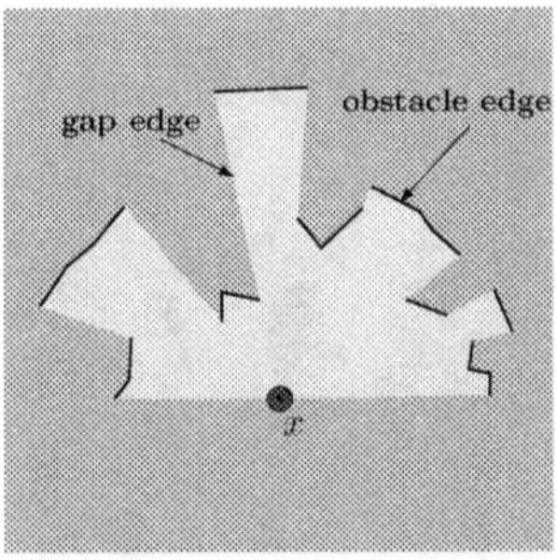

(b)

Fig. 1. A robot tracker operating in a planar environment. (*a*) A robot mounted with a laser range finder tracks another robot. Courtesy of H.H. González-Baños. (*b*) The visibility region from a sensor at x.

pied by obstacles. To simplify the presentation, assume for now that the each robot is a point in $\mathcal{W}$. The extension to the usual cylindrical robots is straightforward.

Both the robot tracker and the target have visual sensors. The sensor's field of view depends on the robot's position. Let $\mathcal{V}(x) \subset \mathcal{W}$ denote the visibility region of a robot located at x. The set $\mathcal{V}(x)$ contains all the points in $\mathcal{W}$ that have a direct line of sight to the sensor, unobstructed by the obstacles in $\mathcal{W}$. The visibility region can be modeled in various ways, depending on sensor capability. Here we assume that the sensor has an omni-directional field of view. Constraints, such as limited viewing angle and range, can be added if necessary. On real robots, the visibility region is obtained by processing sensor data. In simulation, the visibility region can be computed with a rotational plane sweep algorithm [3]. Either way, after the processing, we represent the visibility region as a polygon. The boundary of the visibility polygon consists of two types of edges: *obstacle* edges and *gap* edges (Figure 1*b*). Obstacle edges are part of obstacle boundaries. They directly block the sensor's line of sight. Gap edges result from occlusion of the line of sight elsewhere.

The tracker has two objectives, tracking and stealth. Let x_{s} and x_{t} denote the positions of the tracker and the target, and let $\mathcal{V}_{\mathrm{s}}$ and $\mathcal{V}_{\mathrm{t}}$ be the visibility regions of the tracker at x_{s} and the target at x_{t}, respectively. To track the target, the tracker must keep the target inside the field of view, meaning $x_{\mathrm{t}} \in \mathcal{V}_{\mathrm{s}}$. To maintain stealth, the tracker must stay outside the target's field of view, meaning $x_{\mathrm{s}} \notin \mathcal{V}_{\mathrm{t}}$. If the visual sensor of the tracker is stronger than that of the target, *e.g.*, by having a longer range, then the tracker's task becomes easier. It can stay at a distance from the target to remain outside $\mathcal{V}_{\mathrm{t}}$ while keeping the target inside $\mathcal{V}_{\mathrm{s}}$ at the same time. We assume, however, that the tracker and the target have similar sensors. In this case, the visibility relationship seems symmetric: $x_{\mathrm{t}} \in \mathcal{V}_{\mathrm{s}}$ if and only if $x_{\mathrm{s}} \in \mathcal{V}_{\mathrm{t}}$. Our two objectives are then contradictory and cannot be achieved simultaneously.

The visibility relationship, however, is not exactly symmetric. The target is initially unaware of the tracker and may not detect the tracker even if the tracker is inside the visibility region $\mathcal{V}_{\mathrm{t}}$. So we allow the tracker to operate in a region within a distance L of some gap edge of $\mathcal{V}_{\mathrm{t}}$ (Figure 2). Since there is an upper

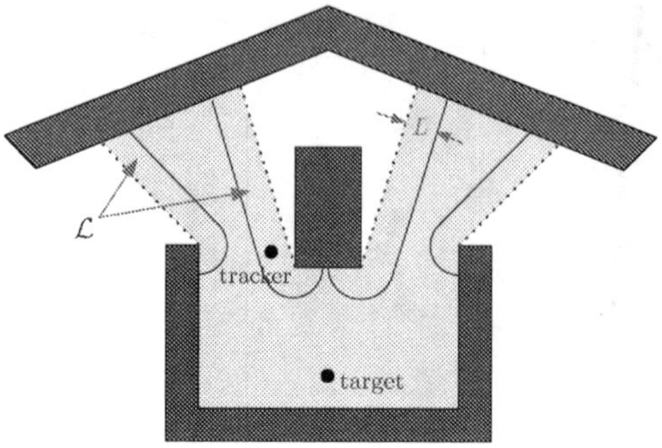

Fig. 2. The lookout region $\mathcal{L}$.

bound on the tracker's maximum velocity, the threshold L reflects the tracker's estimate on the amount of time during which it can remain safely in $\mathcal{V}_t$ without detection, and quickly run out of $\mathcal{V}_t$ via a gap edge if it detects risk of exposure.

More precisely, let the distance from a point $p \in \mathcal{V}_t$ to a gap edge g be the minimum Euclidean distance between p and all points in g that are visible to p via an unobstructed line of sight. We define the *lookout* region of a gap edge g as

$$\mathcal{L}_g = \{p \in \mathcal{V}_t \mid \text{The distance between } p \text{ and } g \text{ is less than } L\}.$$

The total lookout region $\mathcal{L}$ is then the union of $\mathcal{L}_g$ over all gap edges of $\mathcal{V}_t$.

Now the objective of the tracker is to move in $\mathcal{L}$ so that the target remains in the field of view. The motion of the robot tracker and the target is described with a simple discretized model. Let $x(t)$ denote the position of a robot at time t. If it chooses a velocity $v(t)$ at time t, its new position $x(t+1)$ after a fixed time interval Δt is given by

$$x(t+1) = x(t) + v(t)\Delta t.$$

Here we implicitly assume that sensing occurs every Δt time. This discretized model is effective as long as Δt remains small. As we will see, our tracking algorithm is very efficient. Based on the experience of previous work [8], we expect it to run at the rate of 10 Hz, sufficient for keeping Δt small in many common tasks.

There is a maximum bound V on the robot's velocity, but no other constraint. So the robot can reach anywhere inside a circle with radius center $x(t)$ and $V\Delta t$, unless it is obstructed by obstacles. Let V_s and V_t be the maximum velocity bounds of the tracker and the target, respectively. The bounds may be different, but we must have $V_s \geq V_t$. Otherwise, the target can easily escape by simply running straight ahead with maximum velocity, and the tracking problem is uninteresting.

We are now ready to state the problem formally:

Problem 1. A stealth tracking algorithm must compute a sequence of actions—in this case, the velocities $v(t), t = 0, 1, \ldots, T$—for the tracker so that at any time t, $x_t \in \mathcal{V}_s$ and $x_s \in \mathcal{L}(x_t)$, where $\mathcal{L}(x_t)$ is the lookout region for a target at x_t.

For some environments, losing the target is unavoidable. We may then try to keep the target in sight for as long as possible and maximize the target's escape time. In addition, it may be advantageous to choose actions to optimize certain criteria, for example, minimizing the distance travelled by the tracker or minimizing the distance between the tracker and the target.

Our formulation assumes that the target is unpredictable. The tracker does not know the target motion in advance, and does not have a prior map of the environment. It knows only $\mathcal{V}_s$, the part of the environment visible to its sensor. It also knows the position of the target in $\mathcal{V}_s$ and thus can compute the subset of the target's visibility region within $\mathcal{V}_s$. These assumptions certainly limit the tracker's ability to make the best decisions. It must rely on purely local information and choose the necessary motion in an on-line fashion. However, the assumptions are realistic in many applications. They may also simplify the problem in some respects: if the tracker does not rely on a global map, it does not need to localize with respect to the map, a difficult problem in itself [17].

Our formulation also assumes that the tracker can estimate the risk of exposure and take evasive action if necessary. In some applications, the target has directional sensors and must turn to look in the direction of the tracker. Upon observing this, the tracker can then take evasive action. If the target has omni-directional sensors, the best action for the tracker is nevertheless to stay close the boundary of the target's visibility region, but it is more difficult to decide when to take the evasive action. One possible solution is to learn a model of the target's typical movement patterns. Deviation from the model may indicate that the target is trying to look for the tracker. This is an interesting problem in itself, but is beyond the scope of this paper.

4 The Stealth Tracking Algorithm

Our stealth tracking algorithm uses a local greedy strategy. It defines a function that measures the risk for the target to escape. At each step, the tracker moves to a location that minimizes the risk function, subject to the stealth constraint.

4.1 The Risk Function

In our visibility model, the target can only escape from the tracker's visibility region through one of its gap edges. Let us focus on a single gap edge g for now (see Figure 3). If the target plans to escape through g, the best action is to move towards g along a straight-line path perpendicular to g. The length of this path gives the minimum distance d_{esc} that the target must travel to escape through g. Since the target's velocity is bounded, this also gives an estimate on the minimum time to escape. To prevent the target from exiting g, the tracker should move to maximize d_{esc}. If ℓ is the occlusion line containing g and w is the obstacle

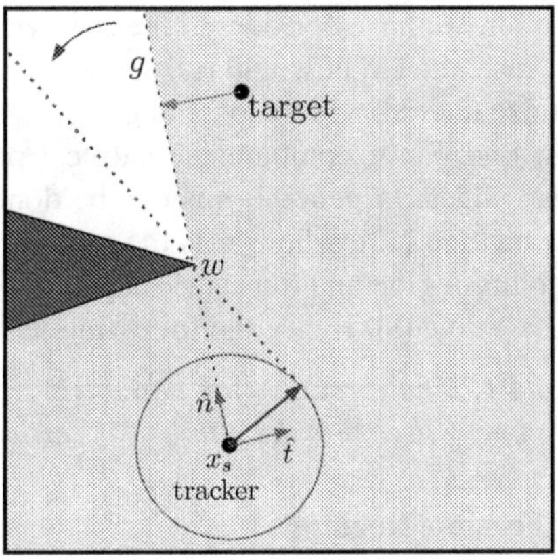

Fig. 3. The tracker's best move in one time step.

vertex abutting g, we can think of such a move as rotating ℓ about w so that the g is as far away from the target as possible. During a time step of length Δt, the tracker can only move inside a circle of radius $V\Delta t$, where V is the tracker's maximum velocity bound. Clearly the best move for the tracker is to go until ℓ is tangent to this circle (Figure 3). This move can be decomposed into two orthogonal directions $\hat{n}$ and $\hat{t}$, where $\hat{n}$ is a unit vector pointing from x_s to w and $\hat{t}$ is a unit vector orthogonal to $\hat{n}$. Both directions contribute to increased d_{esc}. Moving along $\hat{t}$ gives the maximum instantaneous increase of d_{esc}. Moving along $\hat{n}$ brings the tracker much closer to the pivot vertex w so that a future step along $\hat{t}$ gives even larger increase of d_{esc}. Intuitively $\hat{t}$ gives the direction to reduce current escape risk of the target, and $\hat{n}$ gives the direction to reduce future escape risk. These considerations lead to the risk function

$$f = (1/2)(r/e)^2 \tag{1}$$

where r is the distance between w and the tracker position x_s, and e is the target's minimum escape distance with respect to g. The risk f grows, as r increases and the tracker is further away from w, or as e decreases and the target is closer to the gap edge g. The gradient of f with respect to the tracker position x_s gives the direction to increase f at the maximum rate:

$$\nabla f = (r/e^2)\nabla r - (r^2/e^3)\nabla e. \tag{2}$$

The vectors ∇r and ∇e are oriented along $\hat{n}$ and $\hat{t}$, respectively. So the combined direction ∇f balances the current and the future escape risk. There are certainly alternative ways of combining r and e to form the risk function. They basically change the relative weight of ∇r and ∇e in ∇f. It is not clear which choice is the best. So we have decided to use the simplest one.

Given the tracker position and a gap edge, ∇f can be computed in constant time. To minimize f, the tracker follows the negated gradient $-\nabla f$.

So far, we have considered the risk with respect to a particular gap edge. In the worst case, the target would run towards the nearest gap edge $\hat{g}$. We must find $\hat{g}$ and minimize the risk with respect to $\hat{g}$.

To find $\hat{g}$, we compute the shortest path in $\mathcal{V}_s$ that goes from x_t to each of the gap edges. In general, this can be done in $O(n^2 \lg n)$ time using the visibility graph method [12], where n is the number of vertices in $\mathcal{V}_s$. However, since $\mathcal{V}_s$ is a visibility region and thus star-shaped, a better algorithm is available and achieves the $O(n)$ time [8]. Hence the following lemma.

Lemma 1. *The direction for minimizing the worst-case risk can be computed in $O(n)$ time, where n is the number of vertices in the tracker's visibility region $\mathcal{V}_s$.*

4.2 Feasible Regions

Having determined the direction of the tracker's motion $-\nabla f$, we now consider how much the tracker should move along $-\nabla f$. Suppose that the tracker's current position is x_s. The tracker's new position $x_s{}'$ in one time step must satisfy three conditions:

- $x_s' \in R$, where R is a disc with radius $V\Delta t$ and center x_s, because of the maximum velocity bound.
- $x_s' \in \mathcal{V}_s$. The new position must lie inside the tracker's current visibility region $\mathcal{V}_s$, because the tracker has only a local map from the sensor and has no information outside $\mathcal{V}_s$. This condition also guarantees that x_s' is collision-free, $x_s' \in \mathcal{F}$.
- $x_s' \in \mathcal{L}(x_t)$, where x_t is the target's current position. The tracker's new position must remain in the lookout region to maintain stealth.

So x_s' lies in the intersection $R \cap \mathcal{V}_s \cap \mathcal{L}(x_t)$, which we call the *feasible region*.

To obtain the feasible region, we first compute the intersection $I = \mathcal{V}_s \cap R$. Let $p_i, i = 1, 2, \ldots, n$ be the vertices of $\mathcal{V}_s$ in counter-clockwise order. Since $\mathcal{V}_s$ is star-shaped, we represent it as a list of triangles $\Delta x_s p_i p_{i+1}$ for $i = 1, 2, n - 1$. We then intersect each triangle with R and obtain a new convex shape in constant time. So I can be computed in $O(n)$ time.

The set I is basically the visibility region $\mathcal{V}_s$ clipped by the boundary circle of R. Usually the time step Δt is small. Thus R is also small and contains only a small constant number of obstacle vertices and edges, if any. By merging consecutive convex shapes in I whenever possible, the resulting I has only constant size.

Next, we compute the target's visibility region in order to get $\mathcal{L}(x_t)$. The tracker has no information outside of its own visibility region $\mathcal{V}_s$. It can only compute the subset of the target's visibility region within $\mathcal{V}_s$. Since there is no confusion, we use $\mathcal{V}_t$ to denote this subset from now on. Given x_t, there is a simple algorithm that computes $\mathcal{V}_t$ in in $O(n)$ time, because $\mathcal{V}_s$ is star-shaped. We defer the discussion of this algorithm until Section 5.

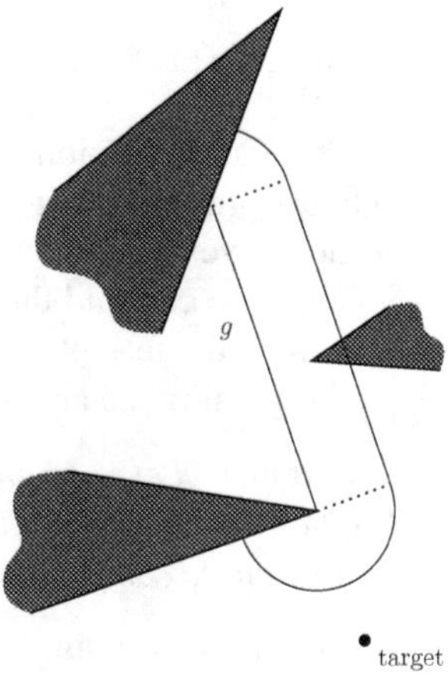

Fig. 4. Comparing $\mathcal{L}_g$ and $\mathcal{L}'_g$. The obstacle in the middle right is ignored in $\mathcal{L}'_g$.

For each gap edge g of $\mathcal{V}_t$, there is a lookout region $\mathcal{L}_g$, which may be quite complex if many obstacles are close together. We compute a simpler set $\mathcal{L}'_g$, which consists of a possibly clipped rectangle adjacent to g and two circular sectors that cap the rectangle (Figure 4). The width of the rectangle and the radius of the circular sectors are both L, which is the distance threshold for satisfying the stealth requirement. Compared with $\mathcal{L}_g$, the set $\mathcal{L}'_g$ ignores all the obstacles except the two at the end points of g. The set $\mathcal{L}'_g$ is sufficient for computing the feasible region, because $\mathcal{L}_g \cap I = \mathcal{L}'_g \cap I$. A point in the obstacle is certainly outside $\mathcal{V}_s \supset I$.

It is important to observe that $\mathcal{L}'_g$ can be represented as a union of at most three convex shapes, each of which has constant size. Assuming I has constant size, we can intersect every convex shape in I with $\mathcal{L}'_g$ in constant time. There are at most $O(n)$ gap edges, and thus we can compute the feasible region in $O(n)$ time. The feasible region is represented as a list of convex shapes approximated by convex polygons, all having constant sizes. Again, if we assume that R is small enough, then R intersects only a constant number of lookout regions, and the feasible region has a constant size. We summarize the result in the lemma below.

Lemma 2. *The feasible region for the tracker's new position can be computed in $O(n)$ time, where n is the number of vertices in the tracker's visibility region $\mathcal{V}_s$. Assuming that the time step Δt is small enough, the feasible region has $O(1)$ size.*

4.3 Locally Optimal Action

To choose the best move, the tracker should minimize the risk f over the feasible region. Consider the linear approximation of f at x_s:

$$f(x_s + \Delta x) \approx f(x_s) + \nabla f(x_s) \cdot \Delta x. \tag{3}$$

Minimizing function (3) is equivalent to

$$\min_{\Delta x} \nabla f(x_s) \cdot \Delta x \qquad \text{subject to } \Delta x \text{ lying in the feasible region.}$$

As the feasible region consists of a list of convex polygons, the problem reduces to linear programming [3]. The minimum solution $\Delta\hat{x}$ must lie at the vertex of one of the convex polygons. By projecting all the vertices in the feasible region along $\nabla f(x_s)$, we can find the minimum in $O(c)$ time, where c is the number of vertices describing the feasible region. We now set the tracker's new position $x_s{}' = x_s + \Delta\hat{x}$.

Combining the above result with Lemmas 1 and 2 gives the following theorem.

Theorem 1. *Let n be the number of vertices in the tracker's visibility region $\mathcal{V}_s$. The best action for the tracker to minimize the risk function f within the feasible region can be computed in $O(n)$ time at each time step.*

Without preprocessing, we cannot hope to achieve running time better than $O(n)$. However, as the tracker's visibility region changes at each time step, there is little opportunity for preprocessing.

4.4 Emergency Actions

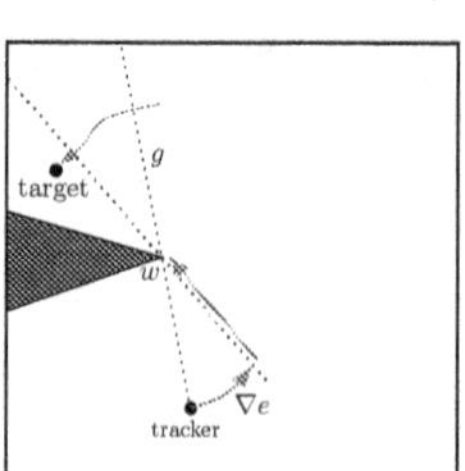

Fig. 5. Emergency action.

Our tracking algorithm also handles two emergency situations. In the first case, the tracker loses the target. Suppose that the target exits a gap edge g and that the obstacle vertex w abuts g (Figure 5). When the tracker detects that the target has exited g, it tries to rotate the occlusion line ℓ that contains g as fast as possible by moving along ∇e for a fixed number of steps, and hopes to regain the target quickly. If this fails, the tracker then goes directly towards w in order to eliminate the occlusion line ℓ. In the second case, the tracker is exposed. It computes a shortest path to each of the lookout regions and goes towards the nearest one.

5 Computing the Target's Visibility Region

In this section, we describe how to compute $\mathcal{V}_t$, the subset of the target's visibility region inside the tracker's visibility region $\mathcal{V}_s$. Since $\mathcal{V}_s$ is a simple polygon, $\mathcal{V}_t$ can be computed in optimal $O(n)$ time [7,14,11]. However, a simpler algorithm is possible here, because $\mathcal{V}_s$ is star-shaped.

Our basic idea is to walk along the boundary of $\mathcal{V}_s$ and compute the vertices of $\mathcal{V}_t$ incrementally. To initialize, we use the line that goes through x_s and x_t to divide $\mathcal{V}_s$ into two halves, P and P' (see Figure 6). We now describe the algorithm for P, the left half.

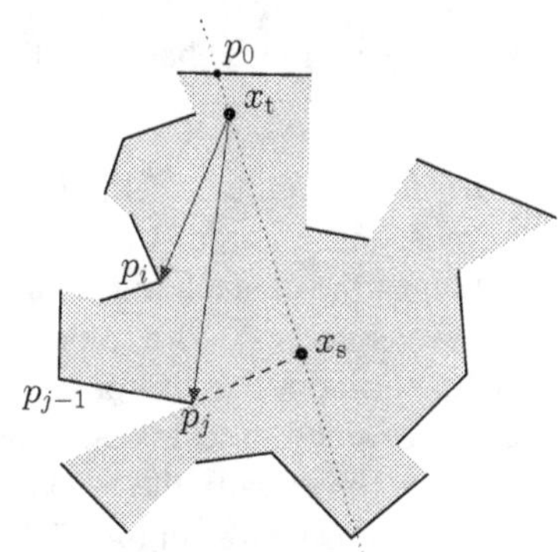

Fig. 6. Computing the target's visibility region.

Let p_0 the point where the ray $\overrightarrow{x_s x_t}$ intersects the boundary of $\mathcal{V}_s$. We number the vertices $p_0, p_1, p_2, \ldots$ of P in counter-clockwise order, starting from p_0. Every vertex of P is visible to x_s, because P is a subset of the visibility region of x_s. The first vertex of the target's visibility region $\mathcal{V}_t$ is p_0, which is visible to x_t because x_t lies on the line segment $\overline{x_s p_0}$. Now we walk along the boundary of P and visit the vertices $p_1, p_2, \ldots$ in this order. Let p_i be the latest vertex of P that is visible to x_t. For every new vertex p_j encountered, where $j > i$, if the ray $\overrightarrow{x_t p_j}$ lies to the right of $\overrightarrow{x_t p_i}$, we simply move to p_{j+1}, because p_j must be blocked by edges adjacent to p_i. If $\overrightarrow{x_t p_j}$ lies to the left of $\overrightarrow{x_t p_i}$, we claim that p_j is visible to x_t. To see this, we have to show that no boundary edge of $\mathcal{V}_s$ intersects $\overline{x_t p_j}$. Suppose, for the purpose of contradiction, that some edges of P intersect $\overline{x_t p_j}$. Let e be the intersecting edge closest to x_t along $\overline{x_t p_j}$. If e belongs to the polygonal chain between p_0 and p_i, the chain must cross $\overline{x_t p_i}$. This is impossible, because p_i is visible to x_t. If e belongs to the polygonal chain between p_i and p_j, then some vertex on the chain (excluding p_i and p_j) must be visible to x_t. This contradicts the fact that p_i is the latest vertex visible to x_t. If e belongs to the polygonal chain after p_j, the chain must cross $\overline{x_s p_j}$. This is also impossible, because p_j is visible to x_s. Of course, none of the edges in P' can intersect $\overline{x_t p_i}$. Otherwise, they would block the visibility line through x_s and x_t. Hence, p_j is visible to x_t. We compute the intersection of the ray $\overrightarrow{x_t p_i}$ and the edge $\overline{p_{j-1} p_j}$ and add both the intersection point and p_j to $\mathcal{V}_t$ as new vertices of $\mathcal{V}_t$. We then move to p_{j+1} and continue until all vertices are visited.

The right half of $\mathcal{V}_s$ can be processed similarly, except that now, we walk along the boundary in clockwise order. We then merger the two halves of $\mathcal{V}_t$.

Our initialization step takes $O(n)$ time, where n is the number of vertices in $\mathcal{V}_s$. When walking along the boundary of $\mathcal{V}_s$, we encounter each vertex exactly once and process it in constant time. Hence the following theorem.

Theorem 2. *Let n be the number of vertices of the tracker's visibility region $\mathcal{V}_s$. Given a target position x_t, the visibility region of x_t within $\mathcal{V}_s$ can be computed in $O(n)$ time.*

6 Experimental Results

To test the effectiveness of our tracking algorithm, we implemented it and ran it with different environments in simulation. In Figures 7–11, we show five representative experiments to illustrate the behavior of the algorithm. In these figures, dark blue regions indicate obstacles. Red crosses mark the target's trajectory. Blue boxes mark the tracker's trajectory. Two circles mark the target's and the tracker's current positions. The tracker's visibility region is marked with thick blue lines. The target's visibility region is shaded in light red. The lookout region is marked with thin black lines.

Corridor (Figure 7) This example shows the tracker's behavior when the target turns round a corner in a corridor. Initially the tracker stays near the lower right corner of the obstacle inside a lookout region to maintain stealth (Figure 7*a*). When the target makes a turn, a new lookout region develops, and the line of sight pivots about the upper right corner of the obstacle. The tracker follows the gradient of the risk function to take advantage of this. It swings out to improve its visibility (Figure 7*b*). Finally the tracker goes towards the upper right corner and stays there to maintain stealth (Figure 7*c*).

The tracker's success in this environment depends on the target's turning radius. If the target makes a sharp turn, the line of sight may pivot too fast for the tracker to follow, and the tracker loses the target. In this case, the tracker executes the emergency action (Section 4.4) to try regaining the target.

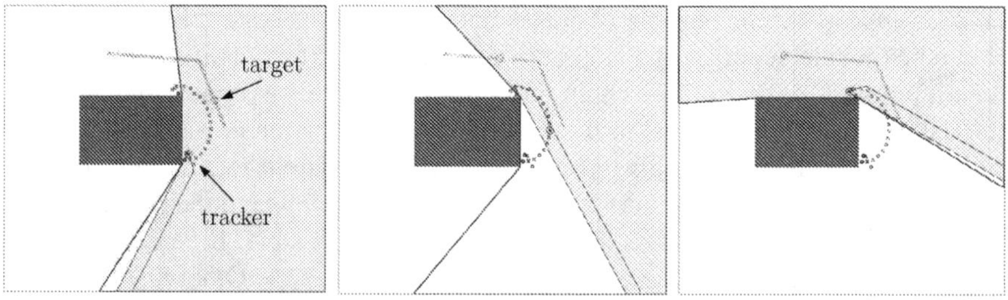

(*a*) (*b*) (*c*)

Fig. 7. The target turns around a corner.

Forest (Figure 8) Imagine a target going straight along a road passing through dense forests. If the tracker follows behind the target on the road, it risks exposure. The figures show the path that the tracker chooses when faced with such a situation. In general, the tracker stays on side of the road near the obstacles to avoid the risk of exposure. It tries to trail the target as closely as possible, given the constraint that it must stay inside the lookout region of some gap edge (Figure 8*a*). When the

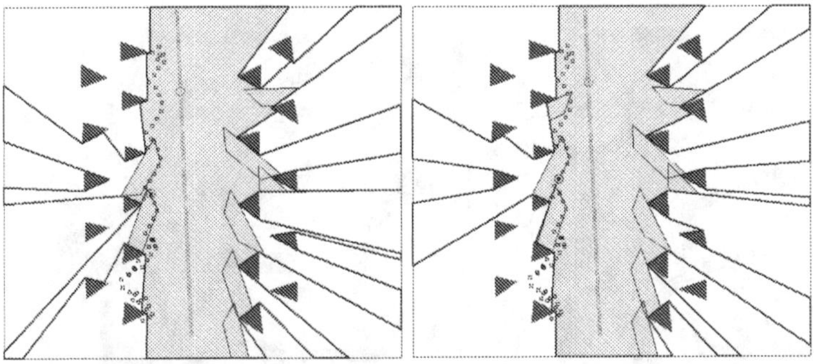

(*a*) (*b*)

Fig. 8. The tracker switches lookout regions.

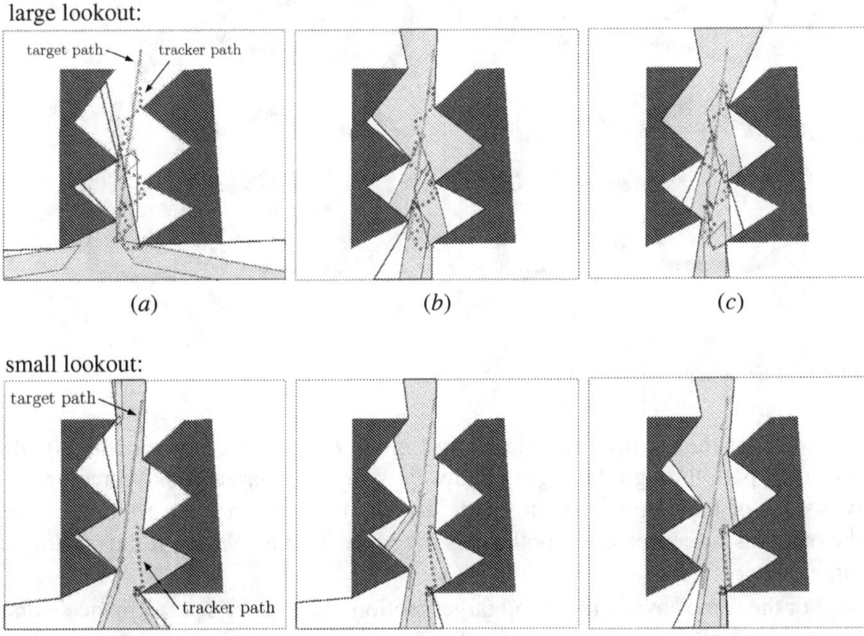

Fig. 9. The tracker's behavior changes due to different sizes of lookout regions.

lookout regions of two gap edges merge, the tracker immediately switches to the new lookout region to further reduce the risk function (Figure 8*b*).

Zigzag Pathway (Figure 9) The tracker's behavior and success may change drastically, depending on how much risk of exposure that it is willing to take. Recall that the distance threshold L controls the size of lookout regions and reflects the tracker's estimate or willingness to take risk of exposure. If L is large, the tracker has more room to maneuver, usually resulting in more successful tracking.

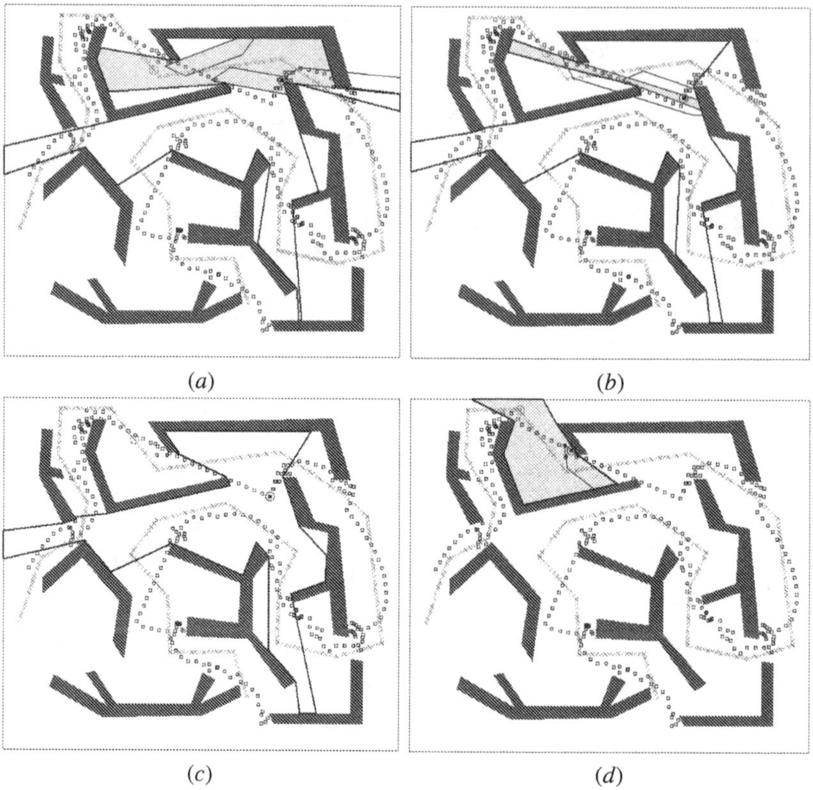

Fig. 10. Losing and regaining the target.

Consider the environment shown in Figure 9. The target moves roughly along a straight path through the zigzag pathway. When L is large, lookout regions from two sides of the pathway often merge and the tracker can jump from one side to the other, using obstacles on both sides for cover (Figures 9*a–c*). It thus follows the target very closely.

For the same environment and target motion, the tracker's performance worsens when L is small. It is unable to move from one side to the other, because the lookout regions are small and do not merge. It is stuck in one lookout region and has to wait for another lookout region to come close as a result of target motion (Figures 9*d–e*). However, by then, it falls behind the target by a large distance (Figure 9*f*). Although the tracker does not lose the target here, the target can escape easily by making a turn at the end of its straight-line motion. No simple emergency action can recover the target, because the distance between the target and the tracker is too large.

Maze (Figure 10) The geometry of this environment is more complex than the others. The target also takes along a long and winding path. The tracker is able to follow the target till the end, with the help of emergency actions. In Figure 10*a*–10*b*, the tracker moves almost entirely in the direction ∇e, because the target is

Table 1. Tracking performance.

Environment	Total No. Steps	No. Steps Visible	No. Times Lost & Regained
Corridor	102	102	0
Forest	214	214	0
Zigzag (large L)	130	130	0
Zigzag (small L)	130	130	0
Maze	730	663	8
City Blocks	468	466	2

very close to the gap edge and the current escape risk is high. At this moment, the tracker does not have the luxury to move along ∇r to reduce the future escape risk. Unfortunately the tracker still loses the target (Figure 10*c*). Taking the emergency action (see Section 4.4), the tracker runs to the vertex that abuts the gap edge from which that the target has escaped and regains the target (Figure 10*d*). The tracker loses the target eight times during the tracking, but every time it regains the target after taking the emergency action.

City Blocks (Figure 11) This environment resembles urban areas with roads and housing blocks and gives another example with many obstacles. The target takes a long path, but the tracker successfully follows the target till the end, with a little help from emergency actions.

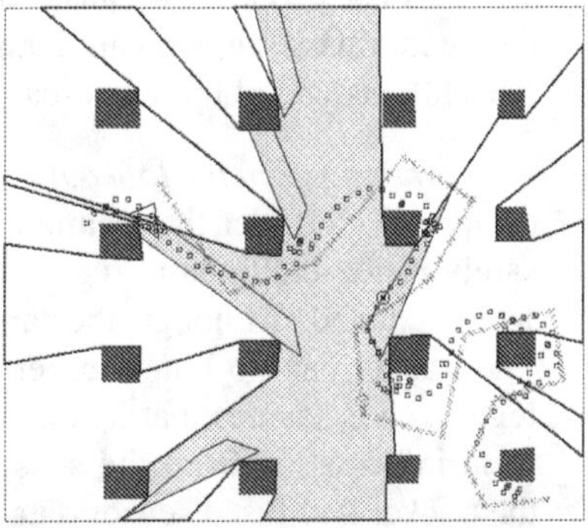

Fig. 11. Tracking a target among many obstacles.

Table 1 shows the performance statistics of the tracker in the above environments. Column 2 lists the total time length of target motion. Column 3 lists the length for which the target is visible to the tracker. Column 4 shows how many times that the target is lost and regained. From the table, we see that the tracker loses the target in the two environments where there are many obstacles. In the City Blocks environment, the tracker loses the target two times, for a duration of one step each. Every time that the tracker loses the target, it recovers it in the next step by taking the emergency action. In the Maze environment, which is more complex, the tracker loses the target eight times. It always recovers the target, but it takes more time than that needed in City Block. When there are many obstacles causing occlusions, the tracker is more likely to lose the target. The emergency action, despite it simplicity, seems quite effective in such environments.

7 Discussion

We now discuss some limitations of our approach and possible solutions.

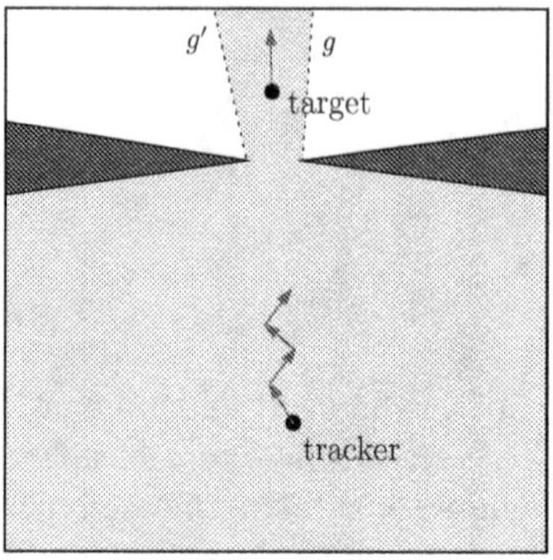

Fig. 12. Changes in the escape edge causing the robot tracker to follow an oscillatory path.

Discontinuity in the risk function. Our tracking algorithm currently minimizes the worst-case risk. The risk function f is defined with respect to the gap edge closest to the target, which we call the *escape edge*. The escape edge may suddenly change, leading to a discontinuity in f. Figure 12 shows an example. The target is moving forward about equally distant to two gap edges g and g'. Initially g is the escape edge. To minimize the escape risk with respect to g, the tracker takes a step to the left. This step increases the target's distance to g, but reduces the target's distance to g'. As the result, g' becomes the escape edge. Now the tracker takes a step to the right, which reduces the distance to g', but increases the distance to g. The escape edge switches back to g. This causes the robot tracker to oscillate and is undesirable. To deal with this issue, we can define the overall risk as the average of all gap edges' risks [8], instead of the worst-case risk.

"Discontinuity" in the lookout regions. Sometimes a lookout region may suddenly disappear. Consider the example shown in Figure 13. Initially the tracker is at x_s, safely inside the lookout region of a gap edge g. As it moves to the new position, $\mathcal{V}_s$ is reduced. Although the target has not moved at all, g now lies outside $\mathcal{V}_s$. The tracker can no longer ascertain the existence of g and the associated lookout region, and assumes that it is exposed. This happens, because the tracker uses only local information from the sensor and knows only the environment within $\mathcal{V}_s$. If the tracker had fused sensor data through history to produce a global map, it would know that it is inside the lookout region of g. The use of only local information is clearly a limitation here.

This situation occurs only if the tracker is close to the obstacle vertex w that supports the line ℓ containing the gap edge g (Figure 13). If the tracker crosses ℓ, the gap edge disappears from $\mathcal{V}_s$. To avoid this situation, we truncate the lookout region $\mathcal{L}_g$ so that the tracker does not cross ℓ.

Nonholonomic constraints. Our current model of the tracker's motion is very simple. Often the tracker robot is wheeled and is subject to nonholonomic or dynamic constraints. To deal with these additional constraints, we can use a control system to describe the robot's motion [1,4]. We no longer use the disc R for computing the feasible region, but replace it with a reachable set, the set of points reachable from the tracker's current position within time Δt according to the motion model. In fact, the disc R is simply the reachable set for our simple motion model. We

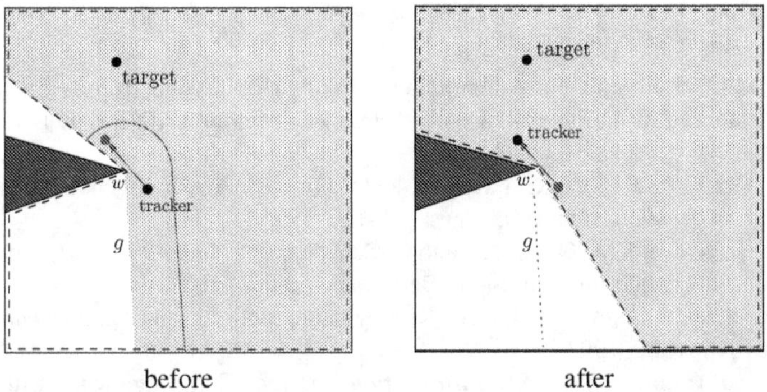

Fig. 13. The target's visibility regions before and after the tracker's move. The dashed lines indicate $\mathcal{V}_s$. The shaded region indicates $\mathcal{V}_t$.

can pre-compute the reachable set and approximate it using convex polygons. The computation of the feasible region can then proceed as before.

8 Conclusion and Future Work

We have introduced the stealth tracking problem, in which a robot tries to track a moving target among obstacles and remain hidden from the target at the same time. Our tracking algorithm uses only local information from the tracker's visual sensors and assumes no prior knowledge of target motion or a global map of the environment. It defines a function that measures the target's escape risk and tries to minimize the risk function, subject to the stealth constraint, in order to achieve the dual objectives of tracking and stealth. The algorithm is efficient, taking $O(n)$ time at each time step, where n is the complexity of the tracker's visibility region. Experiments in simulation show that the algorithm performs well in difficult environments.

For future work, we plan to test our algorithm on real robots to further validate the approach. We also plan to introduce a limited amount of global information to improve tracking performance, due to the limitations of purely local information, as we have discussed in Section 7. With global information, more sophisticated tracking strategies can be applied. For example, instead of following a target in a long corridor, which leads to high risk of exposure, the tracker can simply move to the other end of the corridor and wait there for the target. Global information also allows us to use multiple trackers to further improve performance.

Acknowledgements We thank Ho-Lun Cheng and Jean-Claude Latombe for helpful discussions and thank Hector González-Baños for providing the image in Figure 1.

References

1. J. Barraquand and J.C. Latombe. Nonholonomic multibody mobile robots: Controllability and motion planning in the presence of obstacles. *Algorithmica*, 10(2-4):121–155, 1993.
2. C. Becker, H.H. González-Baños, J.C. Latombe, and C. Tomasi. An intelligent observer. In *Int. Symp. on Experimental Robotics*, pages 153–160, 1995.
3. M. de Berg, M. van Kreveld, M. Overmars, and O. Schwarzkopf. *Computational Geometry: Algorithms and Applications*. Springer, Berlin, 2nd edition, 2000.
4. B.R. Donald, P. Xavier, J. Canny, and J. Reif. Kinodynamic motion planning. *Journal of the ACM*, 40(5):1048–1066, 1993.
5. K. Donald. *Optimal Control Theory: An Introduction*. Prentice Hall, Englewood Cliffs, NJ, 1970.
6. A. Efrat, H.H. González-Baños, S.G. Kobourov, and L. Palaniappan. Optimal strategies to track and capture a predictable target. In *Proc. IEEE Int. Conf. on Robotics & Automation*, pages 3789–3796, 2003.
7. H. ElGindy and D. Avis. A linear algorithm for computing the visibility of polygon from a point. *J. Algorithms*, 2:186–197, 1981.
8. H.H. González-Baños, C.Y. Lee, and J.C. Latombe. Real-time combinatorial tracking of a target moving unpredictably among obstacles. In *Proc. IEEE Int. Conf. on Robotics & Automation*, pages 1683–1690, 2002.
9. S.A. Hutchinson, G.D. Hager, and P.I. Corke. A tutorial on visual servo control. *IEEE Trans. on Robotics & Automation*, 12(5):651–670, 1996.
10. M. Isard and A. Blake. Condensation—conditional density propagation for visual tracking. *Int. J. Computer Vision*, 29(1):5–28, 1998.
11. B. Joe and R.B. Simposon. Corrections to Lee's visibility polygon algorithm. *BIT*, 27:458–473, 1987.
12. J.C. Latombe. *Robot Motion Planning*. Kluwer Academic Publishers, Boston, MA, 1991.
13. S.M. LaValle, H.H. González-Baños, C. Becker, and J.C. Latombe. Motion strategies for maintaining visibility of a moving target. In *Proc. IEEE Int. Conf. on Robotics & Automation*, pages 731–736, 1997.
14. D.T. Lee. Visibility of a simple polygon. *Computer Vision, Graphics, & Image Processing*, 22:207–221, 1983.
15. T.Y. Li and T.H. Yu. Planning tracking motions for an intelligent virtual camera. In *Proc. IEEE Int. Conf. on Robotics & Automation*, pages 1353–1358, 1999.
16. R. Murrieta-Cid, González-Baños, and B. Tovar. A reactive motion planner to maintain visibility of unpredictable targets. In *Proc. IEEE Int. Conf. on Robotics & Automation*, pages 4242–4248, 2002.
17. S. Thrun. Probabilistic algorithms in robotics. *AI Magazine*, 21(4):93–109, 2000.

Multi-Step Motion Planning for Free-Climbing Robots

Tim Bretl[1], Sanjay Lall[1], Jean-Claude Latombe[2], and Stephen Rock[1]

[1] Department of Aeronautics and Astronautics, Stanford University
{tbretl,lall,rock}@stanford.edu

[2] Computer Science Department, Stanford University
latombe@cs.stanford.edu

Abstract. This paper studies non-gaited, multi-step motion planning, to enable limbed robots to free-climb vertical rock. The application of a multi-step planner to a real free-climbing robot is described. This planner processes each of the many underlying one-step motion queries using an incremental, sample-based technique. However, experimental results point toward a better approach, incorporating the ability to detect when one-step motions are infeasible (i.e., to prove disconnection). Current work on a general method for doing this, based on recent advances in computational real algebra, is also presented.

1 Introduction

1.1 Broad context

The Probabilistic-Roadmap (PRM) motion planning approach [19], or one of its variants, is widely used for planning paths through high-dimensional configuration spaces subject to multiple constraints. PRM planners, which capture connectivity by a network of local paths joining sampled configurations, can often construct feasible paths very quickly. However, the approach lacks a formal stopping criterion. Typically, planners are allowed to search for a fixed length of time, after which a motion query is declared infeasible. Because PRM planners often have a fast convergence rate (i.e., the probability that they find a path when one exists goes quickly to 1), this heuristic works fine for processing a small number of queries. But imagine we want to process many queries, each in a different configuration space subject to different constraints. Then, a fundamental question arises: how much time should be spent on each query? Too much, and time is wasted on infeasible queries; too little, and feasible queries – which could be critical – may be falsely declared infeasible.

Multi-step planning for climbing robots (as well as general multi-limbed locomotion and manipulation planning) presents exactly this challenge. By applying a basic multi-step planner to a real climbing robot (Section 4), we establish the importance of incorporating a "disconnection planner" that rejects most infeasible queries. A longer time could then be allowed for remaining queries (knowing that most feasible problems are solved quickly)

M. Erdmann et al. (Eds.): Algorithmic Foundations of Robotics VI, STAR 17, pp. 59–74, 2005.

and a much smaller amount of time would be wasted on infeasible queries. A portion of this paper (Section 5) describes ongoing work on this topic – at this stage, we still do not have a completely satisfactory solution.

1.2 The climbing problem

Our goal is to enable a wide class of multi-limbed robots to climb vertical rock using only natural features and friction for upward progress (i.e., to *free-climb*). Such robots need not incorporate special fixtures or tools, nor need they be designed specifically for climbing, as have previous climbing robots (e.g., [2,3,11,16,17,25]). Their flexibility could benefit several application areas, including search-and-rescue, surveillance, and planetary exploration.

We consider robots with a small number of articulated limbs. We do not distinguish between limbs, and call the end-point of each one a *hand*. To climb vertical terrain, a robot must go through a continuous sequence of configurations satisfying certain constraints (e.g., equilibrium, collision, joint-torque limits). At each configuration, some of the robot's hands are in contact with the terrain – a surface with small, arbitrarily distributed features (e.g., protrusions or holes) called *holds*. During a *one-step* motion, the robot brings one hand to a new hold while using frictional contacts at other hands and internal degrees of freedom (DOF's) to maintain equilibrium. A *multi-step* motion is a sequence of one-step motions.

This paper presents a new framework for computing multi-step climbing moves. We view the multi-step motion space of the robot as a discrete graph. Corresponding to each association of hands to holds (i.e., each *stance*) is a set of feasible configurations (satisfying all constraints), that might have one or more components. Each node in the graph represents one of these components. Each edge is a one-step motion.

It is computationally practical neither to explicitly construct the feasible space at every stance, nor to exactly compute one-step motions to neighboring stances. Instead, we use an augmented PRM approach, based on our previously developed one-step planner [8,9]. Consequently, as described above, we are faced with the challenge of deciding how much time should be spent searching for each one-step motion.

2 Related Work

2.1 Multi-step planning

Multi-step motion planning for climbing robots is related to general multi-limbed locomotion and manipulation planning. Just as for climbing robots, multi-limbed locomotion is characterized by a sequence of continuous one-step movements in which feet are placed on or removed from the terrain. Likewise, manipulation paths are composed of sequences of one-step motions either of a grasped object and its manipulator or of the manipulator alone, respectively called "transfer" and "transit" motions [1].

Three main approaches to multi-step motion planning have been used:

Continuous approach. Here, the motion space is viewed as a single, high-dimensional, continuous space. The configuration of the robot moves on successive manifolds embedded in this space. Each manifold corresponds to a different set of contact constraints. This type of representation (called "stratified" [14,15] or "hybrid mechanical" [10]) is convenient for a controllability-based analysis. In particular, it leads to techniques that transform nominal paths to feasible, gaited trajectories, similar to methods used for smooth, nonholonomic systems [22,24]. However, the computational cost of these methods grows quickly with the dimensionality, geometric complexity, and number of embedded manifolds. They seem impractical for free-climbing robots.

Constraint relaxation. These methods generate an initial path considering only some of the constraints and iteratively refine this path into a feasible one by incorporating additional constraints [13]. Foot or finger placement concepts may be explicitly used in the construction of the paths. One technique proposed for manipulation planning is to first plan a trajectory for the grasped object ignoring manipulators, then generate manipulator trajectories to achieve necessary re-grasps [20]. A similar strategy for multi-limbed robots on uneven terrain is to first generate a feasible path for the center of mass (ignoring foot placement), then construct specific foot placement and limb motions that keep the center of mass stable along this path [12]. This approach can generate multi-step motions quickly for some systems. However, it is not guaranteed to find a path if one exists, and generally requires tuned heuristics to perform well. In addition, it is based on having some way of prioritizing the constraints. So, it does not handle interdependent constraints well (e.g., self-collision, equilibrium, and torque limits in climbing motions).

Foot-placement or grasp-based. These methods, like our own, carefully consider where a robot or manipulator should place its feet, hands, or fingers. Many are based on the framework for manipulation planning proposed by [1], which (as we do) views the multi-step motion space as a graph. This approach has two distinct advantages for systems such as free-climbing robots. First, by viewing the manifold associated with each set of contact constraints as a separate configuration space with its own lower-dimensional parameterization, the dimension of each one-step motion query is reduced. Second, by taking advantage of an intermediate discretization of the environment into specific foot/hand/finger placements, regions of useless contact locations are eliminated, again reducing overall problem complexity. On the other hand, the challenge of this approach is to efficiently compute the structure of the multi-step graph.

For specific systems (e.g., "spider-robots" either walking on horizontal terrain [7] or with fixed gaits in planar tunnels [29,32], and 3-limbed, planar, climbing robots subject to equilibrium constraints only [8]), it is possible to quickly compute the exact structure of this graph. For more general systems (including the climbing robots considered in this paper), it seems impractical to compute the structure of the multi-step motion space exactly. Heuristic search techniques have been used in cases where a predefined set of one-step moves is available [5,21]. If no such set has been predefined, continuous exploration is needed [26,30]. The "fuzzy PRM" approach of [26] is particularly interesting, since it seeks to address, statistically, the problem of where to allocate planning time over possible one-step motions. However, it still does not solve the fundamental problem of when to stop searching for a particular one-step motion.

2.2 Disconnection proofs

The idea of proving that two configurations are not path-connected is not new. This idea is central to the concept of "complete" or exact motion planning, in which a feasible connecting path is returned if and only if one exists [31]. But while work on exact motion planning has provided important insight and bounds on problem complexity, it has not been practical for real applications. Instead, approximate methods such as PRM [19] are used to construct feasible paths quickly, whenever possible.

Recently, more attention has been paid to developing "disconnection" planners [4], which search for proofs that no path exists. Establishing this type of proof is difficult in general, since it involves searching for disconnecting (n-1)-D manifolds in n-D space, rather than for 1-D connecting paths. The work of [4] presents one approach, for a polyhedral robot moving among polyhedral obstacles in 3-D workspace, based on bounding volumes. The proposed technique requires knowledge of critical (i.e., narrow) regions of the workspace, and is not applicable to more general constraints. Another approach is to compute both under- and over-approximations to the connectivity of simpler sub-problems, and use these approximations to prove disconnection [18]. This strategy requires a smart method of decomposition and model-reduction, and the implementation of a complete planner.

Two ideas are common to both approaches. First, both seek to relax the original feasibility problem, so that conservative proofs of infeasibility are found in a simpler space. Second, both integrate the search for disconnection proofs with a search for connecting paths (loosely, this can be described as a primal-dual method). In Section 5, we describe a systematic method of problem relaxation and a general framework for integrating "primal" and "dual" search when an over-approximation to the feasible space can be expressed as a semialgebraic set.

3 Free-Climbing Robots

3.1 Description of robot

We have focused our work on a real free-climbing robot, LEMUR IIb, developed by the Mechanical and Robotic Technologies Group at the Jet Propulsion Laboratory [9]. This robot was designed to have a number of capabilities in addition to climbing (e.g., assembly, inspection, maintenance, transport, intervention) and to be able to traverse a variety of other types of terrain (e.g., roads, talus, dirt, urban rubble).

LEMUR IIb consists of four identical limbs attached to a circular chassis, with a total mass of 7 kg (Fig. 1). Each limb contains three revolute joints, providing two in-plane (yaw) and one out-of-plane (pitch) DOF's. Each joint has an identical drive-train, capable of a maximum continuous torque of 5.0 N-m and a maximum speed of 45 deg/s. Each end-effector is a single peg wrapped in high-friction rubber. LEMUR IIb can be field-operated, with on-board batteries, processing, and sensors (including a swiveling stereo camera pair, a 6-axis force/torque sensor at each shoulder, a 3-axis accelerometer, and joint angle encoders).

The terrain climbed by LEMUR IIb in our tests is an indoor, near-vertical, planar surface. This surface is covered with small, artificial rock features exactly as are indoor "climbing gyms" for human climbers. In order to focus on the planning algorithm, the location and friction characteristics of each feature are identified manually and pre-surveyed. As a further simplification, in our current implementation we maintain LEMUR IIb's chassis parallel to and at a fixed distance from the climbing surface, and use the out-of-plane DOF in each limb only to make or break contact with features.

Finally, at this stage in our research, the robot's motion is slow enough to be assumed quasi-static.

Fig. 1. LEMUR IIb, a multi-use robot capable of free-climbing (Mechanical and Robotic Technologies Group, Jet Propulsion Laboratory) [9].

3.2 Model and notation

We call the robot's circular chassis the *pelvis*. In each limb, the first joint (nearest the pelvis) is called the *shoulder*, the second joint is called the *elbow*, and the third (out-of-plane) joint is called the *wrist*. Assuming that the pelvis moves at a fixed distance parallel to the wall, any configuration of the robot is defined by 15 parameters: the position/orientation (x_p, y_p, θ_p) of the pelvis and the joint angles $(\theta_1, \theta_2, \theta_3)$ of each limb.

Any point on the terrain is a potential contact – either on the contour of a continuous rock feature, or on the planar climbing surface. We assume that a discrete number of useful contacts have been identified; these points are called *holds*. For LEMUR IIb, all holds lie on an inclined plane but can have arbitrary orientation, so each is defined by a 2-D point (x_i, y_i) and a 3-D direction $\boldsymbol{\nu}_i$. The endpoint of each limb is a *hand*. Holds at which hands are in contact are the *supporting holds*. We assume that supporting holds are point contacts, where friction is modeled by Coulomb's law.

When climbing, LEMUR IIb always maintains either three or four supporting holds. The set of supporting holds is a *stance*, denoted σ – to differentiate between 3-hold and 4-hold stances, we write $\sigma 3$ and $\sigma 4$. The linkage between the supporting holds – containing the pelvis and either three or four limbs – is called the *contact chain*. When only three supporting holds are used, the fourth limb is the *free limb*.

Because of the closed-chain constraint, the robot's continuous motion with four supporting holds occurs on a 3-D manifold $C_{\sigma 4}$ in the robot's configuration space. With three supporting holds, motion occurs on a 6-D manifold $C_{\sigma 3}$. This motion is subject to four additional constraints: quasi-static equilibrium, joint angle limits, joint torque limits, and (self-)collision. The *feasible space* at a stance σ is the subset F_σ of C_σ satisfying each of these constraints. The limbs have non-negligible mass, so their motion affects the robot's equilibrium. If two points in F_σ are connected by a continuous path in F_σ, we say they are in the same *component* of F_σ. (Henceforth, a "continuous path" will always be taken to mean a continuous motion of the robot at some fixed stance, i.e., with fixed holds.)

3.3 Climbing motions

To climb upward, the robot must switch between 3-hold and 4-hold stances. Two stances $\sigma 3$ and $\sigma 4$ are *adjacent* if $\sigma 4 = \sigma 3 \cup \{i\}$ for some hold i. The robot can only switch between adjacent stances σ and σ' (i.e., place or remove a hand) at points $q_t \in F_\sigma \cap F_{\sigma'}$. We call such points *transition points*. Given a start configuration $q_s \in F_\sigma$ at a stance σ, we say that a component of the feasible space $F_{\sigma'}$ at an adjacent stance σ' is *reachable* if there is a continuous path connecting q_s to a transition point in that component. This path is a *one-step motion*. Examples of one-step motions are shown in Figs. 2 and 3.

Fig. 2. A one-step motion with a 4-hold stance, to remove the bottom right hand.

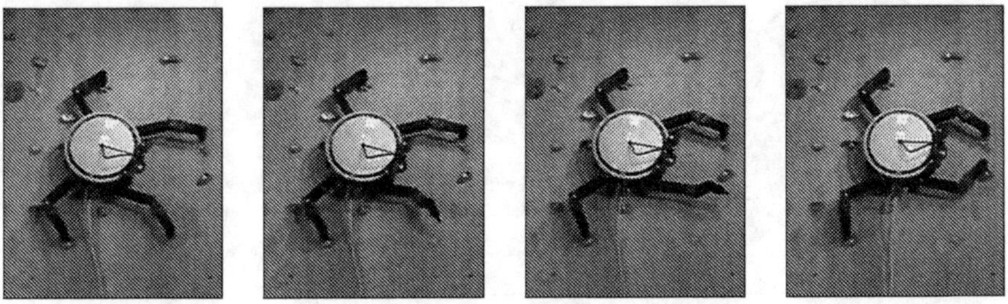

Fig. 3. A one-step motion with a 3-hold stance, to place the bottom right hand.

4 Multi-Step Planning

4.1 Basic concepts

Given a stance σ, a start configuration $q_s \in F_\sigma$, and a goal hold g, the multi-step planning problem is to construct a sequence of one-step motions that will bring the robot to a stance σ_g that contains g. (There may be a number of different valid stances σ_g.) This problem is a search through a graph. Each node in this graph is a component of F_σ at some stance σ. Each edge is a one-step motion. The sequence of one-step motions is not assumed to be periodic (i.e., a gait) – the order in which limbs are placed and removed from holds is arbitrary, and is selected as part of the graph search.

In practice, only local sensing will be available, so long multi-step plans will not be computed. Instead, short (4-10 one-step motions) plans will be generated repeatedly – with graph searches of high breadth but low depth – as the robot explores more of the terrain. However, in our current implementation the terrain has been pre-surveyed, and consequently our tests are for multi-step searches of lower breadth and higher depth.

The concept that nodes in the graph are, fundamentally, components of feasible spaces and not particular configurations is an important one. For example, one-step motions that take a configuration $q_s \in F_\sigma$ to two transition points q_{t_1}, q_{t_2} in the same component of an adjacent feasible space $F_{\sigma'}$ are equivalent. Keeping both one-step motions, and considering both points

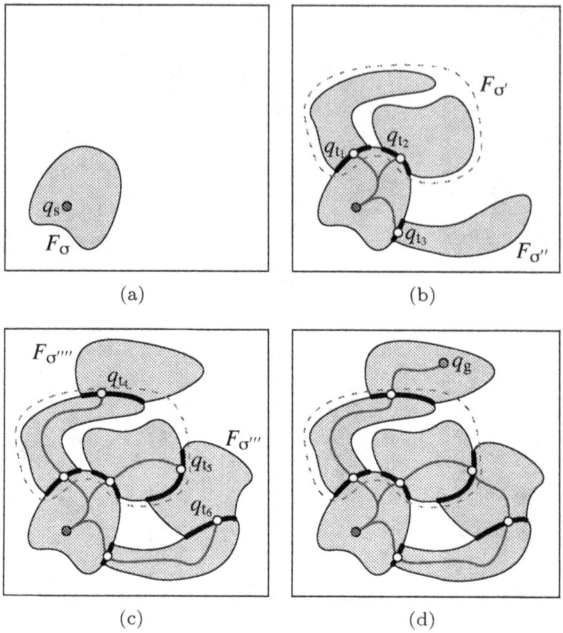

Fig. 4. Schematic of a multi-step search. Gray regions are feasible spaces at different stances (a dotted line shows the two components of $F_{\sigma'}$). Bold arcs are intersections between feasible spaces at adjacent stances. Thin lines are one-step motions.

q_{t_1} and q_{t_2} as distinct, adds redundant nodes and edges to the graph. This increases the overall planning time of a multi-step search. So although it might be convenient to index nodes by individual configurations and edges by point-to-point paths, these always represent entire components and the transitions between them.

Figure 4 shows a schematic of how a multi-step search might proceed. At each stage, transition points are found to adjacent stances at which components of the feasible space $F_{\sigma'}$ intersect the base feasible space F_{σ}. Note that the feasible space at stance σ' has two components, each of which admits one-step motions to different adjacent stances. Eventually, a path is constructed to a configuration q_g in the interior of the feasible space $F_{\sigma''''}$ at a stance containing the goal hold.

4.2 Consequences of an incremental sample-based approach

In general, the graph is much larger – and many more one-step queries must be made – than in the previous example. It is computationally impractical to determine exactly which adjacent stances are reachable from a start configuration $q_s \in F_{\sigma}$. Therefore, we use an approximate method: for each adjacent stance σ', first we sample transition points $q_t \in F_{\sigma} \cap F_{\sigma'}$, then we try to construct a continuous path from q_s to each q_t, using an augmented PRM planner [9]. Similar techniques are used in other multi-step planning algorithms (e.g., [21,26,30]).

However, this approach has two consequences. First, since the structure of each feasible space F_σ is never exactly known, redundant nodes and edges may be introduced. As noted in Section 4.1, this redundancy increases overall planning time. Second, since transition points are sampled and feasible paths are found only with some probability, nodes and edges may be missed. This reduces the overall probability of success and tends to increase the resulting path length (if a path is found).

4.3 Baseline results

Long multi-step paths. We applied our one-step planner to generate climbing motions for LEMUR IIb, which were subsequently executed by the real robot. In these initial tests, only one-step moves were computed by the planner. To get multi-step paths, the user specified which hold to grab or release at each step. Snapshots from one climb, taking the robot from bottom to top of the climbing surface, are shown in Fig. 5. The one-step planner generated, on-line, each of 88 one-step motion trajectories forming the multi-step path.

The distribution of planning times for this example is shown in Fig. 6. Using our sampling scheme, most one-step moves were planned very quickly (less than 0.5 s). However, several difficult moves took more time (more than 5.0 s), even as much as 17.3 s. This illustrates the important issue raised in Section 1.1. Many one-step motion queries are made in a multi-step search; how should we choose the amount of time T_{max} after which the one-step

Fig. 5. Snapshots of a multi-step climb.

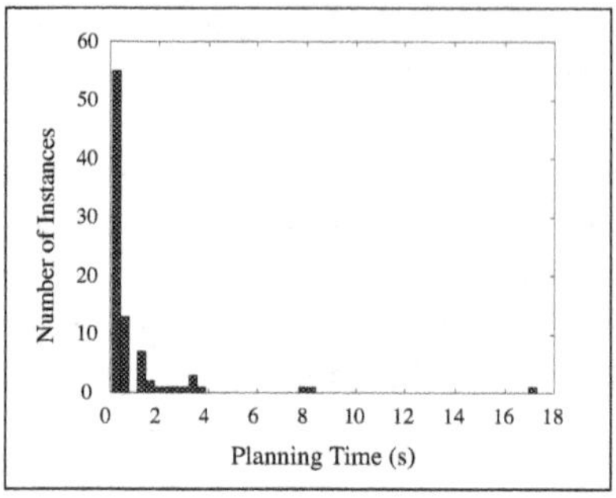

Fig. 6. The distribution of one-step planning times along a multi-step path consisting of 88 one-step motions. Minimum planning time was 0.09 s, maximum was 17.3 s. Mean planning time was 1.02 s, but over 75% of the one-step motions were computed quicker than average.

planner returns failure? In this example, suppose $T_{max} = 2.0\ s$ (twice the mean value) – infeasible one-step motions are rejected quickly, but several difficult moves (15% of the multi-step path) likely are not found. Alternatively, suppose $T_{max} = 20.0\ s$ (greater than the maximum value) – now, all feasible one-step motions likely are found, but every infeasible query takes ten times longer, drastically increasing total search time.

In this example, all one-step motions were computed along a single, user-defined, multi-step path. In a general search, many different multi-step paths may exist between start and goal stances. However, a "critical passage" in the multi-step graph is not necessarily a single critical edge (i.e., a single, difficult, one-step move). If the graph can be divided into two components – one containing the start node, the other containing the goal – such that each edge between these two components represents a one-step motion that is difficult to compute, then this group of edges represents a critical passage.

Performance of a basic multi-step planning algorithm. We have also implemented a fully automatic, multi-step planner for LEMUR IIb, incorporating the same one-step planning algorithm. This planner does not yet exploit search heuristics or other augmentations (e.g., our disconnection proof of Section 5). We have tested our planner in simulation, to understand the effects of varying the amount of time T_{max} it is allowed to search for each one-step motion. Figure 7 shows the results of one experiment. In this example, a goal hold was selected at a distance of eight one-step motions from the robot's initial stance. Results of multi-step planning were averaged over 100 runs at each T_{max}. (In our experiments, it was convenient to bound T_{max} implicitly, e.g., by bounding numbers of iterations – therefore, results are given with respect to mean planning time rather than T_{max} in Fig. 7.)

Both total path length and the ratio of edges to nodes in the graph asymptotically reach minimum and maximum values, respectively, as T_{max} is increased. This trend arises because the structure of the graph is approximated increasingly well (i.e., components of each F_σ are better explored, and feasible one-step motions are found with higher probability). It also indicates why total planning time approaches linear growth with T_{max} – the number of nodes and edges explored in the multi-step search becomes constant.

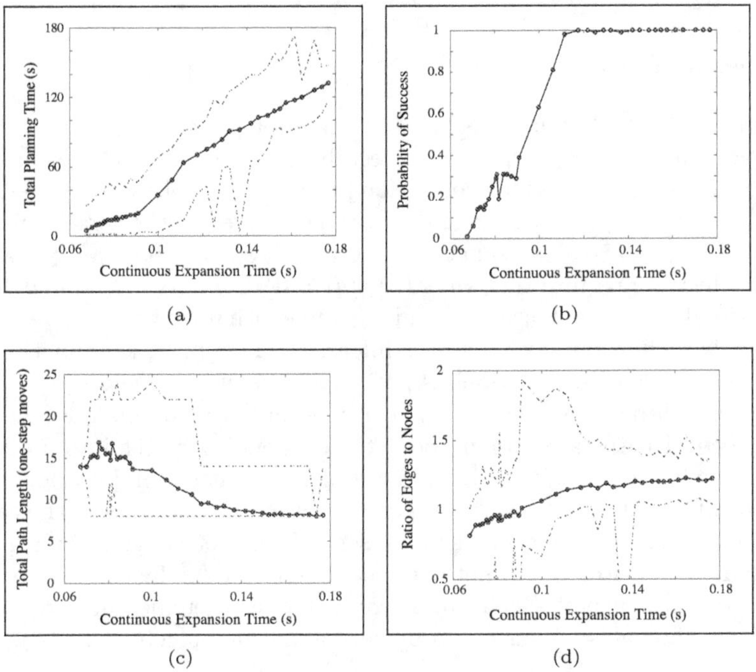

Fig. 7. Performance of a basic multi-step search using PRM, as the average time allowed for continuous expansion is varied. Mean values (over many runs) are shown; bounds are dotted lines. (**a**) Total planning time. (**b**) Probability of success. (**c**) Total number of one-step motions in the final path. (**d**) Ratio of edges to nodes.

4.4 Proposed modifications

There are a number of important modifications we intend to make to our planner in order to improve its performance (e.g., the use of graph search heuristics, and a "lazy" approach to checking the feasibility of edges). However, as shown by our baseline results, a fundamental problem that still must be addressed is to determine when one-step motions are infeasible. We propose a possible, general method for doing so in the following section.

We intend to apply our "disconnection planner" as follows: given each one-step motion query, our sample-based planner will be allowed to run for a short length of time, say 1.0 s. As shown in Fig. 6, many feasible one-step motions will be found immediately. When the one-step planner is unable to find a solution, we will attempt to prove that the motion is infeasible. This will eliminate a large number of infeasible queries quickly. Finally, if no disconnection proof is found, we allow our one-step planner a further amount of time T_{max} sufficient to find feasible one-step motions with high probability. In this way, the fraction of queries processed for time T_{max} will be significantly reduced.

5 Proving One-Step Disconnection

5.1 General approach

Our goal is to show that no feasible, continuous path exists between two configurations $q_s, q_g \in F_\sigma$. One way of establishing such a "disconnection proof" is to construct a manifold embedded in the infeasible space $C_\sigma - F_\sigma$ that separates q_s and q_g. Such a manifold can be defined as the zero-level-set of a continuous, scalar-valued function $g : C_\sigma \to \mathbb{R}$ such that $g(q_s) > 0$ and $g(q_g) < 0$, where we require that $\{q \in F_\sigma | \, g(q) = 0\}$ is empty. The value of g must be zero at some configuration q_i along any continuous path between q_s and q_g, but by definition any such q_i is infeasible. Therefore, the function g is a proof that no feasible, continuous path exists between q_s and q_g.

This approach need not involve the construction of a complete roadmap for F_σ. Indeed, it turns a question about the existence of paths into a question about the feasibility of sets. In certain cases, the latter computation can be done quickly. In this section, we propose a method that applies to the case where F_σ is expressed as a semialgebraic set. We do not suggest the use of exact arithmetic – instead, our method is based on a hierarchy of *Positivstellensatz* refutations, each of which can be computed as a semidefinite program (a result from the fields of convex optimization and real algebra [27,28]).

5.2 Theoretical foundation

Problem statement. Assume the feasible space F_σ can be represented as a *semialgebraic set*. (An over-approximation $\hat{F}_\sigma$ of F_σ is sufficient. If $F_\sigma \subset \hat{F}_\sigma$, then empty $\{q \in \hat{F}_\sigma | \, g(q) = 0\}$ implies empty $\{q \in F_\sigma | \, g(q) = 0\}$.) So, it can be represented as the following subset of an n-dimensional Euclidean space $\mathbb{R}^n$, where all $f_i, h_j \in \mathbb{R}[x_1, \ldots, x_n]$ are polynomials:

$$F_\sigma = \left\{ x \in \mathbb{R}^n \,\middle|\, \begin{array}{ll} f_i(x) \geq 0 & i = 1, \ldots, m_f \\ h_j(x) = 0 & j = 1, \ldots, m_h \end{array} \right\} \quad (1)$$

The disconnection problem is to find a polynomial function $g \in \mathbb{R}[x_1, \ldots, x_n]$ such that $g(q_s) > 0$, $g(q_g) < 0$, and the following set $P_{\text{cut}}(g)$ is empty:

$$P_{\text{cut}}(g) = \left\{ x \in \mathbb{R}^n \,\middle|\, \begin{array}{ll} f_i(x) \geq 0 & i = 1, \ldots, m_f \\ h_j(x) = 0 & j = 1, \ldots, m_h \\ g(x) = 0 & \end{array} \right\} \quad (2)$$

This problem involves two basic challenges. First, given a function g, we must be able to check that $P_{\text{cut}}(g)$ is empty. Second, we must be able to search for such a function g.

Proving that $P_{\text{cut}}(g)$ is empty. A general strategy for showing the infeasibility of $P_{\text{cut}}(g)$ is to assume feasibility and derive a contradiction. When F_σ is semialgebraic, it is natural to begin with the hypothesis that there exists $x_0 \in P_{\text{cut}}(g)$ and try to infer a false conclusion such as $-1 \geq 0$. This is *automated proof by contradiction*, using algebra to perform inference.

For example, consider a feasible space F_σ defined by the single inequality $f_1(x) = x^2 - 1$, so $F_\sigma = (-\infty, -1] \cup [1, \infty) \subset \mathbb{R}$. To separate points $q_s = 2$, $q_g = -2$, we could use the function $g(x) = x$, which defines a manifold that is the single point $\{0\}$. We have $g(q_s) = 2 > 0$ and $g(q_g) = -2 < 0$, so all that remains is to show that $P_{\text{cut}}(g)$ is empty. But by definition, $f_1 \geq 0$ and $g = 0$ on $P_{\text{cut}}(g)$, so by inference $-1 = f_1 + (-x) \cdot g = f_1 + (-x) \cdot 0 = f_1 \geq 0$ on $P_{\text{cut}}(g)$, so if $P_{\text{cut}}(g)$ is nonempty we have a contradiction.

In fact, there is a systematic way to search for such a contradiction. The set of all polynomials that can be inferred from $f_i \geq 0\ \forall i$ to be nonnegative on $P_{\text{cut}}(g)$) is called the *cone* generated by $f_1, \ldots, f_{m_f}$, and is denoted $\mathbf{cone}\{f_1, \ldots, f_{m_f}\}$. Equivalently, the set of all polynomials that can be inferred from $h_j = 0\ \forall j$ and $g = 0$ to be zero on $P_{\text{cut}}(g)$ is called the *ideal* generated by $h_1, \ldots, h_{m_h}, g$, and is denoted $\mathbf{ideal}\{h_1, \ldots, h_{m_h}, g\}$. The *Positivstellensatz* [33] states that $P_{\text{cut}}(g)$ is empty if and only if

$$-1 \in \mathbf{cone}\{f_1, \ldots, f_{m_f}\} + \mathbf{ideal}\{h_1, \ldots, h_{m_h}, g\} \tag{3}$$

As described in [6], the cone and the ideal can be parameterized so (3) is exactly equivalent to the existence of polynomials r and t_i and sum-of-squares polynomials $s_0, s_i, s_{ij}, \ldots,$ (all of arbitrary degree) such that the following algebraic equation holds (note that the series does terminate):

$$-1 = \left(s_0 + \sum_{i=1}^{m_f} s_i f_i + \sum_{i=1}^{m_f} \sum_{j=1}^{m_f} s_{ij} f_i f_j + \ldots \right) + \left(\sum_{i=1}^{m_h} t_i h_i + r g \right) \tag{4}$$

Recently, it has been shown that for fixed degree r, t_i, and $s_0, s_i, s_{ij}, \ldots,$ the feasibility of (4) can be checked by solving an SDP [27]. The basic idea is to search over the coefficients of these polynomials, restricting the coefficients of $s_0, s_i, s_{ij}, \ldots,$ to lie within the positive semidefinite cone. This approach allows the use of floating-point computation, rather than exact algebraic manipulation. A hierarchy of checks is also possible, with polynomial multipliers of higher and higher degree, in which each SDP is a better approximation of the "only if" condition of (4). However, the SDP's become significantly larger and more costly to compute. In practice, a degree is usually fixed (in our experiments, low degree certificates have been sufficient).

Searching for a function g. In some cases good candidate functions g are already available. In fact, this is true for the climbing robot problem, where F_σ is often disconnected across manifolds corresponding to $\theta_2 = 0$ in one limb (i.e., a straight-limb configuration). If candidates are not available, directly incorporating the search for g into the SDP computation (4) is problematic.

However, there is a duality theory associated with the *Positivstellensatz* and its corresponding SDP's [23]. This means that if both primal (connection) and dual (disconnection) searches are conducted in parallel, information from each can be used to help the other (also as suggested in [4]). For example, if $P_{\text{cut}}(g)$ is not empty, there must exist some feasible configuration $q_f \in P_{\text{cut}}(g)$. This can be used to direct the primal search – if the primal algorithm is sample-based, samples can be drawn from a distribution around nonempty $P_{\text{cut}}(g)$. Likewise, candidate functions g can be postulated by fitting hypersurfaces to infeasible sampled configurations $q_i \notin F_\sigma$.

5.3 Preliminary application

To test how this algebraic method might be applied, we used it to prove that the feasible space at the 4-hold stance σ in the last frame of Fig. 3 has at least two components. In this example, F_σ can be separated along a manifold corresponding to $\theta_2 = 0$ for the bottom-left limb. So we can prove that no continuous path exists between any q_s at which $\theta_2 > 0$ and any q_g at which $\theta_2 < 0$ in this limb.

The feasible space F_σ is over-approximated by considering only the closed-chain and joint-angle constraints. There are a variety of possible parameterizations to express F_σ as a semialgebraic set; the choice is important, since it may result in different numbers of variables, numbers of constraints, or total polynomial degrees of the constraints. In this case it is convenient to use variables x_{mi}, y_{mi} and x_{si}, y_{si} defining the locations of the elbow and shoulder joints, respectively, of each limb i.

Equality constraints h_j are introduced to enforce the spatial relationship between joints and the closed-chain constraint between the supporting holds. Inequality constraints f_i are used to enforce the joint-angle limits, which can be represented as minimum distances between joints. The function g can be expressed implicitly by constraining the shoulder of the bottom-left limb to be a fixed distance (corresponding to $\theta_2 = 0$) from its supporting hold. After eliminating redundancy, the set $P_{\text{cut}}(g)$ can be defined by 8 equality constraints and 11 inequality constraints, in 10 variables.

A certificate of the form (4) with fixed total degree 4 (i.e., a proof that no path exists in F_σ that crosses $\theta_2 = 0$ in the bottom-left limb) was found in 90.0 s on a 1GHz PowerPC G4. Our current implementation is still not fast enough to be practical (e.g., compare with the planning times we presented in Section 4.3). However, we have generated much faster (less than 1.0 s) disconnection proofs for a variety of simpler systems, including a 3-limbed planar robot. We hope to reduce the running time for this application by better exploiting the sparsity of the SDP's (e.g., [28]). Many other such improvements may be possible – this is ongoing work.

6 Future Work

This paper presented a framework for non-gaited, multi-step motion planning for free-climbing robots. Based on experimental results, it showed the need to detect when one-step motions are infeasible (i.e., to prove disconnection). It also presented current work on the development of a general method for doing so, based on advances in computational real algebra.

Other challenging problems remain to be addressed – sensing, grasping, and control are prominent among them. For example, here a pre-surveyed model of the terrain was assumed. With only local sensing, the robot will have to explore the environment incrementally, and plan based on incomplete or uncertain information. Also, there are many unanswered questions raised by our treatment of the disconnection problem. Work on exact motion planning has provided important insight, but to date, few practical algorithms. The use of floating-point computational tools, rather than exact arithmetic, could make it possible to implement some of these algorithms more efficiently.

Acknowledgments. The authors would like thank E. Baumgartner, B. Kennedy, and H. Aghazarian of the Mechanical and Robotic Technologies Group at JPL for their help and support.

References

1. R. Alami, J.-P. Laumond, and T. Simeon. Two manipulation planning algorithms. In K. Goldberg, D. Halperin, J.-C. Latombe, and R. Wilson, editors, *Alg. Found. Rob.*, pages 109–125. A K Peters, Wellesley, MA, 1995.
2. M. Almonacid, R. Saltarén, R. Aracil, and O. Reinoso. Motion planning of a climbing parallel robot. *IEEE Tr. Rob. Aut.*, 19(3):485–489, 2003.
3. C. Balaguer, A. Giménez, J. Pastor, V. Padrón, and M. Abderrahim. A climbing autonomous robot for inspection applications in 3d complex environments. *Robotica*, 18:287–297, 2000.
4. J. Basch, L. Guibas, D. Hsu, and A. T. Nguyen. Disconnection proofs for motion planning. In *IEEE Int. Conf. Rob. Aut.*, pages 1765–1772, Seoul, Korea, 2001.
5. D. Bevly, S. Farritor, and S. Dubowsky. Action module planning and its application to an experimental climbing robot. In *IEEE Int. Conf. Rob. Aut.*, volume 4, pages 4009–4014, 2000.
6. J. Bochnak, M. Coste, and M.-F. Roy. *Real Algebraic Geometry.* Springer, 1998.
7. J.-D. Boissonnat, O. Devillers, and S. Lazard. Motion planning of legged robots. *SIAM J. Computing*, 30(1):218–246, 2000.
8. T. Bretl, J.-C. Latombe, and S. Rock. Toward autonomous free-climbing robots. In *Int. Symp. Rob. Res.*, Siena, Italy, 2003.
9. T. Bretl, S. Rock, J.-C. Latombe, B. Kennedy, and H. Aghazarian. Free-climbing with a multi-use robot. In *Int. Symp. Exp. Rob.*, Singapore, 2004.
10. F. Bullo and M. Žefran. Modeling and controllability for a class of hybrid mechanical systems. *IEEE Tr. Rob. Aut.*, 18(4):563–573, 2002.
11. H. Dulimarta and R. L. Tummala. Design and control of miniature climbing robots with nonholonomic constraints. In *WCICA*, Shanghai, P.R.China, 2002.
12. C. Eldershaw and M. Yim. Motion planning of legged vehicles in an unstructured environment. In *IEEE Int. Conf. Rob. Aut.*, Seoul, South Korea, 2001.

13. P. Ferbach and J. Barraquand. A method of progressive constraints for manipulation planning. *IEEE Tr. Rob. Aut.*, 13(4):473–485, 1997.
14. B. Goodwine and J. Burdick. Controllability of kinematic control systems on stratified configuration spaces. *IEEE Tr. Aut. Cont.*, 46(3):358–368, 2001.
15. B. Goodwine and J. Burdick. Motion planning for kinematic stratified systems with application to quasi-static legged locomotion and finger gaiting. *IEEE Tr. Aut. Cont.*, 18(2):209–222, 2002.
16. J. C. Grieco, M. Prieto, M. Armada, and P. G. de Santos. A six-legged climbing robot for high payloads. In *IEEE Int. Conf. Cont. App.*, Trieste, Italy, 1998.
17. S. Hirose, A. Nagabuko, and R. Toyama. Machine that can walk and climb on floors, walls, and ceilings. In *ICAR*, pages 753–758, Pisa, Italy, 1991.
18. S. Hirsch and D. Halperin. Hybrid motion planning: Coordinating two discs moving among polygonal obstacles in the plane. In *WAFR*, pages 225–241, Nice, France, 2002.
19. L. E. Kavraki, P. Svetska, J.-C. Latombe, and M. Overmars. Probabilistic roadmaps for path planning in high-dimensional configuration spaces. *IEEE Tr. Rob. Aut.*, 12(4):566–580, 1996.
20. Y. Koga and J.-C. Latombe. On multi-arm manipulation planning. In *IEEE Int. Conf. Rob. Aut.*, volume 2, pages 945–952, San Diego, CA, 1994.
21. J. J. Kuffner, Jr., K. Nishiwaki, S. Kagami, M. Inaba, and H. Inoue. Motion planning for humanoid robots. In *Int. Symp. Rob. Res.*, Siena, Italy, 2003.
22. G. Lafferriere and H. Sussmann. A differential geometric approach to motion planning. In Z. Li and J. Canny, editors, *Nonholonomic Motion Planning*, pages 235–270. Kluwer, 1993.
23. J. Lasserre. Global optimization with polynomials and the problem of moments. *SIAM J. Opt.*, 11(3):796–817, 2001.
24. J. Laumond, P. Jacobs, M. Taix, and R. Murray. A motion planner for nonholonomic mobile robots. *IEEE Tr. Rob. Aut.*, 10(5):577–593, 1994.
25. W. Neubauer. A spider-like robot that climbs vertically in ducts or pipes. In *IROS*, pages 1178–1185, Munich, Germany, 1994.
26. C. L. Nielsen and L. E. Kavraki. A two level fuzzy prm for manipulation planning. In *IEEE/RSJ IROS*, pages 1716–1721, Takamatsu, Japan, 2000.
27. P. A. Parrilo. *Structured Semidefinite Programs and Semialgebraic Geometry Methods in Robustness and Optimization.* PhD thesis, California Institute of Technology, Pasadena, CA, 2000.
28. P. A. Parrilo and S. Lall. Semidefinite programming relaxations and algebraic optimization in control. *European Journal of Control*, 9(2-3):307–321, 2003.
29. E. Rimon, S. Shoval, and A. Shapiro. Design of a quadruped robot for motion with quasistatic force constraints. *Autonomous Robots*, 10:279–296, 2001.
30. A. Sahbani, J. Cortés, and T. Siméon. A probabilistic algorithm for manipulation planning under continuous grasps and placements. In *IEEE/RSJ IROS*, pages 1560–1565, Lausanne, Switzerland, 2002.
31. J. Schwartz and M. Sharir. On the 'piano movers' problem II: General techniques for computing topological properties of real algebraic manifolds. *Advances in Applied Mathematics*, 4(1):298–351, 1983.
32. A. Shapiro and E. Rimon. PCG: A foothold selection algorithm for spider robot locomotion in 2d tunnels. In *IEEE Int. Conf. Rob. Aut.*, pages 2966–2972, Taipei, Taiwan, 2003.
33. G. Stengle. A nullstellensatz and a positivstellensatz in semialgebraic geometry. *Mathematische Annalen*, 207:87–97, 1974.

Sampling-Based Motion Planning under Kinematic Loop-Closure Constraints

Juan Cortés and Thierry Siméon

LAAS-CNRS, Toulouse, France

Abstract. Kinematic loop-closure constraints significantly increase the difficulty of motion planning for articulated mechanisms. Configurations of closed-chain mechanisms do not form a single manifold, easy to parameterize, as the configurations of open kinematic chains. In general, they are grouped into several subsets with complex and a priori unknown topology. Sampling-based motion planning algorithms cannot be directly applied to such closed-chain systems. This paper describes our recent work [7] on the extension of sampling-based planners to treat this kind of mechanisms.

1 Introduction

Robot motion planning has led to active research over the two last decades [13]. More recently, several sampling-based approaches (e.g. [12,16]) have been proposed and successfully applied to challenging problems that remained out of scope for previously existing techniques. They allow today to handle practical motion planning problems arising in such diverse fields as robotics, graphic animation, virtual prototyping or computational biology [14].

In this paper, we consider motion planning for closed-chain mechanisms. We present an extended formulation of the motion planning problem in presence of kinematic loop-closure constraints and we introduce a framework for the development of sampling-based algorithms (Sects. 2 and 4). The additional difficulty of this instance of the problem is that feasible configurations form lower-dimensional subsets in the search-space with no available representation. The performance of the very few approaches proposed for closed-chain mechanisms [15,11] (Sect. 3) significantly degrades for reasonably complex systems, mainly due to the difficulty of generating random samples in such subsets. We propose a general and simple geometric algorithm, called Random Loop Generator (RLG), for sampling random configurations satisfying loop-closure constraints (Sect. 5). RLG enables virtually any sampling-based planning algorithm to be extended to closed-chain mechanisms. We have implemented and experimented with its integration within PRM-based and RRT-based planners, obtaining very good results (Sect. 6).

Complex articulated mechanisms with closed kinematic chains appear in all the domains where motion planning techniques can be applied. Figure 1 illustrates an example of coordinated manipulation of an object handled by several robots. The generality and the practical efficiency of the extended planners incorporating RLG allow to tackle such kind of problems as well as problems involving parallel robots, or problems arising in computational

M. Erdmann et al. (Eds.): Algorithmic Foundations of Robotics VI, STAR 17, pp. 75–90, 2005.

Fig. 1. A "closed-chain" version of the piano mover's problem. The piano is moved by three cooperating mobile manipulators, creating multiple closed kinematic chains.

biology for the structural analysis of protein loops. All these applications are commented in Sect. 7. We conclude devising new directions for future research (Sect. 8).

2 Problem Formulation

The motion planning problem consists in finding a path between two given locations of a mobile system that satisfies intrinsic constraints as well as constraints that arise from the environment. Basically, motion constraints are due to the kinematic structure of the mechanism and to collision avoidance. Under these constraints, the problem can be posed and solved in the *configuration-space* $\mathcal{C}$ [19]. Then, it is reduced to explore the connectivity of the subset $\mathcal{C}_{free}$ of the collision-free configurations.

The problem has been clearly formulated for articulated mechanisms without kinematic loops (see [13] for a detailed formulation). In this case, $\mathcal{C}$ corresponds to the space of the joint variables $\mathcal{Q}$, called the *joint-space*. Topologically, $\mathcal{Q}$ is a smooth manifold, with a simple parameterization [2]. For an articulated mechanism with m independent joint variables defined in real intervals, $\mathcal{Q}$ can be seen as a m-dimensional hypercube. If the articulated mechanism contains closed kinematic chains, then some joint variables are related by loop-closure equations [20]. A general expression of loop-closure constraints is: $f(\mathcal{Q})=\mathbf{I}$, where $f(\mathcal{Q})$ is a system of non-linear equations and $\mathbf{I}$ is the identity displacement. The configuration-space $\mathcal{C}$ of a closed-chain mechanism is the subset of $\mathcal{Q}$ satisfying such equations. The stratification of $\mathcal{C}$ leads to several ρ-dimensional manifolds $\mathcal{M}_i$ which can be connected through sets of lower dimension $\mathcal{S}_k$ [3,27]. Note that ρ corresponds to the global mobility of the mechanism. The $\mathcal{M}_i$ are called *self-motion manifolds* and the $\mathcal{S}_k$ are sets of singular configurations. The number of self-motion manifolds is bounded, and it tends to decrease as ρ increases [3].

Figure 2 illustrates a fictive example with three joint variables $\{\theta_1, \theta_2, \theta_3\}$ and $\rho = 2$. Let us consider a function of the form $f(\theta_1, \theta_2, \theta_3)=0$, representing loop-closure constraints. This function maps to several surfaces embedded in the joint-space $\mathcal{Q}$. Such surfaces are the different self-motion manifolds $\mathcal{M}_i$. In this example, $\mathcal{M}_1$ and $\mathcal{M}_2$ intersect at a singular set $\mathcal{S}$. We have represented the obstacle region $\mathcal{Q}_{obst}$ in the joint-space. $\mathcal{Q}_{free}$ is the complementary subset: $\mathcal{Q}_{free} = \mathcal{Q} \setminus \mathcal{Q}_{obst}$. $\mathcal{C}_{free}$ is the intersection of $\mathcal{Q}_{free}$ with the different $\mathcal{M}_i$.

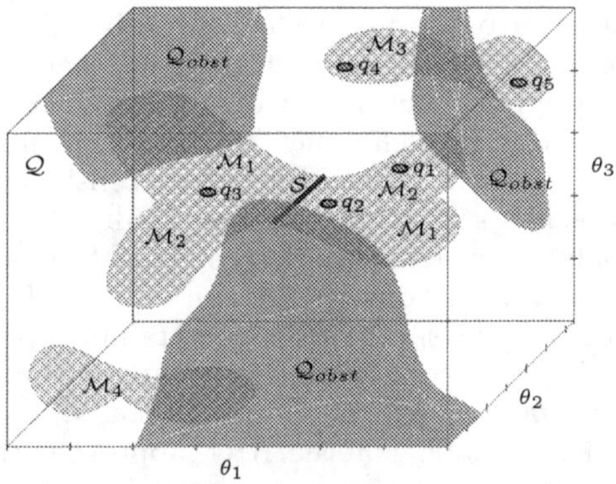

Fig. 2. Illustration for the formulation of the motion planning problem under kinematic loop-closure constraints. Configurations are grouped into several subsets embedded in the joint-space.

Several situations can rise in motion planning queries within this example. The best case is a query for a path between q_1 and q_2. These configurations lie on the same self-motion manifold and in the same connected component of $\mathcal{C}_{free}$, thus there is a free path between them. A path is also feasible between q_1 and q_3, even if it contains singular configurations. However, for q_4 and q_5, the presence of obstacles makes these configurations cannot be connected by a free path. Finally, a case that does not appear for open kinematic chains can rise under closure constraints: q_1 and q_4 lie on the same connected component of $\mathcal{Q}_{free}$ but on different components of $\mathcal{C}$!

3 Available Techniques

Only a few exact motion planning approaches have been proposed that treat kinematic loop-closure and collision avoidance simultaneously, including these two types of constraints within algebraic expressions (e.g. [4,1]). The computational complexity of these approaches makes them unpractical. Normally, different techniques have to be combined for planning motions under kinematic loop-closure constraints. First, loop-closure equations must be solved to obtain the configuration-space $\mathcal{C}$. Then, motion planning algorithms can be applied to compute paths in the collision-free subset $\mathcal{C}_{free}$.

Techniques that provide a complete solution of loop-closure equations are very limited in practice. Currently, they can be applied to non-redundant mechanisms, single loops with only a few (two or three) degrees of redundancy or particular classes of parallel mechanisms [22,21,23]. For more complex closed-chain mechanisms, only discrete points in $\mathcal{C}$ can be obtained. The use of a grid for globally representing $\mathcal{C}$ is not applicable to high-dimensional spaces. The remaining possibility is then to use sampling techniques combined with numerical or algebraic techniques to obtain single configurations satisfying loop-closure equations. This fact restrains the choice of motion planning algorithms to those based on sampling.

Sampling-based planners have demonstrated their efficacy for solving difficult problems in high-dimensional spaces. The *Probabilistic RoadMap* (PRM) [12] and the *Rapidly-exploring Random Trees* (RRT) [16] are two approaches that have had a particular success. However, only two attempts had been made to extend such sampling-based planners to closed-chain mechanisms [15,11].

The first PRM-based approach able to handle mechanisms with closed chains was presented in [15]. The problem is formulated in the joint-space $\mathcal{Q}$. Closure constraints are expressed by error functions involving distances in the Euclidean space. Numerical optimization techniques are used to sample and to connect configurations in the subset of $\mathcal{Q}$ satisfying these constraints within a given tolerance. The approach is general but suffers the drawbacks of optimization-based methods to solve inverse kinematic problems: they are exposed to the local minima problem and the convergence can be very slow. A technique to randomly sample the tangent space of the constraints is proposed that increases the efficiency of the process to connect sampled configurations. More details of the method and the extension of RRT-based algorithms were subsequently published in [29,30]. For the RRT approach, the random configurations used to bias the exploration are generated ignoring closure constraints. The argument is that computing closed configurations is too expensive and does not provide appreciable benefit. We will show in Sect. 6 that this last assertion is not totally right.

The approach described in [11] treats closed kinematic chains within a PRM-based planner. Each loop in the mechanism is broken into two subchains. For computing nodes, uniform random sampling is used to generate the configuration of one of the subchains (called *active subchain*) and then an inverse kinematics problem is solved to obtain the configuration of the remaining part of the loop (called *passive subchain*) in order to force closure. For computing edges, the local planner is limited to act on the active configuration parameters and the corresponding passive variables are computed for each intermediate configuration along the local path. For the efficiency of the roadmap computation, the passive subchain of each loop must be a non-redundant mechanism with closed-form inverse kinematics solution. As the authors admit, this means an important drawback when the approach is applied to a highly-redundant loop. The probability of randomly generating configurations of a long active subchain for which a configuration of the passive chain satisfying closure constraints exists is very low. The performance of the algorithm drops off significantly due to this fact.

Our approach shares some ideas used for the extension of PRM-based planners in [11]. One of our contributions is to resolve the main drawback of the referred technique by the integration of the RLG sampling technique (explained in Sect. 5).

4 Sampling-Based Planning and Closed-Chain Mechanisms

This section presents a general framework to extend sampling-based approaches for planning the motions of general closed-chain mechanisms. The use of sampling-based planners is strongly justified since, for the kind of problems we address (see Sect. 7), there is no available technique providing a complete, exact representation of $\mathcal{C}_{free}$. However, there are important difficulties for sampling and for checking the connectivity of configurations of closed-chain mechanisms. Next, we discuss how to deal with these difficulties.

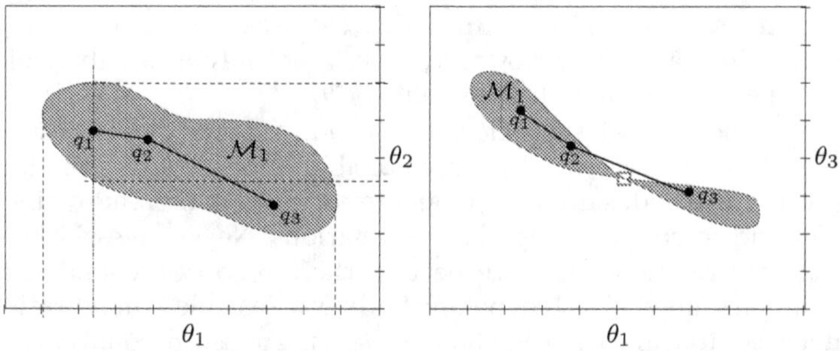

Fig. 3. Projection of $\mathcal{M}_1$ on the planes $\theta_1\theta_2$ and $\theta_1\theta_3$.

On the Parameterization of $\mathcal{C}$: Configurations of a closed-chain mechanism are grouped into a finite number of lower-dimensional manifolds $\mathcal{M}_i$ embedded in the joint-space $\mathcal{Q}$ (the search-space in our problem). These manifolds can be parameterized, at least locally, by a set of ρ independent parameters, selected from the joint variables. Points in the different manifolds can be generated by sweeping the ρ parameters through their range and evaluating the loop-closure equations that provide the value of the other (dependent) variables. An *atlas* of each one of these manifolds $\mathcal{M}_i$ can be constituted by a finite number (in general not exceeding the dimension of $\mathcal{Q}$, m) of *local charts* considering different combinations of ρ parameters. Without loss of generality, sets of ρ consecutive joint variables in a kinematic chain can be chosen as local coordinates [27].

Following a terminology also used in [11], we call *active variables* q^a the set of the ρ joint variables chosen as parameters of a local chart and *passive variables* q^p the remaining set of the $m - \rho$ dependent joint variables, so that $\{q^a, q^p\} = q \in \mathcal{C} \subset \mathcal{Q}$.

Main Principle: The core of our approach is to explore the connectivity of $\mathcal{C}_{free}$ by sampling configurations and by testing feasible connections through local parameterizations of $\mathcal{C}$. Motion planning algorithms are applied on the local parameters q^a. Using a roadmap method such as PRM, the nodes are generated by sampling q^a and local paths are obtained by applying a local planner (also called *steering method*) to these parameters. In a similar way, q^a are the configuration parameters directly handled by incremental search methods like RRT. Obviously, for each computed value of the parameters, loop-closure equations must be solved for obtaining the whole configuration of the mechanism $q \in \mathcal{C}$. Therefore, the efficiency of the planner partially relies on the efficient solution of these equations.

Configuration Sampling: Given a set of active variables q^a, loop-closure equations have real solutions only for a range of values of each joint variable, that we call the *closure range*. Besides, the closure range of a parameter depends on the value of the other parameters. The last assertion is illustrated in Fig. 3, that shows the projection of the manifold $\mathcal{M}_1$ of Fig. 2 on the planes $\theta_1\theta_2$ and $\theta_1\theta_3$. Let us consider $q^a = \{\theta_1, \theta_2\}$ (the left image). If we sample first θ_1 and then θ_2 for generating a configuration $q_1 \in \mathcal{M}_1$, then θ_1 can be sampled in its whole closure range (i.e. the feasible range for any value of the other joint variables). However, θ_2 is valid only in a subset of its whole closure range, determined by the value of θ_1.

There is no general and efficient method to define closure ranges of joint variables. Thus, in practice, the only possibility for sampling configurations is to use a trial method: sampling parameters q^a in the intervals defined for joint variables and solving the loop-closure equations. Nevertheless, when a closure range is very restricted with respect to the interval of a joint variable, too many samples maybe tested before finding a feasible configuration. Hence, too much computing time is spent in solving closure equations leading to imaginary values. This is an important drawback for the efficiency of motion planners, and mainly for those using a roadmap approach, such as the planner in [11]. The RLG algorithm, further discussed in Sect. 5, resolves this problem using simple geometrical operations.

Computing Local Path: Sampling-based roadmap methods check the connectivity of nearby configurations by local paths. Incremental search methods try to expand a configuration toward a local goal, producing also a sort of local path. Such paths can be computed by explicitly variating the local parameters q^a and solving loop-closure equations with a given resolution for obtaining q^p.

Without a topological characterization of $\mathcal{C}$, the number of different sets of parameters q^a required to determine if two configurations are connected or not is a priori unknown. This fact can easily be understood on Fig. 3. Let us consider that a linear interpolation of the parameters q^a produces a kinematically feasible path (i.e. there are no differential constraints). Then, any two configurations lying on $\mathcal{M}_1$ can be connected (directly or indirectly) using $q^a = \{\theta_1, \theta_2\}$ as local parameters. On the contrary, choosing $q^a = \{\theta_1, \theta_3\}$ leads to a non-complete solution of motion planning queries within $\mathcal{M}_1$. A solution path between configurations q_1 and q_2 can be immediately obtained since they are directly connected by a local path. However, a path between q_1 and q_3 can not be found using sampling-based techniques. The point indicated by the small square is a singularity of this parameterization. The probability of generating this point by sampling values of $q^a = \{\theta_1, \theta_3\}$ is null, as well as the probability of sampling two points on a line (local path) passing through this singularity. Thus, finding a feasible path between q_1 and q_3 requires another set of local parameters than $\{\theta_1, \theta_3\}$.

Local paths have to satisfy other motion constraints besides loop-closure, such as collision avoidance. In this paper, we do not talk about these other constraints, that can be checked along local paths by techniques (e.g. [18]) similar to those used for open-chain mechanisms.

Dealing with Kinematic Singularities: Up to now we have limited our discussion to the case of a single manifold. However, $\mathcal{C}$ may be composed of several manifolds. These manifolds are either disjoint, or they intersect at lower-dimensional subsets corresponding to kinematically singular configurations. Therefore, exploring the connectivity of $\mathcal{C}_{free}$ requires to deal with these singularities.

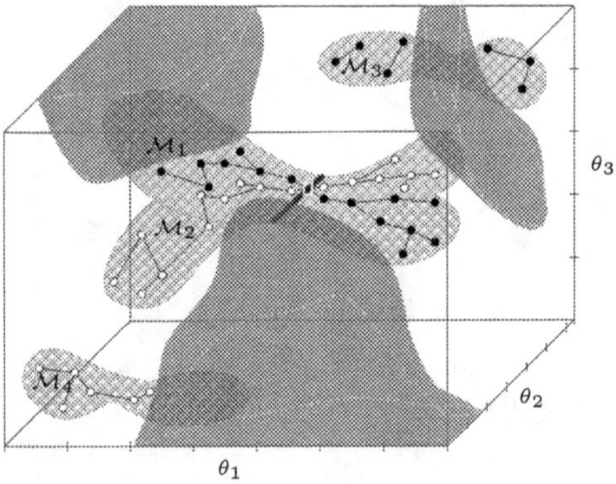

Fig. 4. Probabilistic roadmap capturing the connectivity of $\mathcal{C}_{free}$ for the fictive motion planning problem involving kinematic closure constraints of Sect. 2.

The roadmap represented in Fig. 4 has been built using $q^a = \{\theta_1, \theta_2\}$ as only set of parameters. Each point $\{\theta_1, \theta_2\}$ maps to two configurations (i.e. there are two solutions for $q^p = \theta_3$) on different manifolds for all the domain under $\mathcal{M}_1$ and $\mathcal{M}_2$ except for the singular set $\mathcal{S}$ where they intersect, so that configurations in these manifolds are connected by this singular set. This kinematic singularity corresponds to the singularity of the parameterization $q^a = \{\theta_1, \theta_3\}$ commented above. It can be traversed by a local path on the current parameterization. When a local path is computed between two configurations at different sides of $\mathcal{S}$, the bifurcation of the solution can be easily detected when checking the local path validity, and the singular configuration (marked by the small rectangle in Fig. 4) connecting $\mathcal{M}_1$ and $\mathcal{M}_2$ can then be identified.

Let us consider now a more difficult case where $\mathcal{M}_1$ and $\mathcal{M}_2$ do not intersect along a line but meet at a point. None of the three possible parameterizations would allow to identify such a singular point exactly. The difference in this case is that the singular set has dimension $\rho - 2$ instead of $\rho - 1$. In theory, sets of kinematically singular configurations can have dimension from $\rho - 1$ to zero. Using sampling techniques for generating configurations on $\mathcal{C}$ and steering methods on subsets of configuration parameters q^a, our approach has only the guarantee (if we do not admit a tolerance) to find connections through singular sets of dimension $\rho - 1$. The other singular sets, from dimension $\rho - 2$ to isolated singularities, must be identified by other methods. The general treatment of such singularities goes beyond the scope of this paper. As far as we know, techniques able to globally characterize singular configurations have been proposed only for particular classes of mechanisms (e.g. [10,28]).

5 The RLG Algorithm

We have developed an algorithm, that we call Random Loop Generator (RLG), for sampling configurations of closed-chain mechanisms. The general approach was presented in [5]. Then, a variant that treats more efficiently parallel mechanisms was introduced in [6]. More details can be found in [7]. In this section, we present an overview containing main ideas.

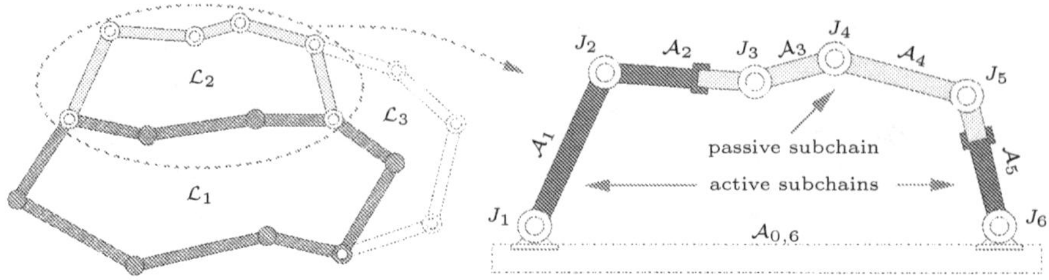

Fig. 5. Decomposition of a multi-loop in single loops, and a possible choice of passive and active subchains in loop $\mathcal{L}_2$.

5.1 Mechanical System Decomposition

General Case: RLG is based on a decomposition of the mechanism into open kinematic chains. In the general case, single loops are handled separately, in a determined order. For each loop, sets of active and passive joint variables are defined consecutively such that they correspond to segments of the kinematic chain. We call *passive subchain* the segment involving the passive variables and *active subchains* to the other segments. There can be one or two active subchains depending on the placement of the passive subchain. The passive subchain is a non-redundant mechanism whose end-frame can span full-rank subsets of the workspace. In general, this requires three joint variables for a planar mechanism and six for a spatial mechanism. Efficient methods to solve inverse kinematics problems for such mechanisms are usually available [22].

In multi-loops, some joint variables are involved in the configuration of several individual loops. Their value is computed for the single loop treated first, and then, these common portions of the mechanism become rigid bodies when treating the other loops. Figure 5 shows an example of a planar multi-loop mechanism. The individual loops are designated by $\mathcal{L}_i$, where the index i indicates the order for the treatment. The figure also illustrates the decomposition of $\mathcal{L}_2$. This $6R$ planar linkage has mobility $\rho = 3$. Thus, q^a and q^p contain three joint variables each. In this illustration we have chosen θ_3, θ_4 and θ_5 (the variables associated with joints J_3, J_4 and J_5) to be the passive variables. Then, active variables can be seen as configuration parameters of two open chains rooted at a (fictive) link $\mathcal{A}_{0,6}$.

Parallel Mechanisms: A parallel mechanism is an articulated multi-loop structure in which a solid, the end-effector or platform $\mathcal{P}$, is connected to the base $\mathcal{A}_0$ by at least two independent kinematic chains $\mathcal{K}_i$. The pose of $\mathcal{P}$ is defined by a vector $q_\mathcal{P} = \{x_\mathcal{P}, y_\mathcal{P}, z_\mathcal{P}, \gamma_\mathcal{P}, \beta_\mathcal{P}, \alpha_\mathcal{P}\}$. The three first elements represent the position of $F_\mathcal{P}$ relative to $F_{\mathcal{A}_0}$, the frames associated with $\mathcal{P}$ and $\mathcal{A}_0$ respectively. The orientation is given by three consecutive rotations around the coordinate axes of $F_\mathcal{P}$ [1]. We consider the configuration q of a parallel mechanism is defined by the platform pose and the configuration of the chains $\mathcal{K}_i$: $q = \{q_\mathcal{P}, q_{\mathcal{K}_1}, \ldots, q_{\mathcal{K}_{n_k}}\}$. The parameters defining the platform

[1] RLG could also handle other parameterizations (e.g. Euler angles).

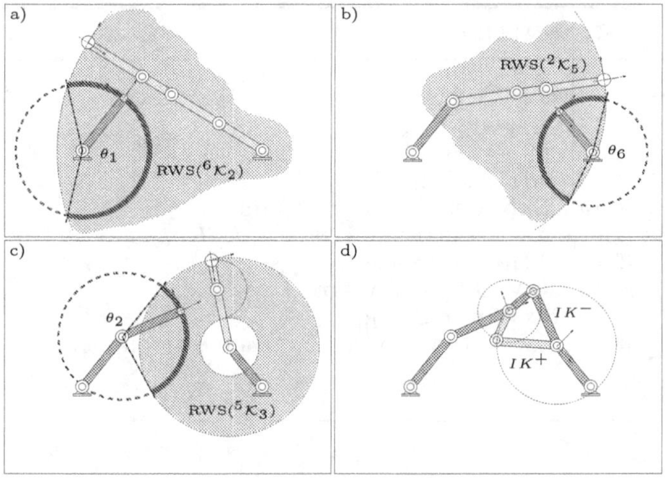

Fig. 6. Steps of the RLG algorithm performing on a $6R$ planar linkage.

pose $q_{\mathcal{P}}$ are selected as active variables. For a given platform pose, the base-frame and the end-frame of each chains $\mathcal{K}_i$ have fixed relative location. Thus, they have to be treated as closed kinematic chains. If a chain $\mathcal{K}_i$ is redundant, it is decomposed in active and passive subchains as explained above for a single loop. Thus, we can separate joint variables as follows: $q_{\mathcal{K}_i} = \{q^a_{\mathcal{K}_i}, q^p_{\mathcal{K}_i}\}$. Hence, variables defining the configuration q of a parallel mechanism are divided into active and passive such that:

$$q^a = \{q_{\mathcal{P}}, q^a_{\mathcal{K}_1}, \ldots, q^a_{\mathcal{K}_{n_k}}\} \ , \quad q^p = \{q^p_{\mathcal{K}_1}, \ldots, q^p_{\mathcal{K}_{n_k}}\}$$

5.2 Sampling process

RLG performs a "guided"-random sampling for q^a that notably increases the probability of obtaining real solutions for q^p when solving the loop-closure equations. Figure 6 illustrates the process on a planar $6R$ linkage (the loop $\mathcal{L}_2$ in Fig. 5). The active joint variables are computed sequentially by the function SAMPLE_q^a detailed in Algorithm 1. The two active subchains are treated alternately. The idea of the algorithm is to progressively decrease the complexity of the closed chain until only the configuration of the passive subchain, q^p, remains to be solved (by inverse kinematics).

At each iteration, the function COMPUTECLOSURERANGE returns a set of intervals I_c which approximate the closure range of a joint variable, for a fixed configuration of the portions of loop previously generated. The approximation must be conservative in the sense that no region of $\mathcal{C}$ is excluded for the sampling. This is required in order to guarantee any form of sampling-based completeness (e.g. probabilistic completeness) of motion planning algorithms. The value of the joint variable is randomly sampled inside the intervals I_c.

The problem of computing the closure range of a joint variable can be formulated as follows. Given a closed kinematic chain $^b\mathcal{K}_e$ involving joints from J_b to J_e (we consider $b < e$ in this explanation), two open kinematic chains

Algorithm 1: SAMPLE_q^a

input : the loop $\mathcal{L}$
output : the parameters q^a
begin
1 $(J_b, J_e) \leftarrow$ INITSAMPLER($\mathcal{L}$);
while not ENDACTIVECHAIN($\mathcal{L}, J_b$) **do**
  $I_c \leftarrow$ COMPUTECLOSURERANGE($\mathcal{L}, J_b, J_e$);
  if $I_c = \emptyset$ **then goto** line 1;
  SETJOINTVALUE(J_b, RANDOM(I_c));
  $J_b \leftarrow$ NEXTJOINT($\mathcal{L}, J_b$);
  if not ENDACTIVECHAIN($\mathcal{L}, J_e$) **then** SWITCH(J_b, J_e);
end

are obtained by breaking the link $\mathcal{A}_b$ between J_b and J_{b+1}. A suitable break-point is the physical placement of J_{b+1}, but any other point can be chosen. A frame F_c associated with this break-point can be seen as the end-frame of both open chains. The closure range of the joint variable corresponding to J_b is the subset of values making F_c reachable by the chain ${}^e\mathcal{K}_{b+1}$. Solving such a problem requires to represent the workspace of this chain, which is in general very complicated. For our purpose, a simple and fast method is preferred versus a more accurate but slower one. RLG handles simple volumes that bound the reachable workspace (i.e. only considering positional reachability). They are denoted by RWS in Fig. 6. In general, a reasonable choice for the RWS is a spherical shell with external and internal radii corresponding respectively to an upper bound of the maximum extension and a lower bound of the minimum extensions of the chain. Once defined RWS(${}^e\mathcal{K}_{b+1}$), computing the approximation of the closure intervals for J_b is very simple. If J_b is a revolute joint, then the origin of F_c describes a circle around its axis. If J_b is a prismatic joint, the origin of F_c moves on a straight-line segment. Then, I_c is obtained from the intersection of a circle or a line with a simple volume RWS.

The function to sample $q_{\mathcal{P}}$ for a parallel mechanism is based on the same principle that SAMPLE_q^a. The parameters are sampled progressively from the computed closure range approximations. The main difference is that the closure range depends now on the satisfaction of closure constraints that simultaneously involve several individual loops.

6 Performance of RLG

RLG is a general technique, applicable to any sampling-based planner. In this section, we show examples of motion planning problems solved by PRM and RRT planners extended to handle loop-closure constraints. The goal of the experiments is to compare the performance of the planners with and without incorporating RLG. The tests have been made with the software Move3D [25], in which our algorithms have been implemented.

Fig. 7. Motion planning problem for four robotic arms manipulating an object. Start and goal configurations (left) and trace of the solution path (right).

6.1 Results with a PRM-based Planner

PRM-based algorithms build a graph (the roadmap) whose nodes are randomly sampled configurations that satisfy motion constraints. For a closed-chain mechanism, only the parameters q^a are sampled, and loop-closure equations have to be solved to determine if the sample yields a valid configuration or not. We have compared the performance of the planner using an uniform random sampling [2] or RLG for sampling q^a.

Figure 7 shows four manipulators handling an object. The whole system can be seen as an articulated parallel structure. If the grasps are modeled as fixed attachments, the composed mechanism involves $m = 30$ joint variables (6 for each manipulator and 6 for the movable object), and the mobility is $\rho = 6$. Since the manipulators are non-redundant, $q^a = q_{\mathcal{P}}$: the parameters defining the location of the object. Almost all (more than 90%) the configurations $q_{\mathcal{P}}$ generated by RLG make the object simultaneously reachable by the four manipulators. Using a uniform random sampling, the bounds of the parameters $q_{\mathcal{P}}$ have first to be adjusted "by hand" with relation to the workspace of the manipulators (this is not necessary for RLG). Even with a good setting of the bounds, less than 0.05% of the samples yield valid (closed) configurations. Let us see now the repercussion when solving motion planning problems using the PRM approach. In the problem illustrated in the figure, the manipulators have to unhook an object and to insert it into the cylindrical axis. For computing a roadmap containing the solution path to this problem, millions of poses generated by uniform random sampling were necessary and the process took more than 20 minutes [3]. Using RLG for sampling configurations, less than 500 random platform poses were generated and the roadmap was built in less than 20 seconds. This result show that RLG avoids an enormous number of futile operations (i.e. calls to inverse kinematics functions) which drop off the performance of the planner.

[2] Implemented using the `rand()` function of the GNU C Library.

[3] Tests were performed using a Sun Blade 100 Workstation with a 500-MHz UltraSPARC-IIe processor.

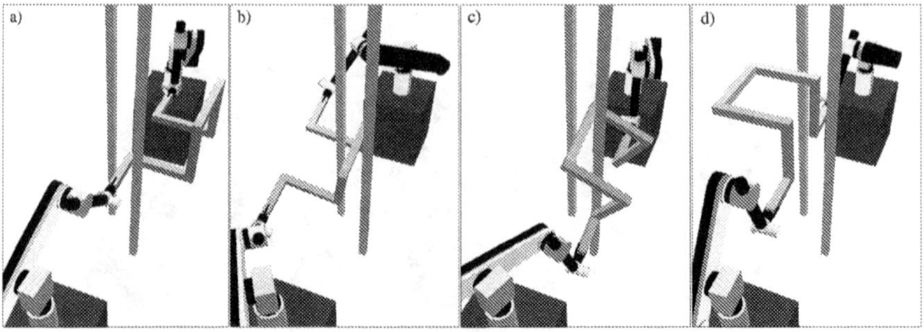

Fig. 8. Sequence of the solution to a difficult motion planning problem for two robotic arms manipulating an object among obstacles.

6.2 Results with an RRT-based Planner

The principle of RRT is to use uniformly distributed configurations q_{rand}, sampled at random in $\mathcal{C}$, to bias the expansion of search trees in $\mathcal{C}_{free}$. Under loop-closure constraints, the use of samples in $\mathcal{Q}$ was proposed in [29] in order to evade the cost of sampling $\mathcal{C}$. We have compared the performance of a uniform random sampling in $\mathcal{Q}$ versus RLG for generating q^a_{rand}.

In the example illustrated in Fig. 8, two robotic arms coordinate for manipulating a twisted bar among two vertical bars that restrict its motion. The goal is to solve a motion planning query between configurations in Fig. 8.a and Fig. 8.d. The figure shows a sequence of intermediate configurations of the solution of this puzzle-like problem. The difficulty of this problem depends on the distance between the vertical bars d_{bars}. We have made tests with three settings: $d_{bars} = \{150, 175, 200\}$. In this example, q^a corresponds to the configuration of one of the arms grasping the bar. The next table shows averaged results of tests. **N** is the number of iterations for expanding the search trees. **T** is the computing time. These results show the importance of an appropriate sampling considering the presence of kinematic loops. The gain obtained using the RLG sampling technique increases with the difficulty of the motion planning problem.

d_{bars}	Uniform		With RLG		
	N	**T**	**N**	**T**	**gain T**
200	2719	42.26	118	3.23	× 13
175	4995	102.92	424	6.30	× 16
150	9761	312.76	615	9.14	× 34

7 Applications of Closed-Chain Motion Planning

We have studied some of the possible applications of algorithms for motion planning under loop-closure constraints. One of them, coordinated manipulation planning, has been illustrated with the two examples in Sec. 6 and the problem shown in Fig. 1. This last problem combines several types of difficulty. First, the virtual structure composed by the three mobile manipulators grasping the piano can be seen as a parallel mechanism with redundant legs

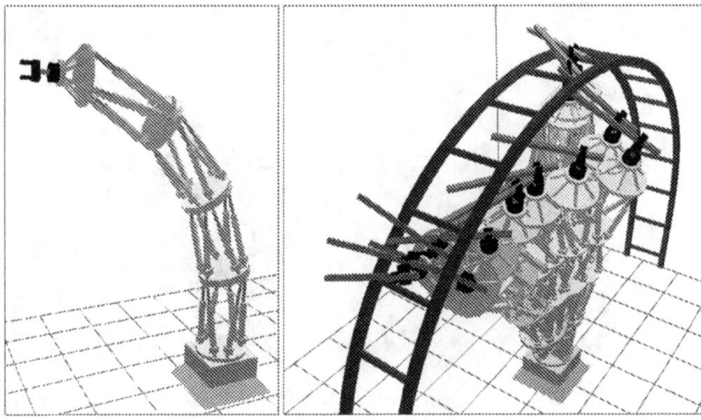

Fig. 9. Model of the robot Logabex-LX4, composed of four Gough-Stewart platforms connected in series, and trace of a collision-free path.

(the chains $\mathcal{K}_i$ correspond to the mobile manipulators). And second, the geometric complexity of the scene makes collision checking very hard. Besides, obstacles are strategically placed in order to hinder the motion of robots for changing the orientation of the piano. A roadmap that permits to solve most possible queries in this scene was computed using the an extended version of the Visibility-PRM algorithm [24] in 5 minutes.

Applied to parallel robots, motion planning algorithms can help designers of these mechanisms, or can provide useful data for real-time trajectory planning. Our work on parallel mechanisms [6] represents the first effective application of sampling-based planners to this kind of articulated structures. The generality of our approach is demonstrated by the complexity of the systems that it is able to treat, such as the model of the Logabex-LX4 (Fig. 9), whose configuration-space is a 25-dimensional variety embedded in a 97-dimensional joint-space. Planning queries for moving the manipulator with the grasped bar from one to another opening of the bridge were solved by the extended RRT-based algorithm in less than one minute.

The above expounded closed-chain planners can also be used as a key component of a novel manipulation planning approach described in [26]. The clever idea is to explore the connectivity of the subset where the manipulation sub-paths (i.e. *transit* and *transfer* paths) meet via a virtual closed-chain mechanism consisting of the robot grasping the movable object placed at a stable position. Our manipulation planner automatically generates, among continuous sets, the grasps and the intermediate placements of the movable object required to solve complicated problems. It is the first general manipulation planner with this capability.

We also began to investigate applications out of the field of robotics. Motion planning techniques can be used as new tools to help the resolution of important open problems in computational biology [9]. We discuss in [8] the application of closed-chain planning techniques to the structural analysis of protein loops. The algorithms that we have presented can act as efficient filters for conformational search methods by making a geometric treatment of strong energetic constraints: maintaining the backbone integrity (i.e. loop-

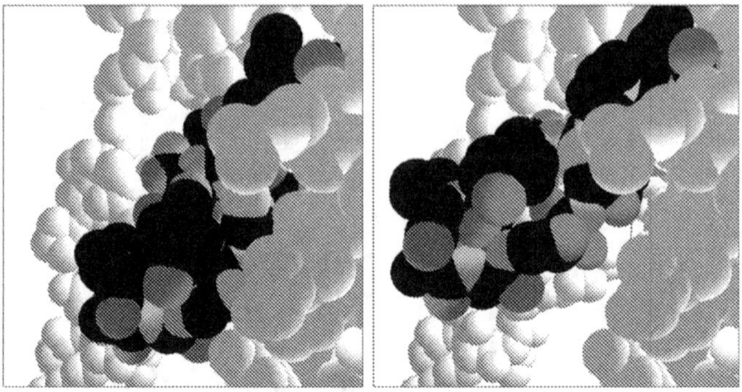

Fig. 10. Simulated conformational gating of loop 7 in amylosucrase. Crystallographic conformation (left) and geometrically feasible change (right).

closure) and avoiding steric classes. We have developed a conformational sampling algorithm that provides random conformations achieving such geometric constraints. The algorithm combines RLG for generating the backbone conformation with other sampling and collision detection techniques to obtain feasible conformations of the side-chains. We also propose a new conformational search technique, inspired by the RRT approach, for studying geometrically feasible loop motions. Preliminary results are promising. They demonstrate the capacity of the techniques to handle protein models with long loops that involve several dozens of degrees of freedom in the backbone and in the side-chains. Figure 10 illustrates an example of such a long loop in the model of amylosucrase from *Neisseria polysaccharea*.

8 Conclusions and Future Work

We have introduced sampling-based motion planning algorithms into an extended formulation of the motion planning problem under kinematic loop-closure constraints. We have tried to give general directives without focusing on a particular implementation. The RLG algorithm allows to overcome the challenge of sampling random configurations for general closed-chain mechanisms. The results obtained, in different domains of application, when solving difficult problems with PRM-based and RRT-based extended planners, demonstrate the efficacy and the generality of the approach.

Several points remain for future research. Some of them concern the RLG sampling algorithm. RLG provided good results in all our experiments. Nevertheless, a deeper analytical work is necessary in order to characterize its performance. Also, studying new forms of sampling, using quasi-random sequences or multi-resolution grids recently proposed [17], seems to be an interesting way to follow. Another improvement involves the selection of active and passive subchains. A general automatic method, based on an analysis of kinematic diagrams of mechanisms, remains to be devised.

A general methodology for the treatment of singularities within sampling-based motion planning algorithms remains an open topic to be further investigated. Recent interval methods [21,23] appear to be another matter to study for the improvement of closed-chain motion planning techniques. Interval methods can provide a complete approximated representation of the whole set of configurations satisfying loop closure equations. They compute a set of boxes that contain the continuum of solutions. Such a representation is very suitable for the application of sampling-based motion planning algorithms.

Another direction for future research concerns the application of closed-chain motion planning to computational biology. We intend to further study problems in structural biology that require the development of new techniques for the conformational analysis of protein loops. A first goal is to get efficient algorithms to capture the whole subset of the geometrically feasible conformations of one or several loops in the same protein. The next goal is to handle several proteins that interact while changing the conformation of loops on their surfaces. Applications to structural biology are attractive for the algorithmic development because of the kinematic complexity of molecular models. Obviously, robotic applications could also benefit from improvements achieved in this field.

Acknowledgments.
This work has been supported by the European project IST-37185 MOVIE and the project BioMove3D of the CNRS Bioinformatics Research Program.

References

1. Basu S., Pollack R., Roy M.-F. (2000). Computing Roadmaps of Semi-algebraic Sets on a Variety. J. Am. Math. Soc. **13(1)**, 55–82
2. Burdick J.W. (1988). Kinematic Analysis and Design of Redundant Robot Manipulators. PhD Thesis, Stanford University.
3. Burdick J.W. (1989). On the Inverse Kinematics of Redundant Manipulators: Characterization of the Self-Motion Manifold. Proc. IEEE Int. Conf. Rob. & Autom., 264–270
4. Canny J.F. (1988). The Complexity of Robot Motion Planning. MIT Press, Cambridge
5. Cortés J., Siméon T., Laumond J.-P. (2002). A Random Loop Generator for Planning the Motions of Closed Kinematic Chains using PRM Methods. Proc. IEEE Int. Conf. Rob. & Autom., 2141–2146
6. Cortés J., Siméon T. (2003). Probabilistic Motion Planning for Parallel Mechanisms. Proc. IEEE Int. Conf. Rob. & Autom., 4354–4359
7. Cortés J. (2003). Motion Planning Algorithms for General Closed-Chain Mechanisms. PhD Thesis, Institut National Polytechnique de Toulouse
8. Cortés J., Siméon T., Remaud-Siméon M., Tran V. (2004). Geometric Algorithms for the Conformational Analysis of Long Protein Loops. J. Comp. Chem. **25(7)**, 956–967
9. Finn P.W., Kavraki L.E., Latombe J.-C. et al. (1998). RAPID: Randomized Pharmacophore Identification. Comp. Geom.: Theory & Appl. **10(4)**, 263–272
10. Gosselin C. (1988). Kinematic Analysis, Optimization and Programming of Parallel Robotic Manipulators. PhD Thesis, McGill University

11. Han L., Amato N.M. (2001). A Kinematics-Based Probabilistic Roadmap Method for Closed Kinematic Chains. In: Donald B.R., Lynch K.M., Rus D. (Eds.) Algorithmic and Computational Robotics: New Directions (WAFR2000). A.K. Peters, Boston, 233–245
12. Kavraki L.E., Svestka P., Latombe J.-C., Overmars M.H. (1996). Probabilistic Roadmaps for Path Planning in High-Dimensional Configuration Spaces. IEEE Trans. Rob. & Autom. **12(4)**, 566–580
13. Latombe J.-C. (1991). Robot Motion Planning. Kluwer Academic Publishers
14. Latombe J.-C. (1999). Motion Planning: A Journey of Robots, Molecules, Digital Actors, and Other Artifacts. Int. J. Rob. Res. **18(11)**, 1119–1128
15. LaValle S.M., Yakey J.H., Kavraki L.E. (1999). A Probabilistic Roadmap Approach for Systems with Closed Kinematic Chains. Proc. IEEE Int. Conf. Rob. & Autom., 473–479
16. LaValle S.M., Kuffner, J.J. (2001). Rapidly-Exploring Random Trees: Progress and Prospects. In: Donald B.R., Lynch K.M., Rus D. (Eds.) Algorithmic and Computational Robotics: New Directions (WAFR2000). A.K. Peters, Boston, 293–308
17. LaValle S.M., Branicky M.S. (2004). On the Relationship Between Classical Grid Search and Probabilistic Roadmaps. In: Boissonnat J.-D., Burdick J., Goldberg K., Hutchinson S. (Eds.). Algorithmic Foundations of Roadmaps V (WAFR2002). Springer-Verlag, Berlin, 59–75
18. Lin M.C., Manocha D. (2004). Collision and Proximity Queries. In: Handbook of Discrete and Computational Geometry: Collision Detection. To appear. Available at `http://www.cs.unc.edu/~dm/`
19. Lozano-Pérez, T. (1983). Spatial Planning: A Configuration Space Approach. IEEE Trans. Computers **32(2)**, 108–120
20. McCarthy J.M. (2000). Geometric Design of Linkages. Springer-Verlag, New York
21. Merlet J.-P. (2001). A Parser for the Interval Evaluation of Analytical Functions and its Applications to Engineering Problems. J. Symb. Comp. **31**, 475–486
22. Nielsen J., Roth B. (1997). Formulation and Solution for the Direct and Inverse Kinematics Problem for Mechanisms and Mechatronic Systems. Proc. NATO Adv. Study Inst. on Comp. Meth. in Mech. **1**, 233–252
23. Porta J.M., Ros L., Thomas F., Torras C. (2003). A Branch-and-Prune Algorithm for Solving Systems of Distance Constraint. Proc. IEEE Int. Conf. Rob. & Autom., 342–348
24. Siméon T., Laumond J.-P., Nissoux C. (2000). Visibility-Based Probabilistic Roadmaps for Motion Planning. Adv. Rob. J. **14(6)**, 477–494
25. Siméon T., Laumond J.-P., Lamiraux F. (2001). Move3D: a Generic Platform for Path Planning. Proc. IEEE Int. Symp. Assembly & Task Planning, 25–30
26. Siméon T., Laumond J.-P., Cortés J., Sahbani A. (2004). Manipulation Planning with Probabilistic Roadmaps. Int. J. Rob. Res., *in press*
27. Thomas F. (1993). The Self-Motion Manifold of the N-bar Mechanism. In: Angeles J., Hommel G., Kovács P. (Eds.) Computational Kinematics. Kluwer Academic Publishers, Dordrecht, 95–107
28. Trinkle J.C., Milgram R.J. (2002). Complete Path Planning for Closed Kinematic Chains with Spherical Joints. Int. J. Rob. Res. **21(9)**, 773–789
29. Yakey J.H. (2000). Randomized Path Planning for Linkages with Closed Kinematic Chains. MA Thesis, Iowa State University
30. Yakey J.H., LaValle S.M., Kavraki L.E. (2001). Randomized Path Planning for Linkages with Closed Kinematic Chains. IEEE Trans. Rob. & Autom. **17(6)**, 951–958

Composing Navigation Functions on Cartesian Products of Manifolds with Boundary

Noah J. Cowan

Department of Mechanical Engineering
Johns Hopkins University; Baltimore, MD 21218

Abstract. Given two compact, simply connected manifolds with boundary, and a navigation function (NF) on each manifold, this paper presents a simple composition law that yields a new NF on the cross product space. The method provides tunable "hooks" for shaping the new potential function while still guaranteeing obstacle avoidance and essentially global convergence. The composition law is associative, and successive compositions fold into a single, computational simple expression, enabling the practical construction of NFs on the Cartesian product of several manifolds.

1 Introduction

The gradient vector field of a properly designed artificial potential function can steer a robot to a goal, while avoiding obstacles along the way. Adding a damping term to flush out any unwanted kinetic energy generalizes this approach to the second-order setting, since total energy always decreases in damped mechanical systems [14]. Of course, if a well-tuned robot control system is already in place, the potential field may be used in a more traditional first-order manner to generate trajectories via the gradient flow (in the case of a position controller) or field itself (in the case of a velocity controller).

The problem of spurious minima and safety for second-order plants lead Rimon and Koditschek to introduce navigation functions (NFs), a refined notion of artificial potential functions [16]. They showed that every smooth compact connected manifold with boundary, $\mathcal{M}$, admits a smooth NF, $\varphi : \mathcal{M} \to \mathbb{R}$. Thus, given a fully actuated Lagrangian system evolving on a such a configuration manifold, the machinery of NFs "solves" the global dynamical control and obstacle avoidance problem [20]. *Caveat emptor:* constructing NFs for arbitrary manifolds remains an art: each new model space requires a handcrafted NF. Example constructions for special model spaces include "sphere" worlds [16] and "star" worlds [19].

While a general methodology for constructing NFs seems unlikely, an "NF designers toolbox" for configuration spaces of practical interest seems plausible. Cartesian product spaces provide a compelling starting point since they imbue engineering applications: a serial-link robotic manipulator constructed of lower pairs between links comes to mind. Mathematical abstractions of

M. Erdmann et al. (Eds.): Algorithmic Foundations of Robotics VI, STAR 17, pp. 91–106, 2005.

physical phenomena, such as rigid motions ($\mathrm{SE}(n) = \mathrm{SO}(n) \times \mathbb{R}^n$), also highlight this point. Other application specific examples include the "occlusion-free configuration space" in visual servoing (see Sec. 4) [7, 8]. In each case, the systems are *topologically decoupled*, but *mechanically coupled.* Thus, a systematic set of tools for building NFs on Cartesian product spaces adds a key engineering tool to the toolbox.

1.1 Contribution

This paper presents a mathematically simple method by which to compose two NFs on two respective manifolds with boundary, to generate a new NF on the Cartesian product space. The composition law provides a control system designer with a set of gains to shape the resultant potential function, without imperiling the formal convergence and obstacle avoidance guarantees afforded by the NF methodology. Thus if the topology of the free configuration space can be expressed as the Cartesian product of several lower dimensional manifolds for which known NFs exists, a new NF for the free configuration space may be easily constructed from the constituent components.[1]

1.2 Related literature

Potential Fields and Energy Methods. Khatib introduced artificial potential functions for robot control in his Ph.D. dissertation over 20 years ago [11], and later employed potentials for dynamical obstacle avoidance [12]. Since 1980, several researchers have tried to repair inadequacies present in the early application of potential functions to robot control, most notably, the problem of local minima. In addition to addressing local minima, Koditschek and Rimon's NFs "lift" to second order dynamical settings while still ensuring essentially global convergence and obstacle avoidance [14, 16, 19, 20]. Lopes and Koditschek recently extended the use of NFs to the control of nonholonomic systems, in the context of perceptual constraints [17].

Other energy methods also round out the toolbox. For example, gyroscopic forces between multiple agents, each negotiating the same potential field, avoids mutual collisions [5], providing an elegant compliment to NFs. Also, the method of controlled Lagrangians [1,2] shapes the input to a system so that the closed loop behavior is that of a system with a different, desired Lagrangian; this is, in a sense, a generalization of potential shaping.

Composition. This paper presents a technique for controller composition: building up a large, integrated control systems based on relatively simple components. Burridge *et al.* describe a means by which to compose controllers sequentially [4]; roughly speaking, given a set of controllers, with known attractors and known domains of attraction, the resulting composition yields

[1] The constituent manifolds can be arbitrarily complicated, so long as an NF has been designed for each one.

a controller whose domain is the union of the constituent controllers. Their method applies to any dynamical system for which a set of constituent controllers (possessing certain technical requirements) exists. Sequential composition sits at a level above NFs, so the two may be integrated synergistically. For example, Conner et al. applied sequential composition to a collection of NF-like potentials for global robot navigation [6].

Parallel composition aims to allocate a small number of actuated degrees of freedom to manage the energy of multiple, independent dynamical systems. Early attempts at parallel composition, applied to robot juggling, succeeded empirically [3], leading to recent formal methods for parallel and interleaving composition by Klavins and Koditschek [13]. Their method abstracts each of n independent dynamical systems to its phase, which evolves on a copy of S^1. The total system evolves on the n-fold Cartesian product of S^1, T^n, where they construct a flow derived in part from a potential field. The systems they consider are topologically *and* (at least piecewise) mechanically "decoupled" (save the mechanical coupling of a shared actuation resource, such as the paddle in a juggling system [21]). An interesting extension of the current work would be to consider the parallel control of mechanically coupled systems evolving on "topologically decoupled" (i.e. cross product) spaces.

2 Control via Navigation Functions

This section summarizes the work on NFs developed by Rimon and Koditschek [14, 16, 19, 20].

2.1 Plant model

Consider a holonomically constrained, fully actuated mechanical system with known kinematics and suppose the configuration space is modeled by a compact, connected n-dimensional manifold with boundary, $\mathcal{Q}$. Let $(q, \dot{q}) \in T\mathcal{Q}$ denote the generalized positions and velocities on $\mathcal{Q}$. The equations of motion may be found using Lagrange's equations (see, for example, [9, 18]), namely

$$\frac{d}{dt}\frac{\partial}{\partial \dot{q}}L(q,\dot{q}) - \frac{\partial}{\partial q}L(q,\dot{q}) = u \tag{1}$$

where $L(q, \dot{q})$ is the Lagrangian, and u is a set of generalized force inputs. Assume that any external potentials (such as gravity) or non-viscous forces can be canceled by an appropriate feed-forward control term.

2.2 Task specification

Assume that any obstacles in the workspace are accounted for by the construction of $\mathcal{Q}$, so that for obstacle avoidance, trajectories must avoid crossing

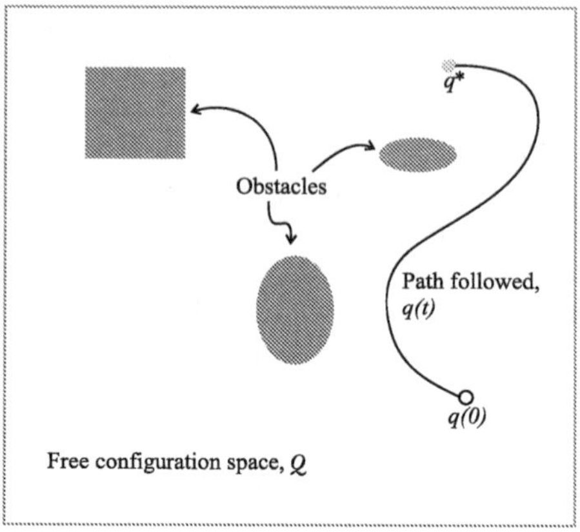

Fig. 1. An example of a free configuration space, $\mathcal{Q}$. In this case, $\mathcal{Q} \subset \mathbb{R}^2$ is a "square with three holes".

the boundary $\partial\mathcal{Q}$, for all $t \geq 0$ (see Fig. 1). The positioning objective is described in terms of a *goal*, q^*, in the interior of the domain, $\mathring{\mathcal{Q}}$. The task is to drive q to q^* asymptotically subject to (1) by an appropriate choice of u while remaining in $\mathcal{Q}$. Moreover, the basin of attraction $\mathcal{E}$ must include a dense subset of the zero velocity section of $T\mathcal{Q}$; this guarantees convergence from almost every initial zero velocity state $(q(0), \dot{q}(0)) = (q_0, 0)$ whose position component lies in $q_0 \in \mathcal{Q}$.

2.3 First order gradient systems

Let $\mathcal{Q}$ be an n-dimensional manifold with boundary, and consider the kinematic control system given by $\dot{q} = u$ where $u \in T\mathcal{Q}$ is a generalized velocity input. One possible control strategy involves following the gradient of a potential function $\varphi : \mathcal{Q} \to \mathbb{R}$, namely

$$\dot{q} = -\nabla\varphi(q). \tag{2}$$

A smooth scalar valued function whose Hessian matrix is non-singular at every critical point is called a *Morse* function [10]. Potential field controllers (2) arising from Morse functions impose global steady state properties that are particularly easy to characterize, as summarized in the following result.

Theorem 1 (Koditschek, [14]). *Let φ be a twice continuously differentiable Morse function on a compact Riemannian manifold, $\mathcal{Q}$. Suppose that $\nabla\varphi$ is transverse and directed away from the interior of $\mathcal{Q}$ on any boundary of that set. Then the negative gradient flow has the following properties:*

1. *$\mathcal{Q}$ is a positive invariant set;*
2. *the positive limit set of $\mathcal{Q}$ consists of the critical points of φ*
3. *there is a dense open set $\widetilde{\mathcal{Q}} \subset \mathcal{Q}$ whose limit set consists of the local minima of φ.*

2.4 Second order, damped gradient systems

The first order dynamical convergence results above do not directly apply to Lagrangian systems. This section reviews machinery to "lift" the gradient vector field controller (2) to one appropriate for the second order plant (1). Adding a linear damping term yields a nonlinear "PD" style feedback,[2]

$$u = -\nabla\varphi(q) - K_d\,\dot{q}, \tag{3}$$

and it follows that the total energy,

$$\eta(q,\dot{q}) = \varphi(q) + \kappa(q,\dot{q}),$$

where κ is the kinetic energy functional, is non-increasing [14].

Note that if the total initial energy exceeds the potential energy at some point on the boundary $\partial\mathcal{Q}$, a trajectory beginning within $\mathcal{Q}$ may intersect $\partial\mathcal{Q}$. Fortunately, further refining the class of potential functions will enable the construction of controllers for which the basin of attraction contains a dense subset of the zero velocity section of $T\mathcal{Q}$.

Definition 1. Let $\mathcal{Q}$ be an n-dimensional compact, simply connected manifold with boundary, and let $q^* \in \mathcal{Q}$ be distinct point. Let $\mathcal{C} \subset \partial\mathcal{Q}$, called the "corners" of $\mathcal{Q}$, be a nowhere dense subset. A functional $\varphi : \mathcal{Q} \to [0,1]$ is a *navigation function*, if it

1. is continuous on $\mathcal{Q}$ and C^2 on $\mathcal{Q} - \mathcal{C}$;
2. achieves its unique minimum of 0 at $q^* \in \mathring{\mathcal{Q}}$;
3. attains its maximal value of 1 uniformly on $\partial\mathcal{Q}$, the boundary of $\mathcal{Q}$ (assuming $\partial\mathcal{Q} \neq \emptyset$);
4. is Morse on $\mathcal{Q} - \mathcal{C}$.

This definition was adapted from [14,15,19]. In particular, [19] addresses the issue of manifolds with smooth interiors but possibly "sharp corners." For example, the square, $\mathcal{Q} = [-1,1] \times [-1,1]$, which has four sharp corners, admits an NF as will be seen in Sec. 4.

This notion of an NF, together with the observation that total energy decreases in a damped mechanical system, now yields the desired convergence result for the Lagrangian system (1).

Theorem 2 (Koditschek, [14]). *Given the system described by (1) subject to the control (3), almost every initial condition within the set*

$$\mathcal{E} = \{(q,\dot{q}) \in T\mathcal{Q} : \eta(q,\dot{q}) \le 1\} \tag{4}$$

converges to $(q^*,0)$ asymptotically. Furthermore, transients remain within $\mathcal{Q}$, namely $q(t) \in \mathcal{Q}$ for all $t \ge 0$.

Theorem 2 generalizes the kinematic global convergence of Theorem 1. Starting within $\mathcal{E}$ imposes a "speed limit" as well as a positional limit, since the total energy must be initially bounded [15].

[2] The allusion to PD control derives from the fact that near a minimum of φ, (3) reduces to $u \approx -K_p(q-q^*) - K_d\dot{q}$, where $K_p = \frac{\partial^2\varphi}{\partial q^2}(q^*)$ is the Hessian of φ.

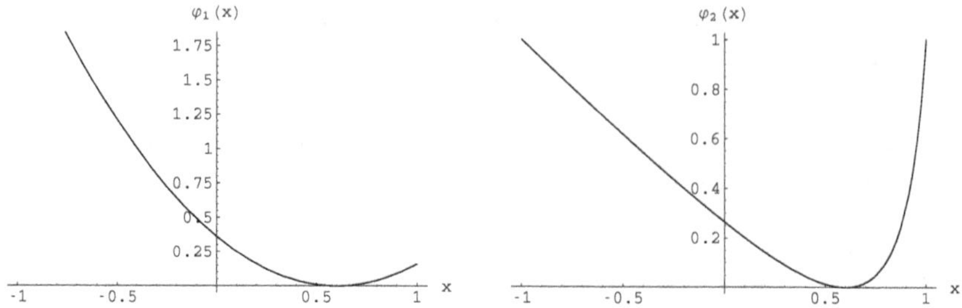

Fig. 2. An NF should evaluate uniformly at the boundary. In this case, $\mathcal{X} = [-1, 1]$, and φ_1 does not obtain the same value at $x = \pm 1$, whereas φ_2 does.

2.5 Invariance under diffeomorphism

A key ingredient in the mix of geometry and dynamics involves the realization that NFs are invariant to diffeomorphism [14]. This affords the introduction of geometrically simple model spaces and correspondingly simple NFs.

2.6 Why uniformly maximal at the boundary?

The requirement that an NF uniformly evaluate to a constant on the boundary is often overlooked. To illustrate this point, let $\mathcal{X} = [-1, 1]$, be the configuration manifold, and let $x^* = 0.6$ be the goal. The boundary, $\partial\mathcal{X}$, consists of two points, -1 and $+1$. As a candidate NF, one may naively consider

$$\varphi_1(x) = (x - 0.6)^2 \tag{5}$$

and note that $\nabla\varphi_1(x) = 2(x - 0.6)$, which has no local minima on $[-1, 1]$ except at the goal of $x^* = 0.6$. If the system begins at the left edge of the configuration manifold, it will have a total initial potential of $1.6^2 = 2.56$, but the potential barrier on the right edge is only 0.16. If the system is not highly over-damped, there are no guarantees that the right boundary will be avoided for all zero-velocity initial conditions within the domain $\mathcal{X}$. Thus, *second order safety* requires *boundary uniformity.* The NF

$$\varphi_2(x) = \frac{(x - 0.6)^2}{(1 - x^2) + (x - 0.6)^2}, \tag{6}$$

satisfies this requirement, as illustrated in Fig. 2.

3 Composing NFs on Cross Product Topologies

3.1 Construction

Consider two compact manifolds with boundary, $\mathcal{X}_1$ and $\mathcal{X}_2$, together with associated NFs, φ_1 and φ_2. Denote the Cartesian product space, $\mathcal{Q} = \mathcal{X}_1 \times \mathcal{X}_2$.

As a naive attempt construct a navigation function on $\mathcal{Q}$, consider the sum of NFs, $\varphi : \mathcal{Q} \to [0,1]$, given by

$$\varphi(q) = \tfrac{1}{2}(\varphi_1(x_1) + \varphi_2(x_2)),$$

where $q = (x_1, x_2)$. Such a function is Morse with unique minimum at $q^* = (x_1^*, x_2^*) \in \mathcal{Q}$, where $x_1^* \in \mathcal{X}_1$, $x_2^* \in \mathcal{X}_2$ are the respective goals of each NF. However, φ is *not* an NF on $\mathcal{Q}$. The problem arises because φ is not uniformly maximal on the boundary of $\mathcal{Q}$, given by the disjoint union of three components

$$\partial\mathcal{Q} = (\partial\mathcal{X}_1 \times \partial\mathcal{X}_2) \cup (\partial\mathcal{X}_1 \times \mathring{\mathcal{X}}_2) \cup (\mathring{\mathcal{X}}_1 \times \partial\mathcal{X}_2)\,.$$

For example, $\varphi(\cdot) < 1$ on $\partial\mathcal{X}_1 \times \mathring{\mathcal{X}}_2$, since $\varphi_2(\cdot) < 1$ on $\mathring{\mathcal{X}}_2$. Thus, the sum of two NFs, though possibly adequate for first-order kinematic systems, does not ensure safety with respect to the boundary when lifted to the second-order dynamic setting as described in Sec. 2.6.

Note that the function given by $q \mapsto (\varphi_1(x_1), \varphi_2(x_2))$, maps the boundary, $\partial\mathcal{Q}$, to the "top and right edges" of $[0,1]\times[0,1]$, namely $\{(1,\cdot)\}\cup\{(\cdot,1)\}$ and encodes the goal q^* at the point $(0,0)$. This motivates the following definition.

Definition 2. A *composition function* is a functional $\vartheta : [0,1]\times[0,1] \to [0,1]$ such that

1. $\vartheta(0,0) = 0$;
2. $\vartheta(1,\cdot) = \vartheta(\cdot,1) = 1$;
3. ϑ is C^2 everywhere but at the point $(1,1)$, where it is C^0.
4. ϑ is monotone increasing in both variables, i.e. $\frac{\partial\vartheta(z_1,z_2)}{\partial z_i} > 0$, $i = 1, 2$.

We are now ready to compose two NFs.

Proposition 1 (Navigation Function Product). *Consider two compact manifolds with boundary, $\mathcal{X}_1$ and $\mathcal{X}_2$, together with associated NFs, φ_1 and φ_2, and a composition function, ϑ. The* navigation function product

$$\varphi = \varphi_1 \vee \varphi_2,$$

given by $\varphi(q) = \vartheta(\varphi_1(x_1), \varphi_2(x_2))$, is a navigation function on $\mathcal{Q} = \mathcal{X}_1 \times \mathcal{X}_2$, with unique global minimum at $q^* = (x_1^*, x_2^*) \in \mathcal{Q}$.

Proof. (Uniformly maximal) Since ϑ is a composition function, φ achieves the value of 1 exactly when either φ_1 and/or φ_2 achieve a value of 1, which comprises the entire boundary $\partial\mathcal{Q}$.

(Corners are nowhere dense) Let $\mathcal{C}_i \subset \partial\mathcal{X}_i$, $i = 1, 2$ be the sets of corners associated with each respective manifold. Consider the set $\mathcal{C} \subset \partial\mathcal{Q}$, comprised of three disjoint components

$$\mathcal{C} = \mathcal{J}_1 \cup \mathcal{J}_2 \cup \mathcal{J}_3, \text{ where } \begin{cases} \mathcal{J}_1 = \mathcal{C}_1 \times \mathcal{X}_2 \\ \mathcal{J}_2 = \mathcal{X}_1 \times \mathcal{C}_2 \\ \mathcal{J}_3 = \partial\mathcal{X}_1 \times \partial\mathcal{X}_2\,. \end{cases}$$

This set comprises the set of corners of $\mathcal{Q}$, and as we will see, is the exact set on which $\varphi = \varphi_1 \vee \varphi_2$ is not smooth. Note that even if $\mathcal{X}_1$ and $\mathcal{X}_2$ are both corner-free, the Cartesian product $\mathcal{X}_1 \times \mathcal{X}_2$ has corners $\mathcal{J}_3$.

By hypothesis, $\mathcal{C}_i \subset \partial\mathcal{X}_i$, $i = 1, 2$ are nowhere dense, therefore $\mathcal{C}_1 \times \mathcal{X}_2 \subset \partial\mathcal{Q}$ (and vice versa) are nowhere dense. Moreover, $\partial\mathcal{X}_1 \times \partial\mathcal{X}_2$ is nowhere dense on $\partial\mathcal{Q}$. The finite union of nowhere dense sets is nowhere dense.

(Continuity) Note that φ_i is C^2 on $\mathcal{X}_i - \mathcal{C}_i$, $i = 1, 2$ and ϑ is C^2 on $[0,1] \times [0,1] - \{(1,1)\}$. By construction, $\mathcal{C}$ comprises exactly the "bad regions" where these functions are not smooth, and thus φ is C^2 on $\mathcal{Q} - \mathcal{C}$. Moreover, φ is the composition of continuous functions and is therefore continuous (C^0) everywhere on $\mathcal{Q}$.

(Morse) Note that

$$\nabla\varphi = \begin{bmatrix} \frac{\partial\vartheta(z_1,z_2)}{\partial z_1}\nabla\varphi_1 \\ \frac{\partial\vartheta(z_1,z_2)}{\partial z_2}\nabla\varphi_2 \end{bmatrix}\Bigg|_{z_1=\varphi_1, z_2=\varphi_2}$$

which is well defined everywhere except on $\mathcal{C}$. By Definition 2, $\frac{\partial\vartheta(z_1,z_2)}{\partial z_i} > 0$, $i = 1, 2$, so on $\mathcal{Q} - \mathcal{C}$ we have

$$\nabla\varphi(q) = 0 \iff \nabla\varphi_1(x_1) = 0 \text{ and } \nabla\varphi_2(x_2) = 0\,.$$

Thus, the critical points of φ are given simply by all combinations of critical points of φ_1 and φ_2. The Hessian at a critical point[3] is given by

$$\frac{\partial^2\varphi}{\partial q^2} = \begin{bmatrix} \frac{\partial\vartheta(z_1,z_2)}{\partial z_1}\frac{\partial^2\varphi_1}{\partial x_1^2} & 0 \\ 0 & \frac{\partial\vartheta(z_1,z_2)}{\partial z_2}\frac{\partial^2\varphi_2}{\partial x_2^2} \end{bmatrix}\Bigg|_{z_1=\varphi_1, z_2=\varphi_2}. \tag{7}$$

Thus, the Hessian matrix evaluated at a critical point (7) is block diagonal with the positively scaled Hessians of each constituent NF on the diagonal. Thus, since the constituent Hessians are nondegenerate at a critical point, then φ is also nondegenerate.

(Unique minimum) The critical point $q^* = (x_1^*, x_2^*)$ corresponding to a minimum of both constituent potential functions is also a minimum. Moreover, it is the global minimum since $\varphi(q) = 0$ iff $\varphi_i(x_i) = 0$, $i = 1, 2$, which is only true at q^*. By inspection of the Hessian, all other others critical points are saddles and maxima. □

3.2 Designing composition functions

The composition functions presented below, while by no means exhaustive, have certain convenient properties, such as associativity, tunability, and successive compositions reduce to a single, computationally simple expression.

[3] This expression is not valid away from the critical points, since it explicitly uses the fact that $\nabla\varphi_i = 0$, $i = 1, 2$.

Consider the "squashing diffeomorphism" $\sigma : [0, \infty) \to [0, 1)$, defined by

$$\sigma(\alpha) := \frac{\alpha}{1 + \alpha} \tag{8}$$

and the function, $\chi : [0, 1) \times [0, 1) \to [0, \infty)$, given by

$$\chi(z_1, z_2) = \frac{z_1}{1 - z_1} + \frac{z_2}{1 - z_2}. \tag{9}$$

Finally, define $\vartheta : [0, 1] \times [0, 1] \to [0, 1]$ by

$$\vartheta(z_1, z_2) := \begin{cases} \sigma \circ \chi(z_1, z_2), & \text{when } z_1 < 1 \text{ and } z_2 < 1 \\ 1, & \text{otherwise.} \end{cases} \tag{10}$$

To see that this function satisfies Definition 2 note that

$$\vartheta(z_1, z_2) = \frac{z_1 + z_2 - 2z_1 z_2}{1 - z_1 z_2} \tag{11}$$

everywhere on $[0, 1] \times [0, 1]$, except at $z_1 = z_2 = 1$ (i.e. the "upper right corner"). In particular, note that the above expression evaluates to 1 when either $z_1 = 1$ or $z_2 = 1$ (but the expression is not well defined for $z_1 = z_2 = 1$). Nevertheless, the limit

$$\lim_{z_1, z_2 \to 1} \frac{z_1 + z_2 - 2z_1 z_2}{1 - z_1 z_2} = 1$$

and therefore ϑ is continuous, and well defined on its domain $[0, 1] \times [0, 1]$. Furthermore, the Jacobian of ϑ,

$$\frac{\partial \vartheta(z_1, z_2)}{\partial z} = \frac{1}{(1 - z_1 z_2)^2} \left[(1 - z_2)^2 \;\; (1 - z_1)^2 \right],$$

is well defined except when $z_1 = 1$ or $z_2 = 1$, and can be easily extended by taking the limit as $z_1 \to 1$ or $z_2 \to 1$ (but not both) to yield

$$\lim_{z_1 \to 1} \left(\frac{\partial \vartheta(z_1, z_2)}{\partial z} \right) = [1 \; 0], \quad \lim_{z_2 \to 1} \left(\frac{\partial \vartheta(z_1, z_2)}{\partial z} \right) = [0 \; 1].$$

As can be seen, the Jacobian is discontinuous at the point $z = (1, 1)$. It is also straight forward to compute the Hessian matrix, which is well defined everywhere on the domain $[0, 1] \times [0, 1]$ except at $z = (1, 1)$.

Remark 1 (Semigroup property). By Proposition 1, if φ_1 and φ_2 are, respectively, NFs on compact manifolds with boundary, $\mathcal{X}_1$ and $\mathcal{X}_2$, then $\varphi_1 \vee \varphi_2$ is an NF on $\mathcal{X}_1 \times \mathcal{X}_2$, and thus $\vee$ is closed. Moreover, Cartesian products of spaces are associative, i.e.

$$(\mathcal{X}_1 \times \mathcal{X}_2) \times \mathcal{X}_3 = \mathcal{X}_1 \times (\mathcal{X}_2 \times \mathcal{X}_3).$$

Finally, the navigation product is associative because ϑ is associative, namely $\vartheta(z_1, \vartheta(z_2, z_3)) = \vartheta(\vartheta(z_1, z_2), z_3)$. This can be verified by direct algebraic substitution. Thus, the set NFs on compact manifolds with boundary forms a semigroup. □

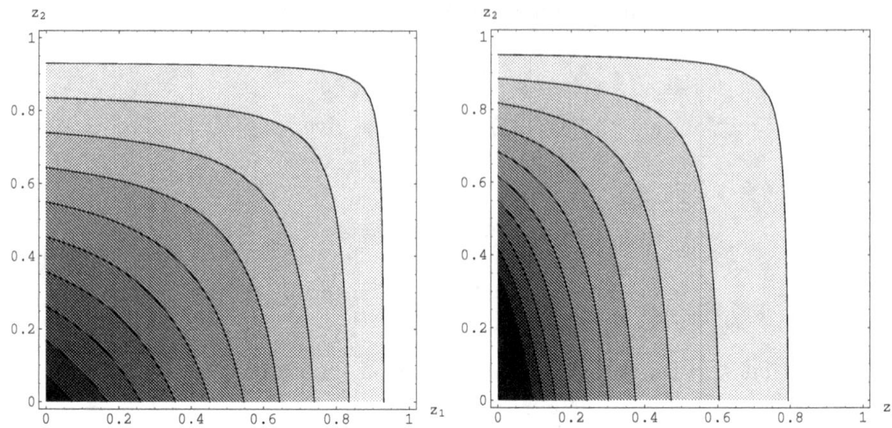

Fig. 3. Contour plots for different composition functions. *Left:* untuned ϑ in (10). *Right:* tuned $\vartheta_{\rm k}$ in (12), with $k = (1, 5)$.

Remark 2 (Fold). Let $\mathcal{X}_i$, $i = 1, \ldots, n$ be n manifolds with boundary and let φ_i be their respective NFs with goals x_i, $i = 1, \ldots, n$. Using the composition function ϑ, let $\varphi = \varphi_1 \vee \varphi_2 \cdots \vee \varphi_n$, $\mathcal{Q} = \mathcal{X}_1 \times \mathcal{X}_2 \times \cdots \mathcal{X}_n$, and let $q = (x_1, x_2, \ldots, x_n) \in \mathcal{Q}$. Then[4]

$$\varphi(q) = \sigma\left(\sum_{i=1}^{n} \frac{\varphi_i(x_i)}{1 - \varphi_i(x_i)} \right).$$

(Proof by induction). By definition, this is true for $n = 2$. Suppose it is true for $n = k$, and let $\varphi^k := \varphi_1 \vee \varphi_2 \cdots \vee \varphi_k$ be the navigation product of the first k functions. Dropping the explicit dependence on the x_i's, from (11) we have that

$$\varphi^{k+1} = \vartheta(\varphi^k, \varphi_{k+1}) = \frac{\varphi_{k+1} + \sigma\left(\sum_{i=1}^{k} \frac{\varphi_i}{1-\varphi_i}\right) - 2\varphi_{k+1}\sigma\left(\sum_{i=1}^{k} \frac{\varphi_i}{1-\varphi_i}\right)}{1 - \varphi_{k+1}\sigma\left(\sum_{i=1}^{k} \frac{\varphi_i}{1-\varphi_i}\right)}$$

which, upon multiplying numerator and denominator by $1 + \sum_{i=1}^{k} \frac{\varphi_i}{1-\varphi_i}$, and simplifying yields the desired result. □

Tunable composition function. While the NF framework affords certain formal guarantees, such as dynamical obstacle avoidance and essentially global convergence, one of the practical weaknesses of the existing NF literature concerns the tunability of NF-based controllers. In (3), the damping gain, K_d, comprises a set of a free design parameters, but there are no explicit hooks for

[4] One must use care on $\partial \mathcal{Q}$, e.g. when $\varphi_i = 1$, but at these points $\varphi(q) = 1$.

tuning the potential function, unless the designer builds such hooks directly into the NF. The composition function formed by replacing χ (9) with

$$\chi_k(z_1, z_2) = k_1 \frac{z_1}{1 - z_1} + k_2 \frac{z_2}{1 - z_2}$$

introduces a set of tunable gains $k = (k_1, k_2)$. As long as $k_1, k_2 > 0$, it is readily shown that the resulting composition function

$$\vartheta_k(z_1, z_2) := \begin{cases} \sigma \circ \chi_k(z_1, z_2), & \text{when } z_1 < 1 \text{ and } z_2 < 1 \\ 1, & \text{otherwise.} \end{cases} \tag{12}$$

satisfies Definition 2.

With appropriate choice of gains at each stage of composition, the $\vee$ is associative as before. For example

$$\vartheta_{(k_1,1)}\big(z_1, \vartheta_{(k_2,k_3)}(z_2, z_3)\big) = \vartheta_{(1,k_3)}\big(\vartheta_{(k_1,k_2)}(z_1, z_2), z_3\big),$$

and $\varphi = \varphi_1 \vee \varphi_2 \cdots \vee \varphi_n$ is given by

$$\varphi(q) = \sigma\left(\sum_{i=1}^{n} k_i \frac{\varphi_i(x_i)}{1 - \varphi_i(x_i)}\right).$$

4 Examples

We consider a few simple examples of the navigation product.

4.1 Cross product of a circle and an interval

Note that $\varphi_1(\theta) = (1 - \cos\theta)/3$ is an NF on S^1 with a goal at $\theta = 0$. The maximum value of φ_1 is chosen to be $2/3$, rather 1, since there is no boundary on S^1. The function $\varphi_2(x) = x^2$ is an NF on $[-1, 1]$, with a goal at $x = 0$. Let $\mathcal{Q} = S^1 \times [-1, 1]$, and let $q = (\theta, x) \in \mathcal{Q}$. The navigation product $\varphi = \varphi_1 \vee \varphi_2 : \mathcal{Q} \to [0, 1]$ is given by

$$\varphi(q) = (\varphi_1 \vee \varphi_2)(\theta, x) = \frac{1 + x^2 + \left(-1 + 2\,x^2\right)\cos\theta}{3 - x^2 + x^2\cos\theta}.$$

A 3D plot is shown in Figure 4.

4.2 Cross product of intervals

Let $\mathcal{X}_i = [\alpha_i, \beta_i]$, for some $\alpha_i, \beta_i \in \mathbb{R}$, be two intervals and consider a different NF, $\varphi_i : \mathcal{X}_i \to [0, 1]$, for each

$$\varphi_i(x) = \frac{(x - q_i^*)^2}{(\beta_i - x)(x - \alpha_i) + (x - q_i^*)^2}, \quad i = 1, 2, \tag{13}$$

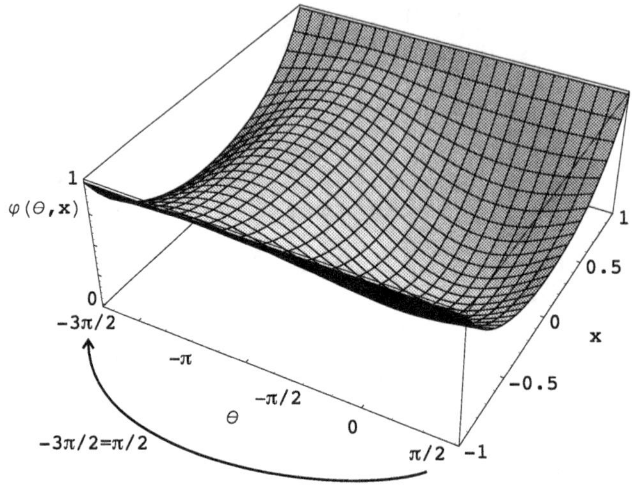

Fig. 4. A 3D plot of an NF on $S^1 \times [-1, 1]$ derived from two simple NFs on S^1 and $[-1, 1]$, respectively. The endpoints of the θ axis are identified. There are two critical points, a saddle and a minimum.

with goals at q_i^*, i=1,2. Let $\mathcal{Q} = \mathcal{X}_1 \times \mathcal{X}_2$, and let $q = (q_1, q_2) \in \mathcal{Q}$. Consider the naive NF candidate given by

$$\varphi_{\text{naive}}(y) = \tfrac{1}{2}(\varphi_1(q_1) + \varphi_1(q_2)).$$

As can be seen from Fig. 6, this naive construction is not an NF, whereas the navigation product $\varphi = \varphi_1 \vee \varphi_1$ is an NF. In the example plot, the configuration space is $\mathcal{Q} = [-\frac{\pi}{4}, \frac{\pi}{4}]^2$, with the goal location $q^* = (-\frac{3\pi}{20}, 0)$.

Note that the corners of $\mathcal{Q}$ are the four points $(-\frac{\pi}{4}, -\frac{\pi}{4})$, $(-\frac{\pi}{4}, \frac{\pi}{4})$, $(\frac{\pi}{4}, \frac{\pi}{4})$, $(\frac{\pi}{4}, -\frac{\pi}{4})$. Similarly, for the Cartesian product of three intervals, e.g. $\mathcal{Q} = [-1, 1]^3$, the corners are the edges of the cube.

4.3 Double pendulum

We simulated a two link, revolute-revolute mechanical system, as shown in Fig. 5, to illustrate the consequences of naive artificial potential function design. We assumed joint limits of $\pm\frac{\pi}{4}$, and thus $\mathcal{Q} = [-\frac{\pi}{4}, \frac{\pi}{4}]^2$. We chose simulated link lengths of $\ell_1 = \ell_2 = 1$ and masses of $m_1 = m_2 = 1$ located at the end of each respective link. Finally, we chose a diagonal damping matrix with diagonal elements given by $(1.5, 0.5)$.

For each DOF, we computed the NF, φ_i, $i = 1, 2$, as given in (13), where the goals were chosen to be $(q_1^*, q_2^*) = (-\frac{3\pi}{20}, 0)$. We applied the control law given by (3) for both the "naive" potential function, $\varphi_{\text{naive}} = \frac{1}{2}(\varphi_1 + \varphi_2)$, and a true NF given by $\varphi = \varphi_1 \vee \varphi_2$.

If the system were kinematic (2), or the links were mechanically uncoupled, then control based on, φ_{naive}, would guarantee that all zero-velocity

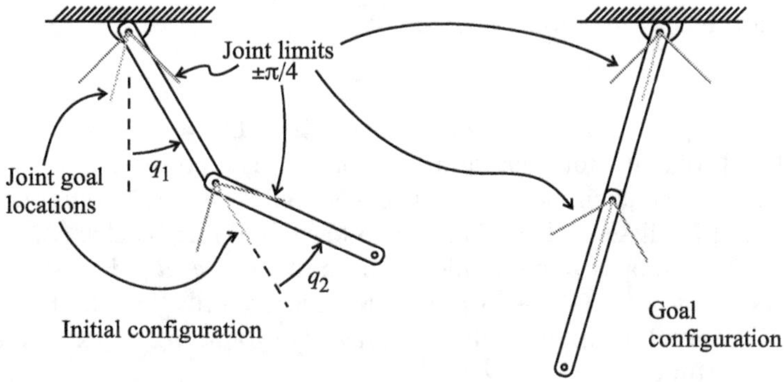

Fig. 5. A "double pendulum."

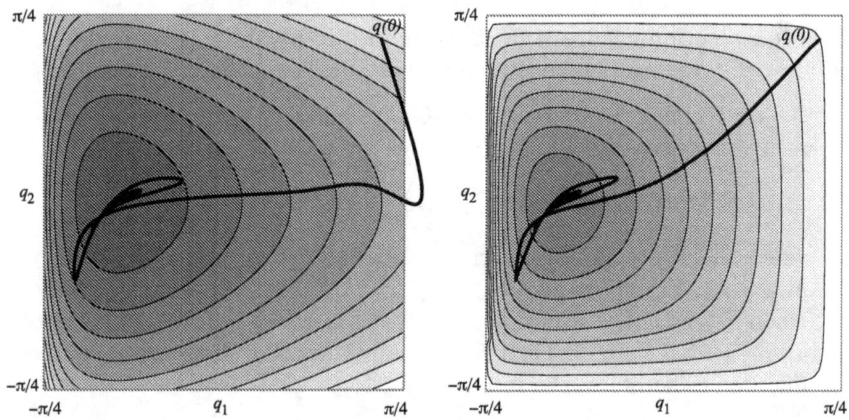

Fig. 6. Contour plots of two candidate NFs on the space $\mathcal{Q} = [-\frac{\pi}{4}, \frac{\pi}{4}]^2$. The configuration space trajectories resulting from a zero velocity initial condition,$(q(0), 0) \in T\mathcal{Q}$, of the double pendulum system shown in Fig. 5, subject to the "gradient + damping" feedback in (3) are shown by the bold curves. *Left.* The potential function, $\varphi_{\text{naive}}(q_1, q_2) = \frac{1}{2}(\varphi_1(q_1) + \varphi_2(q_2))$, is not an NF, because it is not uniformly maximal on the boundary, and thus the trajectory crosses the boundary. *Right.* The navigation product, $\varphi = \varphi_1 \vee \varphi_2$, from Proposition 1 is an NF, thus ensuring safety with respect to the boundary.

initial conditions within the free configuration space, $\mathcal{Q}$, safely converge. However, the mechanical coupling between the links renders the behavior undesirable since the first link violates its joint limit, as shown in Fig. 6.

4.4 Occlusion-Free visual servoing

We turn to a slightly more complex example. Cowan and Chang [7] showed that the set of all configurations of a perspective projection camera with a limited field of view, that keeps a specific visual target completely in view, is diffeomorphic to the model space

$$\mathcal{Q} = \mathcal{P} \times [-1,1] \times \mathcal{U} \tag{14}$$

where $\mathcal{P} := \{R \in \mathrm{SO}(3) : r_1 \cdot e_1 \geq \cos\psi\}$, $\mathcal{U} = \overline{\mathbf{D}^2} = \{u \in \mathbb{R}^2 : u_1^2 + u_2^2 \leq 1\}$ is a closed, planar unit disk, and $\psi \in (0, \pi/2)$ is a constant.

The basic construction of NFs on each configuration component follows Rimon and Koditschek [16]. Define two basic building blocks: a *goal function*, $\gamma : \mathcal{X} \to [0,\infty)$, possessing a unique minimum at a *goal point*, $x^* \in \mathcal{X}$, and an *obstacle function*, $\beta : \mathcal{X} \to [0,\infty)$ that vanishes (only) on the boundary $\partial\mathcal{X}$, which is treated as an obstacle set. These two building blocks are assembled to create a function $\varphi : \mathcal{X} \to [0,1]$:

$$\varphi := \frac{\gamma}{\gamma + \beta}. \tag{15}$$

This construction ensures that φ is uniformly maximal on $\partial\mathcal{X}$. The critical point structure must be verified on a case-by-case basis.

1. *NF on* $\mathcal{P}$. Let $R = \begin{bmatrix} r_1 \; r_2 \; r_3 \end{bmatrix} \in \mathrm{SO}(3)$. Define goal and obstacle functions, respectively, as

$$\gamma_1(R) := \mathrm{trace}\left((I - RR^{*T})\Lambda\right),\ \beta_1(R) := r_1^T e_1 - \cos\psi \tag{16}$$

   where Λ is a diagonal matrix with three distinct positive eigen values, $\lambda_1, \lambda_2, \lambda_3$. Let $\varphi_1 := \gamma_1/(\gamma_1 + \beta_1)$ as described above. As shown in [7], when $R^* = I$, the function $\varphi_1 : \mathcal{P} \to \mathbb{R}$, has a unique global minimum at $R = I$, and any other critical points are non-degenerate saddles and maxima. Therefore, the function φ_1, for a goal location of $R^* = I$, is an NF.
2. *NF on* $[-1,1]$. Define goal and obstacle functions, respectively, as

$$\gamma_2(\zeta) := (\zeta - \zeta^*)^2,\ \beta_2(\zeta) := 1 - \zeta^2, \tag{17}$$

   where $\zeta^* \in (-1,1)$ is the goal point. One can show that φ_2, defined in (15) is an NF. (This is equivalent to (13) with $\alpha = -1$ and $\beta = 1$.)
3. *NF on* $\mathcal{U}$. The manifold $\mathcal{U} := \{u \in \mathbb{R}^2 : \|u\| \leq 1\}$ is the simplest form of a "sphere world," as defined by Koditschek and Rimon [16]. Define goal and obstacle functions, respectively, as

$$\gamma_3(u) := \|u - u^*\|^2,\ \beta_3(u) := 1 - u_1^2 - u_2^2, \tag{18}$$

   where $u^* \in \mathbf{D}^2$ (the interior of $\mathcal{U}$). The function φ_3, as defined in (15), is an NF [16].

Let $q = (R, \zeta, u)$. Given the NFs φ_1, φ_2 and φ_3 defined on their respective manifolds $\mathcal{P}$, $[-1,1]$ and $\mathcal{U}$, then the function $\varphi = \varphi_1 \vee \varphi_2 \vee \varphi_3$, given by

$$\varphi(q) := \sigma\left(k_1 \frac{\varphi_1(R)}{1 - \varphi_1(R)} + k_2 \frac{\varphi_2(\zeta)}{1 - \varphi_1(\zeta)} + k_3 \frac{\varphi_3(u)}{1 - \varphi_3(u)}\right), \tag{19}$$

is an NF on the cross product space, $\mathcal{Q} = \mathcal{P} \times [-1,1] \times \mathcal{U}$.

5 Conclusions

The goal of this paper is to render the machinery of NFs more useful. Typically, each new space – even if just a simple Cartesian product of other spaces with known NFs – has required the construction of a new NF. This paper enables a designer to construct an obstacle avoiding "spring law" for each separate configuration component, and then the navigation product, $\vee$, stitches the spring laws together so that dynamical obstacle avoidance is maintained even for coupled mechanical systems. The technique makes no extra assumptions on the topology or geometry of the underlying spaces than those already required for the NF literature. Moreover, the composition law has the virtue of being associative, tunable, and computational simple even for large numbers of successive cross products.

For real systems, a significant amount of "hand" design may still be necessary, but the design process is vastly simplified by decomposing the problem into (topologically) separate pieces, and then combining the results together with the navigation function product.

Acknowledgment

The author would like to thank the reviewers for their many useful comments and suggestions. In addition, thanks to Andrew Lamperski, Richard Groff, Dong Eui Chang, Elon Rimon, Rob Ghrist, Eric Klavins and Jacopo Piazzi, for providing very useful feedback.

References

1. Anthony M. Bloch, Dong E. Chang, Naomi E. Leonard, and Jerrold E. Marsden, *Controlled Lagrangians and the stabilization of mechanical systems II: Potential shaping*, IEEE Transactions on Automatic Control **46** (2001), no. 1, 1556–1571.
2. Anthony M. Bloch, Naomi E. Leanard, and Jerald E. Marsden, *Controlled Lagrangians and the stabilization of mechanical systems I: The first matching theorem*, IEEE Transactions on Automatic Control (2000), 2253–2270.
3. Martin Buehler, Daniel E. Koditschek, and Peter J. Kindlmann, *Planning and control of robotic juggling and catching tasks*, International Journal of Robotics Research **13** (1994), no. 2, 101–118.
4. Robert R. Burridge, Alfred A. Rizzi, and Daniel E. Koditschek, *Sequential composition of dynamically dexterous robot behaviors*, Int. J. Rob. Res. **18** (1999), no. 6, 534–555.
5. Dong E. Chang, Shawn C. Shadden, Jerrold E. Marsden, and Reza Olfati-Saber, *Collision avoidance for multiple agent systems*, Proc. Conference on Decision and Control (Hawaii), vol. 1, IEEE, 2003, pp. 539–543.
6. David C. Conner, Alfred Rizzi, and Howie Choset, *Composition of local potential functions for global robot control and navigation*, International Conference on Intelligent Robots and Systems, vol. 4, IEEE/RSJ, Las Vegas, NV, 2003, pp. 3546– 3551.

7. Noah J. Cowan and Dong Eui Chang, *Geometric visual servoing*, IEEE Transactions on Robotics (2004), Submitted.
8. Noah J. Cowan, Omid Shakernia, Rene Vidal, and Shankar Sastry, *Vision-based follow-the-leader*, Intelligent Robots and Systems (Las Vegas, NV), vol. 2, IEEE, October 2003, pp. 1796– 1801.
9. John J. Craig, *Introduction to robotics*, Addison-Wesley, Reading, Mass., 1986.
10. Morris W. Hirsch, *Differential topology*, Springer-Verlag, 1976.
11. Oussama Khatib, *Commande dynamique dans l'espace operationnel des robots manipulateurs en presence d'obstacles*, Ph.D. thesis, L'Ecole Nationale Superieure de l'Aeronautique et de l'Espace, 1980.
12. ______, *Real-time obstacle avoidance for manipulators and mobile robots*, International Journal of Robotics Research **5** (1987), no. 1, 90–98.
13. Eric Klavins and Daniel E. Koditschek, *Phase regulation of decentralized cyclic robotic systems*, International Journal of Robotics and Research **21** (2002), no. 3, 257–276.
14. Daniel E. Koditschek, *The application of total energy as a Lyapunov function for mechanical control systems*, Dynamics and control of multibody systems (Brunswick, ME, 1988), Amer. Math. Soc., Providence, RI, 1989, pp. 131–157.
15. ______, *The control of natural motion in mechanical systems.*, ASME Journal of Dynamic Systems, Measurement, and Control **113** (1991), no. 4, 547–551.
16. Daniel E. Koditschek and Elon Rimon, *Robot navigation functions on manifolds with boundary*, Advances in Applied Mathematics **11** (1990), 412–442.
17. Gabriel A. D. Lopes and Daniel E. Koditschek, *Level sets and stable manifold approximations for perceptually driven nonholonomically constrained navigation*, Intelligent Robots and Systems, IEEE, 2003, submitted.
18. Richard M. Murray, Zexiang Li, and S. Shankar Sastry, *A mathematical introduction to robotic manipulation*, CRC Press, Boca Raton, FL, 1994.
19. Elon Rimon and Daniel E. Koditschek, *The construction of analytic diffeomorphisms for exact robot navigation on star worlds*, Transactions of the American Mathematical Society **327** (1991), no. 1, 71–115.
20. ______, *Exact robot navigation using artificial potential fields*, IEEE Transactions on Robotics and Automation **8** (1992), no. 5, 501–518.
21. Alfred A. Rizzi, Louis L. Whitcomb, and Daniel E. Koditschek, *Distributed real-time control of a spatial robot juggler*, IEEE Computer **25** (1992), no. 5, 12–24.

Adaptive RRTs for Validating Hybrid Robotic Control Systems

Joel M. Esposito[1], Jongwoo Kim[2], and Vijay Kumar[2]

[1] WSE Department, US Naval Academy, Annapolis, MD 21403, USA
[2] MEAM Department, University of Pennsylvania, Philadelphia, PA 19104
esposito@usna.edu, jwk,kumar@grasp.cis.upenn.edu

Summary. Most robot control and planning algorithms are complex, involving a combination of reactive controllers, behavior-based controllers, and deliberative controllers. The switching between different behaviors or controllers makes such systems hybrid, *i.e.* combining discrete and continuous dynamics. While proofs of convergence, robustness and stability are often available for simple controllers under a carefully crafted set of operating conditions, there is no systematic approach to experimenting with, testing, and validating the performance of complex hybrid control systems. In this paper we address the problem of generating sets of conditions (inputs, disturbances, and parameters) that might be used to "test" a given hybrid system. We use the method of Rapidly exploring Random Trees (RRTs) to obtain test inputs. We extend the traditional RRT, which only searches over continuous inputs, to a new algorithm, called the Rapidly exploring Random Forest of Trees (RRFT), which can also search over time invariant parameters by growing a set of trees for each parameter value choice. We introduce new measures for coverage and tree growth that allows us to dynamically allocate our resources among the set of trees and to plant new trees when the growth rate of existing ones slows to an unacceptable level. We demonstrate the application of RRFT to testing and validation of aerial robotic control systems.

1 Introduction

Hybrid systems provide mathematical models of discrete/continuous dynamic systems. Many robotic systems including walking robots, grasping and manipulation, or logic-based software controlled robots can be modelled under this framework. In fact most robot control and planning algorithms are complex, involving a combination of reactive controllers [2], behavior-based controllers [27], and deliberative controllers [13, 19]. While it is possible to analyze each controller in isolation it is well known that the interaction between discrete and continuous time dynamics of such systems can produce rich and often unexpected behavior. For this reason, as these systems grow in complexity and sophistication, the need for automated design tools increases. The focus to date in the literature has been on the formal verification of safe operation, via the solution of the reachability problem, initially through symbolic methods [31, 20] and later through numerical techniques [3, 1, 5, 28, 9].

M. Erdmann et al. (Eds.): Algorithmic Foundations of Robotics VI, STAR 17, pp. 107–121, 2005.

However, it soon became apparent that the class of hybrid control systems for which the reachability problem was decidable is quite limited in both expressiveness and dimensionality [15, 21, 33]. Interestingly motion planning research experienced a similar evolution from exact (symbolic) methods [32], followed by the result that the problem was fundamentally hard [6], to a shift toward approximate methods that worked well in practice [29]. Most recently research activity focuses on *randomized* approaches to the problem which have been shown to scale well with dimension.

Test generation – a well established concept for software design – is a relatively new approach to analyzing control systems. Rather than prove controller safety exhaustively, our approach is to try to generate a set of test scenarios, using Rapidly Exploring Random Trees (RRTs), that cause the system to fail. In addition to considering traditional continuous inputs, we have extended the method to consider uncertain time invariant parameters. We call our algorithm the Rapidly exploring Random Forrest of Trees (RRFT). The merit of this approach as compared with reachability analysis is that decidability issues do not come into play because we are not attempting to represent or manipulate a reachable set. The drawback of the approach is that it is a semi-decision method, meaning that we can only *disprove* system safety by counter-example – safety cannot be proved. Despite this draw back we feel that randomized approaches hold the most promise for addressing complex nonlinear real-world problems for which trial and error testing is not sufficient; and formal analysis is intractable. Similar work has recently appeared which uses genetic algorithms [30]. Some works have used RRTs as a synthesis tool for hybrid control systems [12, 4]; but the idea of using RRTs to explore a system's faults has only appeared in passing [18, 8].

Our approach consists of drawing a parallel to, and using methods from, motion planning. Informally the *Motion Planning Problem* is: given a robot with dynamics and constraints (obstacles), find a path (if one exists) from the starting configuration to the goal configuration in some complex high dimensional configuration space. Similarly, the goal of test generation is to find a sequence of inputs (or disturbances) and parameters (if one exists) which will take a hybrid system from an initial state to some unsafe set in the hybrid state space.

The primary differences between motion planning and testing lie in the types of systems considered. For example, motion planning approaches do not traditionally consider hybrid systems (though recent work has [12]). Another difference is that in motion planning problems the state space is not simple connected, in the geometric sense, due to the presence of obstacles, necessitating the use of sophisticated collision detection algorithms. For hybrid system the state space is usually simply connected with a given mode. Perhaps most importantly, robotic systems are almost always output controllable (by design), so the reachable space is the entire output space. Therefore a solution usually exists, unlike testing problems. As a result considerations of when to stop growing the tree are rarely discussed.

The contributions and outline of this paper are as follows

- Formally introduce the Test Generation problem for complex control systems, point out the similarities to motion planning (Sections 2-3).

- Define new coverage criteria for RRTs (Section 4).
- Introduce the RRFT algorithm which is an capable of searching both over traditional continuous inputs (like the RRT) *and uncertain time invariant parameters*. RRFT is an adaptive algorithm which modifies its search strategy based on the run time coverage estimates (Section 5).

Incidentally, the algorithm can be used for motion planning or testing. Finally we apply the algorithms to validate two controllers for robotic unmanned aerial vehicles in Section 6.

2 Problem Statement

Definition 1. *We define a* **Finite Time Hybrid Control System** *with unknown parameters dependencies (modified from the Hybrid Automata, see [26]) as a tuple* $H = (X, Q, U, T, P, Init, f, Inv, E, G, R)$ *where*

- $X \subset \mathbb{R}^N$ *is a set of continuous variables;*
- $Q \subset \mathbb{N}$ *is a set of discrete variables which index the system modes;*
- $U \subset \mathbb{R}^m$ *is a compact set of continuous input values;*
- $T = [t_0, t_f] \subset \mathbb{R}$ *is a compact time interval the system evolves over;*
- $P \subset \mathbb{R}^p$ *is a compact set of uncertain, time invariant parameters;*
- $Init \subset Q \times X$ *is a set of possible initial conditions;*
- $f : Q \times X \times U \times P \to \mathbb{R}^N$ *is a vector field which prescribes the time derivative of the continuous variables* (i.e., $\dot{x} = f(q, x, u; p)$)*;*
- $Inv : Q \to 2^X$ *assigns to each* $q \in Q$ *an invariant set;*
- $E \subset Q \times Q$ *is a collection of edges describing the possible discrete transition (a.k.a.- mode switches);*
- $G : E \to 2^{X \times P}$ *assigns to each* $e = (q, q') \in E$ *a guard; and*
- $R : E \times X \to 2^X$ *assigns to each* $e = (q, q') \in E$ *a reset relation.*

Throughout this paper we refer to (x, q) as the state of the hybrid system. Note that we use the term "input signal" in the most general sense in that it can include yet unspecified feedback control inputs, human in the loop type inputs, disturbances, etc. Informally, the evolution of such a system consist of concatenations of continuous flows and discrete switches. Given an initial state (x^0, q^0), the derivative of the continuous state is prescribed by the map f as long as the state remains within the invariant set Inv. Once the continuous state leaves this set time stops, and the discrete q state will change to some new state q' along some edge e as determined by which guard g condition is true. If a reset relation is activated, the continuous state may jump as well. Time and the continuous state begin to evolve once again in this new discrete state. Again, many robotic system can be modelled in this way (see Sect. 1).

Note that the uncertain parameters p can affect the continuous or discrete dynamics. Examples of P could include a control gain, the initial condition of an adversarial agent, or the width of a narrow passage.

Definition 2. *The* **Testing Problem** TP *is specified as a tuple* $(H, x^0, q^0, s, \bar{U}, \delta t)$ *where*

- H *is a finite time hybrid control system as described above;*
- $x^0, q^0 \in Init;$
- $\bar{U}$ *is user defined discretization of* U*;*
- δt *is the fixed time period for which a constant* $u \in \bar{U}$ *is applied such that* $(t_f - t_o)/\delta t = k$ *is an integer;*
- $s : X \times Q \times P \to \mathbb{R}$ *is a specification;*

Given an initial state and a particular control function $u(t)$, s can be used to define a set. If $s(x, q; p) > 0$, then (x, q) is an acceptable state inside the specification set for p; otherwise it is unacceptable.

Problem 1. Given an initial state, (x^0, q^0), the **Testing Problem** is to determine $\{u_1, u_2, \ldots u_k\} \in \bar{U}$, which define a piecewise constant control sequence

$$u(t) = u_i \;\; if \;\; (i-1) \cdot \delta t \leq t < i \cdot \delta t \tag{1}$$

for $i = 1, \ldots, k$, and a $p \in P$, if they exist, such that $\exists t \in T$ for which $s(x(t), q(t); p) \leq 0$.

In words, the goal of the test generator is to determine a counter-example – an piecewise constant input sequence and a value of the time invariant uncertain parameters which will cause the system to fail – if one exists. Note the similarity to trajectory planning (for example see the problem statement in [7]).

3 Testing Through Rapidly Exploring Random Trees

The similarities between the Testing Problem and the motion planning problem, suggest the use of a randomized methods such as Probabilistic Road Maps [16] or Rapidly Exploring Random Trees (RRTs) [23]. We choose the RRT primarily because it works directly with the set of admissible inputs and is therefore directly applicable to systems with complex dynamics. This algorithm has experienced widespread success in solving a variety of high dimensional and nonlinear problems in motion planning and has recently been applied to controller synthesis problems for hybrid systems [12, 18, 4]. Figure 1 illustrates the concepts and a very basic algorithm is given in Algorithms 1 and 2. Note that ρ is a suitable metric function; and the notation $(x, q) + \int^{\delta t} H(u)dt$ means: using x, q as an initial condition, simulate the evolution of the hybrid system for δt seconds using $u(t)$ as the control input. Simulating hybrid systems is numerically sensitive and a specialize event detection algorithm must be used such as [11]. Various versions of the algorithm can be generated using different metrics [10], or random distributions. In Sect. 5 we focus on stopping criteria if no solution is found.

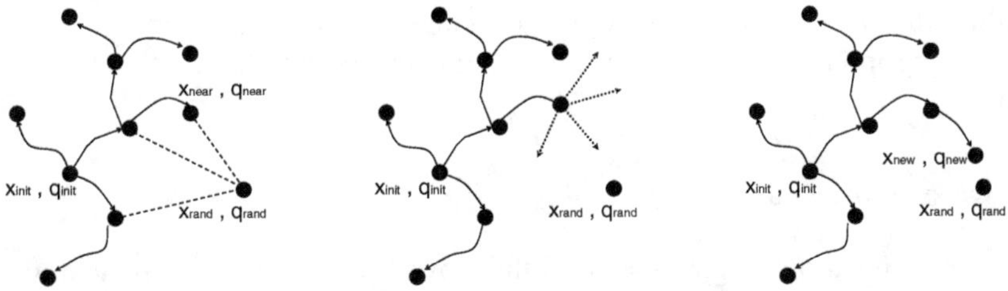

Fig. 1. The Testing algorithm (inspired by the RRT [23]). The test set is represented as a tree $\mathcal{T}$ with nodes as states (x, q) and edges as inputs $u \in \bar{U}$. First a new state is generated at random, x_{rand}, q_{rand}. The algorithm then determines the closest state, x_{near}, q_{near} in the tree to the random state (*left*). It determines which $u \in \bar{U}$ brings x_{near}, q_{near} closest to x_{rand}, q_{rand} (*center*). u_{new} is applied for a duration δt and the new node x_{new}, q_{new} and edge u_{new} are added to the tree (*right*).

Algorithm 1 Grow Test Set T

Initialize RRT: $\mathcal{T}$.addvertex(x^0, q^0)
while $\nexists(x, q) \in \mathcal{T}$ such that $s(x, q) \leq 0$ **do**
  Extend($\mathcal{T}$)
end while

Algorithm 2 Extend($\mathcal{T}$)

$x_{rand}, q_{rand} \leftarrow$ random()
$x_{near}, q_{near} \leftarrow$ nearestNeighbor($\mathcal{T}$, (x_{rand}, q_{rand}))
$u_{new} = \arg\min_{u \in \bar{U}} \{\rho((x_{rand}, q_{rand}), (x_{near}, q_{near}) + \int^{\delta t} H(u)dt)\}$
$(x_{new}, q_{new}) = (x_{near}, q_{near}) + \int^{\delta t} H(u_{new}(t))dt$
$\mathcal{T}$.add vertex(x_{new}, q_{new})
$\mathcal{T}$.addEdge$(u_{new}, (x_{near}, q_{near}) \rightarrow (x_{new}, q_{new}))$

4 Coverage Measures

It has been shown that, for a controllable system, the RRT will ultimately cover the entire state space as the number of sample points goes to infinity [24]. Unfortunately, because many of the systems we consider are not controllable with respect to the test inputs, the reachable set is not the entire space. It is very difficult in practice to estimate coverage quality because the reachable set is not known a priori. It is therefore important for us to estimate coverage for two reasons.

- Many testing problems may have no solution, meaning that there is not a counter-example to be found. In such a case we must decide when to terminate Algorithm 1.
- It is possible for Algorithm 1 to get stuck in "local minima" due to its greedy strategy [8] or to slow down because a tree is fully grown.

Regarding the second point, we can use coverage measures to determine when it might be appropriate to alter the search strategy. Indeed we explore this further in Section 5.

4.1 Coverage: previous work

It has been pointed out many times [22] that coverage of X by $\mathcal{T}$ is related to the *Voronoi Diagram* of the vertices of the tree. While this connection is useful for theoretical analysis the major problem is that it is impractical to compute Voronoi diagrams in high dimensions. The *Discrepancy* (a concept from the Monte Carlo literature) is also mentioned in [22] but it too is very difficult to compute. Another appealing idea to measure the growth or coverage of the RRT is to compute the volume of the convex hull. Unfortunately the convex hull is more indicative of the distribution of the points than it is of the coverage. For example, in Figure 2 the left and right panels represent two sample sets whose convex hulls are identical. Obviously the sample shown to the right covers the state space better. In [25] a variant of the convex hull is explored. Rather than compute the hull of all tree vertices, vertices are grouped according to their depth from parent nodes. The union of these hulls clearly provides a better approximation however the selection of the grouping is somewhat arbitrary. It is not clear how to relate the union to coverage due to possible overlaps.

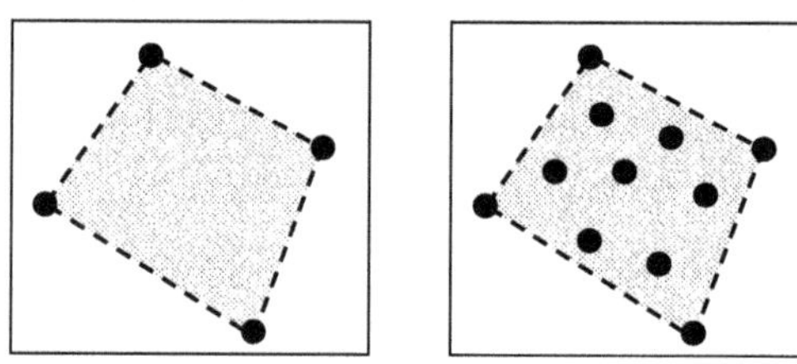

Fig. 2. Two sample sets which have the same convex hull. The set on the left clearly has inferior coverage to the set on right.

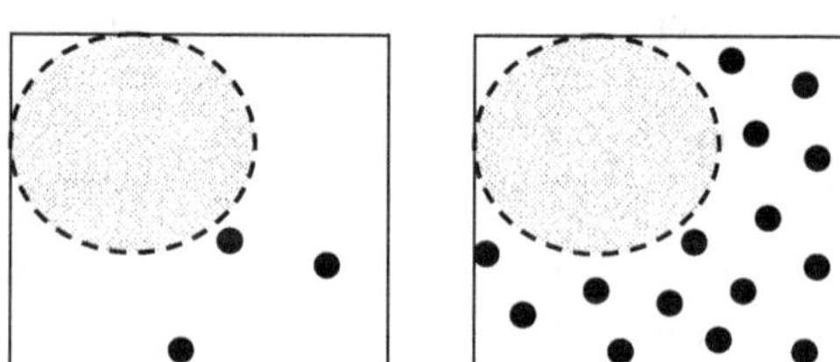

Fig. 3. Two sample sets which have the same dispersion (the size of the largest empty ball, drawn with dashed line). The set on the left clearly has inferior coverage to the set on right.

It appears that the most accepted measure to date is the *Dispersion* (see [14] or more recently [22]). Assuming we have a sample set $\bar{X}$, which contains N points, over the space X, it is defined as

$$\mu(\bar{X}, \rho) = \sup_{x \in X} \min_{\bar{x} \in \bar{X}} \rho(x, \bar{x}) \tag{2}$$

and can be thought of as the radius of the largest empty ball in X and obviously depends greatly on the choice of metric ρ. While its use has been advocated for analyzing planners we reject it for computation on two grounds: (1) it is impractical to compute in high dimensions; and, (2) it is an overly conservative coverage measure because it only considers the *largest* ball. For example, in Figure 3 the left and right panels represent two sample sets with the same dispersion. Obviously the sample shown on the right covers the state better.

4.2 New Coverage Measures

We have three goals: to measure the coverage of the state space X by the tree $\mathcal{T}$; to measure the growth of the tree $\mathcal{T}$ and to measure the coverage of the set of time invariant parameters P by a finite set of values $\bar{P}$.

RRT Coverage

We begin by overlaying a grid of n_g points and spacing δ on the state space. We calculate the minimum distance from each grid point j to the set of nodes in the tree, d_j. The quantity $\min(d_j, \delta)$ may be thought of as the radius of the largest ball centered at each grid point which does not contain a tree node or adjacent grid points (see Figure 4). Given a tree $\mathcal{T}$ we define its coverage, $c(\mathcal{T})$, as

$$c(\mathcal{T}) = \frac{1}{\delta} \sum_{j=1}^{n_g} \frac{\min(d_j, \delta)}{n_g} \tag{3}$$

which is the average of all the node distances, normalized by the grid spacing.

Our measure is similar to an approximation of an "average" dispersion, but far less conservative and faster to compute. Clearly this measure is a monotonically decreasing function. If it goes to zero on a given grid it tells us that any set whose distance along its smallest dimension is greater than the grid spacing has been entered. Said another way, the state space is covered up to a resolution equal to the grid spacing. Overall one of the advantages of this measure is that, the grid size can be as fine or coarse as one chooses. Finer grids will require more distance queries but are more accurate indications of coverage. From a computational point of view it should be stressed that this list of distances can be updated incrementally as new tree nodes are added, since the affect of each new node is local.

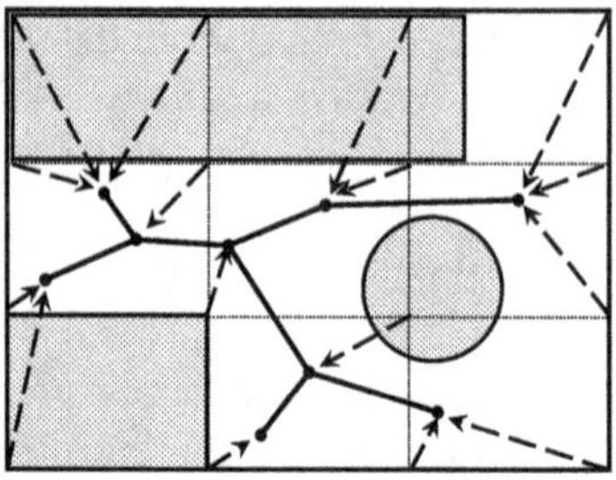

Fig. 4. A grid is super imposed on the state space. The shaded regions indicate unreachable sets. The distances from the grid points to the closest nodes are d_j(shown as dashed arrows) and the grid spacing is δ.

RRT Growth

The derivative of $c(\mathcal{T})$ with respect to the number of vertices in the tree, n_v, indicates the growth. Therefore

$$g(\mathcal{T}) = -\Delta c(\mathcal{T})/\Delta n_v \tag{4}$$

In practice the derivative is actually a finite difference and we may choose to look at the change in c over the course of adding several new vertices to $\mathcal{T}$.

Time invariant parameter set coverage

The set of parameters over which we may test the system may be generated directly since they are not subject to differential constraints and therefore the sample set can be engineered to have a certain coverage of P. Halton sequences [14] have naturally low dispersions and are cheap to compute. Therefore we define the coverage of P using the dispersion defined in eq.(2) normalized by μ_{max}.

5 Forest of Random Trees Algorithm

The original RRT given in Algorithm 1 only addresses time varying inputs such as $u(t)$. Recall that the evolution of our system is characterized by time invariant parameters, $p \in P$ as well. In our RRFT algorithm (see Algorithm 3), the repeated application of the RRT algorithm results in a tree for every choice $p \in P$ (called the seed value). Accordingly, we need to consider a *set of trees* (a forest) that rapidly explore the state space. We call a RRT grown (or rooted) at $p_i \in \bar{P}$, $\mathcal{T}_{p_i}$. Initially we plant trees at $\bar{P} = \{p_1, \ldots, p_{n_t}\}$, where n_t is the maximum number of active trees we can consider concurrently. At any point if a counter example is found (a state and parameter $s(x, q; p_i) \leq 0$ for which $(x, q) \in \mathcal{T}_{p_i}$) the algorithm terminates.

Because we have limited computational resources, we must decide how to allocate them in growing the trees – choosing which to grow and which to terminate. As the RRT algorithm progresses, we monitor the progress of each tree. If at any point the growth of one of the trees as measured by $g(\mathcal{T}_{p^i})$ drops below a threshold $\bar{g}$, the tree is considered no longer actively growing; or, if the coverage $c(\mathcal{T}_{p^i})$ is less than a threshold $\bar{c}$, the tree is considered fully grown. In either case the tree is terminated. Provided the set P is not adequately covered with seeds (as measured by the dispersion) a new "seed" is planted and a new tree is initiated. The process of planting and growing new trees continues until a counter example is discovered, or until P is sufficiently covered ($\mu(\bar{P}) \leq \bar{\mu}$) with seed values, whose trees have stopped growing.

One key component of this approach is that each RRT can be computed in parallel on a different CPU's, therefore we assume a fixed computational resource that will dictate the number of trees that can be simultaneously computed in parallel.

Algorithm 3 Test Generation: $\mathcal{T}_p$ = RRFT$(H, x^0, q^0, s, \bar{U}, \bar{g}, \bar{c}, \bar{\mu}, n_t)$

Generate initial seed set $\bar{P} = \{p_1, \ldots, p_{n_t}\}$ where $p_i \in P$
for $i = 1, \ldots n_t$ **do**
  Initialize RRT: $\mathcal{T}_{p_i}$.addvertex(x^0, q^0)
end for
while $n_t \neq 0$ **do**
  for $i = 1, \ldots n_t$ **do**
    Extend$(\mathcal{T}_{p_i})$
    if $\exists (x, q) \in \mathcal{T}_{p_i}$ such that $s(x, q; p) \leq 0$ **then**
      return $\mathcal{T}_{p_i}$
      break (test case found)
    else
      if $g(\mathcal{T}_{p_i}) \leq \bar{g}$, OR, $c(\mathcal{T}_{p_i}) \leq \bar{c}$ **then**
        terminate $\mathcal{T}_{p_i}$
        $n_t \leftarrow n_t - 1$
        if $\mu(\bar{P}) > \bar{\mu}$ **then**
          $n_t \leftarrow n_t + 1$
          Generate new $p_i \in P$ via Halton sequence and append to $\bar{P}$
          Initialize RRT: $\mathcal{T}_{p_i}$.addvertex(x^0, q^0)
        end if
      end if
    end if
  end for
end while

6 Examples

We demonstrate the algorithm on two examples involving Robotic Unmanned Arial Vehicles.

6.1 Example 1: Aircraft conflict resolution

As a first example we test an aircraft collision avoidance protocol proposed in [28]. We test over a continuous input $u(t)$, (a wind disturbance) and a constant parameter p, (the minimum separation distance) and consider a scenario involving 5 aircraft. The problem has 15 states, which is considerably larger than problems which have been considered in the literature on reachability and verification.

Each aircraft has three states, $X_i = (x, y, \theta)$ and there are 5 aircraft so the continuous state space is $X = X_1 \times \ldots X_5$. The continuous dynamics $f : Q \times X \times U \times P$ are

$$\dot{x}_i = v\cos(\theta_i) + (-d_1 \sin(\theta_i) + d_2 \cos(\theta_i))(-\sin(\theta_i)) \tag{5}$$
$$\dot{y}_i = v\sin(\theta_i) + (-d_1 \sin(\theta_i) + d_2 \cos(\theta_i))(\cos(\theta_i)) \tag{6}$$
$$\dot{\theta}_i = \omega(q; p) \tag{7}$$

Where v is a constant forward velocity; $u = [d_1 \;\; d_2]^T \in [-w, w] \times [-w, w]$ is a wind disturbance whose normal component to the planes alters their dynamics (this is the main difference versus [28]). Note that q and p do not explicitly appear in the dynamics but rather determine ω, the preset yaw rate control law. The control law was designed to bring each plane from $Init$ to its own predetermined final destination (x_i^g, y_i^g) without colliding. The function ω switches depending on the mode. At the start positions, the aircrafts are in $q = 1$, (*heading* mode) and rotate until pointing toward their goal positions, so $\omega(1) = \theta_{goal} - \theta_i$. Once they reach the desired heading, they switch to $q = 2$ (*cruise mode*), $\omega(2) = 0$, and cruise straight toward the goals. If two aircrafts get within a distance p km of each other, each of the two aircraft enters $q = 3$ (*avoid* mode) and makes instantaneous -90° turns, then it follows a half circle with angular velocity $\omega(3) = c$. After finishing the circular turn, they make instantaneous turns again until pointing to their own goal positions and return to cruise mode. In case the aircraft sees another aircraft within p km during the avoid mode, it makes -90° turn again and executes the same operation as above. This is illustrated in (see Figure 5 *left*). The specification is the minimum distance between all pairs of planes.

When $\|u\| \leq 0.03 km/sec$ and $p = 5.25 km$ a collision among the aircraft was discovered (see Figure 5 *right*) after about 8,600 nodes and 5 parameters were explored. A uniform distribution was used to generate samples, and a simple metric based on a weighted Euclidean distance is utilized

$$\rho = d + w_a |\Delta\theta| \tag{8}$$

where d is a Euclidean distance between two (x, y) positions, $\Delta\theta \in (-\pi, \; \pi]$ is a heading difference, and w_a is a weight factor. Figure 6 shows $c(\mathcal{T})$ and $g(\mathcal{T})$ for the trees. Figure 7 shows the coverage of the seed set, $\mu(\bar{P})$ as new seeds are generated. Three initial seeds are planted and two new seeds are generated until solution trajectory is found. Therefore a total of 5 seeds are tried to obtain the trajectory.

6.2 Example 2: Unmanned blimp control law

In this section, we consider the validation of a feedback control algorithm for waypoint to waypoint navigation of an unmanned outdoor blimp under unpredictable but bounded disturbances. The blimp has a 12-dimensional state space. Closed loop control laws use proportional inertial feedback to keep the blimp at the desired altitude with the target speed and to move from one inertial waypoint to the next. Waypoints are generated in the 3-dimensional space $(x,\ y,\ z)$. A change of waypoints can be considered a mode change. We bound the input, $u(t)$, (a wind disturbance) by limiting the magnitude of the wind gust and the rate of change of wind velocity. The maximum change rate of wind direction is $18^o/m$ and the magnitude is bounded as $\|u\| \leq 0.03km/sec$. For a detailed description of the feedback control law, sampling strategy, metric design, and the bounded wind disturbance, refer to [18]. In this case the time invariant parameter we will test over is the position of the waypoint. The waypoints are specified by a high level planner which does not account for the blimp's dynamics. We would like to see if it is possible for this planner to specify a waypoint which would cause the blimp to collide with the obstacle.

Figure 8 shows the trajectories of the blimp. The initial forward velocity is $0.5m/sec$. The target forward velocity is $1m/sec$. The starting waypoint is $[0\ 0\ -5]^T$ and the goal waypoint is $[150\ \ -150\ \ -10]^T$. We assume the high level planner can generate intermediate waypoints $p \in [80, 100] \times [-80, -60] \times (-10)$ to avoid a collision. We assume the navigation plan is achieved if the blimp can reach within 20m of the goal waypoint avoiding the obstacle under the wind disturbance. A counter-example is discovered with the intermediate waypoint at $[95\ \ -62.2\ \ -10]^T$ after exploring about 9,000 nodes. The solution requires 315 minutes of computation time on 1.4GHz PC. Coverage criteria are shown in Figure 9 and Figure 10. In this application, the RRFT analysis technique allows the designer to efficiently explore the safeness of the blimp's closed loop flight control laws in the presence of obstacles.

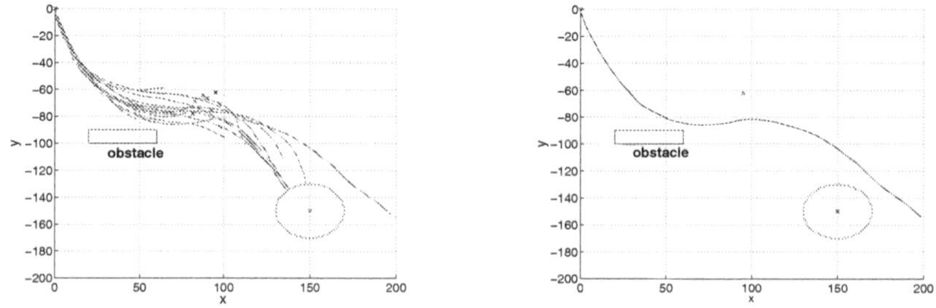

Fig. 8. The RRFT of the blimp under a wind disturbance $\|u\| \leq 0.03km/sec$ and uncertain intermediate waypoints (left). A solution trajectory is obtained after exploring about 9000 nodes (right)

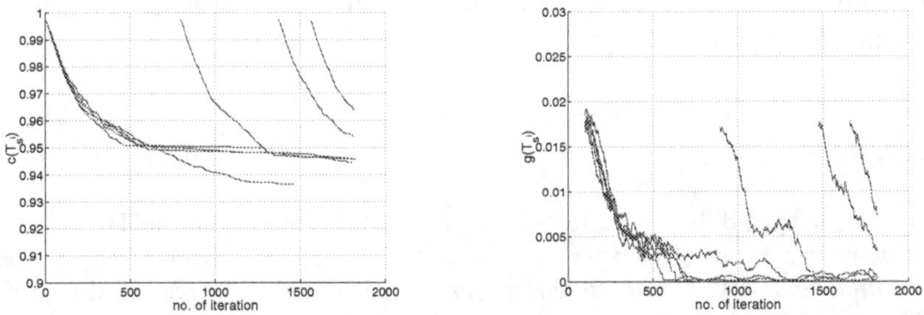

Fig. 9. The coverage of the trees. New trees are started when the growth rate slows below a specified threshold ($\bar{g} = 1 \times 10^{-10}$ is used in this example). A solution is discovered in one of the initial seeds.

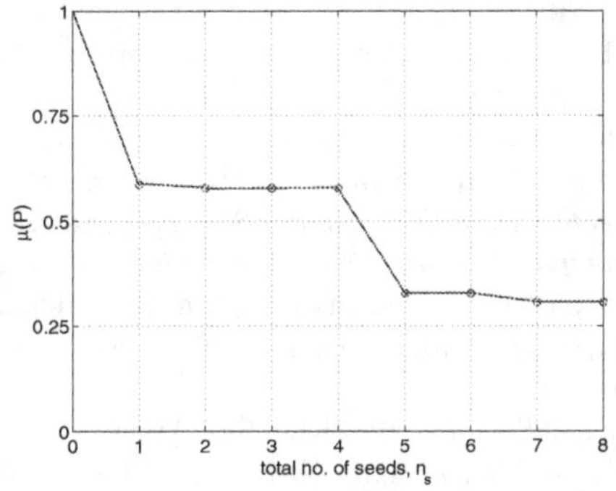

Fig. 10. The coverage improves ($\mu(\bar{P})$ decreases) as new trees are seeded. A solution is discovered at n_s=8.

7 Conclusion

The RRT method is a powerful technique to explore high-dimensional configuration spaces and find motion plans for systems with kinematic and dynamic constraints. In this paper, we presented two enhancements to this method and a novel application. First, we showed how sets of time-invariant parametric uncertainties can be explored with this method to generate a forest of trees. Second, we developed an on-line measure of dispersion that allows us to allocate computational resources among the individual trees based on their growth rates. We presented the application of both methods to the testing and validation of hybrid robot control systems, systems that do not lend themselves to proofs of convergence and stability. In both these examples, because the controller is fixed, the resulting trees do not expand to fill the configuration space. Instead, they fill a "tube" of configuration space that is defined by allowable disturbances and external inputs. The first example showed the ability to analyze multiple-agent systems with uncertainties, while the second example addressed the generation of worst case disturbances for the analysis of full dynamic models of aerial vehicles. The adaptation of the growth of the individual tree to coverage in configuration or state space is a direction of current research and is reported in a forthcoming publication [17].

References

1. T.A. Henzinger, B. Horowitz, R. Majumdar, and H. Wong-Toi. Beyond HyTech: Hybrid systems analysis using interval numerical methods. In *Hybrid Systems : Computation and Control*, volume 1790 of *Lecture Notes in Computer Science*, pages 130–144. Springer Verlag, 2000.
2. R. Arkin and T. Balch. *Artificial Intelligence and Mobile Robots*, chapter Cooperative Multiagent Robot Systems. MIT Press, 1998.
3. E. Asarin, T. Dang, and O. Maler. d/dt: A tool for reachability analysis of continuous and hybrid systems. In *5th IFAC Symposium on Nonlinear Control Systems*, St. Petersburg, Russia, 2001.
4. M.S. Branicky, M.M. Curtiss, J. Levine, and S. Morgan. RRTs for nonlinear, discrete, and hybrid planning and control. December 2003.
5. O. Butchkarev and S. Tripakis. Verification of hyrbid systems with linear differential inclusions using elipsoidal roximations. In *Hybrid Systems : Computation and Control*, volume 1790 of *Lecture Notes in Computer Science*, pages 73–88. Springer Verlag, 2000.
6. J. F. Canny. *The complexity of robot motion planning*. MIT Press, Cambridge, MA, 1988.
7. P. Cheng and S. M. LaValle. Reducing metric sensitivity in randomized trajectory design. In *Proceedings of IEEE/RSJ International Conference on Intelligent Robots and Systems*, pages 43–48, 2001.
8. P. Cheng, Z. Shen, and S.M. LaValle. RRT-Based trajectory design for autonomous vehicles and spacecraft. *Archives of control science*, 11(3-4):51 –78, 2001.
9. A. Chutinam and B. Krogh. Computational techniques for hybrid system verification. *IEEE transactions on automatic control*, 48(1):64–75, 2003.
10. J.M. Esposito. Randomized test case generation for hybrid systems: metric selection. In *IEEE Southeastern Symposium on Systems Theory*, Atlanta, GA.
11. J.M. Esposito and V. Kumar. An asynchronous integration and event detection algorithm for simulating multi-agent hybrid systems. To Appear in ACM Transactions on Computer Modeling and Simulation, 2004.
12. E. Frazzoli, M.A. Dahleh, and E. Feron. Real-time motion planning for agile autonomous vehicles. In *AIAA Confernce on guidance, navigation and control*, August 2000.
13. R.W. Ghrist and D.E. Koditschek. Safe cooperative robotic patterns via dynamics on graphs. In *Proceedings of the 8th International Symposium on Robotics Research*, Shonan Village, Japan, oct 1997.
14. J.H. Halton. On the efficiency of certain quasi-random sequences of points in evaluating multi-dimensional integrals. *Numer. Math.*, (2):84–90, 1960.
15. T.A. Henzinger, P.W. Kopke, A. Pujri, and P. Varaiya. What is decidable about hybrid automata? In *Proceedings of the 27th annual ACM Symposia on the Theory of Computation STOC*, pages 373–382, May 1995.
16. L.E. Kavraki, P. Svestka, J.C. Latombe, and M.H. Overmars. Probabilistic roadmaps for path planning in high dimensional configuration space. *International Transactions on Robotics and Automation*, 12:566–580, 1996.
17. J. Kim and J.M. Esposito. Adaptive biasing of RRTs for testing hybrid systems. Submitted to *IEEE Confernce on Decision and Control*, 2004

18. J. Kim, J. Keller, and V. Kumar. Design and verification of controllers for airships. *IEEE/RSJ International Conference on Intelligent Robots and Systems*, Oct 2003.
19. V. Kumar, M. Zefran, and J. Ostrowski. Intelligent motion planning and control. In *Handbook of Industrial Robotics*. John Wiley and Sons, 1999.
20. G. Lafferriere, G.J. Pappas, and S. Yovine. Symbolic reachability for families of linear vector fields. *Journal of Symbolic Computation*, 32:231–253, 2001.
21. G. Lafferriere, G.J. Pappas, and S. Yovine. A new class of decidable hybrid systems. In *Hybrid Systems : Computation and Control*, volume 1590 of *Lecture Notes in Computer Science*, pages 137–151. Springer Verlag, 1999.
22. S.M. LaValle and M.S. Branicky. On the relationship between classical grid search and probabilistic roadmaps. In *Workshop on the Algorithmic Foundation of Robotics (WAFR)*.
23. S.M. LaValle and J.J. Kuffner. Randomized kinodynamic planning. *International Journal of Robotics Research*, 20(5):378–400, 2001.
24. S.M. LaValle and J.J. Kuffner. Rapidly exploring random trees: Progress and prospects. In B. Donald, K. Lynch, and D.Rus, editors, *Algorithmic and computational robotics: new directions*, pages 293–308. A.K. Peters, Wellesley, MA, 2001.
25. J.A. Levine. Sampling-based planning for hybrid systems. Master's thesis, Case Western Reserve University, September 2003.
26. J. Lygeros and G. Pappas. A tutorial on hybrid systems: Modeling, analysis and control. 14th IEEE International Symposium on Intelligent Control/Intelligent Systems and Semiotics, September 1999.
27. M. Mataric. Issues and approaches in the design of collective autonomous agents. *Robotics and Autonoumous Systems*, 16(2-4):321–331, Dec 1995.
28. I. Mitchell and C. Tomlin. Level set methods for computation in hybrid systems. In B. Krogh and N. Lynch, editors, *Hybrid Systems : Computation and Control*, volume 1790 of *Lecture Notes in Computer Science*, pages 310–323. Springer Verlag, 2000.
29. P. Chen and Y.K. Hwang. SANDROS: A dynamic graph search algorithm for motion planning. *IEEE Transactions on Robotics and Automation*, 14(3):390–403, June 1996.
30. Q. Zhao, B.H. Krogh, and P. Hubbard. Generating test inputs for embedded control systems. *IEEE Control Systems Magazine*, 23(4):49–57, 2003.
31. R. Alur, T.A. Henzinger, and P.-H Ho. Automatic symbolic verification of embedded systems. *IEEE Transactions on Software Engineering*, 22:181–201, 1996.
32. J.T. Schwartz and M. Sharir. On the 'Piano Movers' problem: II General techniques for computing topological properties of real algebraic manifolds. *Advances in applied mathematics*, 4:298–351, 1983.
33. P. Tabuada and G.J. Pappas. Model checking LTL over controlable linear systems is decidable. In *Hybrid Systems : Computation and Control*, volume 2623 of *Lecture Notes in Computer Science*. Springer Verlag, 2003.

Collision Free Motion Planning on Graphs

Michael Farber

University of Durham, Durham DH1 3LE, UK

Abstract. A topological theory initiated in [4], [5] uses methods of algebraic topology to estimate numerically the character of instabilities arising in motion planning algorithms. The present paper studies *random* motion planning algorithms and reveals how the topology of the robot's configuration space influences their structure. We prove that the topological complexity of motion planning $\mathbf{TC}(X)$ coincides with the minimal n such that there exists an n-valued random motion planning algorithm for the system; here X denotes the configuration space. We study in detail the problem of collision free motion of several objects on a graph Γ. We describe an explicit motion planning algorithm for this problem. We prove that if Γ is a tree and if the number of objects is large enough, then the topological complexity of this motion planning problem equals $2m(\Gamma) + 1$ where $m(\Gamma)$ is the number of the essential vertices of Γ. It turns out (in contrast with the results on the collision free control of many objects in *space* [7]) that the topological complexity is independent of the number of particles.

1 Introduction

Algorithmic motion planning in robotics is a well established discipline which provides (1) a wide variety of general-purpose efficient theoretical algorithms, (2) more efficient algorithms designed for a number of special situations and (3) some practical solutions which work reasonably well in fairly involved scenarios. We refer to [14] for a recent survey and to [11] for a comprehensive textbook.

In general, one is given a moving system R with k degrees of freedom and a two or three-dimensional workspace V. The geometry of R and of V is given in advance which determines the configuration space of the system, X. The latter is a subset of $\mathbf{R}^k$ consisting of all placements (or configurations) of the system R, each represented by a tuple of k real parameters, such that in this placement R lies fully in V. For simplicity we may restrict our attention to a single connected component of R, the one containing a prescribed initial placement of R.

Being a subset of the Euclidean space $\mathbf{R}^k$, the configuration space X naturally inherits its *topology*. Many questions of control theory depend solely on the configuration space X viewed as a topological space. One of the advantages of this approach is that different control problems could be treated simultaneously for all systems having homeomorphic configuration spaces. It is well-known that any real analytic manifold can be realized as the configuration space of a simple mechanical system (linkage). Therefore topological

M. Erdmann et al. (Eds.): Algorithmic Foundations of Robotics VI, STAR 17, pp. 123–138, 2005.

questions of robotics lead to interesting new topological invariants of abstract manifolds.

We are interested in motion planning algorithms which work as follows: the algorithm gets as its input the present and the desired states of the system and it produces as the output a continuous motion of the system from its current state to the desired state. It turns out that the topology of the configuration space of the system imposes important restrictions on the discontinuities of the robot motion as a function of the input data. We emphasize that these are not discontinuities of the robot motion as a function of time. The discontinuities which we study here are in the way the decision (the whole motion) depends on the input data.

The approach to the motion planning problem described in [4], [5] was initially inspired by my discussions with Dan Halperin and Micha Sharir in 2000. It was also influenced by the well-known previous work of S. Smale [15] on algorithms of finding roots of algebraic equations.

Our results on topological robotics were published in mathematical journals [4], [5], [6] and in [7]. They led to an interesting new topological invariant $\mathbf{TC}(X)$ of topological spaces. In robotics applications, the number $\mathbf{TC}(X)$ has at least three different appearances. Firstly, it is the minimal number of domains of continuity of any motion planning algorithm for a system having X as its configuration space. Secondly, it is the minimal *order of instability* (see [5]) which have motion planning algorithms in X. The third interpretation (see §6 below) allows to measure $\mathbf{TC}(X)$ while relying on *random* motion planning algorithms: we show that $\mathbf{TC}(X)$ is the minimal integer n such that there exists an n valued random motion planning algorithm for the system.

The main part of the paper is devoted to a very specific motion planning problem: simultaneous control of many objects whose motion is restricted by a graph and the goal is to construct a motion planning algorithm avoiding collisions between the objects. This problem was initially studied by R. Ghrist, D. Koditschek and A. Abrams [8], [9], [1]. We calculate here the topological complexity of the problem and describe an explicit motion planning algorithm solving it.

For convenience of the reader we have included sections §§2 - 5 which give a brief description of some results of [4], [5] used later in this paper.

2 Motion Planning Algorithms

Consider a mechanical system (robot) controlled by a motion planning algorithm. It is supposed to function as follows: an operator introduces into the computer of the system the current and the desired states of the system and the motion planning algorithm determines a continuous motion of the system from its current state to the desired state.

Let X be the configuration space of the system. We will always assume that X is path connected, i.e. any pair of points $A, B \in X$ may be joined by a continuous path γ in X. This means that it is possible to bring our system, by a continuous movement, from any given configuration A to any given configuration B. This assumption does not represent a restriction since in practical situations when the natural configuration space of a given system has several connected components, we may simply restrict our attention to one of them.

Given two points $A, B \in X$, one wants to connect them by a path in X; this path represents a continuous motion of the system from one state to the other. A motion planning algorithm is a rule (algorithm) that takes pairs of configurations $(A, B) \in X \times X$ as an input and produces a continuous path in X from A to B as an output.

Let PX denote the space of all continuous paths $\gamma : [0,1] \to X$, equipped with the compact-open topology, and let $\pi : PX \to X \times X$ be the map assigning the end points to a path: $\pi(\gamma) = (\gamma(0), \gamma(1))$. Rephrasing the above discussion we see that a motion planning algorithm is a *section* of this fibration

$$s : X \times X \to PX, \quad \pi \circ s = 1_{X \times X}. \tag{1}$$

Here $1_{X \times X} : X \times X \to X \times X$ denotes the identity map. The algorithm associates the curve $s(A, B)(t)$, where $t \in [0,1]$, with any two given configurations of the system $A, B \in X$.

Given a mechanical system, one asks *does there exist a continuous motion planning algorithm for it*? In other words, whether it is possible to find a continuous section (1). The answer is negative in most cases as the following theorem proven in [4] states:

Theorem 1. *A globally defined continuous motion planning algorithm (1) exists if and only if the configuration space X of the system is contractible.*

3 Topological Complexity of Motion Planning Algorithms

In view of Theorem 1, one expects that in general a motion planning algorithm is only *piecewise continuous*. The following definition describes an important class of piecewise continuous motion planning algorithms having *finitely many* domains of continuity:

Definition 1. Let X be a path-connected topological space. A motion planning algorithm (1) is called *tame* if there exist finitely many subsets

$$F_1, \ldots, F_k \subset X \times X \tag{2}$$

such that the following conditions are satisfied:

(a) the sets $F_1, \ldots, F_k$ are pairwise disjoint $F_i \cap F_j = \emptyset$, $i \neq j$, and cover $X \times X$, i.e. $X \times X = F_1 \cup F_2 \cup \ldots \cup F_k$;

(b) each restriction $s|_{F_j}$ is continuous;
(c) each set F_j is an ENR (see below).

Condition (a) means that the sets $F_1, \dots, F_k$ partition the total space of all possible pairs $X \times X$. Condition (b) is the major continuity assumption. Condition (c) is technical, it allows to avoid pathological (exotic) decompositions. Recall, a topological space Y is called an *Euclidean Neighborhood Retract* (ENR) if it is homeomorphic to a subset of a Euclidean space $Y' \subset \mathbf{R}^n$, such that Y' is a retract of some open neighborhood $Y' \subset U \subset \mathbf{R}^n$; in other words, $U \subset \mathbf{R}^n$ is open and there exists a continuous map $r : U \to Y'$ such that $r(y) = y$ for all $y \in Y'$. Such a continuous map r is called a *retraction.*

It is well-known that all manifolds and polyhedra are ENRs.

For a given algorithm (1) there may exist many different decompositions (2) satisfying the conditions of Definition 1.

Definition 2. *The topological complexity* of a tame motion planning algorithm $s : X \times X \to PX$, $\pi \circ s = 1_{X \times X}$ is defined as the minimal number k of domains of continuity F_j which appear in Definition 1.

Given a concrete mechanical system one wishes to construct motion planning algorithms for it with the minimal possible topological complexity. The problem clearly depends only on the topology of the configuration space X of the system. This leads to the following purely topological notion:

Definition 3. Let X be a path-connected topological space. *The topological complexity* of X is defined as the minimal topological complexity of motion planning algorithms in X.

The topological complexity of a topological space X coincides (for nice spaces X) with the invariant $\mathbf{TC}(X)$ which was introduced in [4]. Its definition (which is more convenient from the purely topological point of view) appears in the following section.

4 Topological Invariant TC(X)

Definition 4. Let X be a path-connected topological space. The number $\mathbf{TC}(X)$ is defined as the minimal integer r such that the Cartesian product $X \times X$ can be covered by r open subsets

$$X \times X = U_1 \cup U_2 \cup \ldots \cup U_r, \tag{3}$$

such that for any $i = 1, 2, \dots, r$ there exists a continuous map

$$s_i : U_i \to PX \quad \text{with} \quad \pi \circ s_i = 1_{U_i}. \tag{4}$$

If no such r exists, we set $\mathbf{TC}(X) = \infty$.

It is shown in [4] that $\mathbf{TC}(X)$ depends only on the homotopy type of X. The invariant $\mathbf{TC}(X)$ admits an upper bound [4]:

$$\mathbf{TC}(X) \leq 2\dim X + 1. \tag{5}$$

The number $\mathbf{TC}(X)$ was computed in [4],[5],[6], [7] for a number of important configuration spaces appearing in robotics.

Theorem 2. *Let X be a connected smooth manifold. Then the topological complexity of X (cf. Definition 3) coincided with* $\mathbf{TC}(X)$ *(cf. Definition 4).*

We refer to [5] for a proof.

A lower bound for $\mathbf{TC}(X)$ is based on the knowledge of the cohomology algebra of X. To describe this result (which will be used later in this paper) we first observe that the singular cohomology $H^*(X;\mathbf{R}) = H^*(X)$ is a graded $\mathbf{R}$-algebra with the multiplication

$$\cup : H^*(X) \otimes H^*(X) \to H^*(X) \tag{6}$$

given by the cup-product, see [3], [16]. The tensor product $H^*(X) \otimes H^*(X)$ is again a graded $\mathbf{R}$-algebra with the multiplication

$$(u_1 \otimes v_1) \cdot (u_2 \otimes v_2) = (-1)^{|v_1|\cdot|u_2|}\, u_1u_2 \otimes v_1v_2. \tag{7}$$

Here $|v_1|$ and $|u_2|$ denote the degrees of cohomology classes v_1 and u_2 correspondingly. The cup-product (6) is an algebra homomorphism.

Definition 5. The kernel of homomorphism (6) is called *the ideal of the zero-divisors* of $H^*(X)$. The *zero-divisors-cup-length* of $H^*(X)$ is the length of the longest nontrivial product in the ideal of the zero-divisors of $H^*(X)$.

The next result is the main cohomological lower bound for the topological complexity.

Theorem 3. $\mathbf{TC}(X)$ *is greater than the zero-divisors-cup-length of the cohomology algebra $H^*(X)$.*

See [4] for a proof.

As an illustration we state the following result from [5] which is relevant for the sequel:

Theorem 4. *Let X be a connected graph. Then*

$$\mathbf{TC}(X) = \begin{cases} 1, \text{ if } b_1(X) = 0, \\ 2, \text{ if } b_1(X) = 1, \\ 3, \text{ if } b_1(X) \geq 2. \end{cases} \tag{8}$$

Here $b_1(X)$ denotes the first Betti number of X.

5 Order of Instability

Besides the number of domains of continuity, the motion planning algorithms could be characterized by their orders of instability.

Let $s : X \times X \to PX$, $\pi \circ s = 1_{X \times X}$ be a tame motion planning algorithm (cf. Definition 1). Let $F_1, \dots, F_k \subset X \times X$ be pairwise disjoint subsets as in Definition 1, i.e. such that $s|_{F_j}$ is continuous, each F_j is an ENR and union of the sets F_j equals $X \times X$.

Definition 6. The *order of instability* of a motion planning algorithm s is defined as the smallest integer r such that the subsets $F_1, \dots, F_k \subset X \times X$ as above could be constructed in such a way that for any sequence of $r+1$ indices $1 \le i_1 < i_2 < \dots < i_{r+1} \le k$ one has

$$\bar{F}_{i_1} \cap \bar{F}_{i_2} \cap \dots \cap \bar{F}_{i_{r+1}} = \emptyset. \tag{9}$$

Here $\bar{F}_j$ denotes the closure of F_j in $X \times X$.

The order of instability represents a very important functional characteristic of a motion planning algorithm. If the order of instability equals r then there exists a pair of initial - final configurations $(A, B) \in F_j$ such that arbitrarily close to (A, B) there exist $r-1$ pairs of configurations (A_1, B_1), (A_2, B_2), ..., (A_{r-1}, B_{r-1}) (which are all distinct small perturbations of (A, B)), belonging to distinct sets F_i, $i \neq j$. This means that *small perturbations of the input data (A, B) may lead to r essentially distinct motions given by the motion planning algorithm s.*

In practical situations one prefers to have motion planning algorithms with order of instability as low as possible.

Theorem 5. *Let X be a connected smooth manifold. Then the order of instability of any tame motion planning algorithm (1) is at least* $\mathbf{TC}(X)$. *Moreover, there exists a motion planning algorithm with the order of instability* $\mathbf{TC}(X)$.

This theorem proven in [5] gives yet another way the topological invariant $\mathbf{TC}(X)$ appears in robotics. In the next section we explain how the topological quantity $\mathbf{TC}(X)$ shows in random algorithms.

6 Random Motion Planning Algorithms

According to Theorem 1, *deterministic* continuous motion planning algorithms $s : X \times X \to PX$ exist only when the configuration space X is contractible. Instead, one may work with *random* algorithms solving the motion planning problem.

Let X be a path-connected topological space. A *random n-valued path* σ in X starting at $A \in X$ and ending at $B \in X$ is given by an ordered sequence of paths $\gamma_1, \dots, \gamma_n \in PX$ and an ordered sequence of real numbers

$p_1, \ldots, p_n \in [0,1]$ such that each $\gamma_j : [0,1] \to X$ is a continuous path in X starting at $A = \gamma_j(0)$ and ending at $B = \gamma_j(1)$, and

$$p_j \geq 0, \quad p_1 + p_2 + \ldots + p_n = 1.$$

One thinks of the paths $\gamma_1, \ldots, \gamma_n$ as of the ***states*** of σ and of the number p_j as being the ***probability*** that the random path σ is in state γ_j. Random path σ as above will be written as a formal linear combination

$$\sigma = p_1\gamma_1 + p_2\gamma_2 + \ldots + p_n\gamma_n. \tag{10}$$

Equality between n-valued random paths is understood as follows: the random path (10) is equivalent to $\sigma' = p'_1\gamma'_1 + p'_2\gamma'_2 + \ldots + p'_n\gamma'_n$ iff $p_j = p'_j$ for all $j = 1, \ldots, n$ and, besides, $\gamma_j = \gamma'_j$ for all indices j with $p_j \neq 0$. In other words the path γ_j which appears with the zero probability $p_j = 0$ could be replaced by any other path starting at A and ending at B.

We denote by P_nX the space of all n-valued random paths in X. The space P_nX has a natural topology: it is a factor-space of a subspace of the Cartesian product of n copies of $PX \times [0,1]$. Note that the space P_1X coincides with PX. The canonical map

$$\pi : P_nX \to X \times X \tag{11}$$

assigns to a random path its initial and end points. Map (11) is continuous.

Definition 7. An *n-valued random motion planning algorithm* is defined as a continuous map

$$s : X \times X \to P_nX \tag{12}$$

such that $\pi \circ s = 1_{X \times X}$.

Given a pair $(A, B) \in X \times X$, the output of the algorithm

$$s(A,B) = p_1\gamma_1 + \ldots + p_n\gamma_n \tag{13}$$

is an ordered probability distribution on the paths between A and B. In other words, the algorithm s produces the motion γ_j with probability p_j where $j = 1, \ldots, n$.

Theorem 6. *Let X be a path-connected metric space. Then the minimal integer n such that there exists an n-valued random motion planning algorithm $s : X \times X \to P_nX$ in X coincides with* $\mathbf{TC}(X)$.

Proof. The following proof is an adjustment of the proof of Proposition 2 from [13]. Assume that there exists an n-valued random motion planning algorithm $s : X \times X \to P_nX$ in X. The right hand side of formula (13) defines continuous real valued functions $p_j : X \times X \to [0,1]$, where $j = 1, \ldots, n$. Let U_j denote the open set $p_j^{-1}(0,1] \subset X \times X$. The sets $U_1, \ldots, U_n$ form an open covering of $X \times X$. Setting $s_j(A,B) = \gamma_j$, one gets a continuous map $s_j : U_j \to PX$ with $\pi \circ s_i = 1_{U_j}$. Hence, $n \geq \mathbf{TC}(X)$ according to the definition of $\mathbf{TC}(X)$.

Conversely, setting $k = \mathbf{TC}(X)$, we obtain that there exists an open cover $U_1, \ldots, U_k \subset X \times X$ and a sequence of continuous maps $s_i : U_i \to PX$ where $\pi \circ s_i = 1_{U_i}$, $i = 1, \ldots, k$. Extend s_i to an arbitrary (possibly discontinuous) mapping

$$S_i : X \times X \to PX$$

satisfying $\pi \circ S_i = 1_{X \times X}$. This can be done without any difficulty; it amounts in making a choice of a connecting path for any pair of points $(A, B) \in X \times X - U_i$. Next, one may find a continuous partition of unity subordinate to the open cover $U_1, \ldots, U_k$. It is a sequence of continuous functions $p_1, \ldots, p_k : X \times X \to [0,1]$ such that for any pair $(A,B) \in X \times X$ one has

$$p_1(A,B) + p_2(A,B) + \ldots + p_k(A,B) = 1$$

and the closure of the set $p_i^{-1}(0,1]$ is contained in U_i. We obtain a continuous k-valued random motion planning algorithm $s : X \times X \to P_n X$ given by the following explicit formula

$$s(A,B) = p_1(A,B)S_1(A,B) + \ldots + p_k(A,B)S_k(A,B). \tag{14}$$

The continuity of s follows from the continuity of the maps S_i restricted to the domains $p_i^{-1}(0,1]$. This completes the proof.

7 Configuration Spaces of Graphs

Let Γ be a connected finite graph. The symbol $F(\Gamma, n)$ denotes the configuration space of n distinct particles on Γ. In other words, $F(\Gamma, n)$ is the subset of the Cartesian product

$$\underbrace{\Gamma \times \Gamma \times \ldots \times \Gamma}_{n \text{ times}} = \Gamma^{\times n}$$

consisting of configurations $(x_1, x_2, \ldots, x_n)$ where $x_i \in \Gamma$ and $x_i \neq x_j$ for $i \neq j$. The topology of $F(\Gamma, n)$ is induced from its embedding into $\Gamma^{\times n}$.

Configuration spaces of graphs were studied by R. Ghrist, D. Koditschek and A. Abrams, see [8], [9], [1].

To illustrate the importance of these configuration spaces for robotics one may mention the control problems where a number of automated guided vehicles (AGV) have to move along a network of floor wires [9]. The motion of the vehicles must be safe: it should be organized so that the collisions do not occur. If n is the number of AGV then the natural configuration space of this problem is $F(\Gamma, n)$ where Γ is a graph. Here we idealize reality by assuming that the vehicles have size 0 (i.e. they are points). Although this assumption simplifies our discussion, it is in fact irrelevant for the topological problems which we study.

The first question to ask is whether the configuration space $F(\Gamma, n)$ is connected. Clearly $F(\Gamma, n)$ is disconnected if $\Gamma = [0, 1]$ is a closed interval (and $n \geq 2$) or if $\Gamma = S^1$ is the circle and $n \geq 3$. These are the only examples of this kind as the following simple lemma claims:

Lemma 1. *Let Γ be a connected finite graph having at least one essential vertex Then the configuration space $F(\Gamma, n)$ is connected.*

An *essential vertex* is a vertex which is incident to 3 or more edges.

One of the main results of topological robotics states that the configuration spaces $F(\Gamma, n)$ are aspherical; see [8].

For the topological complexity of the configuration spaces one has:

Theorem 7. *Let Γ be a connected graph having an essential vertex. Then the topological complexity of $F(\Gamma, n)$ satisfies*

$$\mathbf{TC}(F(\Gamma, n)) \leq 2m(\Gamma) + 1, \tag{15}$$

where $m(\Gamma)$ denotes the number of essential vertices in Γ.

Proof. By a theorem of R. Ghrist [8], the configuration space $F(\Gamma, n)$ has homotopy type of a cell complex of dimension $\leq m(\Gamma)$. Combining this result with the theorem about the homotopy invariance of $\mathbf{TC}(X)$ (see Theorem 3 in [4]) and the with the upper bound (5) one gets (15).

We prove below that equality holds in (15) in many cases.

We shall also see examples where (15) holds as a strict inequality.

8 A Motion Planning Algorithm in $F(\Gamma, n)$

In this section Γ denotes a *tree*[1] having an essential vertex. Fix a univalent

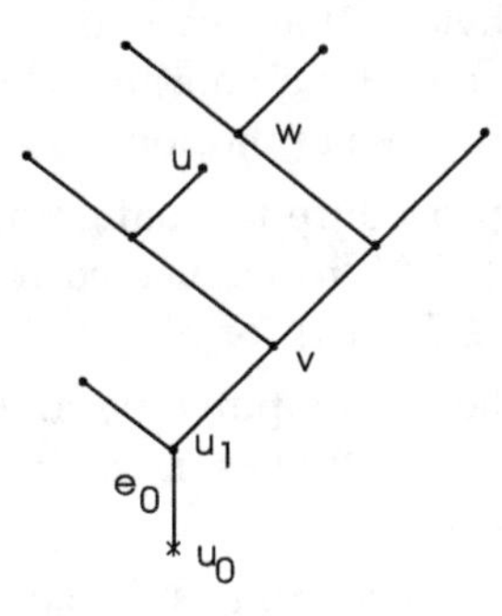

Fig. 1. A partial order between the points of the tree.

[1] Recall that tree is a connected graph with no cycles.

vertex $u_0 \in \Gamma$ which will be called *the root*. Any point in Γ can be connected by a path to the root u_0 and this connecting path is unique up to homotopy. The choice of the root determines a partial order on Γ: we say that $x \succ y$, where $x, y \in \Gamma$ if any path from x to the root u_0 passes through y. Of course, $\succ$ is only a partial order, i.e. there may exist pairs $x, y \in \Gamma$ such that neither $x \succ y$, nor $y \succ x$. On Figure 1 we see $u \succ v$ and $w \succ v$ however the points u and w are not comparable.

Let $e_0 \subset \Gamma$ denote the root edge of Γ.

Fix a configuration $\alpha_0 \in F(\Gamma, n)$ of n distinct points lying on e_0 and a continuous collision free motion connecting (in $F(\Gamma, n)$) any pair of permutations of α_0. Such motions exist as we assume that Γ has an essential vertex and hence the configuration space $F(\Gamma, n)$ is connected (see Lemma 1).

The algorithm works as follows. Let $\alpha = (A_1, A_2, \dots, A_n) \in F(\Gamma, n)$ and $\beta = (B_1, B_2, \dots, B_n) \in F(\Gamma, n)$ be two given configurations of n distinct points on Γ. Let $A_{i_1}, \dots, A_{i_r}$ be all the minimal elements (with respect to the order $\succ$) of the set of points of α. Here the notation is such that $i_1 < i_2 < \dots < i_r$. First we move the point A_{i_1} down to an interior point of the root edge e_0. Next we move A_{i_2} to the root edge e_0 and we continue moving similarly the following points $A_{i_3}, \dots, A_{i_r}$.

As a result, after this first stage of the algorithm, all the minimal points of α are transferred onto the root edge e_0. On the second stage we find the minimal set among the remaining points of α and move them down, one after another, to the edge e_0. Repeating in this way we find continuous collision free motion of all the points of α moving them onto the interior of the root edge e_0. We obtain a configuration of points $\alpha' = (A'_1, \dots, A'_n)$ which all lie in the interior of the root edge e_0, in certain order.

For points lying on the root edge e_0 the partial order $\succ$ is a linear order. Given a configuration $A = (A_1, \dots, A_n)$ where all points A_i lie on the edge e_0, there exists a unique permutation $\alpha : \{1, \dots, n\} \to \{1, \dots, n\}$ such that $A_{\alpha_n} \succ A_{\alpha_{n-1}} \succ \dots \succ A_{\alpha_1}$. This permutation describes the order of the points on the edge. It is obvious that any two configurations having the same permutation can be connected by a continuous collision free path such that no points leave e_0 in the process of motion.

Applying the similar procedure to configuration β we obtain a configuration $\beta' = (B'_1, \dots, B'_n) \in F(\Gamma, n)$ connected to β by a continuous collision free motion, such that $B'_i \in \mathrm{Int}(e_0)$ for all $j = 1, \dots, n$.

Next we apply one of the precomputed permutational motions which takes α' to a configuration α'' which also lies in the interior of e_0 and has the same order as the configuration β'.

Finally, the output of the algorithm is the concatenation of (1) the motion from α to α'; (2) the motion from α' to α''; (3) the obvious motion from α'' to β'; and (4) the reverse of the motion $\beta \to \beta'$. The motion (3) exists since both α'' and β' lie on e_0 and have the same order.

The above algorithm is discontinuous: if one of the points A_j is a vertex then a small perturbation of A_j may lead to a different set of minimal points and hence to a completely different motion. Note that the vertices of T which have valence one or two do not cause discontinuity.

Let $S_i \subset F(\Gamma, 2)$ denote the set of all configurations $\alpha = (A_1, \dots, A_n)$ such that precisely i points among A_j are essential vertices of Γ. If we restrict the above algorithm on the pairs $(A, B) \in S_i \times S_j$ then the result is a continuous function of the input. The sets $S_i \times S_j$ and $S_{i+1} \times S_{j-1}$ are disjoint and each of the sets contain no limit points of the other. This follows since the closure of S_i is contained in the union of the sets S_j with $j \geq i$. We may define

$$Y_k = \bigcup_{i+j=k} S_i \times S_j, \quad k = 0, 1, \dots, 2m(\Gamma)$$

where $m(\Gamma)$ denotes the number of the essential vertices of Γ. The algorithm described above is continuous when restricted on each set Y_k. Hence we obtain:

Corollary 1. *The topological complexity of the algorithm described above is less than or equal to* $2m(\Gamma) + 1$ *where* $m(\Gamma)$ *denotes the number of the essential vertices of* Γ.

9 The Main Result

Our main result which will be proven later in this paper states:

Theorem 8. *Let* Γ *be a tree having an essential vertex. Let* n *be an integer satisfying* $n \geq 2m(\Gamma)$ *where* $m(\Gamma)$ *denotes the number of essential vertices of* Γ. *In the case* $n = 2$ *we will additionally assume that the tree* Γ *is not homeomorphic to the letter* Y *viewed as a subset of the plane* $\mathbf{R}^2$. *Then the upper bound (15) is exact, i.e.*

$$\mathbf{TC}(F(\Gamma, n)) = 2m(\Gamma) + 1. \tag{16}$$

We conjecture that this theorem holds for any connected graph Γ having ≥ 2 essential vertices without assuming that Γ is a tree and for any number of particles $n \geq 2m(\Gamma)$.

If Γ is homeomorphic to the letter Y then $m(\Gamma) = 1$ and $F(\Gamma, 2)$ is homotopy equivalent to the circle S^1. Hence in this case $\mathbf{TC}(F(\Gamma, 2)) = 2$, see [4]. The equality (16) fails in this case.

From Theorem 10 of the next section and Theorem 4 it follows that for any tree Γ one has

$$\mathbf{TC}(F(\Gamma, 2)) = \begin{cases} 2, & \text{if } \Gamma \text{ is homeomorphic to the letter } Y, \\ 3, & \text{otherwise.} \end{cases} \tag{17}$$

This example shows that the assumption $n \geq 2m(\Gamma)$ of Theorem 8 cannot be removed: if Γ is a tree with $m(\Gamma) \geq 2$ then (16) would give $\mathbf{TC}(F(\Gamma, 2)) = 2m(\Gamma) + 1 \geq 5$ contradicting (17).

Here are some more examples. For the graphs K_5 and $K_{3,3}$ (see Figure 2) one has

$$\mathbf{TC}(F(K_5, 2)) = \mathbf{TC}(F(K_{3,3}, 2)) = 5. \tag{18}$$

This follows by combining a theorem of A. Abrams [1] (stating that the spaces

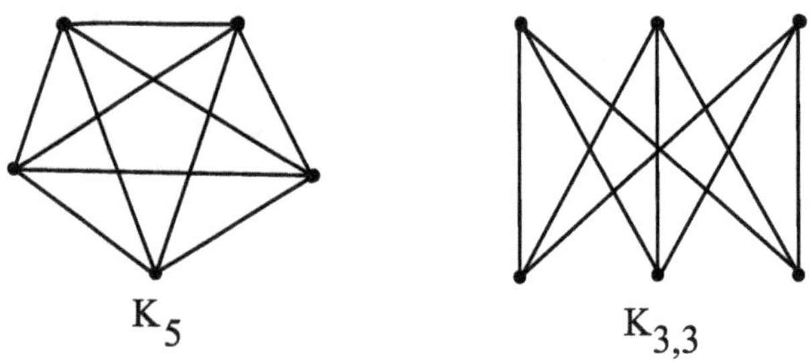

Fig. 2. Graphs K_5 and $K_{3,3}$.

$F(K_5, 2)$ and $F(K_{3,3}, 2)$ are homotopy equivalent to closed orientable surfaces of genus 6 and 4, correspondingly) with Theorem 9 from [4] computing the invariant $\mathbf{TC}$ for the surfaces. In these examples the equality(16) is again violated.

10 Configuration Spaces of Two Particles

In the present section we explicitly describe the topology of the configuration spaces $F(\Gamma, 2)$ and $B(\Gamma, 2) = F(\Gamma, 2)/\mathbf{Z}_2$ where Γ is a tree. The latter space is the quotient of $F(\Gamma, 2)$ with respect to the involution interchanging the particles. The results of this section are used in the proof of Theorem 8.

Theorem 9. *Let Γ be a tree having an essential vertex. Then the space $B(\Gamma, 2)$ is homotopy equivalent to the wedge of*

$$\frac{1}{2}\sum_v (\eta(v) - 1)(\eta(v) - 2) \tag{19}$$

circles where v runs over the vertices of Γ and the symbol $\eta(v)$ denotes the number of edges incident to the vertex v.

Note that only essential vertices v contribute nonzero summands to (19).

In the next theorem we identify the equivariant homotopy type of $F(\Gamma,2)$ with respect to the canonical involution interchanging the labels of the particles. Fix a univalent root vertex u_0 of Γ. Then any vertex $v \neq u_0$ has a well-defined single *descending* edge which is incident to it. It is the edge connecting v with the root vertex: removing the descending edge makes v and u_0 lying in different connected components. The other $\eta(v)-1$ edges incident to the vertex v will be called *ascending*.

Theorem 10. *Let Γ be a tree having an essential vertex. Then the space $F(\Gamma,2)$ is homotopy equivalent to the wedge of*

$$\sum_v (\eta(v)-1)(\eta(v)-2)-1 \tag{20}$$

circles where v runs over the vertices of Γ. Moreover, the $\mathbf{Z}_2$-equivariant homotopy type of $F(\Gamma,2)$ can be described as follows. Consider a 1-dimensional cell complex Y_Γ having two vertices A and B and a number of 1-dimensional cells connecting A to B, each labelled by an ordered pair of distinct ascending edges of Γ incident to an essential vertex $v \in \Gamma$. In total there are $\sum_v(\eta(v)-1)(\eta(v)-2)$ such 1-cells. The complex Y_Γ has an involution $T: Y_\Gamma \to Y_\Gamma$ which takes A to B and which takes the edge with the label (e_i,e_j) to the edge with the label (e_j,e_i). Then $F(\Gamma,2)$ and Y_Γ are $\mathbf{Z}_2$-equivariantly homotopy equivalent.

Figure 3 illustrates the construction of the cell complex Y_Γ.

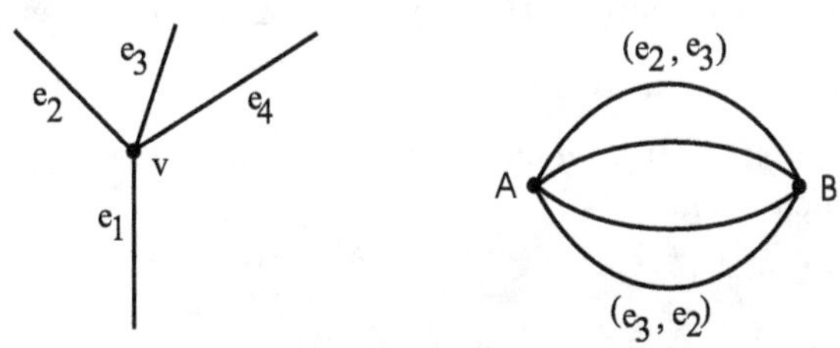

Fig. 3. Descending edge e_1 and ascending edges e_2, e_3, e_4 incident to a vertex v (left) and a part of the cell complex Y_Γ (right).

The complex Y_Γ can also be described as follows. Consider the standard unit sphere $S^2 \subset \mathbf{R}^3$ given by the equation $x^2+y^2+z^2=1$. Let P_ϕ denotes the 2-dimensional plane spanned by the vectors $(0,0,1)$ and $(\cos\phi, \sin\phi, 0)$. The intersection $P_\phi \cap S^2$ is a circle S^1_ϕ containing the North and the South poles $(0,0,\pm 1)$. Set $\phi_i = \frac{i\pi}{n}$ where $i = 0,1,\ldots,n-1$ and

$$n = 1/2 \cdot \sum_v \eta(v)-1)(\eta(v)-2),$$

the sum taken over the vertices of Γ. Then

$$Y_\Gamma = \bigcup_{i=0}^{n-1} S^1_{\phi_i}. \tag{21}$$

The latter space is viewed with the standard antipodal involution.

11 Sketch of Proof of Theorem 8.

First we consider the special case when the tree Γ has a single essential vertex, $m(\Gamma) = 1$. Then, by a theorem of R. Ghrist [8], the configuration space $F(\Gamma, n)$ has homotopy type of a wedge of

$$1 + [(n-1)(\eta-2)-1] \cdot \frac{(n+\eta+2)!}{(\eta-1)!} \tag{22}$$

circles where $\eta = \eta(v)$ denotes the number of edges incident to the essential vertex. Using Theorem 4 we find that the topological complexity of the configuration space $F(\Gamma, n)$ equals either 2 or 3 depending on whether the number of circles in the wedge is 1 or > 1. It is easy to see that (22) equals 1 if and only if $\eta = 3$ and $n = 2$. Since this possibility is excluded by our assumption we find that $\mathbf{TC}(F(\Gamma, n)) = 3$ for $m(\Gamma) = 1$.

The proof in the case $m(\Gamma) > 1$ uses the following lemma:

Lemma 2. *Let X be a topological space and let*

$$u_1, \dots, u_m, w_1, \dots, w_m \in H^1(X) \tag{23}$$

be cohomology classes (where $H^*(X) = H^*(X; \mathbf{k})$ and $\mathbf{k}$ is a field) satisfying

$$u_i w_j = 0, \quad i, j = 1, \dots, m, \tag{24}$$

and such that their cup-products $u_1 \dots u_m$ and $w_1 \dots w_m \in H^m(X)$ are linearly independent. Then

$$\mathbf{TC}(X) \geq 2m + 1. \tag{25}$$

Proof. Consider the classes $\bar{u}_i = 1 \otimes u_i - u_i \otimes 1$ and $\bar{w}_j = 1 \otimes w_j - w_j \otimes 1$ lying in the tensor product $H^*(X) \otimes H^*(X)$. The classes $\bar{u}_i$ and $\bar{w}_j$ are zero-divisors. The product of all these classes in $H^*(X) \otimes H^*(X)$ equals

$$\prod_{i=1}^m \bar{u}_i \cdot \prod_{j=1}^m \bar{w}_j = \pm \prod_{i=1}^m u_i \otimes \prod_{j=1}^m w_j \pm \prod_{j=1}^m w_j \otimes \prod_{i=1}^m u_i. \tag{26}$$

Since the classes $\prod_{i=1}^m u_i$ and $\prod_{j=1}^m w_j$ are linearly independent we may find two linear functionals $\phi, \psi : H^m(X) \to \mathbf{k}$ such that $\phi(\prod_{i=1}^m u_i) \neq 0$ and $\phi(\prod_{j=1}^m w_j) = 0$ while $\psi(\prod_{j=1}^m w_j) \neq 0$ and $\psi(\prod_{i=1}^m u_i) = 0$. Using (26) we see that applying $\phi \otimes \psi$ to the product (26) is nonzero which proves the nontriviality of the product (26). Hence we have a nontrivial product of $2m$ zero divisors. Applying Theorem 3 gives the desired result.

Assume now that Γ is a tree having $m(\Gamma) \geq 2$ essential vertices.

Let $v_1, \ldots, v_m$ be the essential vertices of Γ. The space $F(\Gamma, 2)$ is homotopy equivalent to a wedge of circles (cf. Theorem 10). For any index $i = 1, 2, \ldots, m$ we fix a nonzero cohomology class $\alpha_i \in H^1(F(\Gamma, 2))$ with the only condition that it vanishes on all the circles in the wedge except the circles associated with the vertex v_i, see formula (20).

We define two continuous maps

$$\Phi_i : F(\Gamma, n) \to F(\Gamma, 2), \quad \Psi_i : F(\Gamma, n) \to F(\Gamma, 2),$$

where $i = 1, 2, \ldots, m$ and

$$\Phi_i(x_1, \ldots, x_n) = (x_{2i-1}, x_{2i}),$$

$$\Psi_i(x_1, \ldots, x_n) = \begin{cases} (x_{2i+1}, x_{2i+2}), & \text{for } i = 1, \ldots, m-1, \\ (x_1, x_2), & \text{for } i = m. \end{cases} \tag{27}$$

Finally we denote

$$v_i = \Phi_i^*(\alpha_i) \in H^1(F(\Gamma, n)), \quad w_i = \Psi_i^*(\alpha_i) \in H^1(F(\Gamma, n)). \tag{28}$$

One checks that the conditions of Lemma 2 are satisfied for the constructed cohomology classes, i.e. the cup-products $u_1 \cdots u_m$ and $w_1 \cdots w_m$ are linearly independent and $u_i w_j = 0$. Applying Lemma 2 completes the proof.

The full details will appear elsewhere.

12 Conclusion

The topological invariant $\mathbf{TC}(X)$ imposes important restrictions on the structure of motion planning algorithms for the mechanical systems having X as their configuration space. $\mathbf{TC}(X)$ bounds from below the *order of instability* of deterministic motion planning algorithms. We prove in this paper that the number $\mathbf{TC}(X)$ equals the minimal integer k such that there exists a k-valued *random motion planning algorithm* in X.

The motion planning algorithm in the configuration space $F(\Gamma, n)$ of n distinct points on a tree Γ which is described in §8 has the minimal possible topological complexity, as Theorem 8 states. This algorithm may be used in practical control problems when several objects have to be moved along a tree Γ avoiding collisions.

We observe that for a large number of particles n, the topological complexity of this algorithm depends only on the tree Γ and does not depend on the number of the moving objects n.

This result could be compared with the earlier results of [7] which gives the topological complexity of the motion planning problem of many objects in the space $\mathbf{R}^3$ and on the plane $\mathbf{R}^2$. The topological complexity of the motion planning algorithms in these situations depends linearly on the number of particles; it equals $2n-2$ for $F(\mathbf{R}^2, n)$ and is $2n-1$ for $F(\mathbf{R}^3, n)$.

We obtain: *for a large number of objects which must be simultaneously controlled avoiding collisions, a great simplification can be achieved by restricting the motion of the objects to a one-dimensional net.*

This result may potentially have practical applications in some traffic control problems.

References

1. Abrams A. (2002) Configuration spaces of colored graphs. Geometriae Dedicata **92**, 185 – 194
2. Dold A. (1972) Lectures on Algebraic Topology. Springer - Verlag.
3. Dubrovin B., Novikov S. P. and Fomenko A. T. (1984) Modern Geometry; Methods of the Homology Theory. Springer-Verlag.
4. Farber M. (2003) Topological Complexity of Motion Planning. Discrete and Computational Geometry **29**, 211–221
5. Farber M. (2004) Instabilities of Robot Motion. Topology and its Applications (to appear). Available as preprint cs.RO/0205015.
6. Farber M, Tabachnikov S., Yuzvinsky S. (2003) Topological Robotics: Motion Planning in Projective Spaces. "International Mathematical Research Notices" **34**, 1853–1870
7. Farber M., Yuzvinsky S. (2004) Topological Robotics: Subspace Arrangements and Collision Free Motion Planning. To appear in AMS volume dedicated to S.P.Novikov's 65th birthday. Available as preprint math.AT/0210115
8. Ghrist R. (2001) Configuration spaces and braid groups on graphs in robotics. Knots, braids, and mapping class groups – papers dedicated to Joan S. Birman, AMS/IP Stud. Adv. Math. **24**, Amer. Math. Soc., Providence, 29 – 40
9. Ghrist R., Koditschek D. (2002) Safe cooperative robot dynamics on graphs. SIAM J. Control Optim. **40**, 1556 – 1575
10. Halperin D., Sharir M. (1995) Arrangements and their applications in robotics: recent developments. The Algorithmic Foundations of Robotics. K. Goldberg, D. Halperin, J.C. Latombe and R. Wilson eds., Boston, MA, 495 - 511
11. Latombe J.-C. (1991) Robot Motion Planning. Kluwer Academic Publishers
12. Schwartz J. T., Sharir M. (1983) On the piano movers' problem: II. General techniques for computing topological properties of real algebraic manifolds. Adv. Appl. Math., **4**, 298–351
13. Schwarz A.S. (1966) The genus of a fiber space. Amer. Math. Sci. Transl. **55**, 49- 140
14. Sharir M. (1997) Algorithmic motion planning. Handbook of Discrete and Computational Geometry. J.E. Goodman, J. O'Rourke eds. CRC Press, Boca Raton, FL, 733 - 754
15. Smale S. (1987) On the topology of algorithms, I. J. of Complexity, **3**, 81-89.
16. Spanier E.(1966) Algebraic Topology.

Online Searching with an Autonomous Robot

Sándor P. Fekete[1], Rolf Klein[2], and Andreas Nüchter[3]

[1] Department of Mathematical Optimization, Braunschweig University of Technology, D–38106 Braunschweig, Germany. Email: s.fekete@tu-bs.de
[2] Institute of Computer Science, University of Bonn, D–53117 Bonn, Germany. Email: rolf.klein@uni-bonn.de
[3] Fraunhofer Institute for Autonomous Intelligent Systems, Schloss Birlinghoven, D–53754 Sankt Augustin, Germany. Email: andreas@nuechti.de

Summary. We discuss online strategies for visibility-based searching for an object hidden behind a corner, using Kurt3D, a real autonomous mobile robot. This task is closely related to a number of well-studied problems. Our robot uses a three-dimensional laser scanner in a stop, scan, plan, go fashion for building a virtual three-dimensional environment. Besides planning trajectories and avoiding obstacles, Kurt3D is capable of identifying objects like a chair. We derive a practically useful and asymptotically optimal strategy that guarantees a competitive ratio of 2, which differs remarkably from the well-studied scenario without the need of stopping for surveying the environment. Our strategy is used by Kurt3D, documented in a separate video.

Keywords: Searching, visibility problems, watchman problems, online searching, competitive strategies, autonomous mobile robots, three-dimensional laser scanning, Kurt3D.

1 Introduction

Visibility Problems. Visibility-based problems of surveying, guarding, or searching have a long-standing tradition in the area of computational optimization; they may very well be considered a field of their own. Using stationary positions for guarding a region is the well-known *art gallery problem* [15]. The *watchman problem* [3,18,19] asks for a short tour along which one mobile guard can see the entire region. If the region is unknown in advance, we are faced with the *online watchman problem*. For a simple polygon, Hoffmann et al. [7] achieve a constant competitive ratio of 26.5, while Albers et al. [1] show that no constant competitive factor exists for a region with holes, and unbounded aspect ratio. Kalyanasundaram and Pruhs [12] consider the problem in graphs and give a competitive factor of 16.

M. Erdmann et al. (Eds.): Algorithmic Foundations of Robotics VI, STAR 17, pp. 139–154, 2005.

In the context of geometric searching, a crucial issue is the question of how to look around a corner: Given a starting position, and a known distance to a corner, how should one move in order to see a hidden object (or the other part of the wall) as quickly as possible? This problem was solved by Icking et al. [9, 10] who show that an optimal strategy can be characterized by a differential equation that yields a competitive factor of 1.2121..., which is optimal. Note that actually using this solution requires numerical evaluation.

An Autonomous Mobile Robot. From the practical side, our work is motivated by an actual application in robotics: The Fraunhofer Institute for Autonomous Intelligent Systems (AIS) has developed autonomous mobile robots that can survey their environment by virtue of a high-resolution, 3D laser scanner [17]. By merging several 3D scans acquired in a stop, scan, plan, go fashion, the robot Kurt3D builds a virtual 3D environment that allows it to navigate, avoid obstacles, and detect objects [14]. This makes the visibility problems described above quite practical, as actually using good trajectories is now possible and desirable.

However, while human mobile guards are generally assumed to have full vision at all times, our autonomous robot has to stop and take some time for taking a survey of its environment. This makes the objective function (minimize total time to locate an object or explore a region) a sum of travel time and scan time; a somewhat related problem is searching for an object on a line in the presence of turn cost [5], which turns out to be a generalization of the classical linear search problem. Somewhat surprisingly, scan cost (however small it may be) causes a crucial difference to the well-studied case without scan cost, even in the limit of infinitesimally small scan times.

Independent from our work, the problem of looking around a corner in the presence of scan cost has been studied by Isler et al. [11], who described two deterministic strategies achieving competitive ratios of 3.14 and 2.22, and also considered a probabilistic framework dealing with prior knowledge about the possible values of corners. We improve on these results with a different, asymptotically optimal strategy, and prove a matching lower bound.

Other Related Work. Visbility-based navigation of robots involves a variety of different aspects. For example, Efrat et al. [4] study the task of developing strategies for tracking and capturing a visible target with known trajectory, while maintaining line-of-sight among obstacles. Kutulakos et al. [13] consider the task of vision-guided exploration, where the robot is assumed to move about freely in three dimensions, among various obstacles.

Our Results. The main objective of this paper is to demonstrate that technology has reached the stage of actually applying previous theoretical studies, at the same time triggering new algorithmic research. We hope that this will highlight the need for and the opportunities of closer interaction between theoreticians and practitioners. In particular, we describe the problem of online searching by a real autonomous robot, for an object (a chair) hidden behind a corner, which is at distance d from the robot's starting position. Our mathematical results are as follows:

- We show that for an initial distance of at least $d \geq 1/\delta$ from the corner, a competitive ratio of $2 - \delta$ cannot be achieved. This implies a lower bound of 2 on the competitive ratio by any one strategy, and proves that there is an important distinction from the case without scan cost.
- We describe a heuristic strategy that is fast to evaluate and easy to implement in real life.
- We show that this strategy is asymptotically optimal by proving that for large distances, the competitive ratio converges to 2.
- We give additional numerical evidence showing that the performance of our strategy is within about 2% of the optimum.
- Most importantly, we describe how our strategy can actually be used by Kurt3D, a real mobile autonomous robot.

Further documentation of our work is provided by a video [6] that is also available at the authors' web addresses.

The rest of this paper is organized as follows. In Section 2, we describe the technical details, properties, and capabilities of Kurt3D, an autonomous mobile robot that was used in our experiments. Section 3 provides mathematical results on the problem arising from Kurt searching for a hidden object. Section 4 gives a description of how our results are used in practice. The final Section 5 provides some directions for future research.

2 The Autonomous Mobile Robot

In this section we describe technical details and background of the autonomous mobile robot Kurt3D.

2.1 The Kurt3D Robot Platform

Kurt3D (Figure 1, top left) is a mobile robot platform with a size of 45 cm (length) $\times$ 33 cm (width) $\times$ 26 cm (height) and a weight of 15.6 kg. Equipped with the 3D laser range finder the height increases to 47 cm and the weight to 22.6 kg.[4] Kurt3D's maximum velocity is 5.2 m/s (autonomously controlled 4.0 m/s). Two 90 W motors are used to power the 6 wheels, where the front and rear wheels have no tread pattern to enhance rotating. Kurt3D operates for about 4 hours with one battery (28 NiMH cells, capacity: 4500 mAh) charge. The core of the robot is a Pentium-III-600 MHz with 384 MB RAM. An embedded 16-Bit CMOS microcontroller is used to control the motor.

2.2 The AIS 3D Laser Range Finder

The AIS 3D laser range finder (Figure 1, top right) [16,17] is built on the basis of a 2D range finder by extension with a mount and a standard servo motor.

[4]Videos of the exploration with the autonomous mobile robot can be found at `http://www.ais.fhg.de/ARC/kurt3D/index.html`

The 2D laser range finder is attached in the center of rotation to the mount for achieving a controlled pitch motion. The servo is connected on the left side (Figure 1, top middle). The 3D laser scanner operates up to 5h (Scanner: 17 W, 20 NiMH cells with a capacity of 4500 mAh, Servo: 0.85 W, 4.5 V with batteries of 4500 mAh) on one battery pack.

Fig. 1. Top left: The autonomous mobile robot Kurt3D equipped with the 3D scanner. Top right: The AIS 3D laser range finder. Its technical basis is a SICK 2D laser range finder (LMS-200). Bottom row, left: A scanned scene as depth image. Middle and right: Scanned scenes as point cloud viewed with a camera orientation towards the door.

The area of 180°(h) × 120°(v) is scanned with different horizontal (181, 361, 721 pts.) and vertical (210, 420 pts.) resolutions. A plane with 181 data points is scanned in 13 ms by the 2D laser range finder (rotating mirror device). Planes with more data points, e.g., 361, 721, double or quadruple this time. Thus, a scan with 181 × 210 data points needs 2.8 seconds. In addition to the distance measurement, the 3D laser range finder is capable of quantifying the amount of light returning to the scanner, i.e., reflectance data [14]. Figure 1 (bottom left) shows a scanned scene as depth image, created by off-screen rendering from the 3D data points (Figure 1, bottom middle) by an `OpenGL`-based drawing module.

2.3 Basic 3D Scanner Software

The basis of the scan matching algorithms and the reliable robot control are algorithms for reducing points, line detection, surface extraction and object segmentation. Next we give a brief description of these algorithms. Details can be found in [16].

The scanner emits the laser beams spherically from one center, such that the data points close to the source are more dense. The first step is to reduce the data. Therefore, data points located close together are joined into one point. The number of these so-called *reduced points* is one order of magnitude smaller than the original one.

Second, a simple length comparison is used as a line detection algorithm. Given that the counterclockwise ordered data of the laser range finder (points $a_0, a_1, \ldots, a_n$) are located on a line, the algorithm has to check for a_{j+1} if $\|a_i, a_{j+1}\| \;/\; \sum_{t=i}^{j} \|a_t, a_{t+1}\| < \epsilon(j)$ in order to determine if a_{j+1} is on line with a_j. (Figure 2, left)

The third step is surface detection. Scanning a plane surface, line detection returns a sequence of lines in successive scanned 2D planes approximating the shape of surfaces. Thus a plain surface consists of a set of lines. Surfaces are detected by merging similar oriented and nearby lines. (Figure 2, middle)

The fourth and final step computes occupied space. For this purpose, conglomerations of surfaces and polygons are merged sequentially into objects. Two steps are necessary to find bounding boxes around objects. First a bounding box is placed around each large surface. In the second step objects close to each other are merged together, e.g., one should merge objects closer than the size of the robot, since the robot cannot pass between such objects (Figure 2, right). These bounding boxes are used for avoiding obstacles.

Data reduction, line, surface and object detection are real-time capable and run in parallel to the 3D scanning process.

Fig. 2. Left: Line detection in every scan slice. Middle: Surface segmentation. Right: Bounding boxes of objects superimposing the surfaces

2.4 3D Scan Matching

To create a correct and consistent representation of the environment, the acquired 3D scans have to be merged in one coordinate system. This process is called registration. Due to the robot's sensors, the self-localization is usually erroneous and imprecise, so the geometric structure of overlapping 3D scans has to be considered for registration. The odometry-based robot pose serves as a first estimate and is corrected and updated by the registration process. We use the well-known Iterative Closest Points (ICP) algorithm [2] to compute the transformation, consisting of a rotation $\mathbf{R} \in \mathbb{R}^{3\times3}$ and a translation $\mathbf{t} \in \mathbb{R}^3$. The ICP algorithm computes this transformation in an iterative fashion. In each iteration the algorithm selects the closest points as correspondences and computes the transformation $(\mathbf{R}, \mathbf{t})$ for minimizing

$$E(\mathbf{R}, \mathbf{t}) = \sum_{i=1}^{N_m} \sum_{j=1}^{N_d} w_{i,j} \left|\left|\mathbf{m}_i - (\mathbf{R}\mathbf{d}_j + \mathbf{t})\right|\right|^2 ,$$

where N_m and N_d are the number of points in the model set M, i.e., first 3D scan, or data set D, second 3D scan, respectively, and w_{ji} are the weights for a point match. The weights are assigned as follows: $w_{ji} = 1$, if $\mathbf{m}_i$ is the closest point to $\mathbf{d}_j$ within a close limit, $w_{ji} = 0$ otherwise. It is shown in [2] that the iteration terminates in a minimum. The assumption is that in the last iteration the point correspondences are correct. In each iteration the transformation is computed in a fast closed-form manner by the quaternion-based method of Horn [8]. In addition, point reduction and kD.trees speed up the computation of the point pairs, such that only the time required for scan matching is reduced to roughly one second [17]. Figure 3 shows three iteration steps for 3D scan alignment.

Fig. 3. Three iteration steps of scan alignment process for the two 3D scans presented in Figure 1 (bottom, middle, right).

2.5 3D Object Detection

Fig. 4: Object detection in range images.

Automatic, fast and reliable object detection algorithms are essential for mobile robots used in searching tasks. To perceive objects, we use the 3D laser range and reflectance data. The 3D data is transformed into images by off-screen rendering. To detect objects, a cascade of classifiers, i.e., a linear decision tree, is used. Following the ideas of Viola and Jones, we compose each classifier from several simple classifiers, which in turn contain an edge, line or center surround feature [20]. There exists an effective method for the fast computation of these features using an intermediate representation, namely, integral image. For learning of the object classes, a boosting technique, namely, Ada Boost, is used [20]. The resulting approach for object classification is reliable and real-time capable and combines recent results in computer vision with the emerging technology of 3D laser scanners. For a detailed discussion of object detection in 3D laser range data, refer to [14]. Figure 4 shows an office chair detected by a cascade of classifiers.

3 Algorithmic Approach

Now we turn to algorithmic aspects of the online problem faced by the robot who is trying to look around a corner in the presence of scan cost: Given an initial position at a known distance from a corner or door, and an object that is hidden at an unknown angle behind this obstruction, how should one move in order to see the object as fast as possible? The total time incurred arises from travel at a known maximum velocity, and the total time for stopping, scanning, processing, and re-starting the robot.

When trying to develop a good search strategy, we have to balance theoretical quality with practical applicability. More precisely, we have to keep a close eye on the trade-off between these objectives: An increase in theoretical quality may come at the expense of higher mathematical difficulty, possibly requiring more complicated tools. In an online context, the use of such tools may cause both theoretical and practical difficulties: Complicated solutions may cause computational overhead that can change the solution itself by causing extra delay; on the practical side, actually applying such a solution may be difficult (due to limited accuracy of the robot's motion) and without significant use. To put relevant error bounds into perspective: The largest room available to us is the great hall of Schloss Birlinghoven; even there, the size of Kurt and the object is still in the order of 2% of the room diameter.

On the mathematical side, it should be noted that even in the theoretical paper [7], semi-circles are considered instead of the solution to the differential equation, in order to allow analysis of the resulting trajectories.

In the following, we will start by giving some basic mathematical observations and properties (Section 3.1); this is followed by a discussion of globally optimal strategies (Section 3.2). Section 3.3 describes a natural heuristic solution that is both easy to describe and fast to evaluate; we give a number of computational and empirical results that suggest our heuristic is within 2% of an optimal strategy. Finally, Section 3.4 provides a number of mathematical results, showing that our fast and easy heuristic is asymptotically optimal.

3.1 Basic Observations

First, we introduce some notation that will be used throughout this section.

Let us assume that the corner that hides the object is at distance d from the start. Let x_i denote the distance the robot travels in the i–th step, i.e., on its way from position $i-1$ to position i, from which the i–th scan will be taken. If the object was hidden infinitesimally behind position i, the optimal solution would go perpendicularly to the line L_i that runs from the corner through position i, and then take one scan from there. Let d_i denote the length of this line segment and observe that it meets L_i at a point that lies on the semi-circle spanned by the start and the corner. Then the optimum cost to detect the object would be $1 + d_i$, whereas the robot would only see the object at position $i+1$, having accumulated a cost of

$$i + 1 + \sum_{j=1}^{i+1} x_j.$$

Now suppose that c is the smallest competitive ratio that can be achieved in this setting. By local optimality, for any scan position, the ratio of the solution achieved and the optimal solution must be equal to c. Therefore,

$$x_{i+1} = c(1 + d_i) - (i + 1) - \sum_{j=1}^{i} x_j \tag{1}$$

must hold for $i = 1, 2, \ldots$ In particular, we have $x_1 = c - 1$ for the first step.

3.2 Globally Optimal Strategies

The above recursion can be used for proving a lower bound.

Theorem 1. *There is no global c-competitive strategy with $c < 2$.*

Proof. Assume the claim was false, and there was a c-competitive strategy for $c = 2 - \delta$. We show that $x_i \leq (1-\delta)^i$ holds, making it impossible for the robot to get further than a distance of $1/\delta$ away from the start, a contradiction. Clearly, we have $x_1 = 1 - \delta$ for step 1. Moreover,

$$d_i \leq \sum_{j=1}^{i} x_j$$

holds, because d_i is the shortest path from the start to line L_i, whereas the sum denotes the length of the robot's path. Plugging this into our recursion yields

$$x_{i+1} \leq (1-\delta)(1+\sum_{j=1}^{i} x_j) - i.$$

By induction, we have $x_j \leq (1-\delta)^j$, hence

$$\begin{aligned} x_{i+1} \leq (1-\delta)\frac{1-(1-\delta)^{i+1}}{\delta} - i &\leq (1-\delta)\frac{1-(1-(1+i)\delta)}{\delta} - i \\ &= 1-(1+i)\delta \leq (1-\delta)^{i+1}, \end{aligned}$$

using the Bernoulli inequality $1-(1+i)\delta \leq (1-\delta)^{i+1}$ twice. □

Instead of increasing the distance d we could as well consider a situation where start and corner are a distance 1 apart, but the scan cost is only $1/d$. Now Theorem 1 shows a remarkable discontinuity: Even for a scan cost arbitrarily small, a lower bound of 2 cannot be beaten, whereas for zero scan cost, a factor of 1.212... can be obtained [9].

On the positive side, for n intermediate scan points, Equation (1) provides n optimality conditions. As there are $2n$ degrees of freedom (the coordinates of intermediate scan points), we get an underdetermined nonlinear optimization problem for any given distance d, provided that we know the number of scan points. For $d = 1$, this can be used to derive an optimal competitive factor of 1.808201..., achieved with one intermediate scan point. For larger d (and hence, larger n) one could derive additional geometric optimality conditions and use them in combination with more complex numerical methods. However, this approach appears impractical for real applications, for reasons stated above. As we will see in the following, there is a better approach.

3.3 A Simple Heuristic Strategy

Now we describe a simple strategy for the searching problem that uses trajectories inscribed into a circle. This reduces the degrees of freedom to the point where evaluation is fast and easy. What is more, it works very well in realistic settings, and it is asymptotically optimal for decreasing cost of scanning, or growing size of the environment.

The robot simply follows a polygonal path inscribed into the semi-circle of diameter d, spanned by start and corner. It remains to determine those points where it stops for scanning its environment. This is done by applying the optimality condition derived in Section 3.1. In step j, the robot moves along a chord of length x_j. From the corner, this chord is visible under an angle of $\varphi_j = \arcsin(x_j/d)$. The chord connecting the start to position i is of length

$$d_i = d\ \sin(\sum_{j=1}^{i} \varphi_j),$$

so that the recursion (1) obtained in Section 3.1 turns into

$$x_{i+1} = c\left(1 + d\,\sin\left(\sum_{j=1}^{i} \arcsin\left(\frac{x_j}{d}\right)\right)\right) - (i+1) - \sum_{j=1}^{i} x_j .$$

Given any $c > 1$, we can tentatively compute steps of length x_i by this formula, starting with $x_1 = c - 1$. If the resulting sequence reaches the corner, the ratio of c can indeed be achieved. If it collapses prematurely (by returning negative values) c was too small. (For example, $c = 2.001525...$ is optimal for $d = 40$; see Figure 3.3 for an illustration of upper and lower bounds on this value.)

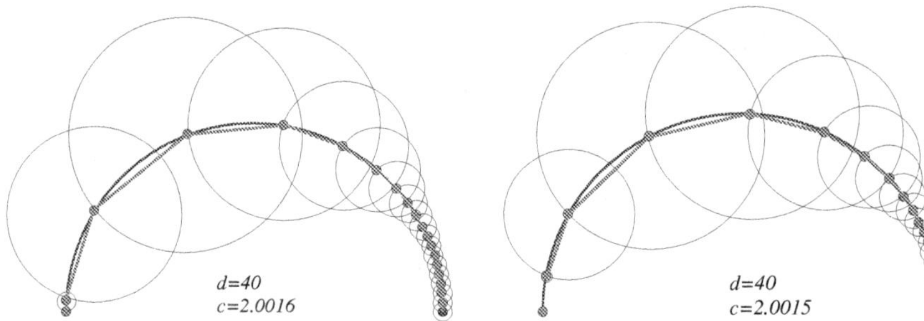

Fig. 5. An example for $d = 40$, with starting point on the right, corner on the left of the semi-circles: (Left) For $c = 2.0016$, the circle sequence reaches the corner, showing that the chosen c can be achieved. (Right) For $c = 2.0015$, the sequence collapses before reaching the corner, showing that the chosen c cannot be achieved. The actual optimum is about 2.001525...

By performing a binary search, the optimal ratio and the necessary step lengths can be computed extremely fast. Moreover, an analysis of the optimal ratio as a function of d shows that a maximum is reached for $d = 4.400875...$ which is precisely at the threshold between three and four necessary scans, with a competitive ratio of 2.168544. (See Table 1 for an overview of the critical values for which the number of scans increases, and Figure 7 for the achievable ratios as a function of the distance.) This is still within about 2% of the global optimum, which appears to be at about 2.12 (see Figure 6.) Moreover, numerical evidence shows that the ratio approaches 2 quite rapidly as d tends to infinity. This is all the more surprising, as the resulting initial step length converges to 1, while a constant step length of 1 yields a competitive ratio of π. In the following Section 3.4 we give a mathematical proof of this observation.

3.4 Asymptotics

As we have seen in Theorem 1, there is a lower bound of 2 on the competitive ratio for all strategies and large d. In the following we will show that for large d, there is a matching upper bound on our circle strategy presented in

Table 1. Threshold values for small numbers of scans, rounded to six digits.

Number of scans	Maximal d	c at upper bound
0	0.618034	1.618034
1	1.530414	2.040287
2	2.799395	2.155363
3	4.400876	2.168544
4	6.316892	2.147994
5	8.514200	2.118498

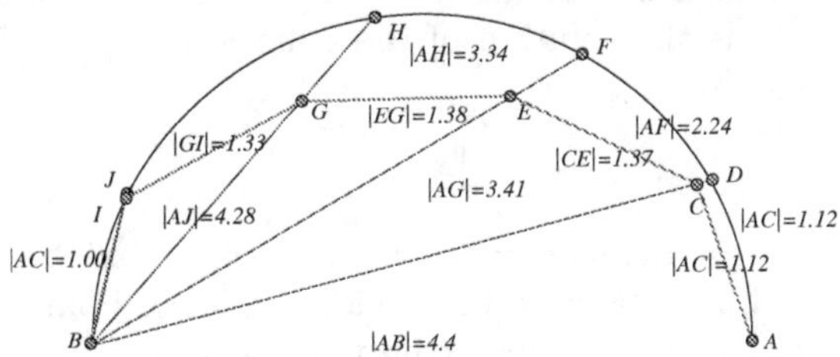

Fig. 6. A solution for $d = 4.4$ that achieves competitive ratio 2.12: The starting position is at A, the corner at B.

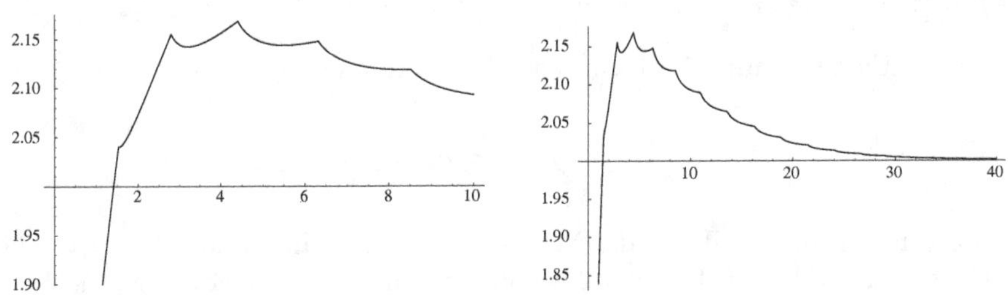

Fig. 7. The competitive ratio as a function of d: (Left) for small values of d. (Right) for larger values of d. Note the cusps at threshold values, the sharp peak at (4.4,2.17), and the clear asymptotic behavior. The first step length, x_1, is given by $c - 1$.

Section 3.3, proving it to be asymptotically optimal. For limited physical distances, it shows that even for arbitrarily small scan times, there is a relatively simple strategy that achieves the optimal ratio of 2.

Our proof of the upper bound proceeds as follows. Let us assume that we are given some fixed $\varepsilon > 0$. We then proceed to show that for $c = 2 + \varepsilon$, the recursion presented in Section 3.3 does not collapse before the corner is reached, if the diameter d of the semi-circle is large enough.

In proving the lower bound stated in Theorem 1, we have used the obvious fact that the length d_n of the optimal path cannot exceed the length of the robot's path. Now we are turning this argument around: The robot's path to position n does not exceed the length of the circular arc leading from the start to position n. As this arc is not much longer than d_n, the length of the chord from the start to n, if the diameter d of the circle is large enough. More precisely, we use the following.

Lemma 1. *(i) There is an upper bound on the total length of the first n steps of the circle strategy that does only depend on n and ε, but not on d.*
(ii) Given any $A > 0$, we can find d_0 such that each arc of length $\leq A$ in a circle of diameter $\geq d_0$ exceeds the length of its chord by at most ε^2.

Proof. Claim (i) can be shown by the same technique as in the proof of Theorem 1. In order to prove claim (ii), let a and c denote the maximum lengths of an arc and its chord in a circle of diameter d satisfying $a \leq b + \varepsilon^2$. Let 2β denote the angle of the arc, as seen from the center, so that $a = d\beta$ and $c = d \sin \beta$ hold. The maximum arc satisfying the condition is of length $a = d\beta_d$ where β_d is the solution of the equation $\beta_d - \sin \beta_d = \varepsilon^2/d$. In the equivalent expression

$$d\beta_d \left(1 - \frac{\sin \beta_d}{\beta_d}\right) = \varepsilon^2$$

the fraction tends to one, so $a = d\beta_d$ must be unbounded. □

These facts will now be used in providing a lower bound for the first steps along the semi-circle, aiming for a competitive ratio of $c = 2 + \varepsilon$.

Lemma 2. *Let $\varepsilon > 0$ and N be given. Then there is a number d_0 such that for each diameter $d \geq d_0$ we have $x_n \geq 1 + (2^n - 1)\,\varepsilon$, for $n \leq N$.*

Proof. Using Lemma 1 we can choose d_0 large enough that

$$\sum_{i=1}^{n} x_i \leq d_n + \varepsilon^2$$

holds for all $n \leq N$ if $d \geq d_0$. Now we proceed by induction. For $x_1 := (1+\varepsilon)$ the claim is fulfilled. For $n = 2$ we observe that $d_1 = x_1$ holds, so the recursive formula (1) yields

$$\begin{aligned} x_2 &= (2+\varepsilon)(1+d_1) - 2 - x_1 \\ &= (2+\varepsilon)^2 - 3 - \varepsilon \geq 1 + 3\varepsilon. \end{aligned}$$

Now assume the claim was true for $x_1, \ldots, x_{n-1}$, where $n \geq 3$, and let d_{n-1} be the $(n-1)$st chord, arising by connecting the start point with the $(n-1)$st scan point. The induction hypothesis implies

$$\sum_{i=1}^{j} x_i \geq \sum_{i=1}^{j} (1 + (2^i - 1)\,\varepsilon) = j + (2^{j+1} - j - 2)\varepsilon.$$

From the recursion we obtain

$$x_n = (2+\varepsilon)(1 + d_{n-1}) - n - \left(\sum_{i=1}^{n-1} x_i\right).$$

As $d \geq d_0$, we have $d_{n-1} \geq \left(\sum_{i=1}^{n-1} x_i\right) - \varepsilon^2$ for $n \leq N$. As $n \geq 3$, we get

$$
\begin{aligned}
x_n &\geq (2+\varepsilon)\left(1+\left(\sum_{i=1}^{n-1} x_i\right) - \varepsilon^2\right) - n - \left(\sum_{i=1}^{n-1} x_i\right) \\
&= (1+\varepsilon)\left(\sum_{i=1}^{n-1} x_i\right) + (2+\varepsilon)(1-\varepsilon^2) - n \\
&= 1 + (2^n-1)\varepsilon + (2^n - n - 3 - \varepsilon)\,\varepsilon^2 \\
&\geq 1 + (2^n - 1)\varepsilon. \quad \square
\end{aligned}
$$

Under the assumptions of Lemma 2 we can now prove the following.

Lemma 3. *For the first N steps of the robot, $\frac{1}{N}\sum_{i=0}^{N-1} x_i \geq 5$ holds.*

Proof. We may assume that

$$x_n \geq 1 + (2^n - 1)\,\varepsilon$$

holds for $n \leq N$. If N is large enough and $n \geq N/2$, we get

$$x_n \geq 1 + \left(\frac{10}{\varepsilon} - 1\right)\varepsilon \geq 10.$$

Thus,

$$\sum_{i=1}^{N-1} x_i \geq \sum_{i=N/2}^{N-1} x_i \geq 5N,$$

as claimed. $\square$

To conclude the proof, we consider a diameter d large enough for Lemma 3 to hold, so we have a lower bound of 5 on the average size for the first N steps. This suffices to show that all following steps are at least of length 5.

Lemma 4. *Assume that for some $N \geq 12$, we have $\sum_{i=1}^{N-1} x_i \geq 5N$. Then $x_n \geq 5$ for all $n \geq N$.*

Proof. Again we proceed by induction and consider

$$x_n = (2+\varepsilon)(1 + d_{n-1}) - n - \left(\sum_{i=1}^{n-1} x_i\right).$$

As all x_i are lengths of chords of the semi-circle with diameter d, we have

$$d_{n-1} \geq \frac{2}{\pi}\sum_{i=1}^{n-1} x_i.$$

By a similar argument as before, we get

$$\begin{aligned} x_n &\geq (2+\varepsilon)\left(1+\frac{2}{\pi}\left(\sum_{i=1}^{n-1} x_i\right)\right) - n - \left(\sum_{i=1}^{n-1} x_i\right) \\ &\geq \left(\frac{4}{\pi}-1\right)\left(\sum_{i=1}^{n-1} x_i\right) - n + 2 \\ &\geq \left(\frac{4}{\pi}-1\right) 5n - n + 2 \geq 5, \end{aligned}$$

since $n \geq 12$, as claimed. □

With the help of these lemmas, we get

Theorem 2. *The circle strategy is asymptotically optimal: For any* $\varepsilon > 0$, *there is a* d_ε, *such that for all* $d \geq d_\varepsilon$, *the strategy is* $(2+\varepsilon)$*-competitive.*

Proof. The preceding Lemmas 2, 3, 4 show that for any large enough d, the sequence will consist of step lengths that are all at least 5. This implies that the sequence will reach the corner in a finite number of steps, showing that a competitive factor of $(2+\varepsilon)$ can be reached. □

4 Practical Application

Our strategy was used in a practical setting, documented in the video [6]. In the great hall of Schloss Birlinghoven, starting about 8 meters from a door ($d = 1$ for the right scanner setting), Kurt follows the trajectory developed in the third part; depending on the position of a hidden object (a chair) he may have to perform a second scan from the corner. The second scenario shows a starting distance of $d = 2$, resulting in two intermediate scan points.

5 Conclusions

We have developed a search strategy that can be used for an actual autonomous robot. Obviously, a number of problems remain. Just like [9] provided a crucial step towards the solution for exploring general simple polygons described in [7], one of the most interesting challenges is to extend our results to more general settings with a larger number of obstacles, or the exploration of a complete region. See Figure 8 for a typical realistic scenario. It should be noted that scan cost (and hence positive step length without vision) can cause theoretical problems in the presence of tiny bottlenecks; even without scan cost, this is the basis of the class of examples in [1] for polygons with holes. However, in a practical setting, lower bounds on the feature size are given by robot size and scanner resolution. Thus, there may be some hope.

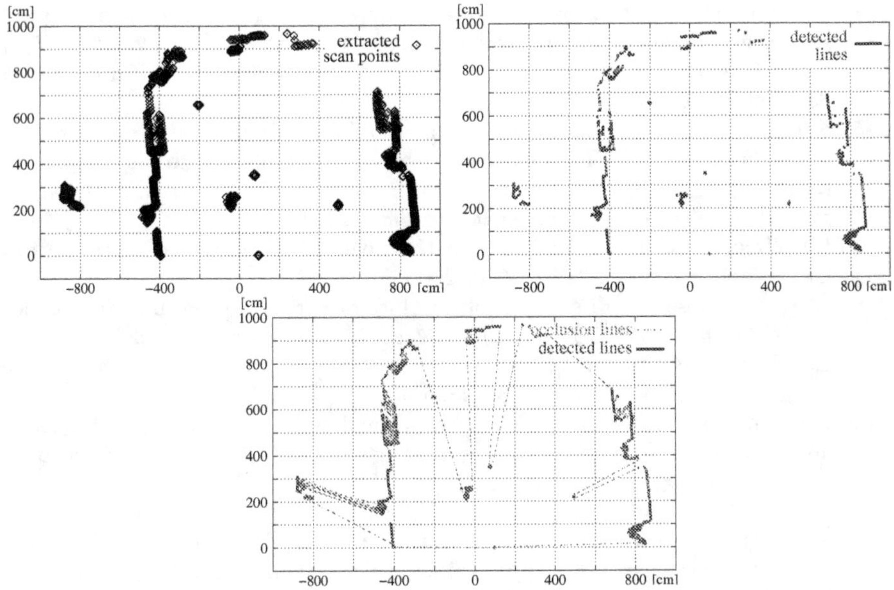

Fig. 8. A typical scenario faced by Kurt3D. Top left: Extracted points at height 75 cm (corresponding to figure 1, bottom middle). Top right: Line detection using Hough transform. Bottom: Automatically generated map with occlusion lines [17].

Acknowledgments

This research was motivated by the Dagstuhl workshop on robot navigation, Dec 7-12, 2003. We thank all other participants for a fruitful atmosphere and motivating discussions. We thank Hartmut Surmann, Joachim Hertzberg, Kai Lingemann, Kai Pervölz, Matthias Hennig, Erik Demaine, Shmuel Gal, Christian Icking, Elmar Langetepe, Lihong Ma for preceding joint research that laid the foundations for this work, and Matthias Hennig, Rolf Mertig, and Jan van der Veen for technical assistance.

References

1. S. Albers, K. Kursawe, and S. Schuierer. Exploring unknown environments with obstacles. In *Proc. 10th ACM-SIAM Sympos. Discrete Algorithms (SODA'99)*, pages 842–843, 1999.
2. P. Besl and N. McKay. A method for registration of 3–D shapes. *IEEE Transactions on PAMI*, 14(2):239–256, 1992.
3. S. Carlsson, H. Jonsson, and B. J. Nilsson. Finding the shortest watchman route in a simple polygon. *Disc. Comput. Geom.*, 22:377–402, 1999.

4. A. Efrat, H. González-Baños, S. G. Koburov, and L. Palaniappan. Optimal strategies to track and capture a predictable target. In *Proc. 2003 IEEE Int. Conf. Robotics and Automation (ICRA 2003)*, pages 3789–3796, Taipei, Taiwan, September 2003. IEEE.
5. S. P. Fekete, S. Gal, and E. D. Demaine. Online searching with turn cost, 2004.
6. S. P. Fekete, R. Klein, and A. Nüchter. Searching with an autonomous robot (video). In *13th Ann. Video Review of Comp. Geom.*, available at `http://give-lab.cs.uu.nl/socg04video/`, 2004.
7. F.Hoffmann, C.Icking, R.Klein, and K.Kriegel. The polygon exploration problem. *SIAM J. Comp.*, 31:577–600, 2001.
8. B. Horn. Closed–form solution of absolute orientation using unit quaternions. *Journal of the Optical Society of America A*, 4(4):629–642, April 1987.
9. C. Icking, R. Klein, and L. Ma. How to look around a corner. In *Proc. 5th Can. Conf. Comp. Geom.*, pages 443–448, 1993.
10. C. Icking, R. Klein, and L. Ma. An optimal competitive strategy for looking around a corner. Technical Report 167, Department of Computer Science, FernUniversität Hagen, Germany, 1994.
11. V. Isler, S. Kannan, and K. Daniilidis. Local exploration: Online algorithms and a probabilistic framework. In *Proc. IEEE Int. Conf. Robotics and Automation (ICRA 2003)*, pages 1913–1920, Taipei, Taiwan, September 2003. IEEE.
12. B. Kalyanasundaram and K. Pruhs. Constructing competitive tours from local information. *Theoret. Comput. Sci.*, 130:125–138, 1994.
13. K. N. Kutulakos, Charles R. Dyer, and V. J. Lumelsky. Provable strategies for vision-guided exploration in three dimensions. In *Proc. 1994 IEEE Int. Conf. Robotics and Automation (ICRA 1994)*, pages 1365–1372. IEEE, 1994.
14. A. Nüchter, H. Surmann, and J. Hertzberg. Automatic classification of objects in 3D laser range scans. In *Proc. 8th Conf. Intelligent Autonomous Systems*, pages 963–970, March 2004.
15. J. O'Rourke. *Art Gallery Theorems and Algorithms.* Internat. Series of Monographs on Computer Science. Oxford University Press, New York, NY, 1987.
16. H. Surmann, K. Lingemann, A. Nüchter, and J. Hertzberg. A 3D laser range finder for autonomous mobile robots. In *Proc. 32nd Int. Symp. on Robotics (ISR '01)*, pages 153–158, Seoul, Korea, April 2001.
17. H. Surmann, A. Nüchter, and J. Hertzberg. An autonomous mobile robot with a 3D laser range finder for 3D exploration and digitalization of indoor environments. *Robotics and Automation*, 45:181–198, 2003.
18. X. H. Tan, T. Hirata, and Y. Inagaki. An incremental algorithm for constructing shortest watchman routes. *Int. J. Comput. Geom. Appl.*, 3(4):351–365, 1993.
19. X. H. Tan, T. Hirata, and Y. Inagaki. Corrigendum to "An incremental algorithm for constructing shortest watchman routes". *Int. J. Comput. Geom. Appl.*, 9(3):319–323, 1999.
20. P. Viola and M. Jones. Robust real-time object detection. In *Proc. 2nd Int. Workshop on Statistical and Computational Theories of Vision – Modeling, Learning, Computing and Sampling*, Vancouver, Canada, July 2001.

Competitive Complexity of Mobile Robot On Line Motion Planning Problems

Yoav Gabriely[1] and Elon Rimon[1]

Technion, Israel Institute of Technology
meeryg,rimon@tx.technion.ac.il

Abstract. *This paper is concerned with on-line problems where a mobile robot of size D has to achieve a task in an unknown planar environment whose geometry is acquired during task execution. The critical parameter in such problems is physical motion time which corresponds to length or cost of the path traveled by the robot. The* competitiveness *of an on-line algorithm measures its performance relative to the optimal off line solution to the problem. While competitiveness usually means constant relative performance, this paper generalizes competitiveness to any functional relationship between on-line performance and optimal off-line solution. Given an on-line task, its* competitive complexity class *is a pair of lower and upper bounds on the competitive performance of all on-line algorithms for the task, such that the two bounds satisfy the same functional relationship. We classify some common on-line motion planning problems into competitive classes. In particular, navigation to a target whose position is either apriori known or recognized only upon arrival belongs to a* quadratic *competitive class. The hardest on-line problems belong to exponential and even non-boundable competitive classes. We present examples of such problems, which involve navigation in unknown variable traversibility environments.*

1 Introduction

This paper is concerned with on-line problems where a mobile robot of size D has to accomplish a task in a planar environment populated by obstacles. The robot has no apriori information about the environment, but may locally acquire this information using on-board sensors. This class of on line problems has a wide range of applications. Examples are area coverage for cleaning supermarkets and train stations, navigation to various targets for mail and material delivery in offices and factories, and planetary exploration and sample acquisition. On the other hand, this paper is not concerned with on line coordination of multiple robots, nor with on line scheduling problems. The most critical parameter in mobile robot motion tasks is physical travel time rather than on-board computation time. Under a uniform velocity assumption, travel time corresponds to path length. Under a uniform power output assumption, travel time corresponds to path traversibility cost. Hence the algorithms discussed in this paper are classified in terms of length or cost of the path traveled by the robot during algorithm execution.

Mobile robot on line algorithms are discussed in the robotics as well as computational geometry literature. Roboticists usually emphasize the type of

M. Erdmann et al. (Eds.): Algorithmic Foundations of Robotics VI, STAR 17, pp. 155–170, 2005.

sensors required to achieve a given task [4]. A notable early paper in this area is by Lumelsky and Stepanov [15]. They describe the $BUG1$ and $BUG2$ algorithms for navigating a point robot in an unknown planar environment using tactile sensors. Their work has been extended to navigation in planar environments using vision and laser sensors, e.g. [14,16]. However, performance of these algorithms is typically characterized in terms of geometric parameters of the environment such as total obstacle perimeter, without any reference to the length of the optimal off-line solution, denoted l_{opt}. For instance, under $BUG1$ the robot may start at S and circumnavigate the entire outer boundary of the environment before reaching the target T, even in cases where l_{opt} is almost the straight line distance from S to T. While the path length of $BUG1$ is always bounded by 1.5 times the total perimeter of the obstacles [15], it may perform unboundedly poorly with respect to l_{opt}.

Computational geometry researchers introduced the notion of competitiveness which compares the performance of an on-line algorithm to the optimal off-line solution to the same problem. In particular, an algorithm for a task P is said to be *competitive* if its solution to every instance of P is bounded by a constant times l_{opt} [13]. A notable early paper in this area is by Baeza-Yates et al. [1]. They investigate navigation of a point robot in a planar unknown environment consisting of m radial corridors to a target whose position is recognized only upon arrival. Their on-line algorithm solves the problem with a path length l satisfying the linear inequality $l \leq \left(1 + 2\frac{m^m}{(m-1)^{m-1}}\right) l_{opt}$. Moreover, up to lower order terms this upper bound is tight. Other notable papers discuss on-line navigation in specific classes of rectangular rooms [2,8] and generalized streets [5]. However, all of these papers strive to identify specific classes of environments in which *constant* competitiveness can be achieved. In contrast, we seek to identify the competitive relationship governing the fully general problems.

The structure of the paper is as follows. In the next section we generalize the traditional notion of competitiveness into *competitive classes.* We define the *competitive complexity class* of a task P as a pair of lower and upper bounds on the competitive performance of all on-line algorithms for P, such that the two bounds satisfy the same functional relationship up to constants. Note that competitive complexity characterizes the task P itself, not of any specific algorithm for P. The notion of competitive complexity was first proposed by Icking and Klein [13]. However, they consider only linear competitiveness (i.e. a constant ratio between on-line and off-line performance), and require that the lower and upper bounds match precisely. We consider a collection of competitive complexity classes, and only require that the two bounds belong to the same class. The remainder of the paper classifies some common mobile robot problems into competitive complexity classes, under a key assumption that *the robot has a physical size D such that $D > 0$.* While this assumption may seem obvious, only few papers make use of this assumption (e.g. [6]). In Section 3 we discuss the well understood area coverage problem. In Section 4 we discuss the celebrated problem of navigation in an unknown planar environment to a target whose position is either apriori known or recognized only upon arrival. In Section 5 we discuss on-line navigation in unknown variable traversibility environments. Finally, in the concluding section we discuss open problems, and consider extension of the results to motion tasks in three-dimensions.

2 Definition of Competitive Classes and Complexity

In this section we specify our basic assumptions and define competitive complexity classes. We consider *geometric* on line problems, where a mobile robot has to execute a task in a planar environment populated by unknown stationary obstacles. The mobile robot is a freely moving planar body of size D, where $D > 0$ is a given constant. One may think of the mobile robot as a disc of diameter D. However, the results reported in this paper hold true for any planar body of minimum width D. The assumption that the robot has a physical size $D > 0$ is well justified in any practical setting. This assumption is implicit in papers which employ on-line discretization of the continuous work area into a grid of D-size cells, e.g. [6,10,12,17]. The following lemma characterizes the number of essentially distinct paths available to a D-size robot in a planar environment.

Lemma 2.1 *Let a planar body $\mathcal{B}$ of size $D > 0$ move in a planar environment populated by stationary obstacles. Let α be a closed curve which encloses $\mathcal{B}$ (the curve may cross obstacles). Then the number of homotopically distinct free paths, or "corridors," along which $\mathcal{B}$ can cross α and move to infinity is bounded by $|\alpha|/D$, where $|\alpha|$ is the length of α.*

While a formal proof of the lemma is omitted, consider a disc robot of size D trying to cross a circle of finite radius R. Since R is finite, the circle is punctured in worst case by finitely many point-size obstacles. Any circular segment bounded by point obstacles whose mutual distance is less than D is entirely blocked for $\mathcal{B}$. Conversely, a circular segment bounded by point obstacles whose mutual distance larger or equal to D forms a passable corridor. Hence the number of such corridors is bounded from above by $2\pi R/D$.

The sensors of the robot are assumed ideal and their specification is as follows. The first sensor measures the robot's position and orientation with respect to a fixed reference frame. This sensor is usually based on odometry measurements indoors and GPS measurements outdoors. However, this sensor usually suffers from measurement errors, and our assumption of perfect readings needs to be revised in future work. The second sensor measures geometric details of the obstacles surrounding the robot within a finite detection range. A zero detection range corresponds to tactile sensors, while a positive detection range can be reasonably accurately implemented with cameras or laser scanners. In addition to sensors the robot has on-board memory in which information on the environment can be accumulated. However, the robot is not allowed to place various environmental markers during task execution

Next we describe the parameters governing the performance of mobile robot tasks. The three most significant parameters are physical travel time, on-board computation time, and on-board memory. In order to simplify the ensuing analysis, we associate physical travel time with either length or traversibility cost of the path traveled by the robot. The ensuing definitions are given in terms of path length denoted l, but the definitions hold for other measures such as path traversibility cost. As for on-board computation time, we limit our discussion to algorithms that take *polynomial time* to compute each physical motion step of the mobile robot. Since the time required to complete a physical motion step is typically several orders of magnitudes longer than the time it takes to execute an on-board computation step, we focus on l as the main performance parameter. Last, we limit the discussion to algorithms whose storage requirement is at most *linear* in the size of the environment. This memory requirement may prove impractical in tasks such as planetary navigation, and may need to be revised in future work.

Thus l denotes length of the path traveled by the robot during algorithm execution, while l_{opt} denotes length of the optimal off-line path for the same problem instance. The following definition generalizes the traditional notion of linear competitiveness to any functional relationship between l and l_{opt}.

Definition 1 (generalized competitiveness) *An on-line algorithm solving a task P is* $f(l_{opt})$**-competitive** *when l is bounded from above by a scalable function $f(l_{opt})$ over all instances of P. The following three functional relationships are of particular interest,*

$$\begin{array}{ll} l \leq c_1 l_{opt} + c_0 & \textit{linear competitiveness} \\ l \leq c_2 l_{opt}^2 + c_1 l_{opt} + c_0 & \textit{quadratic competitiveness} \\ l \leq c_1 \exp(l_{opt}) + c_0 & \textit{exponential competitiveness,} \end{array}$$

where the c_i's are constants that depend on the robot size D.

The meaning of *scalability* is as follows. When performance is measured in physical units such as meters m, one must ensure that both sides of the relationship $l \leq f(l_{opt})$ posses the same units, so that change of scale would not affect the bound [18]. For instance, the coefficient c_2 in the relationship $l \leq c_2 l_{opt}^2 + c_1 l_{opt} + c_0$ must have units of m^{-1}, c_1 must be unitless, and c_0 must have units of m. Note that the definition of $f(l_{opt})$-competitiveness focuses on a particular algorithm solving a task P. However, our objective is to characterize the lowest upper bound that can be achieved over all on-line algorithms for P. This objective requires a universal lower bound on the performance of all on-line algorithms for P. If the lower and upper bounds satisfy the same functional relationship, we associate the functional relationship with P itself. This notion is made formal in the following definition.

Definition 2 (competitive complexity) *A* **universal lower bound** *on the competitiveness of a task P is a scalable lower bound $l \geq g(l_{opt})$ over all on-line algorithms for P.*
If a competitive upper bound $f(l_{opt})$ and a universal lower bound $g(l_{opt})$ for P are the same function up to constants, this function is the **competitive complexity class** *of P.*

The remainder of this paper characterizes the complexity class of some common mobile robot tasks. In particular, we describe problems which belong to the linear, quadratic, and exponential competitive complexity classes.

3 On Line Area Coverage

The area coverage problem is as follows. A mobile robot is equipped with a covering tool which is mounted under the robot platform. The covering tool is typically larger than the robot itself, hence we may identify the robot with the covering tool. Given an unknown planar environment populated by stationary obstacles, the robot has to move the covering tool along a continuous path such that eventually every physical point of the work area is covered, then return to its start position [10,12]. Furthermore, we seek an efficient covering path whose length can be compared with the optimal off-line path. Note that physical area coverage is distinct from sensor coverage, where a mobile robot has to scan with limited visibility sensors an unknown environment for map building purposes. The latter problem is considered below.

We discuss the following discretized version of the area coverage problem. The covering tool is assumed to be a square of size D, and the continuous work area is incrementally discretized into D-size cells, each being identical to the covering tool in shape and size. In each step the robot covers one cell, then moves to a neighboring cell while keeping the covering tool aligned with the grid cells. The problem thus becomes a familiar problem of covering arbitrary planar grids. It is worth noting that the off-line Hamiltonian path problem on planar grids (i.e. the construction of a path that visits every node of a grid precisely once) is NP-complete. The off-line optimal covering of planar grids, being a more general problem, is NP-hard. Given this classification of the off-line problem, it comes as a surprise that the on-line task has a linear competitive complexity.

First we describe a concrete algorithm and discuss its competitive upper bound. The trivial DFS algorithm can cover a general planar grid with a path of length $l = 2nD$ using $O(n)$ memory, where n is the total number of grid cells. Since the off-line optimal solution satisfies $l_{opt} \geq nD$, the DFS algorithm establishes the competitive upper bound $l \leq 2l_{opt}$. On the other hand, Icking et al. [12] and Gabriely et al. [10] have independently demonstrated that for any on-line area coverage algorithm, there exists a grid environment in which the length of the coverage path satisfies $l \geq (2-\epsilon)l_{opt}$. Let us demonstrate this lower bound on a simple corridor whose width is identical to the robot size D. Suppose the robot has a range sensor with a detection range of $1-\delta$, where δ is a small positive parameter. We initially place the robot in an infinite corridor and wait until it covers a length-L portion of the corridor, where $L >> 1$. At that instant we truncate the corridor at a unit distance from both sides of the length-L portion, as shown in Figure 1. Since the robot has no way of knowing about this change in the environment, its on-line behavior would be identical if it starts at the same point in the truncated corridor. (If the algorithm is non-deterministic, one outcome would match this on-line behavior.) The length of the on-line covering path is $l \geq 2L+3$, since the robot must visit both ends of the truncated corridor before coverage is complete. The shortest off-line covering path, achieved by sweeping the corridor from end to end, has length $l_{opt} = L+2$. Thus $l \geq (2-\epsilon)l_{opt}$, and this is the universal lower bound on l in terms of l_{opt}. Using more sophisticated environments, the same lower bound holds for *covering tours*, where the robot must return to its starting point.

Fig. 1. A corridor environment with an on-line covering path depicted.

Since the universal lower bound and the competitive upper bound belong to the same linear competitive complexity class, we have the following theorem.

Theorem 1 (linear complexity) *The path length l generated by any on-line grid coverage algorithm satisfies $(2-\epsilon)l_{opt} \leq l \leq 2l_{opt}$, where l_{opt} is the length of the optimal off-line coverage path. The grid coverage problem thus belongs to the* **linear** *competitive complexity class.*

Let us briefly discuss the sensor coverage problem. In this problem the robot must scan an unknown planar environment with limited visibility sensors, attempting to generate a map of the environment as efficiently as possible [3,7,11]. Local visibility sensors can inspect the environment only along rays emanating from the robot. Hence one can always produce a sufficiently congested environment in which visibility sensors have no advantage over tactile sensors. In such environments sensor coverage becomes area coverage. It follows that sensor coverage has the same universal lower bound of $l \geq (2-\epsilon)l_{opt}$ as area coverage. An upper bound on sensor coverage is known only for simple polygons without internal holes. For such polygons, Hoffmann et al. [11] describe a sensor coverage algorithm which has a competitive upper bound $l \leq 26.5l_{opt}$, where l_{opt} is the length of the optimal off-line watchman route for the given environment. However, a general $f(l_{opt})$-competitive algorithm and the competitive complexity class of this problem are still unknown.

4 On Line Navigation

The on line navigation problem is as follows. A mobile robot initially at a configuration S must find a path to a target configuration T in an unknown planar environment. (The target configuration is assumed to be free from obstacles.) We first describe a navigation algorithm to a target whose position is recognized only upon arrival, then modify the algorithm to the case where the target position is apriori known to the robot. Both algorithms are based on a search-area doubling principle, and are called $SAD1$ and $SAD2$. We obtain a competitive upper bound for each algorithm, then obtain a universal lower bound for each version of the problem. To our knowledge only one general navigation algorithm has a known performance bound in terms of l_{opt}, by Datta and Soundaralakshmi [6]. However, their algorithm has an exponential upper bound in terms of l_{opt}, while it is shown below that $SAD1$ and $SAD2$ have a *quadratic* upper bound in terms of l_{opt}.

4.1 $SAD1$ Algorithm—Navigation to an Unknown Target

The principle idea of $SAD1$ is as follows. The robot selects an initial disc of radius R_0 centered at S, and searches the portion of this disc accessible from S. If the target is detected the algorithm terminates. Otherwise the robot repeats the process on discs of radii $2^{i/2}R_0$ for $i = 1, 2, \ldots$ until the target is detected, or the entire region accessible from S is explored without finding T (Figure 2). The search process in each disc is as follows. The robot imposes an on-line discretization of the continuous area into a grid of D-size cells. The grid consists only of free cells and is surrounded by partially occupied cells. The robot executes a standard area coverage tour on the grid of free cells, while scanning each new cell for the target. Once entering a new cell, the robot additionally scans the neighboring partially occupied cells for T. If the discretization preserves the connectivity of the accessible region (this assumption can be relaxed by a more sophisticated algorithm that monitors local connectivity breakage), then clearly all free and partially occupied cells in the region accessible from S are eventually inspected by the robot. A formal description of the algorithm follows.

SAD1 Algorithm:

Sensors: A position and orientation sensor. An obstacle detection sensor.
Input: A start S, an initial search radius R_0, but no apriori knowledge on T.
Initialization: Set initial search radius $r(1) = R_0$. Set $i = 1$.
Repeat
1. Execute a *coverage tour* on grid contained in disc of radius $r(i)$ and center S.
 1.1 Scan each new free cell and its partially occupied neighboring cells for T.
 1.2 STOP if T is found.
2. If no new free cell is encountered during i^{th} coverage tour:
 STOP, the target is unreachable.
3. Set $r(i+1) = \sqrt{2}r(i)$. Set $i = i + 1$.
(End of repeat loop)

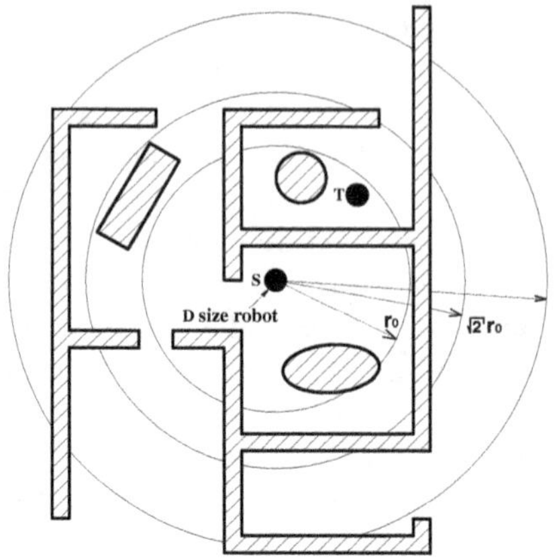

Fig. 2. The area of the disc searched by $SAD1$ doubles in each step.

Rather than give a formal proof of correctness, we make some informal remarks on the algorithm. First, the coverage tour in step 1 can be executed with a trivial DFS algorithm using linear memory. Second, the algorithm's average performance can be improved if the robot covers in each step only those cells which lie in the ring added to the previous disc. Third, if the target is inaccessible from S, the algorithm stops only when it has completely covered the portion of the environment accessible from S. Clearly this is a limitation of the problem itself, which assumes no knowledge of the target location. We now establish an upper bound on the path length of $SAD1$ in terms of l_{opt}. The following proposition establishes a quadratic competitive upper bound on $SAD1$.

Proposition 4.1 *If the target T is reachable from S, $SAD1$ finds the target using a path of length l which satisfies the quadratic inequality,*

$$l \le \frac{8\pi}{D} l_{opt}^2 + \frac{2\pi R_0^2}{D},$$

where D is the robot size, R_0 is the initial search radius, and l_{opt} is the length of the optimal off-line path from S to T.

Note: the upper bound is scalable, since both summands have length units.

Proof: First consider the case where $l_{opt} > R_0$. In this case the initial search disc is expanded at least once before the target is found. Suppose the search disc is expanded k times until the target is found. Since the radius of the disc increases by $\sqrt{2}$ at each step, its area doubles in each step. Let $S_0 = \pi R_0^2$, and let S denote the total area of the regions inspected by the robot inside the k discs. (These areas include free as well as partially occupied cells inspected by the robot.) Then S is bounded by $S \le S_0 + 2S_0 + \cdots + 2^{k-1}S_0 < 2^k S_0$, where we used the formula $2^0 + 2^1 + \cdots + 2^{k-1} = 2^k - 1$. Since the disc of radius l_{opt} already contains at least one path from S to T, $SAD1$ finds the target in a disc of radius at most $\sqrt{2}l_{opt}$. It follows that the area of the kth search disc, $2^{k-1}S_0$, satisfies the inequality $2^{k-1}S_0 \leq 2\pi l_{opt}^2$. Substituting this inequality into the bound on S gives $S < 4\pi l_{opt}^2$. Now recall that $SAD1$ guides the robot only through free grid cells. The total area of the free grid cells is at most S. Hence the total number of grid cells visited by the robot (including repetitive visits), denoted n, is bounded by $n \leq S/D^2$. On line coverage algorithms such as DFS guide the robot along a path whose length in cells is at most twice the total number of grid cells. Since each cell has size D, the total length of the path traveled by the robot is $l \leq 2nD$. Thus, $l \leq 2nD \leq \frac{2}{D}S < \frac{8\pi}{D}l_{opt}^2$, where we substituted the inequality $S < 4\pi l_{opt}^2$. Finally, the constant term $2\pi R_0^2/D$ bounds the path length traveled by the robot in the case where $l_{opt} \leq R_0$. In this case the target is found inside the initial search disc of radius R_0. □

4.2 A Lower Bound on Navigation to an Unknown Target

In this section we establish a universal lower bound on navigation to a target which is recognized only upon arrival. The environment that serves to establish the lower bound is a disc containing D-width corridors that emanate radially from the start point S (Fig. 3(a)). Initially a small disc centered at S is free of obstacles. At a certain distance from S eight equally spaced point-size obstacles appear, such that the distance between the obstacles is D (the number eight has no special meaning here). The eight obstacles extend radially as lines and form the boundary of eight passable corridors for the robot. The width of the eight corridors increases as they stretch radially away from S. When the width of a corridor becomes $2D$, the corridor splits into two D-width corridors separated by a cone-shaped obstacle (Fig. 3(b)). By symmetry all corridors split simultaneously. Hence the cone obstacles that separate corridors just before splitting are truncated at the splitting radius, and become radial lines from this radius onward. A close inspection of Figure 3(b) reveals that the cone obstacles occupy *one third* of the disc's total area. Finally, the tip of each cone obstacle is symmetrically "shaved", so that its tip would lie at a distance D away from the truncated cone lying closer to S (Fig. 3(b)). This shaving allows a D-size robot to enter the two corridors generated by splitting. Note that the shaved-off area becomes negligible relative to the cone's total area as the radius of the environment increases.
The following theorem establishes a quadratic lower bound on the performance of all on-line navigation algorithms to a target whose position is recognized only upon arrival.

Theorem 2 (quadratic complexity) *Let A be any navigation algorithm in an unknown planar environment to a target whose position is recognized only upon arrival. Let l_{opt} be the length of the optimal off-line path. Then the length l of the path generated by A satisfies the quadratic lower bound,*

$$l \geq \frac{4\pi}{3D}(1-\epsilon)l_{opt}^2,$$

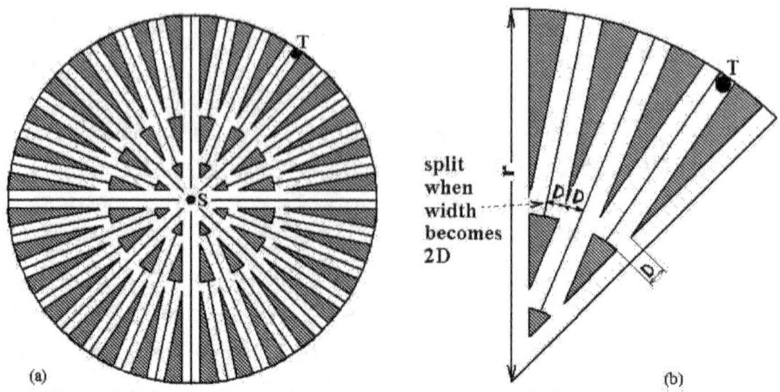

Fig. 3. (a) A D-width radial corridors environment. (b) Close up view.

where D is the robot size and ϵ is an arbitrary small positive parameter. Since the upper bound of $SAD1$ is also quadratic in l_{opt}, this navigation problem belongs to the **quadratic** *competitive complexity class.*

Proof: Consider the radial corridors environment with the target T placed at the end of a distal corridor, at a distance r from S. Since the robot has no knowledge of the environment and has no information where T might lie, it must in worst case inspect every corridor including all distal corridors. (If A is deterministic, we can enforce this worst case scenario by first watching the behavior of A, then placing the target in the last inspected corridor. If A is non-deterministic, we can only guarantee that *one outcome* of the algorithm would match this worst case scenario.) By construction every distal corridor can be approached from S along a simple radial path. Moreover, the robot must eventually move twice through every corridor of the environment—once in order to inspect a distal corridor and once in order to exit the corridor. An exception to this rule is the last corridor which is considered below. The total area of the obstacles in the corridor environment is almost one third of the disc area, with the approximation becoming arbitrary close to one third as the disc's radius increases. The total area inspected by the robot is therefore $2\pi r^2/3$. Since all corridors have a width D which is identical to the robot size, the total length of the path traveled by the robot satisfies in worst case the inequality $l \geq 4\pi r^2/3D - r$, where the substraction of r is due to the last corridor which need not be traced backward. Since T is placed at a radial distance r from S we have that $l_{opt} = (1+\epsilon')r$, where ϵ' is an arbitrary small positive parameter. Substituting for r gives $l \geq cl_{opt}^2/(1+\epsilon')^2 - l_{opt}$, where $c = 4\pi/3D$. We can write the last inequality as $l \geq cl_{opt}^2(1 - \epsilon'' - 1/cl_{opt})$, where we substituted $1/(1+\epsilon')^2 = 1-\epsilon''$. The term $\epsilon = \epsilon''+1/cl_{opt}$ can be made arbitrarily small for sufficiently large environments, hence $l \geq c(1-\epsilon)l_{opt}^2$. □

4.3 *SAD*2 Algorithm—Navigation to a Known Target

Consider now the celebrated problem of finding a path in an unknown environment to a target T whose position is apriori known to the robot. Algorithm $SAD2$ for this problem is based on the same principles as $SAD1$, except that now the robot searches a sequence of expanding ellipses whose focal points are S and T. The specific operation of $SAD2$ is as follows. The robot selects an initial ellipse with area S_0, and searches the portion of this ellipse accessible from S. If the target is found the algorithm terminates. Otherwise the robot repeats the process on ellipses with areas $2^i S_0$ until the target is found, or the entire region accessible from S is explored without finding a path to T (Figure 4). The search process in each ellipse is based on the same on-line discretization approach employed by $SAD1$. The robot executes a standard area coverage tour on the grid of free cells, while inspecting neighboring partially occupied cells for T. Making the assumption that the discretization preserves the connectivity of the accessible region, all free and partially occupied cells in the region accessible from S are eventually inspected by the robot. A formal description of $SAD2$ follows.

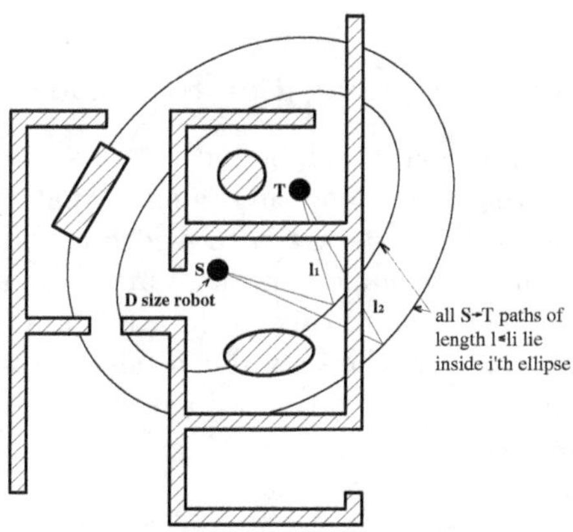

Fig. 4. The area of the ellipse searched by $SAD2$ doubles in each step.

SAD2 Algorithm:

Sensors: A position and orientation sensor. An obstacle detection sensor.
Input: A start S, a target T, an initial ellipse with focal points S,T and area S_0.
Initialization: Set initial search area $S(1) = S_0$. Set $i = 1$.
Repeat
1. Execute a *coverage tour* on grid in ellipse of area $S(i)$ w/ focal points S, T.
   1.1 Scan each new free and its partially occupied neighboring cells for T.
   1.2 STOP if T is found.
2. If no new free cell is encountered during i^{th} coverage tour, or an obstacle encountered during i^{th} coverage tour surrounds the target:
   STOP, the target is unreachable.

3. Set $S(i+1) = 2S(i)$. Set $i = i+1$.
(End of repeat loop)

Here, too, we omit a formal proof of correctness of the algorithm. Note that in step 2 the algorithm terminates the search when it discovers that the target is completely surrounded by an obstacle. We now turn to the task of establishing an upper bound on l in terms of l_{opt}. The following proposition establishes a quadratic upper bound on the path length of $SAD2$.

Proposition 4.3 *If the target T is reachable from S, $SAD2$ finds the target using a path whose length l satisfies the quadratic inequality,*

$$l \leq \frac{2\pi}{D} l_{opt}^2 + \frac{2S_0}{D},$$

where D is the robot size, S_0 is the area of the initial search ellipse, and l_{opt} is the length of the optimal off-line path from S to T.

Note that both summands have length units, so the upper bound is scalable. The proof of the proposition is similar to one presented for Proposition 4.1, and can be found in [9].

4.4 A Lower Bound on Navigation to a Known Target

We now establish a universal lower bound on navigation to a target whose position is known to the robot. The environment that serves to establish the lower bound consists of the same radial D-width corridors used above, with one addition. Now the environment is surrounded by a circular corridor of width D, with a single entry into this corridor (Figure 5). As in the previous environment, the total area of the shaved cone-shaped obstacles is almost *one third* of the extended disc area. The following theorem establishes a quadratic lower bound for this task.

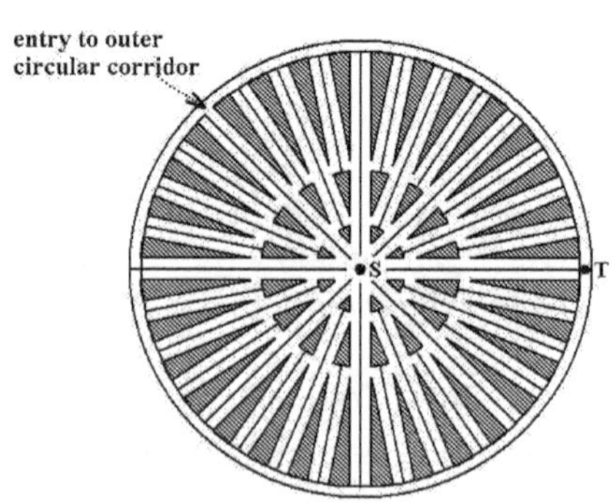

Fig. 5. The same radial corridors environment with an outer circular corridor.

Theorem 3 (quadratic complexity) *Let A be any navigation algorithm in an unknown planar environment to a target whose position is apriori known to the robot. Let l be the length of the path generated by A, and let l_{opt} be the length of the optimal off-line path. Then l satisfies the quadratic lower bound,*

$$l \geq \frac{4\pi}{3(1+\pi)^2 D}(1-\epsilon)l_{opt}^2,$$

where D is the robot size and ϵ is an arbitrary small positive parameter. *Since the upper bound of $SAD2$ is also quadratic in l_{opt}, this navigation problem belongs to the* **quadratic** *competitive complexity class.*

Proof: Suppose the target is placed somewhere in the outer corridor, at a distance r from S. Any on-line algorithm must first find the entry to the outer corridor. In worst case this task requires that the robot move twice through every radial corridor. An exception is the last distal corridor leading to the outer corridor. As argued in the proof of Theorem 4, the total area of the obstacles in the corridor environment is almost one third of the disc area, with the approximation becoming arbitrary close to one third as the disc's radius increases. The total area inspected by the robot is therefore $2\pi r^2/3$. Since all corridors have width D which is identical to the robot size, the total length of the path traveled by the robot satisfies in worst case the inequality $l \geq 4\pi r^2/3D - r$, where the substraction of r is due to the last radial corridor which need not be traced backward. Since T is placed in the circular outer corridor, the optimal off-line path satisfies $l_{opt} \leq (1+\epsilon')(1+\pi)r$, where ϵ' is an arbitrary small parameter. The term πr is due to worst case motion along the circular outer corridor to a target located opposite the entry. It follows that r satisfies the inequality $r \geq l_{opt}/(1+\epsilon')(1+\pi) \geq l_{opt}/(1+\pi) - \epsilon''$. On the other hand, $r \leq l_{opt}$ in the radial corridor environment. Substituting the two inequalities into the lower bound for l gives: $l \geq \frac{4\pi}{3D}r^2 - r \geq \frac{4\pi}{3(1+\pi)^2 D}l_{opt}^2 - l_{opt} - \epsilon''$. Thus $l \geq c(1 - \frac{l_{opt}+\epsilon''}{cl_{opt}^2})l_{opt}^2$, where $c = 4\pi/3(1+\pi)^2 D$. Since the quantity $\epsilon = (l_{opt}+\epsilon'')/cl_{opt}^2$ can be made arbitrarily small for sufficiently large environments, we obtain the lower bound $l \geq c(1-\epsilon)l_{opt}^2$. □

5 Navigation in Variable Traversibility Environments

The key for identifying on-line problems which belong to an exponential competitive complexity class is to break the geometric barrier imposed by the robot size D (Lemma 2.1). In order to break this barrier we introduce the following class of variable traversibility environments.

Definition 3 *Let $\sigma(x)$ denote traversibility per unit length at the obstacle free points of a planar environment. Let $\sigma_0 > 0$ be traversibility at the start point S. Then σ is* **linear-poly** *with respect to S if it satisfies the inequality*

$$\sigma_0(c\|x-s\|+1)^{-1} \leq \sigma(x) \leq \sigma_0 p(\|x-s\|),$$

for some $c > 0$ and monotonically increasing polynomial p s.t. $p(0) = 1$.

Thus we consider environments whose traversibility decays at most linearly and increases at most polynomially with distance from S. Note that σ is defined only at obstacle free points. The traversibility at points occupied by obstacles is assumed to be infinite. Note, too, that σ need not be monotonic with distance from S, only its bounding functions are required to be monotonic. While linear-poly traversibility is not completely general, it captures environments in which on-line navigation belongs to an exponential competitive complexity class. Our first task is to establish an upper bound on the cost of navigating from S to T. We use $SAD1$ as the basic navigation algorithm, and show that the cost of the path generated by this algorithm, denoted ρ, is at most exponential in the optimal off-line path cost, denoted ρ_{opt}.

Proposition 5.1 *$SAD1$ solves both versions of the navigation problem in linear-poly environments with path cost which satisfies the upper bound,*

$$\rho \leq \frac{2\pi}{D} \exp\left(\frac{\rho_{opt}}{\sigma_0}\right),$$

where D is the robot size, ρ_{opt} is the optimal off-line cost of traveling from S to T, and σ_0 is traversibility at S.

It can be verified that the argument of the exponent is unitless, and that the coefficient multiplying the exponent has length units. Since the exponential function itself has traversibility units, the upper bound on ρ is scalable. A proof of the proposition appears in [9].

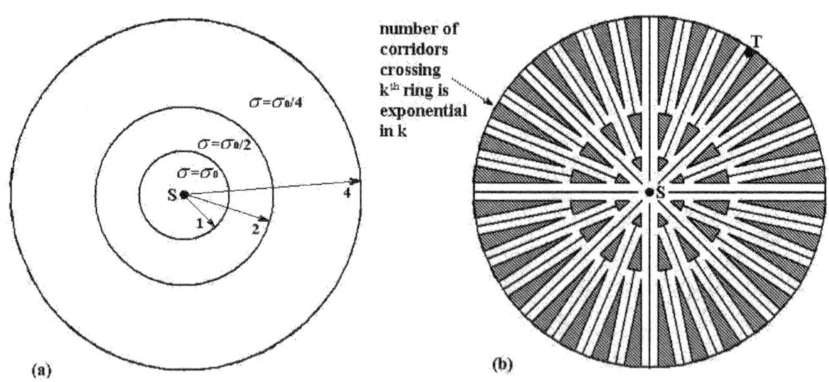

Fig. 6. (a) Travesibility is decaying and piecewise constant on concentric rings. (b) D-width radial corridors are embedded in each concentric ring.

Next we establish a universal lower bound over all on-line navigation algorithms in linear-poly environments. The bound is achieved in the following environment. We select a linearly decaying traversibility function which is piecewise constant on k concentric rings with center at S (Figure 6(a)). This function is defined as $\sigma(x) = \sigma_0$ on the unit disc, and $\sigma(x) = \sigma_0 2^{-i}$ on the rings $2^{i-1} \leq \|x - S\| \leq 2^i$ such that $1 \leq i \leq k$. The resulting environment has the property that the net cost of radially crossing any ring is a constant $\sigma_0/2$, although geometrically the radius of the rings grows exponentially with distance from S. Next we embed in each ring radial corridors of width D (Figure 6(b)). Since the perimeter of the rings doubles in each step, the number of corridors doubles in each ring. The resulting environment is basically an embedding of a binary tree with unit-weight edges as D-width corridors emanating from S. This environment has the property that the number of corridors crossing the kth ring is *exponential* in k, while the cost of reaching this ring is *linear* in k. The following theorem establishes a universal lower bound on navigation in linear-poly environments. A proof of the theorem appears in [9].

Theorem 4 (exponential complexity) *Let A be any navigation algorithm in an unknown planar environment with linear-poly traversibility, where the target is apriori known or recognized upon arrival. Let ρ be the cost of the path generated by A. Then ρ satisfies the exponential lower bound,*

$$\rho \geq \sigma_0 2^{\frac{2(1-\epsilon)}{\sigma_0}\rho_{opt}},$$

where ρ_{opt} is the optimal off-line cost of traveling from S to T, σ_0 is traversibility at S, and ϵ is an arbitrary small positive parameter.
Since the upper bound on this problem is also exponential in ρ_{opt}, navigation in unknown linear-poly environments belongs to the **exponential** *competitive complexity class.*

6 Conclusion

We introduced a new notion of competitive complexity classes and classified some common mobile robot problems into these classes. Let us make some comments regarding future work. First, we classified only a small collection of on-line mobile robot problems. The classification of other problems such as on-line sensory exploration need to be carried out in future work. Second, we assumed linear on-board memory. However, many mobile robot tasks are sufficiently complex as to allow only a constant amount of memory. Given this stricter memory limitation, one must re-explore the competitive complexity class of the basic problems considered in this paper. Third, we have not discussed an important classification criterion of on-line problems which is widely acknowledged but has not been formally treated in the literature. This criterion, tentatively called *on-line disconnection proof,* judges an on-line algorithm according to the efficiency by which it determines that a solution does not exist, relative to the shortest off-line proof. This topic will be discussed in future work. Finally consider extension of the results to three-dimensional environments. Under the assumption that the robot has a physical size $D > 0$, one can adapt the navigation algorithms $SAD1$ and $SAD2$ to three dimensions. The competitive complexity class of these algorithms would be *cubic* in l_{opt}. Another interesting problem is on-line search for the shortest path in an unknown planar environment. This problem belongs to the *quadratic* competitive complexity class [9]. However, in three-dimensions our algorithms for this problem require exponential on-board computation time. Hence our assumption that physical travel time dominates on-board computation time must be re-visited.

References

1. R. Baeza-Yates, J. Culderson, and G. Rawline. Searching in the plane. *J. of Information and Computation*, 106:234–252, 1993.
2. P. Berman, A. Blum, A. Fiat, H. Karloff, A. Rosen, and M. Saks. Randomized robot navigation algorithms. In *SODA*, 75–84, 1996.
3. A. Blum, P. Raghavan, and B. Schieber. Navigating in unfamiliar geometric terrain. *SIAM J. of Comput.*, 26:110–137, 1997.
4. H. Choset and J. W. Burdick. Sensor based planning, part ii: Incremental construction of the generalized voronoi graph. In *ICRA*, 1643–1649, 1995.
5. A. Datta and C. Icking. Competitive searching in a generalized street. In *10th ACM Symp. on Computational Geometry*, 175–182, 1994.
6. A. Datta and S. Soundaralakshmi. Motion planning in an unknown polygonal environment with bounded performance guarantee. In *ICRA*, 1032–1037, 1999.
7. X. Deng and A. Mirzaian. Competitive robot mapping with homogeneous markers. *IEEE Trans. on Robotics and Automation*, 12(4):532–542, 1996.
8. E. Bar Eli, P. Berman, A. Fiat, and P. Yan. Online navigation in a room. *J. of Algorithms*, 17(3):319–341, 1996.
9. Y. Gabriely and E. Rimon. Competitive complexity of mobile robot on line problem. Tech. report, http://robots.technion.ac.il, March 2004.
10. Y. Gabriely and E. Rimon. Competitive on-line coverage of grid environments by a mobile robot. *Comp. Geom.: Theory and Applications* (to appear).
11. F. Hoffmann, C. Icking, R. Klein, and K. Kriegel. The polygon exploration problem. *SIAM J. of Computation*, 2001.
12. C. Icking, T. Kamphans, R. Klein, and E. Langetepe. On the competitive complexity of navigation tasks. In *Sensor Based Intelligent Robots, Lecture Notes in Computer Science*, volume 2238, 245–258. Springer Verlag, 2002.
13. C. Icking and R. Klein. Competitive strategies for autonomous systems. In *Modelling and Planning for Sensor Based Intelligent Robot Systems*, 23–40. World Scientific, Singapore, 1995.
14. I. Kamon, E. Rimon, and E. Rivlin. Tangentbug: A range-sensor based navigation algorithm. *Int. Journal of Robotics Research*, 17(9):934–953, 1998.
15. V. J. Lumelsky and A. Stepanov. Path planning strategies for point automaton moving amidst unknown obstacles. *Algorithmica*, 2:403–430, 1987.
16. B. Tovar, S. M. Lavalle, and R. Murrieta. Optimal navigation and object finding without geometric maps or localization. In *ICRA*, 464–470, 2003.
17. I. A. Wagner, M. Lindenbaum, A. M. Bruckstein. Distributed covering by ant-robots using evaporating traces. *IEEE Trans. on Robotics and Aut.*, 15(5):918–933, 1999.
18. G. Woeginger. Personal communication, Dept. of CS, University of Twente, Netherlands, Dec. 2003.

Pareto Optimal Coordination on Roadmaps*

Robert Ghrist[(1)], Jason M. O'Kane[(2)], and Steven M. LaValle[(2)]

Departments of Mathematics[(1)] and Computer Science[(2)], University of Illinois, Urbana, IL 61801, USA

Summary. Given a collection of robots sharing a common environment, assume that each possesses an individual roadmap for its C-space and a cost function for attaining a goal. Vector-valued (or PARETO) optima for collision-free coordination are by no means unique: in fact, continua of optimal coordinations are possible. However, for CYLINDRICAL obstacles (those defined by pairwise interactions), we prove a finite bound on the number of optimal coordinations. For such systems, we present an exact algorithm for reducing a coordination scheme to its Pareto optimal representative.

1 Introduction

In numerous settings, the coordination of multiple robots remains a basic and challenging research issue. Autonomous guided vehicles (AGVs) are used in a wide variety of industrial settings for problems such as material handling, palletizing, paper roll handling, assembly, conveying, and people moving. Typically, AGVs reliably traverse a fixed roadmap of paths in a complicated factory environment. Although the paths avoid collisions with obstacles in the workspace, the efforts of numerous AGVs may have to be coordinated in a way that avoids collisions between AGVs while at the same time maximizing productivity.

If we wish to coordinate the motions of N robots in a common environment, what is an appropriate notion of optimality? Minimizing the average time robots take to reach their goal? Minimizing the time that the last robot takes? Such approaches are common (e.g., [13, 17, 25]) and may be appropriate in some cases; however, it is important to recognize that scalarization of a vector of N criteria occurs in this process. Each robot has its own cost function, e.g., elapsed time. These N criteria are then converted — often in an arbitrary manner — into a single criterion to be optimized.

*Supported by NSF PECASE DMS-0337713 [RG] and NSF IIS-0296126 [JO,SL].

M. Erdmann et al. (Eds.): Algorithmic Foundations of Robotics VI, STAR 17, pp. 171–186, 2005.

In this paper, we investigate the optimization problem for multiple robot coordination without scalarizing the vector-valued cost function. This centers on the notion of *Pareto optimality* [20, 22], a concept which is widely used in mathematical economics to model individual consumers striving to optimize distinct economic goals. This brand of optimization is more "faithful" in the sense that no data is lost by scalarization. In particular:

1. Any optimum of *any* scalarization of the vector-valued cost function (which is monotone increasing in the components) is in fact one of the Pareto optima: see Lemma 1.
2. For settings in which priorities change over time, knowledge of the Pareto optima allow for easy adaptation. This is particularly relevant in automation settings where repetitive motions are subject to priority changes.
3. By filtering out all of the motion plans that are not worth considering and presenting the user with a small set of the best alternatives, additional criteria, such as priority or the amount of sacrifice one robot makes, can be applied to automatically select a particular motion plan.

Given the desire to filter the space of all possible coordination schemes to a small set of best cases independent of biases on the robots, we are certainly most interested in the cases where this collection of optima is *finite.* A finiteness criterion and an exact algorithm for these optima are the principal results of this paper.

1.1 History

The problem of coordinating robots along fixed roadmaps can be considered as a special case of general motion planning for multiple robots. Previous approaches to multiple-robot motion planning are often categorized as *centralized* or *decoupled.* A centralized approach typically constructs a path in a composite configuration space, which is derived from the Cartesian product of the configuration spaces of the individual robots (e.g., [1, 2, 3, 21, 23]). A decoupled approach typically generates paths for each robot independently, and then considers the interactions between the robots (e.g., [5, 10]). In [4, 6, 19, 24] robot paths are independently determined, and a coordination diagram is used to plan a collision-free trajectory along the paths. In [1, 21], integer linear programming methods are used to determine approximate solutions to the time-minimization problem. In [15, 26], an independent roadmap is computed for each robot, and coordination occurs on the Cartesian product of the roadmap path domains. The suitability of one approach over the other is usually determined by the trade-off between computational complexity associated with a given problem, and the amount of completeness that is lost. In some applications, such as the coordination of AGVs, the roadmap might represent all allowable mobility for each robot.

In this paper, we focus on the multiplicity of Pareto-optimal path coordinations among cylindrical obstacles — those determined by pairwise collisions (see §1.4 below). In [16], an approximate Dijkstra-like algorithm to find Pareto-optimal solutions in this context was given. To our knowledge, the only previously known exact solution is that of [8], which applies only to the case of two robots that translate on acyclic roadmaps.

1.2 Coordination spaces

This paper concerns the coordination of N robots, each having a roadmap Γ_i (a graph within the C-space of the ith robot) precomputed independent of the other robots.

Definition 1. *A* ROADMAP COORDINATION SPACE *of* $\{\Gamma_i\}_1^N$ *is any space of the form* $\mathcal{X} := (\Gamma_1 \times \cdots \times \Gamma_N) - \mathcal{O}$*, where* $\mathcal{O}$ *denotes an (open) obstacle set.*

For the remainder of the paper, all coordination spaces are assumed to be *piecewise linear (PL) manifolds.* This assumption rules out obstacle sets $\mathcal{O}$ which are too intricate to have a nice local structure or for which the system "locks up" in a singular configuration. The results of this paper hold under much weaker technical assumptions (including piecewise analytic or algebraic) [12].

A special example of a roadmap coordination space arises when all of the roadmaps Γ_i are identical, and the obstacle set is an open neighborhood of $\{x_i = x_j : \text{for some } i \neq j\}$. In this case, one can consider the workspace to be the graph itself, and the roadmap coordination space is precisely the configuration space of N labeled objects on the graph.

1.3 Pareto optimality

Pareto optimization refers to optimization of vector-valued functions. In the context of robotics applications, Pareto optimization arises when distinct robots possess distinct goals and cost functions for evaluating performance. Each robot wishes to optimize its cost function independently of the others.

Mathematically, this is characterized as follows. Given a parameterized path $\gamma : [0, T] \to \mathcal{X}$ in a coordination space, each robot executes the projected path $\gamma_i := \text{PROJ}_i \circ \gamma$, where PROJ_i denotes projection onto the ith factor. Given cost functions $\{\tau_i\}$, $i = 1...N$, the cost vector for γ is the vector $\tau(\gamma) := (\tau_i(\gamma_i))_i^N$. The case in which τ_i measures *elapsed time* from start to goal is an important and characteristic example, though more general cost functions are allowed.

We assume that each roadmap Γ_i is outfitted with a metric such that the speed of the ith robot travelling along this graph at maximal speed is exactly one; hence all admissible paths have speeds whose components are bounded above by one. It will be assumed for simplicity that the cost functions τ_i agree with elapsed time.

A path $\gamma : [0, T] \to \mathcal{X}$ is PARETO-OPTIMAL iff $\tau(\gamma)$ is minimal with respect to the partial order on vectors:

$$\tau(\gamma) \leq \tau(\gamma') \Leftrightarrow \tau_i(\gamma_i) \leq \tau_i(\gamma_i') \, \forall \, i. \tag{1}$$

The Pareto optima comprise the set of all optima for all monotone scalarizations (such as, *e.g.*, average time-to-goal and all non-linear generalizations thereof):

Lemma 1. *For any scalarization $f : \mathbb{R}^n \to \mathbb{R}$ with $\partial f / \partial x_i > 0$, all minima of $f \circ \tau$ are Pareto optima.*

Proof: Given any minimal path for $f \circ \tau$ which is not Pareto-optimal, deforming it to a Pareto-optimal path decreases some τ_i without increasing any of the others; hence it decreases the f-value: contradiction. □

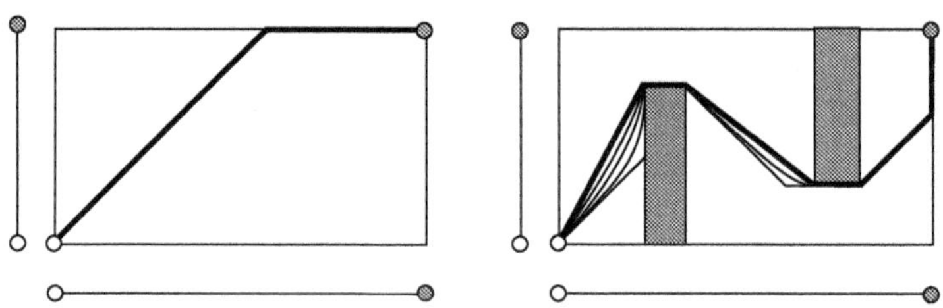

Fig. 1. [left] The unique Pareto optimal path on a rectangle from the lower-left to the upper-right corners; [right] an envelope of Pareto optimal paths weaving through obstacles forms a single equivalence class.

Pareto optimal paths are rarely unique. Two paths γ and γ' are PARETO EQUIVALENT iff they are homotopic through Pareto-optimal paths which are equal in the partial order; *i.e.*, $\tau(\gamma) = \tau(\gamma')$. Fig. 1[right] illustrates a single Pareto-optimal class with many representatives. We show in §2 that certain roadmap coordination spaces admit a continuum of Pareto-optimal *classes*.

1.4 Contributions

For a special class of coordination spaces, we give a **finiteness bound** for the number of Pareto optimal paths and an **exact algorithm** for computing *local* Pareto-optima: paths which are Pareto-optimal among a fixed homotopy class of paths.

We say that the obstacle set $\mathcal{O}$ for $\mathcal{X} = \prod \Gamma_i - \mathcal{O}$ is CYLINDRICAL if $\mathcal{O}$ is of the form

$$\mathcal{O} = \bigcup_{i<j} \left(\Delta_{i,j} \times \prod_{k \neq i,j} \Gamma_k \right), \tag{2}$$

with $\Delta_{i,j} \subset \Gamma_i \times \Gamma_k$. Physically, this means that whenever two robots experience an obstacle (*e.g.*, collision), the states of all other robots are irrelevant:

collisions are pairwise-determined. We require each $\Delta_{i,j}$ to be the interiors of disjoint PL polygonal subsets of $\Gamma_i \times \Gamma_j$.

Main Theorem: Let $\mathcal{X}$ be a cylindrical coordination space:

1. Locally Pareto-optimal path classes are in bijective correspondence with homotopy classes of paths (fixing the endpoints).
2. There is a finite bound on the number of globally Pareto-optimal path classes.

The first statement is that every path in $\mathcal{X}$ is homotopic to a unique class which is Pareto-optimal among paths in its homotopy class; the second statement is that only a finite number of these local optima are in fact global. Most proofs appear in §4; certain technical steps involving techniques from geometric group theory will appear in an extended version of this work [12].

In §5, we present an exact algorithm for determining the local Pareto-optimal representative of a path in $\mathcal{X}$ (under the reasonable assumption that robots do not backtrack). The algorithm is output sensitive with respect to the complexity of the path it computes. Fix the number of robots; let m denote the complexity of the obstacle set $\mathcal{O}$; and let p and p' denote the complexities of the input and output paths respectively. Algorithm 2 executes in $O(p+m\log mp+p'\sqrt{m}\log m)$ time. An extension to arbitrary paths involving backtracking is outlined.

2 Examples

We illustrate a few examples of simple coordination spaces and Pareto optimal path classes.

Example 1. Consider the case where $N = 2$, $\Gamma_1 = \Gamma_2 = [-2,2]$, and $\mathcal{O} = \{(x,y) : x^2 + y^2 < 1\}$, which corresponds to a pair of identical disc-shaped AGV's sliding along interval roadmaps which intersect in the workspace at right angles. There are exactly two Pareto-optimal classes of paths from $(-2,-2)$ to $(2,2)$, as illustrated in Fig. 2[left]. The difference between these two paths lies in which robot decides to pause in order to allow the other to pass through the intersection. Note also that this is a cylindrical coordination space and hence satisfies the Main Theorem above.

Example 2. We modify the previous example by letting $N = 3$ and choosing $\mathcal{O}$ to be a round ball of radius 1 at the origin. By the symmetry of $\mathcal{X}$ about the diagonal of the cube, it is clear that there is a circle's worth of paths which begin at $(-2,-2)$, trace a straight line which is tangent to $\mathcal{O}$, and then exit this sphere tangentially with slope one. The projection of this family of paths to the first two coordinates includes as special cases the distinct Pareto optima of Example 1, as well as a continuum of paths whose goal times interpolate between these two. Hence, there is a continuum of Pareto optimal classes.

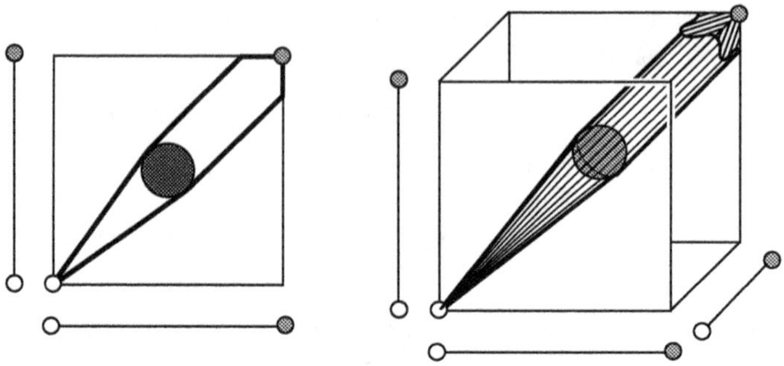

Fig. 2. [left] The case of two robots with a disc obstacle possesses two Pareto-optimal classes; [right] Three robots with a spherical obstacle possesses a continuum of Pareto-optimal classes.

Though neither of these examples is PL, the arguments for the number of Pareto-optimal classes are robust and valid for PL approximations to $\mathcal{O}$.

Example 3. Coordination problems without the fixed path constraint can have continua of optima much more easily than in the fixed path case. Consider the problem of translating two unit squares in the plane in such a way as to exchange their positions. Assume that the squares are centered at $(-1, 0)$ and $(1, 0)$. It is straightforward to show a continuous family of Pareto-optimal classes. Extrema of this family consist of those coordinations for which one square translates exclusively in the horizontal direction, while the other square moves vertically in order to get out of the way. The intermediate coordinations involve one square translating up by an amount $0 < h < 1$ and the other translating down by a total amount $1 - h$. This example works with arbitrary translations, or with translations restricted to coordinate axes directions.

3 Discretized coordination spaces and cube paths

The key step in our analysis of cylindrical coordination spaces is a spatial discretization of them into cubical complexes. This step is inspired by techniques in geometric group theory (*e.g.*, [18]) which have found other applications in robotics contexts [11].

Consider a coordination space $\mathcal{X}$ for roadmaps $\{\Gamma_i\}$. For simplicity, assume that all the edge lengths of the graphs Γ_i are rationally related. Because of this, one may choose a length $\delta > 0$ so that all edge lengths are integer multiples of δ. Denote by $\Gamma_i^{(n)}$ the graph obtained from Γ_i by subdividing it into edges of length $2^{-n}\delta$.

Definition 2. *For each $n \geq 0$, denote by $\mathcal{X}^{(n)} \subset \Gamma_1^{(n)} \times \cdots \times \Gamma_N^{(n)}$ the maximal closed subcomplex of the product cubical complex which does not intersect* *the obstacle set $\mathcal{O}$. This cubical complex is called the* STAGE-n DISCRETIZED COORDINATION SPACE.

These discretized coordination spaces "converge" (in the Hausdorff sense) to the full coordination space $\mathcal{X}$ as $n \to \infty$. The cubical nature of these spaces forms a convenient structure with which to manipulate paths. Our strategy is to approximate Pareto-optimal paths with paths which follow along the diagonals of cubes in $\mathcal{X}^{(n)}$ for n sufficiently large.

Definition 3. *Let X denote a cubical complex. A* CUBE PATH *from vertices v_0 to v_N in X is an ordered sequence of closed cubes $\mathcal{C} = \{C_i\}_1^N$ of X which satisfy (1) $C_i \cap C_{i+1} = v_i$, a vertex; and (2) C_i is the smallest cube of X containing v_{i-1} and v_i. A cube path is said to be* LEFT-GREEDY *if in addition $\forall i$ (3) $C_{i+1} \cap \mathsf{St}(C_i) = v_i$, where $\mathsf{St}(C_i)$ is the* STAR *of C_i (all cells, including C_i, which have C_i as a face).*

To each cube path is associated a PL path in X given by the chain of straight line segments from v_i to v_{i+1} for $i = 0...N-1$. Roughly speaking, a left-greedy cube path is one which uses the highest dimensional cubes as early as possible in the path. This hints at the intuition of left-greedy paths as Pareto-optimal, an intuition which is entirely justified.

4 Topological bounds on Pareto optima

In this section we demonstrate a uniqueness result for Pareto optimal classes. The strategy of our proof is to show that any Pareto-optimal path class contains a representative which is Pareto-equivalent to a unique left-greedy cube path. For the remainder of the paper, $\mathcal{X}$ will denote a cylindrical roadmap coordination space. The cylindrical structure yields uniqueness results for left-greedy cube paths.

Proposition 1. *Any path between vertices of $\mathcal{X}^{(n)}$ is homotopic to a unique left-greedy cube path.*

Proof: Existence of a left-greedy path follows from Algorithm 1 and Lemma 4 following. The techniques of the proof for uniqueness are simple, but lie outside the scope of this paper. Briefly, one uses the cylindrical structure of $\mathcal{X}$ to prove that $\mathcal{X}^{(n)}$ is NONPOSITIVELY CURVED (see, *e.g.*, [11] for a definition). Then, a standard result from geometric group theory (Prop. 3.3 of [18]) implies that left-greedy paths are unique up to homotopy. See [12] for a complete proof. □

Lemma 2. *Any L^∞ geodesic path on $\mathcal{X}$ can be continuously perturbed to an L^∞-close cube path in $\mathcal{X}^{(n)}$ for n sufficiently large.*

Proof: Given any L^∞ geodesic in $\mathcal{X}$, we begin by "pushing" it to an L^∞-approximate path α which lies completely in the interior of $\mathcal{X}$. This step relies crucially on the assumption that $\mathcal{X}$ is a manifold, so that the interior is always accessible from the path.

We next approximate α by a cube path. Since the image of α is compact, there exists an $\epsilon > 0$ so that a tube of radius ϵ about the image of α does not intersect $\partial\mathcal{O}$. Choose n so that the grid size δ is less than ϵ. Modify α so that its endpoints are at vertices of $\mathcal{X}^{(n)}$. Consider the vertex $v^0 := \alpha(0)$ and the L^∞ balls $B_r := B_r^\infty(v^0)$ of radius r about v^0 in $\mathcal{X}^{(n)}$. Let $w^1 := \alpha \cap \partial B_\delta$ and $w^2 := \alpha \cap \partial B_{2\delta}$.

Let $F := B_\delta(v^0) \cap B_\delta(w^2)$. Clearly, $w^1 \in F$ and for any $x \in F$ it holds that the PL path $v^0 \to x \to w^2$ is of the same L^∞ length as the restriction of α to this neighborhood. We claim that F contains a vertex of the grid structure on $\mathcal{X}^{(n)}$. To see this, note that F is a subset of some face of B_δ defined by the set of all points $x = (x_i)$ such that

$$x_i \in [v_i^0 - \delta, v_i^0 + \delta] \cap [w_i^2 - \delta, w_i^2 + \delta] \quad ; \quad i = 1 \dots N \tag{3}$$

Since these intervals have the same length, it follows that either $v_i^0 - \delta$ or $v_i^0 + \delta$ lies within the intersection for each i. Therefore, F contains a gridpoint v^1 of $\mathcal{X}^{(n)}$, and we may replace the segment of the path α with a PL path from v^0 to v^1 to w^2. Repeat the argument inductively, beginning at the vertex v^1 and considering the L^∞ balls about v^1 of radius δ and 2δ.

Marching along yields a PL path passing through vertices of $\mathcal{X}^{(n)}$ which are sequentially of L^∞-distance δ. This is therefore a cube path in $\mathcal{X}^{(n)}$. □

Lemma 3. *Let α be a Pareto-optimal path on $\mathcal{X}$. One can find homotopic cube paths with endpoints close to those of α such that all the goal times approximate those of α.*

Proof: Apply Lemma 2 inductively and use the fact that the restriction of a Pareto-optimal path to any subpath is an L^∞ geodesic. □

A simple algorithm, `LeftGreedy`, converts a cube path $\mathcal{C}$ into a left-greedy cube path by sweeping along $\mathcal{C}$ from left to right, comparing the star of each cube with its neighbor to the right (line 3), and shifting the common directions to the left cube (lines 4-5).

Lemma 4. *Given a Pareto-optimal cube path $\mathcal{C}$,* `LeftGreedy`$(\mathcal{C})$ *is a left-greedy cube path which is Pareto equivalent to $\mathcal{C}$.*

Proof: Since common directions are shifted to the left, each sweep through $\mathcal{C}$ for which $C_{i+1} \cap \mathtt{St}(C_i) \neq \{v_i\}$ for some i strictly decreases the positive integer-valued function $\sum_i i \cdot \dim(C_i)$; hence, convergence.

To argue that `LeftGreedy` can be executed through Pareto-equivalence, note first that the shifting process of lines 4-5 results in a homotopic path,

Algorithm 1 `LeftGreedy`($\mathcal{C}$)

Require: $\mathcal{C} = \{C_i\}_1^m$ is an L^∞-geodesic cube path
1: **while** $\mathcal{C}$ is not left-greedy **do**
2:     **for** $i = 1...m$ **do**
3:         $Z \leftarrow C_{i+1} \cap \mathtt{St}(C_i)$
4:         $C_{i+1} \leftarrow C_{i+1}/Z$
5:         $C_i \leftarrow C_i \times Z$
6:     **end for**
7: **end while**

since all modifications take place within $\mathtt{St}(C_i) \cup C_{i+1}$. If the jth coordinate x_j reaches its goal at vertex v_i of $\mathcal{C}$, then all subsequent vertices v_k $(k \geq i)$ lie within the hyperplane of $\mathcal{X}$ having the same j^{th} coordinate. Hence, in line 3, the common factor Z is trivial in the jth coordinate for v_i and beyond. Thus, the time-to-goal for x_j is unchanged by steps 4-5. □

We now may prove the Main Theorems:

Theorem 1. *Locally Pareto-optimal path classes on $\mathcal{X}$ are in bijective correspondence with homotopy classes of paths (fixing the endpoints).*

Proof: Assume there are two homotopic paths, α and α', which are Pareto optimal within this homotopy class. Approximate these paths by homotopic cube paths $\mathcal{C}$ and $\mathcal{C}'$ on $\mathcal{X}^{(n)}$ for some large n via Lemma 3. Feed these cube paths to `LeftGreedy`. Via Lemma 4, the outputs are left-greedy cube paths which are Pareto equivalent to $\mathcal{C}$ and $\mathcal{C}'$ respectively. By Proposition 1, $\mathcal{C}$ and $\mathcal{C}'$ are the same path and hence have the same output times. Therefore, α and α' have output times which are arbitrarily close. □

It does not automatically follow that there is a bound on the number of Pareto optimal paths, since any cylindrical coordination space $\mathcal{X}$ which is not simply connected must have a countably infinite number of homotopy classes of paths. However, all but a finite set of these can be eliminated as global optima:

Theorem 2. *The number of globally Pareto optimal paths between fixed endpoints on $\mathcal{X}$ is finite.*

Proof: Fix endpoints and consider an infinite collection of Pareto optimal paths, all in different homotopy classes. We induct on K, the dimension of the smallest hyperplane (a subspace obtained by fixing certain coordinates) of $\mathcal{X}$ containing all the paths. The induction hypothesis is that for any such infinite collection of paths, there is an infinite subsequence of paths whose goal times $\{T_i^j\}$ satisfy $T_{i+1}^j \geq T_i^j + 1$ for $i = 1...\infty$ and $j = 1...N$. Assume that $K = 1$. Then, the subspace in which the paths live is a graph, and any infinite collection of homotopy classes must contain representatives which go around a cycle an arbitrarily high number of times, leading to a set of uniformly increasing goal times.

Assuming the hypothesis for systems in a hyperplane of dimension K, consider without loss of generality a system with $K+1$ degrees of freedom. Given an infinite number of Pareto optimal classes of paths $\{\gamma_i\}$, we may, by choosing a subsequence and reindexing, assume that T_i^1 converges to a constant T^1 as a non-increasing sequence and that $T_i^{K+1} \to \infty$. By compactness of the L^∞-ball of radius T^1, we have that the points $\gamma_i(T^1)$ converge to some point $x \in \mathcal{X}$. After reaching the first goal, the remainders of all the paths lie in a hyperplane of dimension K containing the point x.

We know by induction that, after restricting to a subsequence, the goal times T_i^j ($j = 2...K+1$) must march to infinity at a rate bounded from below, since they all have starting point approximately at x. The only way for these paths to be Pareto optimal is to have the times T_i^1 be a *strictly* decreasing sequence. However, one can perform an arbitrarily small perturbation of the paths near x to switch to a low-i path after arriving at x. This makes all the other goal times shorter, contradicting the Pareto assumption. □

It follows easily from the compactness of $\mathcal{X}$ that there is a finite bound on the number of optima which is independent of the endpoints. This completes the proof of the Main Theorem.

5 Algorithms

We now consider the algorithmic problem of computing exact Pareto optimal paths. For concreteness, we restrict attention to coordination problems for N robots $\mathcal{R}_1, \ldots, \mathcal{R}_N$ in which each robot is modelled by a convex polygon translating along a PL path in $\mathbb{R}^2$. As a first step, we develop an algorithm for deforming a path to its Pareto-optimal representative. This is analogous to Algorithm 1 but operates directly on $\mathcal{X}$, rather than using the cube paths in some $\mathcal{X}^{(n)}$. This algorithm is presented in §5.1, followed by an implementation in §5.2 and a discussion in §5.3 of the complexity of computing the set of all Pareto optimal solutions.

5.1 Computing Pareto optimal paths

A path is MONOTONE if it is nondecreasing in each dimension. This is equivalent to the requirement that robots may not "back up" along their paths. For simplicity, we present our algorithm for the case of monotone paths and then outline the changes required to deal with non-monotone paths.

We know from §4 that it suffices to compute γ' to be the unique left greedy path homotopic to γ. Informally, left greedy implies that each robot should move at the fastest possible speed. Each speed change can be characterized in one of four ways: (1) stopping or slowing to avoid an immediate collision

Algorithm 2 ExactLeftGreedy($\mathcal{X}, \gamma$)

Require: γ is a monotone collision-free coordination in $\mathcal{X}$.

$x_0 \leftarrow (0, \dots, 0)$
$w \leftarrow \emptyset$
$\mathcal{O} \leftarrow$ ObstacleRegions($\mathcal{X}$)
$E \leftarrow$ CriticalEvents($\mathcal{O}, \gamma$)
$t \leftarrow 0$
while $x_t \neq x_{goal}$ **do**
  $t \leftarrow t + 1$
  $v \leftarrow$ MaximalLocallyCollisionFreeVelocity($x, w, \mathcal{O}$)
  $e \leftarrow$ NextEvent($x, v, E, \mathcal{O}$)
  $w \leftarrow$ UpdateWaitForSet(p, e, w)
  $x_t \leftarrow x_{t-1} + v \cdot e.\text{time}$
end while
return $(x_0, \dots, x_t)$

(when the path reaches the boundary of a coordination space obstacle); (2) stopping to avoid a future collision (when a projection of the path reaches a horizontal or vertical extremum of some obstacle about which it may not go around); (3) restarting from case 2 when one of the obstacles has been successfully traversed; and (4) stopping when any robot reaches its goal.

The response to reaching a critical event depends on γ. The homotopy class of a monotone path is uniquely determined by assigning to each obstacle in each projection an orientation CW or CCW indicating the direction around which the path passes. As a result, if we build γ' in such a way that it induces the same orientation as γ on each obstacle, we can be certain that γ and γ' are homotopic. This is accomplished by stopping only at vertical critical events for obstacles with CW orientation and horizontal critical events for obstacles with CCW orientation.

Our algorithm builds a path by starting with a single point at $(0, \dots, 0)$ and adding linear path segments sequentially until the goal is reached. We have already classified the critical events at which a robot may need to change speed in a left-greedy coordination. Thus, at every step, we need to compute the appropriate left-greedy velocity v, use v to determine which of these events will occur next, and add a segment to our coordination advancing to this point. We maintain as an invariant that w records the set of robots halted because of a critical event. This method is summarized in Algorithm 2; details follow. In our analysis, N denotes the number of robots, m the total complexity of the pairwise obstacle regions, and p the number of linear segments in γ. We consider N fixed.

The function ObstacleRegions computes the m obstacle regions using a simple collision detection method. In CriticalEvents, we can generate the critical events using a straightforward generalization of the standard vertical decomposition algorithm in [9] in time $O(m \log m)$. To assign an orientation to an obstacle $o \subseteq \mathcal{O}$, we need to determine whether o is above or below

γ. Since γ divides each projection of $\mathcal{X}$ into two regions with PL boundaries, computing the orientation for each obstacle is a planar point location problem easily solved for all obstacles $O(p+m \log p)$ time using recursive triangulation methods [14]. Computing NEXTEVENT involves a ray-shooting query in each projection requiring total time $O(\sqrt{m} \log m)$ to execute the algorithm of [7].

Computing local left greedy gradients

Given a point x in coordination space, we want to compute the maximal velocity $v = (v_1, \dots, v_N)$ that does not result in a local collision. It is straightforward to show that v is the solution to a linear program in $O(N^2)$ constraints with maximization objective $\sum_i v_i$. For fixed N, such a program can be solved in constant time. In practice, the number of constraints is a function of the complexity of obstacle intersections and should be rather small.

Analysis

Our algorithm is output sensitive in the sense that the run time is a function of the complexity p' of the locally Pareto optimal path generated. Each iteration of loop on lines 6-12 adds an additional segment to γ'. Therefore, Algorithm 2 executes in $O(p + m \log mp + p' \sqrt{m} \log m)$ time.

Dealing with nonmonotonicity

We now relax the monotonicity requirement. The primary difficulty to be overcome is that obstacle orientations are no longer sufficient to determine a path's homotopy class. Indeed, one may revisit obstacles numerous times in opposite orientations.

To collate this data, extend rays upward and downward from each obstacle vertex that is a horizontal local extremum and do the same *mutatis mutandis* for vertical extrema. For each ray, assign a symbol a_i to represent the action of crossing this ray in the forward direction and let a_i^{-1} denote crossing the ray backward. The homotopy class of a path is completely determined by the sequence in which it crosses these critical events (extending our earlier usage of the term) and the direction of each crossing. That is, a path's homotopy class is uniquely determined by its representation in the language $L = \{a_i, a_i^{-1}\}^*$.

Conversely, if two paths are homotopic, then by continuously deforming one into the other, the crossing sequence will change by a simple transposition when the deformation passes an intersection point between a pair of critical events or when the order of successive events in different projections is reversed. Thus, we admit commutativity between symbols corresponding to events in different projections and to intersecting events in the same projection. We consider L with these relations as a group under word concatenation.

One can show that the left greedy path in each homotopy class will have a minimal-length crossing sequence in the group L. This dictates, for each direction in each projection, the order in which critical events should be crossed. The straightforward method for doing word reduction in L is quadratic in the length of the word, which itself is $\Theta(pm)$.

In light of this analysis, it is a simple matter to extend Algorithm 2 to optimize non-monotone paths. Having computed the minimal crossing sequence from γ, we can build γ' segment-by-segment, enforcing the sequence of critical event crossings prescribed for each robot. Each robot should stop rather than cross any event other than the next prescribed in the sequence. At each step, the correct direction for each robot (*i.e.*, forward or backward) is determined by the direction required to reach the next event for that robot in the sequence. Aside from the additional preprocessing time required to do word reduction in L, the complexity remains unchanged.

5.2 Simulations

We have implemented a simple version of the algorithm for monotone paths. In particular, we perform both the point-location step and the next event selection using the obvious quadratic time algorithms. Our implementation is in C++ on Linux and experiment times shown are for a 2.55GHz processor. Figure 3 illustrates a simple coordination problem with $n = 3$ for which we found the set of three Pareto optima by exhaustively enumerating monotone homotopy classes. These coordinations took approximately 0.3 seconds to compute. Figure 4 shows two more complex coordination simulations — one in which 8 robots translate left-to-right, and a 'swap' problem in which 20 robots switch sides.

5.3 Complexity of enumeration

Coordination of large numbers of robots under tight constraints is computationally challenging.

Example 4. Consider the generalization of Example 1 with $\Gamma_i = [-2, 2]$ for $i = 1...N$ and having N obstacles, each of the form $x_i^2 + x_j^2 < 1$ for $i < j$. The two Pareto optimal classes from Example 1 generalize (by the cubical symmetry of the space) to yield 2^{N-1} optimal classes between opposite corners in this coordination space.

This type of obstacle set — N intersecting cylinders — can be realized physically by an AGV system with N disc-like robots sliding along intervals, the collection of which have an N-fold intersection at their midpoints. The exponential complexity of this coordination problem is intuitively clear: since everyone cannot pass through the shared center at the same time, they must proceed through one at a time. There are a factorial number of ways to do so.

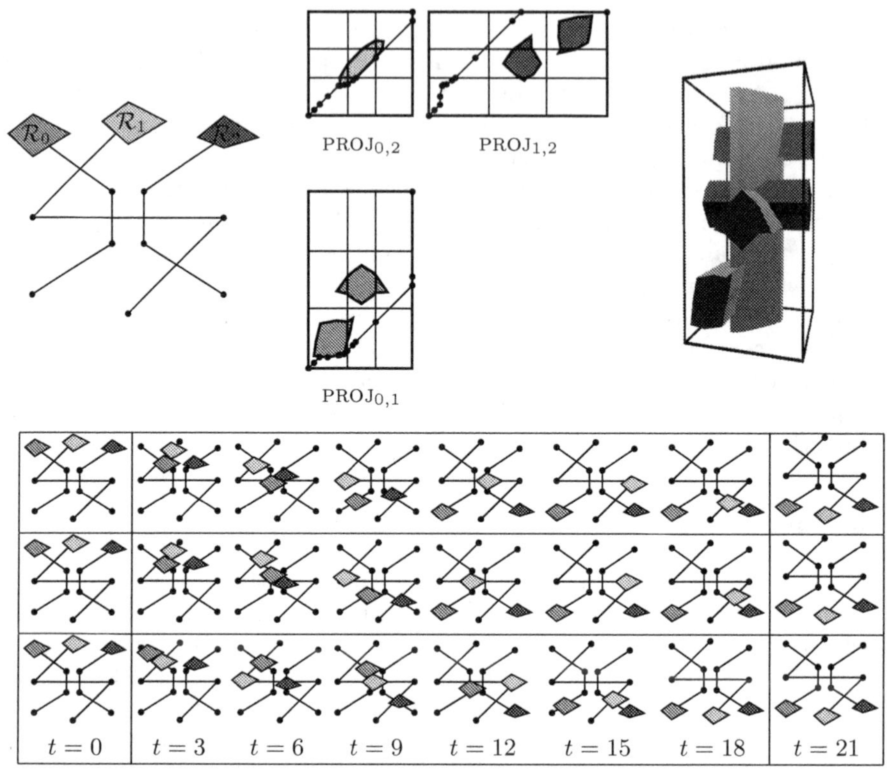

Fig. 3. [top] A coordination problem with 3 robots, its coordination space, and the projections thereof. [bottom] Snapshots of the three Pareto optima.

6 Discussion

It comes as no surprise that the enumeration of all Pareto optima is of high complexity: coordination is inherently difficult, and Lemma 1 implies that such an enumeration is akin to determining all minima for all monotone scalarizations of the cost functions. The algorithms we present are effective enough to be useful in factory AGV systems, where in practice the tracks have relatively low degree of intersection.

What is a surprise is that the set of Pareto optima is finite, the crucial reason being that the cylindrical coordination spaces can be approximated by cubical complexes without any "concave" corners (more specifically, vertices of positive curvature). For such spaces, Pareto-optimal cube path classes are unique up to homotopy. This is not the case for general cube complexes. For example, if one approximates the spaces of Examples 2 and 3 by discretizations $\mathcal{X}^{(n)}$, then, as before, there are a finite number of optimal left-greedy cube paths. However, they are not unique up to homotopy, and the number of

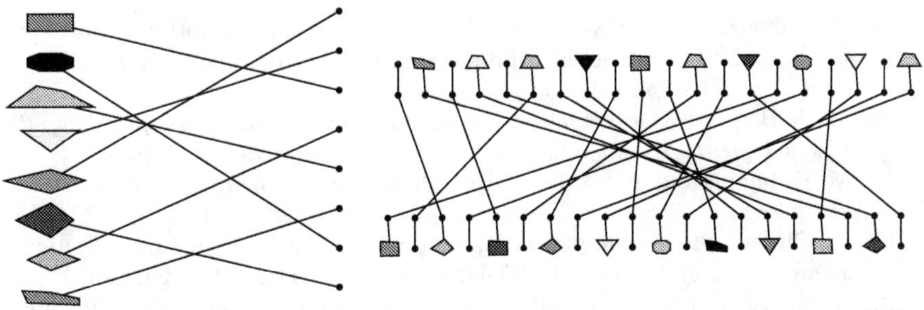

Fig. 4. Two complex coordination problems; [left] for 8 robots, the 37 optimal coordinations were computed in 171.6 seconds by an exhaustive search of monotone classes; [right] for 20 robots, a single optimal coordination was computed in 1.53 seconds.

such paths grows exponentially in the discretization step n, giving rise to the continuum of optima in the limit.

Although our examples and simulations are for convex PL robots sliding by rigid translation in the plane, our setting is much more general. In particular, our coordination spaces are for arbitrary roadmaps (not merely paths) in arbitrary C-spaces. Hence, one can have fully 3-d robots with internal degrees of freedom, nonholonomic constraints, or whatever one wishes, so long as the precomputed roadmap is one-dimensional.

References

1. S. Akella and S. Hutchinson. Coordinating the motions of multiple robots with specified trajectories. In *Proc. IEEE Int. Conf. on Robot. and Autom.*, pages 624–631, 2002.
2. M. D. Ardema and J. M. Skowronski. Dynamic game applied to coordination control of two arm robotic system. In R. P. Hämäläinen and H. K. Ehtamo, editors, *Differential Games - Developments in Modelling and Computation*, pages 118–130. Springer-Verlag, Berlin, 1991.
3. J. Barraquand and J.-C. Latombe. Robot motion planning: A distributed representation approach. *Int. J. Robot. Res.*, 10(6):628–649, December 1991.
4. Z. Bien and J. Lee. A minimum-time trajectory planning method for two robots. *IEEE Trans. Robot. & Autom.*, 8(3):414–418, June 1992.
5. S. J. Buckley. Fast motion planning for multiple moving robots. In *Proc. IEEE Int. Conf. on Robot. and Autom.*, pages 322–326, 1989.
6. C. Chang, M. J. Chung, and B. H. Lee. Collision avoidance of two robot manipulators by minimum delay time. *IEEE Trans. Syst., Man, Cybern.*, 24(3):517–522, 1994.
7. B. Chazelle, H. Edelsbrunner, M. Grigni, L. J. Guibas, J. Hershberger, M. Sharir, and J. Snoeyink. Ray shooting in polygons using geodesic triangulations. *Algorithmica*, 12:54–68, 1994.

8. H. Chitsaz, J. M. O'Kane, and S. M. LaValle. Pareto-optimal coordination of two translating polygonal robots on an acyclic roadmap. In *Proc. IEEE Int. Conf. on Robot. and Autom.*, 2004.
9. M. de Berg, M. van Kreveld, M. Overmars, and O. Schwarzkopf. *Computational Geometry: Algorithms and Applications.* Springer, Berlin, 1997.
10. M. Erdmann and T. Lozano-Perez. On multiple moving objects. In *Proc. IEEE Int. Conf. on Robot. and Autom.*, pages 1419–1424, 1986.
11. R. Ghrist. Shape complexes for metamorphic robot systems. In *Algorithmic Foundations of Robotics V, STAR 7*, Springer-Verlag, 185–201, 2004.
12. R. Ghrist and S. M. LaValle. Nonpositive curvature and Pareto optimal coordination of robots. Preprint, 2004.
13. H. Hu, M. Brady, and P. Probert. Coping with uncertainty in control and planning for a mobile robot. In *IEEE/RSJ Int. Workshop on Intelligent Robots and Systems*, pages 1025–1030, Osaka, Japan, November 1991.
14. D.G. Kirkpatrick. Optimal search in planar subdivisions. *SIAM J. Comput.*, 12(1):28–35, 1983.
15. S. M. LaValle and S. A. Hutchinson. Path selection and coordination of multiple robots via Nash equilibria. In *Proc. IEEE Int. Conf. on Robot. and Autom.*, pages 1847–1852, 1994.
16. S. M. LaValle and S. A. Hutchinson. Optimal motion planning for multiple robots having independent goals. *IEEE Trans. Robot. and Autom.*, 14(6):912–925, December 1998.
17. J. Miura and Y. Shirai. Planning of vision and motion for a mobile robot using a probabilistic model of uncertainty. In *IEEE/RSJ Int. Workshop on Intelligent Robots and Systems*, pages 403–408, Osaka, Japan, May 1991.
18. G. A. Niblo and L. D. Reeves. The geometry of cube complexes and the complexity of their fundamental groups. *Topology*, 37(3):621–633, 1998.
19. P. A. O'Donnell and T. Lozano-Pérez. Deadlock-free and collision-free coordination of two robot manipulators. In *Proc. IEEE Int. Conf. on Robot. and Autom.*, pages 484–489, 1989.
20. L. E. Parker. Cooperative motion control for multi-target observation. In *IEEE/RSJ Int. Conf. on Intelligent Robots & Systems*, pages 1591–1598, 1998.
21. J. Peng and S. Akella. Coordinating multiple robots with kinodynamic constraints along specified paths. In *Algorithmic Foundations of Robotics V, STAR 7*, Springer-Verlag, 221–237, 2004.
22. Y. Sawaragi, H. Nakayama, and T. Tanino. *Theory of Multiobjective Optimization.* Academic Press, New York, NY, 1985.
23. J. T. Schwartz and M. Sharir. On the piano movers' problem: III. Coordinating the motion of several independent bodies. *Int. J. Robot. Res.*, 2(3):97–140, 1983.
24. T. Simeon, S. Leroy, and J.-P. Laumond. Path coordination for multiple mobile robots: a resolution complete algorithm. *IEEE Trans. Robot. & Autom.*, 18(1), February 2002.
25. S.-H. Suh and K. G. Shin. A variational dynamic programming approach to robot-path planning with a distance-safety criterion. *IEEE Trans. Robot. & Autom.*, 4(3):334–349, June 1988.
26. P. Svestka and M. H. Overmars. Coordinated motion planning for multiple car-like robots using probabilistic roadmaps. In *Proc. IEEE Int. Conf. Robot. & Autom.*, pages 1631–1636, 1995.

Automatic Generation of Camera Motion to Track a Moving Guide

Onno Goemans and Mark Overmars

Institute of Information and Computing Sciences, Utrecht University,
P.P. Box 80.089, 3508 TB Utrecht, The Netherlands.
Email: [onno,markov]@cs.uu.nl.

Abstract. Manually following a moving object through a cluttered virtual environment can be challenging for the user. Instead, one would typically rather focus on watching the object and its environment. In this paper, we present an approach to automatically generate a camera motion, such that the user maintains visibility with an object moving along a known path through a virtual environment. To begin, the user specifies the camera placement at the start and end of the object path, and constraints on the camera placement relative to the object. Given this input, the system computes a smooth camera path, satisfying the constraints: Firstly, an initial camera path is generates by applying a single-shot probabilistic road map technique. We augmented this technique with several optimisations, speeding up the path generation considerably. Then, the initial path is smoothed to present the user with a pleasant camera motion. The approach has been implemented and tested: it is fast and computes paths through complicated 2D and 3D environments in less than a second.

1 Introduction

A growing public is coming into contact with virtual environments as a result of developments such as faster and cheaper 3D graphics technology and the increasing use of broadband internet connections. Crucial in all virtual environment applications is effective control of the user's viewpoint, or virtual camera. For novice users, such camera control can be challenging, and when, on top of that, the novice needs to carry out a task within a limited time frame, such a challenge can easily become overwhelming.

A typical task is camera navigating in order to follow a guide (object) moving through the virtual environment. In this paper, we present a technique to generate such camera motion automatically, under the assumption that the guide moves along a known path. Beforehand, the user specifies both the placement of the camera at the start and end of the guide path, and constraints on the camera placement relative to the guide. The generated camera motion must satisfy the user's constraints as well as the "visibility constraint"; the latter demands that the camera maintains visibility with the guide at all times; see Fig. 1(a) and (b).

In applications working with known guide paths, it is possible to optimize a camera path with regard to the given constraints; this optimisation is not

M. Erdmann et al. (Eds.): Algorithmic Foundations of Robotics VI, STAR 17, pp. 187–202, 2005.

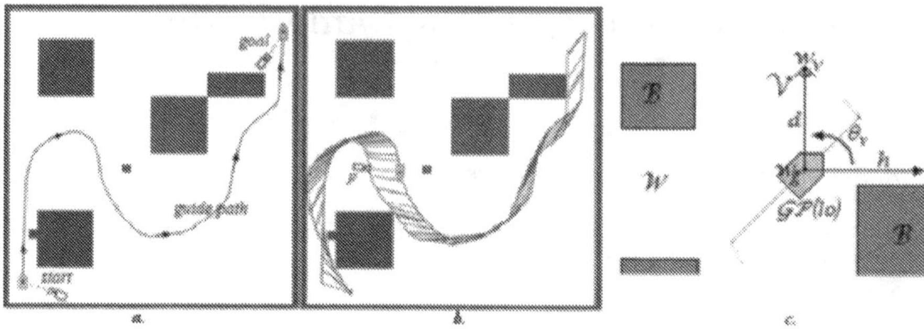

Fig. 1. Figure (**a**) depicts the guide path and the camera placements at the start and end of the guide path. Figure (**b**) shows a possible camera path from the start to the goal. In figure (**c**), the camera configuration v is defined relative to guide configuration $\mathcal{GP}(lo)$ by length offset lo, tracking distance d and tracking angle θ_v.

guaranteed when on-line calculating a camera. Examples of such applications are: a guide showing the architecture of a virtual building or landscape, where the user merely specifies one or more visual goals for the guide to show; or visualization of a robot motion, possibly calculated with an experimental motion planning technique, and so providing more insight into the resulting motion. Moreover, on the fly adaptation of a currently executing camera path can be accomplished by first off-line calculating a new camera path, and then merging this new path with the current one.

In the last decade, a lot of research has been done to assist the user in controlling the camera [9,13,14,22,26] and to control the camera automatically. The latter area of research can be roughly divided into three groups of techniques. To start, the group of techniques that plan camera motion as a sequence of shots. The shots are either based on cinematographic rules [6,15] or calculated by solving a set of constraints [2,8,11]. A second group consists of techniques that generate a collision free camera motion from a start to a goal placement through a virtual environment [4,7,23,24]. And lastly, the group of techniques that generate a camera motion to track a moving guide. Most of these techniques assume the path of the guide to be unknown [10,12], and are therefore not suitable to solve our problem; that is, a smooth motion satisfying the visibility constraint would not be guaranteed.

LaValle *et al.* [18] proposed an approach that describes the guide path as a sequence of configurations in a two-dimensional environment. For each of these configurations, the approach first determines the visibility region, which are then used to calculate an optimal path. We expect that the visibility region approach is not viable in a three-dimensional environment because of large computation times.

Under the assumption of a known guide path, Li *et al.* proposed an approach to plan a camera motion through a two-dimensional environment [20,21]. While maintaining visibility at all time, their approach optimizes several additional camera-specific criteria. Expanding on this approach, a camera configuration is described as a position relative to a guide position; furthermore, a guide position is described as a point on the predefined guide path at some moment in time. All possible camera configurations together form a continuous space, which is approximated with a uniform grid of configurations. To find a collision-free camera path in this configuration-time space, the approach applies a best-first planning algorithm. Li *et al.*'s approach is limited to two-dimensional environments that are not too complicated, as otherwise the grid becomes too large. That is why, although their approach inspired our representation for camera configuration, we use a different planning approach capable of efficiently operating in large 3D environments.

Our aim is to efficiently solve the given problem for two-dimensional as well as three-dimensional environments. The general idea is as follows: All possible placement combinations of camera and guide together form the configuration space. The relevant subspace contains all configurations satisfying both the visibility constraint and the constraints on the camera placement relative to the guide. First, we apply a single-shot probabilistic roadmap method, or PRM [1,3,16,17,25], to construct a roadmap in this relevant subspace. Once the roadmap connects the start and goal configuration, a valid camera motion is found. We then smooth this initial motion in order to eliminate unpleasant camera movement.

The paper is structured as follows. The first four sections address the problem in a 2D workspace, starting with section 2, which provides a more detailed discussion of the guide, camera, and configuration space. Next, we present the basic approach to generate an initial camera path in section 3, along with two improvements. Section 4 then discusses the application we built to test the approach, explains the automation of a PRM-parameter, and provides test results; complex test cases are solved within one second. After this, section 5 discusses the smoothing of the initial camera path. And finally, the extension of the planner to 3D environments is presented in section 6, which is almost straight-forward and solutions are found equally fast.

2 Preliminaries

For the moment, we assume that the guide and camera move in a two-dimensional environment. We specify a guide placement g with three parameters: two for the guide position $\mathcal{W}(g)$ in the workspace, and one for the guide orientation. The guide moves along a known path $\mathcal{GP}$ through the workspace. The normalized length along $\mathcal{GP}$ between configuration $g \in \mathcal{GP}$ and the start configuration is the *length offset lo*, and the mapping from length offset to guide configuration defines the guide path, e.g. $\mathcal{GP}(0)$ and $\mathcal{GP}(1)$ uniquely identify the start and end configuration of $\mathcal{GP}$.

As proposed by Li *et al.* [20,21], we specify the camera position v relative to a given guide configuration g, using the Euclidean distance between v and

g in the workspace, the *tracking distance*, and the orientation of the vector $\overrightarrow{\mathcal{W}(g)\mathcal{W}(v)}$ in the workspace, the *tracking angle*. Furthermore, we choose the camera orientation such that the camera always looks straight at the guide. Together, the length offset, tracking distance and tracking angle uniquely define a camera configuration. See figure 1(c).

We refer to the imaginary line segment connecting the camera and the guide as the *viewing line*. Recall that the user defines constraints on the camera placement relative to the guide. These constraints are a minimum and maximum on both the tracking distance and tracking angle. A note regarding the tracking angle constraints: they are specified relative to the guide orientation, thus if the guide orientation changes, then the allowed tracking angle range changes as well.

Let us now look at the configuration space, in which we will solve our motion planning problem. A configuration describes a combined placement of guide and camera; that is, a guide configuration on the guide path, and a camera configuration relative to this guide configuration. We specify such a configuration with two parameters: the length offset lo, which specifies the guide configuration, and the tracking angle ta, which describes the camera configuration relative to $\mathcal{GP}(lo)$. Note that the tracking distance is not part of the configuration description; this is because we fix the tracking distance at its user-defined minimum. The motivation for this choice is that if no solution exists for the minimum tracking distance, then no solution exists at all; however, if a solution does prove to exist, then the tracking distance can easily be optimized during a post-processing phase. We refer to the configuration space as the *configuration-length offset space*, or $\mathcal{CLO}$-space, and to a path through this space as a *camera path*.

Not all $\mathcal{CLO}$-configurations are valid, because of the visibility constraint and the user's constraints on the tracking angle. We refer to the valid part of the $\mathcal{CLO}$-space as *free* $\mathcal{CLO}$-space, or $\mathcal{CLO}_f$. See figure 2.

We assume that the guide only moves forwards along its path, as well as that the camera only moves when the guide moves. Consequently, a path through the $\mathcal{CLO}$-space must be increasingly monotone in the length offset direction. A camera path is free if it lies entirely in the $\mathcal{CLO}_f$ space, which means that the viewing line never intersects any obstacles in the workspace; again see Fig. 2.

3 Motion Planner

The user specifies the start configuration, q_{start}, and goal configuration, q_{goal}, which are at length offset 0 and 1, respectively. Now the goal is to generate a free camera path $P_{solution}$, connecting q_{start} and q_{goal}. To accomplish this, we apply a single-shot motion planner based on the PRM-approach to construct a graph $G = (V, E)$, or roadmap, in the $\mathcal{CLO}_f$-space. The graph nodes V correspond to free configurations, and edges E to free paths. Now, recall that

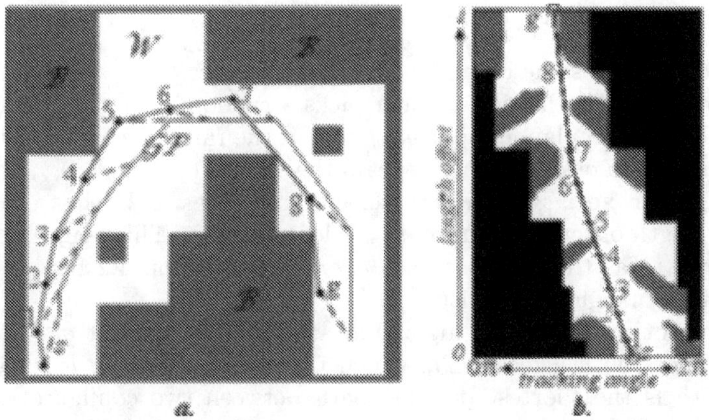

Fig. 2. Figure (**a**) shows a workspace, where the sequence of line segments tagged with $\mathcal{GP}$ is the guide path, and s and g are the camera's start and goal configuration. The sequence of lines segments tagged with numbers is the camera path, and the dotted lines illustrate how the camera looks at the guide.
Figure (**b**) shows the $\mathcal{CLO}$-space. The white area is the $\mathcal{CLO}_f$-space, the gray areas illustrate the collection of occluded configurations, and the black areas illustrate the collection of configurations not allowed because of the tracking angle constraints. The line through $\mathcal{CLO}_f$ is the path connecting the start and goal configuration, and corresponds to the camera path in the workspace (figure 2a)

a path in the $\mathcal{CLO}$-space must be monotone in length offset. This restriction imposes two limitations on the graph: firstly, G is a directed graph and, secondly, for all edges the length offset of the begin configuration must be smaller than that of the end configuration.

The start and goal configuration are the first two nodes added to the graph. Next, the motion planner chooses a random free configuration q_{new} and adds it to the graph. Then, the motion planner tries to connect q_{new} to the graph by attempting to connect q_{new} to nearby nodes (*neighbors*). And finally, the motion planner checks whether the graph connects q_{start} and q_{goal}. These last three steps are repeated in a loop, referred to as the PRM-loop, until $P_{solution}$ is found.

Algorithm 1 *Overview Motion Planner*

```
E ← ∅                                                        [edges]
V ← {q_start, q_goal}                                        [nodes]
loop
    q ← randomly chosen configuration ∈ CLO_f                [Sampling]
    N_q ← set of neighbors of q chosen from V                [Neighbor set]
    Attempt to connect q to its neighbours                   [Connection]
    if G connects q_start and q_goal then return path        [Termination]
    else continue loop
```

3.1 Basic Motion Planner

We need to fill in some details in the basic planner:

Sampling: The motion planner picks a random configuration q_{new} from $\mathcal{CLO}$. The motion planner adds q_{new} to V if the random configuration is free; otherwise, the motion planner repeats the sampling.

Neighbor set: The planner selects the nodes $\in V$ that are within a certain length offset range from q_{new}'s length offset. This length offset range is referred to as the *neighbor set distance*. Subsection 4.2 further discusses the choice of neighbor set distance.

Connection: Next the algorithm attempts to connect q_{new} to each of its neighbors with a *local path*, which is computed by the *local planner*. A local path is the shortest possible path between two configurations in the $\mathcal{CLO}$-space. If a local path is occlusion free, then it is added as an edge to the roadmap.

Termination: The algorithm first checks if q_{start} and q_{goal} are in the same connected graph component. If this test is positive, the algorithm then queries the directed graph to determine whether G connects q_{start} and q_{goal}. If so, the algorithm returns $P_{solution}$; otherwise, the PRM-loop is repeated.

3.2 "Push" Sampling

A narrow passage in the $\mathcal{CLO}_f$-space is a length offset interval for which a relatively small part of the tracking angle interval consists of free configurations. The chance that the sampler randomly generates a free sample in an area of the $\mathcal{CLO}_f$-space is determined by the size of the area. Eventually, enough samples will be generated in a narrow passage to enable the generation of a path though the passage; however, this process is rather time-expensive.

To increase the sample generation in narrow passages, the "push" motion planner pushes a new sample out of the obstacle in case of occlusion, as has also been suggested for PRM application in other motion planning problems [1,25]. The pushing is done by changing the tracking angle value while maintaining the same length offset. Thus, the narrower a passage, the higher the expected number of nodes in the passage.

3.3 "Useful" Nodes

Occlusion checks require collision tests, taking up more than 90% of the motion planner's execution time. Hence, the motion planner should avoid occlusion checks whenever possible. This subsection discusses the concept of usefulness in the framework of our study. We first experimented with growing trees, discussed for example in [19], to reduce useless roadmap growth; however, the gain was only marginal, which is because these techniques, although biased toward the unexplored $\mathcal{CLO}$-space, do not disregard the already explored areas. Hence, we expected that the motion planner should not consider

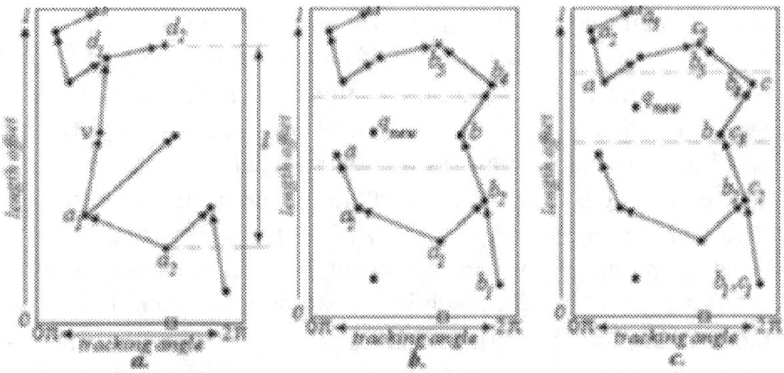

Fig. 3. Figure (**a**): The set of ancestors and descendants of node v are respectively $\{a1, a2\}$ and $\{d1, d2\}$. The longest path $\widehat{P}(v)$ is $P(a2, d2)$ with length l.

explored areas, except when there is a reasonable chance the visit is useful. Due to the monotonicity, the roadmap graph G is a directed acyclic graph. We exploit this roadmap property; as a result, the performance of the motion planner improves drastically.

Let v be a node in G. We refer to the nodes that connect to v as the *ancestors* of v, and the nodes to which v connects as *descendants* of v. We define the length of a path P as the difference in length offset between P's start and goal configuration. Now let av be the ancestor of v with the smallest length offset, and let dv be the descendant of v with the largest length offset. The camera path $P(av, dv)$ is the longest path of which v is a node. If v is not connected to any ancestors and descendants, then the longest path consists only of node v. We denote the longest path containing v by $\widehat{P}(v)$; see Fig. 3(a).

For each graph-node v we store a $\widehat{P}$-record, which describes $\widehat{P}(v)$. The record stores the length offsets of $\widehat{P}(v)$'s begin and end configuration, which are respectively referred to as $\widehat{P}^0(v)$ and $\widehat{P}^1(v)$. Thus for example $\widehat{P}^0(v)$ in Fig. 3(a) is the length offset of node a_2.

To describe the usefulness-criterion, we first introduce an alternative description of the motion planner's goal. Recall that the motion planner expands the roadmap until a path is found. The length offsets of q_{start} and q_{goal} are respectively 0 and 1, hence the length of $P_{solution}$ is 1. All other paths of the roadmap are shorter. Thus we can redefine the goal of the motion planner as the creation a path of length 1. In other words, the motion planner expands the roadmap until $\widehat{P}$ of one of its nodes is of length 1.

Now let us return to the criterion of usefulness. Recall that the PRM-loop starts with the generation of a new sample q_{new}. The next step is the generation of the neighbor set containing all nodes close to q_{new}. One of the three following scenarios occurs after the neighbor set generation: In the first scenario the neighbor set is empty. In the second scenario, the motion planner can connect q_{new} to the roadmap such that $\widehat{P}(q_{new})$ is potentially longer than the longest path of any of its neighbors; see Fig. 3(c): if q_{new} is successfully connected to neighbors a and b, then the longest path $\widehat{P}(q_{new})$ is longer than $\widehat{P}(a)$ and $\widehat{P}(b)$. In the third scenario $\widehat{P}(q_{new})$ does not exceed the length of the longest path of one of its neighbors; see Fig. 3(b): even if q_{new} is successfully connected to neighbors a and b, the longest path $\widehat{P}(q_{new})$ can clearly not become longer than $\widehat{P}(b)$. The sample q_{new} is called useful when one of the first two scenarios occur. We will only add useful nodes. It can be proven that this approach is probabilistically complete.

Once it is decided that the addition of q_{new} is useful, the next step is to connect q_{new} to the roadmap by adding edges. The addition of an edge is only useful, when it increases the length of the current $\widehat{P}(q_{new})$. Let us fill in the details:

Sampling: We can use the basic sampler or the push sampler to generate q_{new}. The difference is that the motion planner does not perform an occlusion check q_{new} before q_{new} is found to be useful.

Neighbor set generation: The motion planner again compiles the neighbor set N_q by taking all nodes $\in V$ within a certain length offset range from q_{new}. We refer to neighbors with a smaller length offset than q_{new} as *backward neighbors*, and to neighbors with a larger length offset than q_{new} as *forward neighbors*. We now determine whether q_{new} is a potentially useful node. If not, q_{new} is discarded and the PRM-loop is repeated.

A longest path of roadmap node v can be divided in two sub paths. The first sub path consists of v and the ancestors of v, and the second sub path consists of v and descendants of v. With this division in mind, we can break the construction of $\widehat{P}(q_{new})$ in two parts as well: the connection of a backward neighbor bn to q_{new}, where $\widehat{P}^0(bn)$ should be as small as possible, and the connection of q_{new} to a forward neighbor fn, where $\widehat{P}^1(fn)$ should be as large as possible.

The motion planner splits the neighbor set N_q into the set of backward neighbors BN_q, and the set of forward neighbors FN_q. The nodes of the first set are sorted on increasing $\widehat{P}_0$ value, and the elements of the latter on decreasing $\widehat{P}^1$ value; see Fig. 4. So we first try the connections that will lead to the longest paths.

Connection: The motion planner attempts to connect the backward neighbors to q_{new}, starting with the first element of BN_q. This connection process continuous until the local planner generates a free local path, or until all backward neighbors are tested and the corresponding local paths are all found occluded. In the first case the motion planner sets $\widehat{P}^0(q_{new})$ as $\widehat{P}^0(nb)$; otherwise $\widehat{P}^0(q_{new})$ is set to q_{new}. Similarly, the motion planner attempts to connect q_{new} to a forward neighbor, and sets $\widehat{P}^1(q_{new})$ accordingly. Again in case of failure $\widehat{P}^1(q_{new})$ is q_{new}. This approach reduces the number of attempted connections considerably.

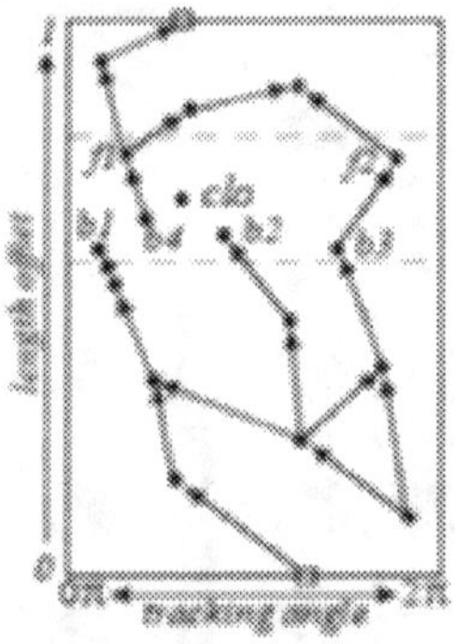

Fig. 4. The sorted backward neighbors set BN_q is $\{b1, b2, b3, b4\}$ and the sorted forward neighbour set is $\{f1, f2\}$.

Once q_{new} is connected to the graph, the $\widehat{P}$-records of q_{new}'s ancestors and descendants must be updated: $\widehat{P}^0(q_{new})$ is propagated through q_{new}'s ancestors, and $\widehat{P}^1(q_{new})$ is propagated through q_{new}'s descendants.

Termination criterion: The goal of the motion planner is to construct a longest path of length 1. This is the case once the length of $\widehat{P}(q_{new})$ is 1.

4 Experimental Results

We implemented the techniques described above in C++. As collision check package we used Solid [5]. All tests are run on a computer with a 2.4 GHz Pentium 4 processor, 512 MB of memory and Windows XP operating system.

4.1 Results

Fig. 5, 6 and 7 illustrate the test cases. Each left picture depicts the workspace containing the guide path, and the start and goal camera configuration at respectively the start and end of the guide path. Each right picture shows a possible camera path. In all experiments we use a maximal tracking angle of $\pi/2$; that is, the camera must stay behind the guide.

We use test case 1 to test the motion planner in an environment with a rather uniform obstacle distribution, and enough room for the camera to maneuver. The purpose of case 2 is to determine the effect of the number of guide path segments on the planner's efficiency. The path is short, but consists of a large number of small segments, increasing the cost of the local planner. Case 3 tests the planner's capability to operate in cluttered environments with a challenging guide path. Below you find the results of the basic planner, the "pusher" planner and the "useful" planner equipped with the push sampling. Recall that the tracking distance is fixed at the most challenging value. The results, expressed in seconds, are averages over 250 runs.

Fig. 5. Test case 1

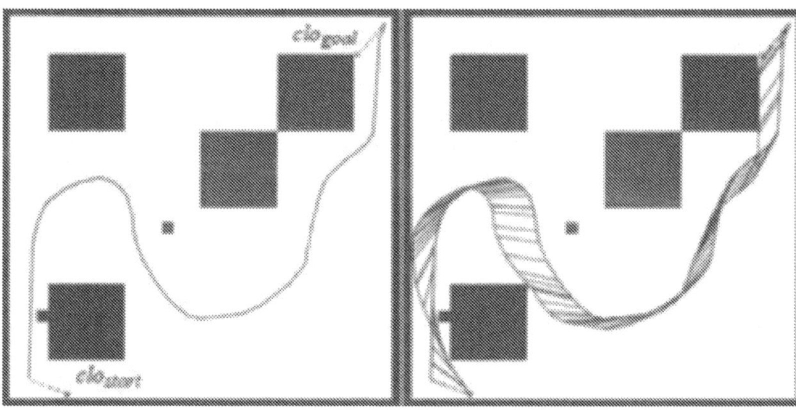

Fig. 6. Test case 2

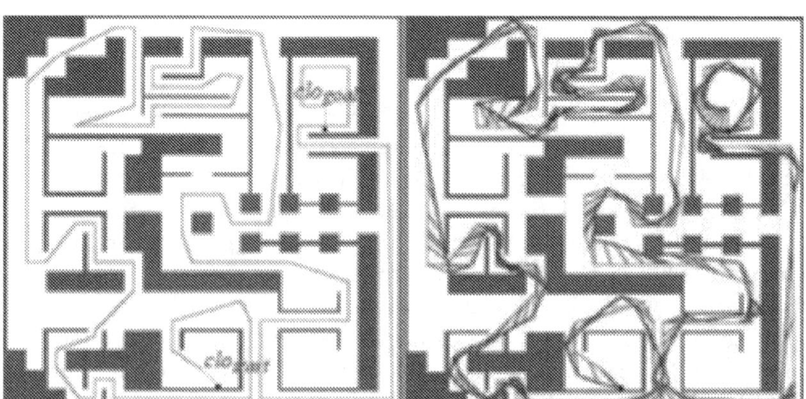

Fig. 7. Test case 3

	Test case 1	Test case 2	Test case 3
Basic planner	1.29	1.02	60+
Push planner	0.91	0.31	45.96
Push + Useful planner	0.22	0.02	0.60

The basic motion planner solves test case 1 and 2 in about one second. However the basic planner is not able to solve the complex test case 3 within a minute. To give an impression of the roadmap size: for test case 3, the roadmap on average consisted of about 60,000 edges and 1,600 nodes.

The "pusher" motion planner clearly performs better overall, although its performance in test case 3 is still not satisfying. For test case 3, the roadmap now consisted of about 43,000 edges and 1,400 nodes.

Adding the usefulness test solved each test case well within a second. The notion of usefulness enables the motion planner to quickly expand the roadmap in uncluttered areas of the $\mathcal{CLO}$-space, and then focus on the cluttered areas, while discarding the already covered easy ones. The roadmap of test case 3 now consists on average of only about 600 edges and 400 nodes.

4.2 Neighbor Set Distance

There is one parameter to fill in: the neighbor set distance (*nsd*). For practical use, it is important that the system determines this value automatically. This subsection describes a technique we refer to as NSD_{auto}.

Let us first define a few roadmap properties. The percentage of all local paths that is tested occlusion free is referred to as the *success ratio*. Furthermore, the percentage of the length offset interval covered by the roadmap G is denoted as the *coverage ratio*.

NSD_{auto} consists of two phases. In the first phase, the learning phase, the roadmap provides too little and too unreliable information to base *nsd* on. Hence in the learning phase, *nsd* is a constant value. Experiments in a variety of scenes show that the constant *nsd* is rather scene independent, somewhere in the range from 0.05 to 0.10 is optimal. The NSD_{auto} goes into the next phase, when the coverage ratio becomes larger than the success ratio. This transition assures on the one hand that the motion planner "learns" enough about the environment, and on the other that the motion planner's efficiency does not decrease because of the relatively large constant *nsd*. During the second phase, we base *nsd* on the average length of the last c free local paths. Experiments show that the results are optimal when c is between about 25 and 100. For test cases of average complexity, NSD_{auto} spends most of its time in the second phase.

The NSD_{auto} as described above causes two undesired side effects, resulting in a decrease of the motion planner's performance. Both problems are caused by the decrease of *nsd* over time and can easily be solved. One problem occurs when the neighbor set of a new sample q_{new} is empty. Fig. 8

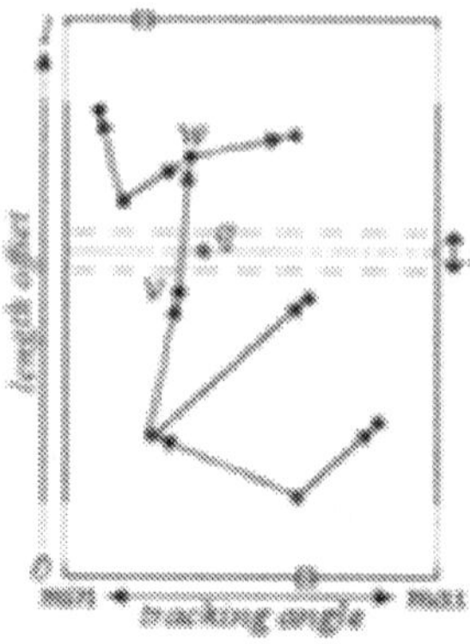

Fig. 8. Illustration of problem occurring because of decreasing *nsd*

illustrates the problem: The neighbor set N of the newly generated sample q is empty, thus the motion planner would mark q as useful, and add q to the roadmap. However, the addition of q is not useful, since the graph already spans the $\mathcal{CLO}$-space covered by the neighbor set N. Note that this situation does not occur when we use a fixed *nsd*, since in that case either node v or w would be in N. To eliminate this problem, the motion planner determines whether the graph already covers the neighbor set length offset interval. If so, then the addition of q_{new} is considered not useful.

The second problem occurs when *nsd* has become small, while a relatively long local path is needed to bridge a gap in the roadmap. We solve this problem by setting a lower boundary for *nsd*; which we base on the tracking distance.

Experiments show that the motion planner performs as well with NSD_{auto} as when the neighbor set distance is manually optimized. The results in the previous section were obtained using NSD_{auto}.

5 Smoothing

Path quality indicates how pleasant the camera images are to watch for a user. The motion planner as described so far disregards this aspect; see Fig. 9(a). Experiments show that the path quality is inversely related to the total tracking angle change. Hence, to improve the path quality, the motion planner has to minimize the total tracking angle change. This smoothing process consists of two phases.

The first phase is active while the motion planner generates the initial camera path: The weight of each roadmap edge e is the tracking angle difference between e's begin and end configuration. When q_{start} and q_{goal} are graph connected, the path with the smallest total weight is taken as initial solution $P_{solution}$.

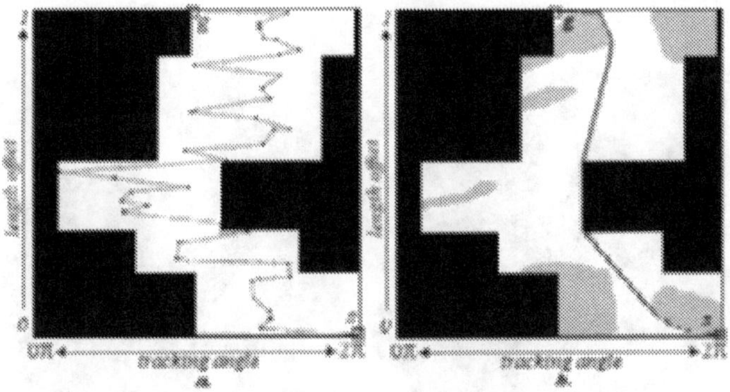

Fig. 9. Figure (**a**): A typical example of an unsmooth camera path in the configuration space. The resulting camera images are unpleasant to watch, because the camera changes its tracking angle frequently. Figure (**b**): The white area is an approximation of $\mathcal{CLO}_f$. The path is an example of a smoothed path.

The second phase applies a standard smoothing approach [24]. The motion planner repeatedly replaces random path segments of $P_{solution}$ with new path segments of higher quality using the local planner, and in doing so increases the overall quality of $P_{solution}$.

The longer this is repeated, the smoother $P_{solution}$ becomes. Fig. 9(b) illustrates a smoothed path for the example test case. Without smoothing, the motion planner generates $P_{solution}$ for the example test case on average in 0.1 second. The (close to) optimally smoothed path, as illustrated in Fig. 9, is on average generated in 0.5 second. However, experiments show that optimal smoothing is not necessary to ensure a satisfying path quality: the motion planner generates a satisfying smooth $P_{solution}$ in on average 0.2 second. The three test cases described in the previous section are all solved and smoothed within a second. We expect that the smoothing phase can easily be improved to reduce its execution time even further.

6 3D Motion Planner

So far, we discussed camera motion generation in a two-dimensional workspace. A virtual environment, however, is three-dimensional and the motions of the guide and camera are not always reducible to a two-dimensional plane. Fortunately, the 2D motion planner is easily extendable to operate efficiently in a three-dimensional workspace. We briefly describe the required changes here.

The guide path remains a map from length offset to guide configuration, where the guide configurations are now described by six parameters. Recall that the camera position is specified relative to a guide configuration. In the two-dimensional workspace we needed the tracking distance and the tracking

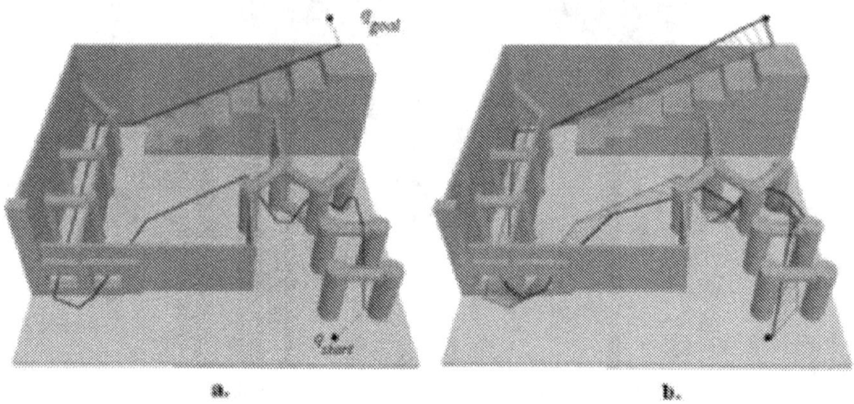

Fig. 10. Figure (**a**) shows the guide path, and the start and goal configurations of the camera as q_{start} and q_{goal}. Figure (**b**) shows the guide path in gray and camera path in black. The short lines between the guide and camera path illustrate viewing directions of the camera.

angle to describe this relative position. In the three-dimensional situation we add a second tracking angle to describe the relative vertical angle. The camera orientation is again discarded: the camera will always look straight at the guide; furthermore, the camera's up-vector is kept equal to the guide's up-vector. The corresponding $\mathcal{CLO}$-space is a three-dimensional space spanned by both tracking angles and the length offset. The motion planner works without adaptation on the expanded $\mathcal{CLO}$-space. We implemented and tested the three-dimensional motion planner. The "push" and "usefulness" techniques again result in a drastic performance improvement.

Fig. 10 shows a typical three-dimensional test case with open spaces, narrow passages and height variations. As for the two-dimensional test cases, the motion planner generates a smooth camera path in less than a second. The resulting roadmap again is relatively small: on average it contains about 150 edges and 100 nodes.

7 Conclusion

We presented a technique to generate a camera motion such that the camera tracks a guide moving through a known environment along a known path. The motion planner applies a single-shot PRM approach to construct a graph in the free configuration space. A substantial performance gain is accomplished by a new technique, which determines whether a node or edge should be added to the roadmap based on its usefulness. The resulting camera path is smoothed to improve the path quality. Experiments show that the planner can operate in three-dimensional environments, and provide the user with a fast responds as is needed for interactive environments.

A number of extensions are possible. Firstly, when the guide is an object rather than a point, we might want to enforce the camera to see the complete guide (and even some part around it). This can easily be achieved by replacing the occlusion check by an intersection test with a visibility cone. Secondly, we can also easily deal with preferred tracking angles by changing the sampling distribution and the smoothing process. And lastly, preferred tracking distance can be achieved in a post processing stage. Finally, additional camera motion properties such as camera speed, camera acceleration, changes in tracking distance and changes in the camera orientation can be incorporated using techniques similar to those proposed in [24].

Acknowledgement This work was partly supported by the Netherlands organization for Scientific Research (NWO), and by the IST Programme of the EU as a Shared-cost RTD (FET Open) Project under Contract No IST-2001-39250 (MOVIE - Motion Planning in Virtual Environments).

References

1. N. Amato, Y. Wu (1996). *"A randomized roadmap method for path and manipulation planning"*. Proc. IEEE Int. Conf. on Robotics and Automation, 1996, pp. 113-120.
2. W.H. Bares, S. Thainimit, S. McDermott (2000). *"A model for constraint-based camera planning"*. Proc. AAAI Spring Symposium on Smart Graphics, pp. 84-91, 2000.
3. J. Barraquand, L. Kavraki, J.-C. Latombe, T.-Y. Li, R. Motwani, P. Raghavan (1997). *"A random sampling scheme for path planning"*. Int. Journal of Robotics Research 16 (1997), pp. 759-774.
4. S. Beckhaus, F. Ritter, T. Strothotte (2000). *"Cubicalpath – Dynamic Potential Fields for Guided Exploration in Virtual Environments"*. Proc. Pacific Graphics 2000, pp. 387-395, 2000.
5. G. van den Bergen (2003), *"Collision Detection in Interactive 3D Environments"*. Morgan Kaufmann 2003.
6. D. Christianson, S. Anderson, L. He, D. Salesin, D. Weld, M. Cohen (1996). *"Declarative camera control for automatic cinematography"*. Thirteenth National Conference on Artificial Intelligence, pp. 148-155.
7. N. Courty, E. Marchand (2001). *"Computer animation: a new application for image-based visual servoing"*. IEEE Int. Conf. on Robotics and Automation, Volume 1, pp. 223-228, 2001.
8. S.M. Drucker, D. Zeltzer (1995). *"CamDroid: A system for implementing intelligent camera control"*.P. Hanrahan and J.Winget, editors, 1995 Symposium on Interactive 3D Graphics, pp. 139-144. ACM SIGGRAPH, 1995.
9. M. Gleicher, A. Witken (1992). *"Through-the-Lens Camera Control"*. Computer Graphics, 26(2), pp. 331-340.
10. H.H. Gonzalez-Banos, C.Y. Lee, and J.-C. Latombe (2002). *"Real-Time Combinatorial Tracking of a Target Moving Unpredictably Among Obstacles"*. Proc. IEEE Int. Conf. on Robotics and Automation, pp. 1683-1690, 2002.

11. N. Halper, P. Olivier (2000). *"Camplan: A Camera Planning Agent"*. Proc. 2000 AAAI Spring Symposium on Smart Graphics, pp. 92-100, 2000.
12. N. Halper, R. Helbing, T.Strothotte (2001). *"Computer Games: A Camera Engine for Computer Games"*. Computer Graphics Forum Volume 20, Issue 3.
13. A. Hanson, E. Wernert (1997). *"Constrained 3D Navigation with 2D Controllers"*. Proc. IEEE Visualisation, pp. 175-182.
14. Hanson, A., E. Wernert, S. Hughes (1999). *"Constrained navigation environments"*. Scientific Visualization: Dagstuhl '97 Proceedings, H. Hagen, G.M. Nielson, and F. Post, Editors. 1999, IEEE Computer Society Press. pp. 95-104.
15. L.-W He, M.F. Cohen, D.H. Salesin (1996). *"The Virtual Cinematographer: a Paradigm for Automatic Real-Time Camera Control and Directing"*. Proceedings of ACM SIGGRAPH'96, pp. 217-224, 1996.
16. L. Kavraki, J.C. Latombe (1994). *"Randomized preprocessing of configuration space for fast path planning"*. Proc. IEEE Int. Conf. on Robotics and Automation, 1994, pp. 2138-2145.
17. L. Kavraki, P. Švestka, J. Latombe, and M. Overmars. (1996). *"Probabilistic Roadmaps for Path Planning in High-Dimensional Configuration Spaces"*. IEEE Transactions on Robotics and Automation, vol. 12, no. 4, pp. 566-580.
18. S.M. LaValle, H.H. Gonzalez-Banos, C. Becker, J.-C. Latombe (1997). *"Motion Strategies for Maintaining Visibility of a Moving Target"*. Proc. IEEE International Conference on Robotics and Automation, pp. 731-736, 1997.
19. S.LaValle, J. Kufner (2001). *"Rapidly-exploring random trees: Progress and prospects"*, in Algorithmic and Computational Robotics: New Directions (B.Donald, K.Lynch, and D.Rus, eds.), pp. 45-59, A.K. Peters, 2001.
20. T.Y. Li (1999). *"An Auto-Navigation System for Virtual Factories"*. Proc. Automation Technology Conference, Taiwan, 1999.
21. T.Y. Li, J.M. Lien, S.Y. Chiu, T.H. Yu (1999). *"Automatically Generating Virtual Guided Tours"*. Proc. Computer Animation Conference 1999, pp. 99-106, 1999.
22. J. Mackinlay, S. Card, G. Robertson (1990). *"Rapid controlled movement through a virtual 3d workspace"*. Computer Graphics (SIGGRAPH '90 Proceedings), pp. 171-176, 1990.
23. E. Marchand, N. Courty (2000). *"Image-based virtual camera motion strategies"*. Graphics Interface Conference, GI2000, pp. 69-76, 2000.
24. D. Nieuwenhuisen, M.H. Overmars (2003). *"Motion Planning for Camera Movements in Virtual Environments"*. Proc. IEEE Int. Conf. on Robotics and Automation 2004, to appear.
25. M.H. Overmars (1992). *"A random approach to motion planning"*. Technical Report RUU-CS-92- 32, Dept. Comput. Sci., Utrecht Univ., Utrecht, the Netherlands, 1992.
26. C.B. Phillips, N.I. Badler, J. Granieri (1992). *"Automatic viewing control for 3D direct manipulation"*. M. Levoy and E. E. Catmull, editors, Proc. Symposium on Interactive 3D Graphics 1992, pp. 71-74, 1992.

Computing Deform Closure Grasps*

K. "Gopal" Gopalakrishnan[1] and Ken Goldberg[2]
[1] IEOR Department, U.C. Berkeley, USA
[2] IEOR and EECS Departments, U.C. Berkeley, USA

Abstract. In prior work, we extended the well-known form closure grasp framework for rigid parts to a class of deformable parts, modeled as frictionless polygons with a finite element mesh and given stiffness matrix. We define the *D-space* (deformation-space) of a part as the C-space of all its mesh nodes and *deform closure* in terms of the work needed to release the part from a set of finger contacts.

In the present paper, we define a measure of grasp quality for two-point deform-closure grasps. This metric is based on balancing the potential energy needed to release the part against the potential energy that would result in plastic deformation. Given two jaw contacts at the perimeter nodes, we give a numerical algorithm to determine the optimal jaw separation based on this metric. For a part with n mesh nodes and p perimeter nodes, the algorithm computes an approximation to the optimal separation in time $O(n^3 p^2 + (p^2/\varepsilon) \log p)$.

1 Introduction

There is a substantial body of research on robotic holding (grasping and fixturing) of rigid objects. Our work is inspired by Rimon and Burdick on rigid body immobility [Rimon96] and by Rimon and Blake [Rimon99] on caging grasps for rigid bodies. We adopt their notation wherever possible.

In [Gopalakrishnan04], we proposed a new model for grasping deformable parts that combines ideas from C-space, FEM meshing, and cage grasping. We define D-space based on the degrees of freedom of mesh nodes and "deform closure" based on local minima in elastic potential energy.

In the present paper, we define a measure of grasp quality for two-point deform-closure grasps. This metric is based on balancing the potential energy needed to release the part against the potential energy that would cause plastic deformation. Given two jaw contacts at the perimeter nodes, we give a numerical algorithm to determine the optimal jaw separation based on this metric.

In our model, the part perimeter behaves like a planar structure whose perimeter is composed of struts of variable length, with hinges at each vertex. We define the *D-space* (deformation-space) of the part as the C-space of all nodes in its mesh. Given the initial configuration q_0 of the mesh in D-space,

* This work was supported in part by the Ford Motor Company URP 2000-403R and NSF Award DMI-0010069. For more information, please contact goldberg@ieor.berkeley.edu or gopal@ieor.berkeley.edu.

M. Erdmann et al. (Eds.): Algorithmic Foundations of Robotics VI, STAR 17, pp. 203–218, 2005.

we define D_T as the set of all configurations with the same mesh topology as q_0. We consider contact with a set of rigid finger bodies $A = \{ A_1, \ldots, A_k \}$. Each finger defines a D-obstacle DA_i, the set of mesh configurations that collide with A_i. The free space D_{free} is the set of feasible configurations, the intersection of D_T with the complement of the union of D-obstacles.

Given part E, mesh M, FEM stiffness matrix K, and contact set A, we can determine the potential energy of any point in D_{free}. Given a candidate configuration q_A, we say that the part can be *released* from A by a sequence of external wrenches if the part can escape from q_A . We define U_A as the minimum work that needs to be done by these wrenches to release the part from A. We say that A holds E in *deform closure* if $U_A > 0$.

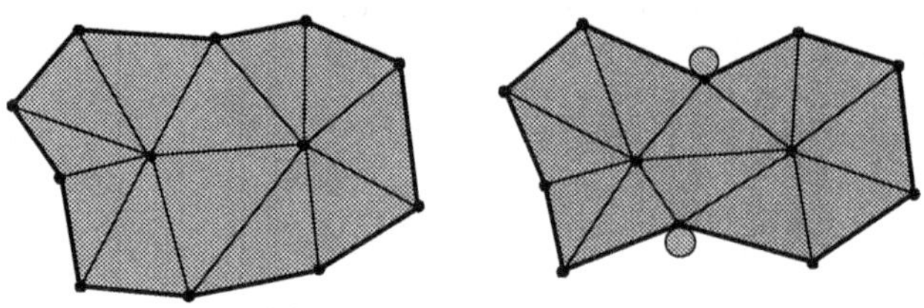

(a) (b)

Fig. 1. (a) Deformable part, modeled as a polygon with a finite element mesh and given stiffness matrix. The part perimeter behaves like a planar structure whose perimeter is composed of struts of variable length, with hinges at each vertex (b) The same part held in *deform closure* by two frictionless point contacts.

As intuition behind our algorithm for computing the optimal jaw separation, consider a "contact graph", where each graph vertex corresponds to a pair of mesh nodes. For each vertex, consider a potential energy function based on varying the distance between the underlying mesh nodes. Given two initial mesh nodes where contact occurs, consider a sequence of vertices -- a path -- that leads to an equilibrium grasp configuration. The upper envelope of the potential functions for all vertices on this path corresponds to the maximal potential energy function for the path. Then consider all possible paths, and the lower envelope of all path potential energy functions. The difference between this lower envelope and the potential function at the initial pair of mesh nodes is the minimum work as a function of jaw separation. The potential energy needed to produce plastic deformation is a quadratic function of jaw separation. The lower envelope of the minimum work function and the plastic deformation function yields the objective function, which defines the optimal jaw separation.

This objective function is continuous and piecewise differentiable, but may have an exponential number of pieces. Our algorithm computes the value of the function by uniform sampling, using a sampling interval that is based on stiffness coefficients and a given approximation bound. At each sample point, we use a dynamic programming algorithm to determine the threshold potential energy and thus the objective function value. The sample point with the maximal value of the objective function provides the approximation to the optimal jaw separation.

2 Related Work

Bicchi and Kumar provide a concise survey of literature on grasping and fixturing in [Bicchi00]. Grasps of rigid bodies can be classified as force or form closure. Form closure (immobility) occurs when any neighboring configuration of the part results in collision with an obstacle. Force-closure occurs if any external wrench can be resisted by applying suitable forces at the contacts [Mason01, Rimon98]. Gripper contacts can be modeled as frictional points, frictionless points or soft contacts [Salisbury82].

The mobility of rigid bodies in contact with frictionless finger bodies was initially studied using first order approximations [Realeaux1876, Somoff1900, Mishra87, Markenscoff90]. The first order theories are based on approximations of part geometry in infinitesimal neighborhoods and part motion of an infinitesimal length. However, these first order approximations of mobility did not always predict immobility correctly.

Rimon and Burdick [Rimon96] give rigorous definitions of first and second order immobility. They express paths of the part in free space using functions $q(t)$ based on scalar parameter t. They consider the distances d_i from the ith C-obstacle surface to $q(t)$. They then consider the first and second order terms in the algebraic expressions of distances as a function of t. The second order terms are needed if first order tests do not show if the distance is increasing or decreasing (the derivative at $t = 0$ is 0). Geometrically speaking, the first order approximation of immobility is equivalent to approximating the trajectory of the part in C-space as a straight line, and the surfaces of the C-obstacles as hyperplanes. Second order immobility is equivalent to approximating the trajectory as an elliptical arc and the C-obstacle surfaces as ellipsoids.

Rimon and Blake [Rimon99] give a method to find caging grasps, configurations of jaws that constrain parts in a bounded region of C-space such that actuating the gripper results in a unique final configuration. They consider the opening parameter of the jaws as a function of the jaws' positions along the perimeter of the part and use stratified Morse theory to find caging grasps by finding limiting cases that occur when the opening parameter is at a saddle point. [Gopalakrishnan02] also uses the distance function to determine immobile grasps of 2D polygonal parts by a pair of vertical cylindrical jaws engaging the part at its concave vertices.

An efficient geometric algorithm to compute all placements of four frictionless point contacts on a polygonal part that ensure form closure is described by van der Stappen et al [vanderStappen99]. Given a set of four edges, they show how to compute critical contact placements in constant time. The time complexity of their algorithm is bounded by the number of such sets, and runs in an expected time of $O(n^2 \log n)$ for n vertices. Cheong et al [Cheong03] give fast algorithms to find immobilizing grasps of 2D polygonal parts with 2 and 3 contacts. The algorithms find sets of contact wrenches that contain the center of mass. Zhu et al [Zhu03] give a grasp metric for 2D and 3D grasps to quantify how firmly the rigid part is held when resisting external forces. Their metric is faster to compute than similar earlier metrics.

Cheong et al [Cheong02] give upper bounds on the number of point contacts needed to hold a chain of polygons hinged at connecting vertices. They consider the centers of rotation for each polygon such that the hinge vertices have the same linear velocities for the pair of polygons they connect. They place contacts such that no such rotation is possible. Milgram and Trinkle [Milgram02] describe the topology of C-spaces of mechanisms that consist of chains of links connected by rotational joints.

[Ramamurti98] gives an excellent review of finite element methods and speed and memory optimizations for numerical solutions of FEM problems. To solve an FEM problem with n elements exactly, $O(n^3)$ time is required. The numerical approximation makes use of the fact that FEM stiffness matrices are sparse. As a result, there are only $O(n)$ non-zero entries in a matrix with n^2 elements. [Corman90] presents several graph algorithms and data structures. Of particular interest to our algorithm is the proof of correctness and time complexity of the modified Dijkstra's algorithm for sparse graphs implemented with binary heaps. For a graph with n vertices and $O(n)$ edges, the shortest path can be computed in $O(n \log n)$ time.

Recent work on fixturing deformable and sheet-metal parts is based on the work of Menassa and De Vries [Menassa91] where they determine the positions of the primary datum (the datum points needed to locate the part in the correct plane) for 3-2-1 fixturing to minimize deformation. They use a finite element model of the part, and determine fixture locations by optimizing an objective that is a function of the deformations at the nodes. Their work is extended by [Rearick93] and [Cai96].

Gopalakrishnan et al [Gopalakrishnan03] propose *unilateral fixtures*, a new class of fixtures for sheet metal parts with holes, where holding elements lie almost completely on one side of the part, maximizing access for welding, assembly, or inspection. Each primary jaw is cylindrical with a conical groove that provides the equivalent of four point contacts and facilitates part alignment. They present a two-phase algorithm for computing unilateral fixtures where the second phase uses a Finite Element Method (FEM) to compute part deformation and to arrange secondary contacts at part edges and interior surfaces. For a given sheet-metal part, given as a 2D surface embedded in 3D with e edges, n concavities and m mesh nodes, Phase I takes $O(e + n^{4/3} \log^{1/3} n + g \log g)$ time to compute a list of g pairs of primary jaws ranked by quality. Phase II computes the location of r secondary contacts in $O(g\, m^3\, r)$ time.

Wakamatsu et al [Wakamatsu96] extend the concept of force closure for rigid parts with unbounded applied forces to deformable parts with bounded applied forces. They consider a candidate grasp and external forces within a bound that can deform and displace the part. They assume that the contact normals do not vary as forces are applied, but allow contacts to be broken. They define "bounded force closure" as grasps that can resist any external force within the bound.

Howard and Bekey [Howard97] model deformable parts using a spring-mass model and use a neural network to control a gripper. They use tactile feedback to learn the properties of the deformable parts, and thus determine the minimum force needed to lift the deformable part. Heinrich and Worn's edited collection of papers [Heinrich00] describes deformation models and control algorithms for the manipulation of deformable objects. These include the modeling of fabrics, continuous modeling of deformation of linear objects such as beams, collaborative manipulation of deformable parts, laying out wires and cables and sewing.

Hirai et al [Hirai01] propose a robust control law for the manipulation of deformable parts. They use tactile feedback from the actuators and video feedback for the positions of selected reference points to control the motion of a deformable part. Li et al [Li02] design fixtures for laser welding using a genetic algorithm within robust design spaces with low to part dimension and jaw location errors. Li et al [Li02b] describe a dexterous part holding mechanism based on vacuum cups and model the elastic deformation of the sheet-metal part using Finite Element Methods and a statistical data model. The results from this model are used to minimize the part's deformation. Shiu et al [Shiu97, Shiu03] give a heuristic algorithm to analyze the deformation of a sheet metal part by decoupling it into beams based on the part's features, and give an algorithm to allocate tolerances to each feature.

Path planning for elastically deformable parts has been studied using probabilistic roadmaps (PRM). Holleman, Kavraki and Warren [Holleman98] give a path planning algorithm for a flexible surface patch. They use a Bezier approximation and an approximate energy function to model deformation of the part. They present experimental results of paths planned for parts generated by a search graph using PRM. Guibas et al [Guibas99] improve on the PRM methods for path planning for a surface patch by studying the medial axis of the workspace. Minimum energy configurations of the part are then computed for positions along the axis and connected by quasi-static paths. Lamiraux et al [Lamiraux99] generate a path for a thin rectangular elastic metal plate represented by a Bezier when it is manipulated by constraining the positions and orientations of two opposite edges. Given the controlling constraints, the shape of the plate is computed by minimizing the potential energy due to the deformations. If there is a collision between the plate and an obstacle, the configuration is discarded. Paths are generated using PRM on the configurations of the edges being controlled. Moll and Kavraki [Moll04] have extended this approach to systems such as surgical sutures and snake robots by computing configurations using constrained minimization of potential energy with the length of the curve fixed.

As noted in the introduction, the present paper extends [Gopalakrishnan04] with a new metric for two-point deformation grasps and an algorithm for computing the optimal jaw separation.

3 Deformation-Space and Deform Closure

3.1 Deformation-Space

In this section, we summarize the relevant definitions and theorems from [Gopalakrishnan04]. We consider a linearly elastic deformable polygonal part E possibly containing polygonal holes. E is discretized into a triangular FEM mesh M with linear interpolation. The stiffness matrix for E and an initial undeformed configuration q_0 of E are also specified.

We define a Deformation Space (D-Space) D of the part. Each of the n nodes in M has two translational degrees of freedom (DOF). The D-space (deformation space) of the mesh is the smooth manifold $D = \mathbf{R}^{2n}$, with each coordinate axis corresponding to one DOF of an FEM node. Any deformed translated or rotated shape of this part can be represented as a unique point in D-space. For any point q in D, we denote the corresponding configuration of the part by $E(q)$.

We define $D_T(q_0)$ as the set of all q that have the same topology as initial configuration q_0.

We consider frictionless contacts between E and a set of finger bodies A_i. Given a finger body A_i, its corresponding D-obstacle DA_i is the set of all configurations q such that $E(q)$ intersects A_i.

We define $D_{free} = \left(\bigcup_{i=1..k} DA_i \right)^C \cap D_T$ as the set of all feasible configurations.

D^C_{free} is the complement of D_{free}.

3.2 Potential Energy in D-Space

The nodal displacement vector X is the vector of the displacements of all nodes from their initial positions in configuration q_0, expressed in the global coordinate frame used to define the FEM model (M, K, X). X can be considered as a special case of expressing q when each node's degrees of freedom are expressed relative to a reference frame whose origin coincides with the mesh node at configuration q_0, and whose axes parallel to the axes of the global coordinate frame. For linear FEM with a stiffness matrix K, the potential energy at configuration q in D_{free} is given by:

$$U(q) = (1/2)\, X^T K\, X.$$

3.3 Deform Closure

We consider a configuration q_A in which the part is held by A.

Definition: An *equilibrium configuration* is any configuration at a local minimum of the potential energy in D_{free}. In the absence of friction and inertial forces, the part comes to rest at an equilibrium configuration.

Definition: We consider a set of wrenches that act on $E\ (q_A)$ for any equilibrium configuration q_A. If after the wrenches are applied and then removed, the part may not return to q_A, the wrenches are said to *release* E.

Definition: If $E(q_A)$ is held by A such that no sequence of wrenches that increases the deformation potential energy of E by at most U can release E from A, we say q_A satisfies the property *stable (U)*.

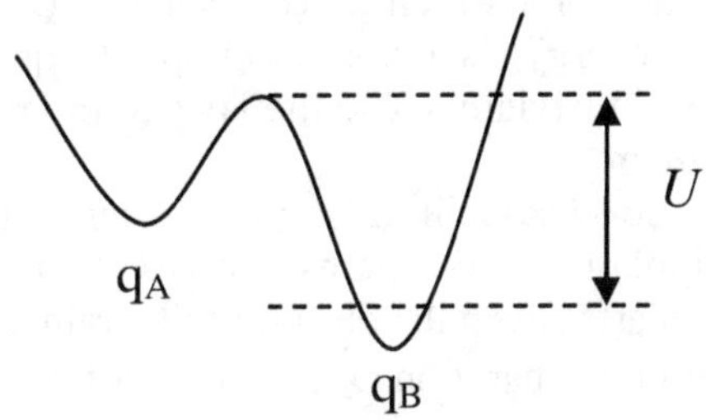

Fig. 2. A slice of the Potential Energy surface. q_A and q_B are stable equilibria. But for the shown value of U, only q_B satisfies stable (U).

Definition: The *threshold potential energy* U_A for A holding E in configuration q_A is defined as 1) $U_A(q_A) = \sup\{\, U \mid \underline{q_A} \text{ satisfies stable } (U)\}$ if q_A is an equilibrium configuration, and 2) 0 otherwise.

Definition: A holds $E(q_A)$ in *deform closure* if $U_A(q_A) > 0$.

Definition: A configuration with potential energy $U(q_A) + U_A(q_A)$ that E assumes when a sequence of wrenches releases it is called an *escape configuration.*

Given a rigid part, we call an undeformed deformable part with the same shape and configuration of the rigid part its equivalent deformable part.

We also prove the following theorems in [Gopalakrishnan04]:

Theorem 1: If A holds a rigid part in form closure, A will hold the equivalent deformable part in deform closure.

Theorem 2: Frame Invariance.

The definition of deform closure is frame invariant, i.e. it does not depend on the global reference frame used.

In the proof of theorem 2, we also show that the potential energy is frame-invariant.

4 Two-Point Deform-Closure Grasps: Optimal Jaw Separation

4.1 Quality Metric

As in the previous section, we consider the same deformable part model, Here, we consider two-point grasps, where there are exactly two contacting jaws (finger bodies). The initial placement of these jaws is specified as two mesh nodes n_0 and n_1. The jaw separation, σ, is the Euclidean distance between the jaw contact points. In what follows, we describe our approach in terms of contracting grasps (where the contacts exert forces towards each other). An analogous approach can be developed for expanding grasps.

We define a quality metric Q to evaluate deform-closure grasps. We define σ^* as the jaw separation for which $Q(\sigma^*) = \max\{Q(\sigma)\}$. Note that U_A, the threshold potential energy, is not an appropriate quality metric because to maximize it we would trivially close the jaws to the point where there is zero distance between them.

We introduce the elastic limit of part material to define a metric that balances the gripping force against forces that will result in plastic deformation of the part. Given the elastic limit strain e of the part material, we consider the maximum separation σ_L at which at least one mesh node reaches the elastic limit when squeezed by the contacts at n_0 and n_1. We define $U_L(\sigma)$ as the difference between the potential energies at separations σ and σ_L. We define:

$$Q(\sigma) = \min\{U_A(\sigma), U_L(\sigma)\}$$

At the optimal jaw separation σ^*, the potential energy needed to release the part equals the potential energy that would be required to squeeze it to its elastic limit.

4.2 Problem Description

Given linearly elastic polygonal part E, triangular mesh M, stiffness matrix K, initial undeformed configuration q_0 of E, elastic limit strain e_L, and a pair of mesh nodes n_0 and n_1 where a part is held by two frictionless point contacts, we give an algorithm that numerically determines contact separation σ_ε as an approximation to σ^*. The maximum allowed error in $Q(\sigma_\varepsilon)$ is specified as ε.

Input: Polygonal part E, mesh M with n mesh nodes and p perimeter nodes, stiffness matrix K, initial undeformed configuration q_0, initial contact mesh nodes n_0 and n_1, and parameters e_L and ε.

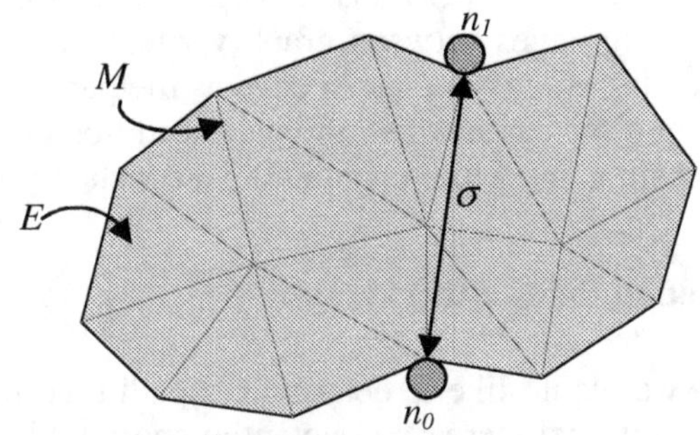

Fig. 3. Part E, mesh M, mesh nodes of contact (n_0, n_1) and stiffness matrix K are given as input to the algorithm. The algorithm numerically maximizes quality metric $Q(\sigma)$ where σ is the contact separation.

Output: A value of σ_ε such that $Q(\sigma_\varepsilon) \geq Q(\sigma^*) - \varepsilon$.

We make the following assumptions:

1. The perimeter of the part does not collide with itself intermediate configurations as it deforms from the deform-closure grasp to the escape configuration.
2. Mesh triangles do not become degenerate. In FEM practice, when mesh triangles become degenerate, a new, denser mesh is automatically generated. We assume that a sufficiently dense mesh that does not result in degeneracy is given initially.
3. If one contact is placed on the perimeter and the other is moved along the perimeter at a fixed separation from the first contact, the local maxima and minima of potential energy occur when the second contact is at a mesh node. Strictly speaking, given a particular mesh, the actual potential energy may be a maximum or a minimum even the contact is at an edge of the mesh, with the line joining the contacts being perpendicular to the edge. However, given a sufficiently dense mesh (with more mesh nodes near the contact), the difference between the maximum potential energy and the potential energy with the second contact at a vertex will be less than the error in potential energy predicted due to the FEM approximation of the part.

Our approach to solve this problem is detailed in the following sections. In brief, we make use of assumption 3 above to identify paths of interest in D-Space through which the part may be release from the grasp. We need not study all paths that release the part as long as we study one where the peak potential energy is at a minimum. Thus, due to assumption 3, we need only consider the paths where at least one contact is at a mesh node at any time. Among these paths, we focus on the points where both contacts are at mesh nodes, since the potential energy varies monotonically between these points. This is a set of discrete configurations that we are interested in. We create a "*contact graph*" whose vertices these configurations and describe all paths of interest in D-Space as paths (traversal of edges) in this contact graph. This is done by constructing edges between graph vertices that differ in exactly one contact's position by exactly one mesh edge as the part slides out of the grasp. We add more edges for cases where the part need not slide out, but can move without contact with a jaw while maintaining a constant potential energy.

4.3 Constructing the Contact Graph

To compute σ_ε, we would like to determine a path that the part can follow to be released with minimal increase in potential energy. Since it is inefficient to consider every possible path in D-space, we create a graph to represent key configurations as graph vertices.

We use the following notation: q_A is the initial configuration of the part when grasped by to contacts at (0, 0) and (σ, 0) engaging the part at n_0 and n_1 respectively. $\sigma(n_a, n_b)$ is the distance between mesh nodes n_a and n_b. $\sigma_0(n_a, n_b)$ is the distance between mesh nodes n_a and n_b at configuration q_0.

To release the part from *A*, we move it along a path in D-space starting at q_A and ending at an escape configuration. To determine the threshold potential energy, we are interested in an escape configuration with least potential energy. From assumption 3, if one contact were fixed on the part's perimeter and the other jaw is allowed to slide along the perimeter, the potential energy as a function of the position of the second jaw is monotonic (increasing or decreasing) in every interval corresponding to one mesh edge. This implies: 1) there exists such an escape configuration q_p with both contacts at mesh nodes, and 2) there exists a path in D-space from q_A to q_p such that one of the contacts is always at an FEM node such that q_p has the maximum potential energy among all configurations on the path. It is sufficient to consider the paths where one contact is always at a mesh perimeter node.

We construct a "contact graph" *G* to represent such paths. The set of vertices in *G* is *V*. Each graph vertex in *V* represents a configuration as the part is released from the point contacts. Thus, every graph vertex $v(n_a, n_b)$ corresponds to point contacts present at a pair of perimeter mesh nodes n_a and n_b. Vertices $v(n_a, n_b)$ and $v(n_a, n_c)$ are connected by an edge if: 1) n_a and n_b are connected by a mesh edge on the part's perimeter, or 2) an arc centered on n_a and having radius σ and lying on the exterior of the part has one end point n_b and the other end-point on a mesh edge flanking n_c. The second case is when the part can be rotated about a mesh node till the contact reaches the perimeter near another mesh node.

4.4 Potential Energies at Graph Vertices

At each graph vertex, we are interested in the minimum possible potential energy when the contacts engage the part at the corresponding mesh nodes n_a and n_b, with separation σ. This is determined by solving the FEM model with boundary conditions restricting n_a at (0, 0) and n_b at $(\sigma, 0)$.

We note that we use a linear FEM system to model the part. Hence, nodal deformations are computed by solving a system of linear equations subject to the boundary conditions. As a result, for any graph vertex when the two contacts perform a contracting grasp at (n_a, n_b), the potential energy associated with the cooresponding vertex is a quadratic function of σ given by $U(n_a, n_b, \sigma) = \frac{1}{2} k_{ab} (\sigma - \sigma_0(n_a, n_b))^2$, where k_{ab} is the x component of the force applied by the contacts towards each other when $\sigma_0(n_a, n_b) - \sigma = 1$ unit. However, for $\sigma > \sigma_0(n_a, n_b)$, the potential energy for contracting grasps is 0.

We first compute the value of k_{ab} for all $v(n_a, n_b)$ by solving the FEM model with boundary conditions n_a is at (0, 0) and n_b is at $(\sigma_0(n_a, n_b) - 1, 0)$.

For the graph vertex $v(n_0, n_1)$ alone, we also compute the strains at each mesh node. We find the maximum strain over all nodes as e_{max}.

Thus each graph vertex $v(n_a, n_b)$ is associated with a potential energy $U(n_a, n_b, \sigma) = \frac{1}{2} k_{ab} (\sigma - \sigma_0(n_a, n_b))^2$, where k_{ab} are known. Some of these vertices correspond to equilibrium configurations. A vertex $v(n_a, n_b)$ is in equilibrium if for every neighbor $v(n_c, n_d)$ of $v(n_a, n_b)$,

$$U(n_c, n_d, \sigma) \geq U(n_a, n_b, \sigma).$$

4.5 Computing σ_ε

Consider a path from $v(n_a, n_b)$ to an equilibrium vertex. The path is a sequence of vertices. Each graph vertex $v(n_c, n_d)$ on the path has a potential energy function $U(n_c, n_d, \sigma)$. The potential energy needed to release the part through this path is given by the upper envelope of all these potential energy functions. There are many equilibrium nodes, and many paths to each equilibrium node. Each path has an upper envelope that gives the potential energy needed. If we consider all the upper envelopes over all the paths and take their lower envelope, this gives the minimum potential energy needed to release the part as a function of contact separation. The difference between this lower envelope and function $U(n_0, n_1, \sigma)$ is the threshold potential energy $U_A(\sigma)$ of the deform-closure grasp.

Recall from the previous section that we computed e_{max}, the maximum strain for a unit reduction of jaw separation from the undeformed configuration with the contacts at n_0 and n_1. We use this to determine σ_L as $\sigma_0(n_0, n_1) - (\varepsilon / e_{max})$. The additional potential energy needed for the contacts to squeeze the part to the elastic limit is given by $U_L(\sigma) = ½\, k_{01} (\sigma_L^2 - \sigma^2)$. The quality metric is given by $Q(\sigma) = \min\{U_A(\sigma), U_L(\sigma)\}$.

$Q(\sigma)$ is a continuous, piecewise differentiable function. However, the number of pieces may be exponential in the number of graph vertices. Hence, computing the exact function and maximizing the metric would require exponential time. Instead, we numerically sample the interval $[0, \sigma_0(n_0, n_1)]$ to maximize $Q(\sigma)$. We define $\lambda = \max\{k_{ab}\, \sigma_0(a, b)\}$. With uniform sampling at intervals of size $\xi = \varepsilon/\lambda$, it can be shown that the error in the optimal metric is at most ε.

For all $\sigma_i = i\ \varepsilon/\lambda$, i = 0, … , $\lceil (\sigma_0(n_0, n_1)\lambda/\varepsilon) \rceil$, we find the threshold potential energy using the algorithm described below.

4.6 Computing Threshold Potential Energy

We are given a value σ_i of contact separation. As in the previous section, consider a path from $v(n_0, n_1)$ to equilibrium configuration $v(n_a, n_b)$. Let U_{max} be the maximum potential energy over all nodes traversed by this path. There are many equilibrium vertices and many such paths. Let U_p be the minimum of the U_{max} values over all such paths. The threshold potential energy $U_A(\sigma_i)$ is given by $U_p - U(n_o, n_1, \sigma_\iota)$.

We determine $U_A(\sigma_i)$ using the contact graph G by an algorithm inspired by Dijkstra's shortest path algorithm. We maintain a set Ω of vertices for whom we known minimum work needed to get to, and a list U_{min} of these minimum works. We also compute a list Ψ of estimated minima of work needed to reach vertices in $V - \Omega$ that are adjacent to vertices in Ω. The estimate is made assuming that in the path used to reach the node, the previous node lies in Ω. We then update Ω by adding the graph vertex with minimum Ψ.

Sub-procedure $U_A(\sigma_\iota)$:

1. Initialize $\Omega \leftarrow \{v(n_0, n_1)\}$, $U_{min}(v(n_0, n_1)) = 0$, $\Psi(v(n_a, n_b)) = U(n_0, n_1, \sigma_\iota) - U(n_0, n_1, \sigma_\iota)$ for all vertices $v(n_a, n_b)$ adjacent to $v(n_0, n_1)$, $\Psi(v(n_a, n_b)) = \infty$ for others.
2. While $\Omega \neq V$
3. Do $v(n_c, n_d) \leftarrow arg\_\min\{\Psi(v(n_a, n_b))\}$
4. $\Omega \leftarrow \Omega \cup \{v(n_c, n_d)\}$
5. $U_{min}(v(n_c, n_d)) = \Psi(v(n_c, n_d))$
6. If $v(n_c, n_d)$ is an equilibrium vertex, return $U_{min}(v(n_c, n_d))$ and stop.
7. For each $v(n_a, n_b)$ in $V - \Omega$ adjacent to $v(n_c, n_d)$, update $\Psi(v(n_a, n_b)) \leftarrow \min\{\max\{\Psi(v(n_c, n_d)), U(n_a, n_b, \sigma_\iota) - U(n_0, n_1, \sigma_\iota)\}, \Psi(v(n_a, n_b))\}$

We can prove that U_{min} always contains the minimum increase in potential needed to reach any element in Ω as follows: Initially, when $\Omega = \{ v (n_0, n_1)\}$, this is true. When a vertex $v (n_c, n_d)$ is added to Ω, assume that U_{min} is larger than the minimum work needed to reach $v (n_c, n_d)$. This implies that in the path that needs to be taken to reach $v (n_c, n_d)$ to minimize the additional potential energy needed, the graph vertex preceding $v (n_c, n_d)$ is in $V - \Omega$. In the sequence of graph vertices represented by the path, consider the first graph vertex $v (n_a, n_b)$ in $V - \Omega$. $\Psi(v (n_a, n_b)) \geq \Psi(v (n_c, n_d))$. Thus, $U (v (n_c, n_d)) \geq \Psi(v (n_c, n_d))$. This is against the assumption that $U_{min} (v (n_c, n_d))$ is not the minimum possible.

4.7 Numerical Example

We consider a part shown in Figure 4. The part material is rubber with the standard Young's modulus of 2 MPa, and Poisson ration of 0.1. The strain at the elastic limit was taken as 0.8. The undeformed distance between the mesh nodes shown is 10 mm. We modeled the part using I-deas and ANSYS and determined σ_ε for the mesh shown. Each FEM problem was solved in about 0.28 seconds. The optimal contact separation is 5.6 mm.

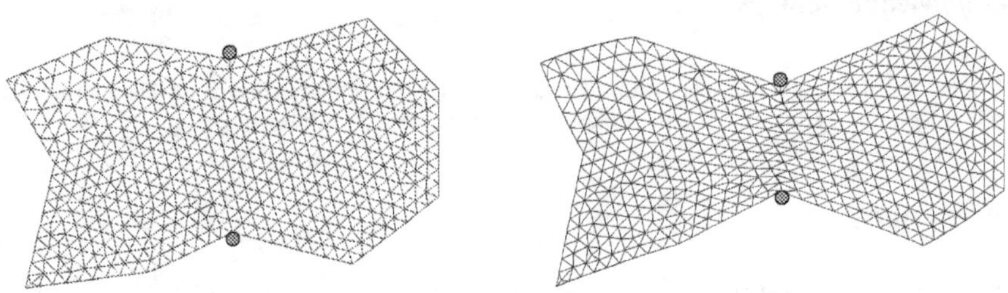

(a) (b)

Fig. 4. (a) Example part with initial jaw separation 10mm for the perimeter mesh nodes shown. (b) The same part at optimal jaw separation at optimal jaw separation 5.6 mm.

5 Discussion

5.1 Complexity

As shown by [Ramamurti98], FEM systems can be solved exactly in $O(n^3)$ time for a part with n mesh nodes. For a part with n mesh nodes and p perimeter nodes, the graph used for the algorithm has $O(p^2)$ nodes and $O(p^2)$ edges. Thus, pre-computing the k_{ab} values for each graph vertex requires $O(n^3 p^2)$ time. Computing U_A for a known separation requires $O(p^2 \log p)$ time when the graph is implemented as a binary heap (this complexity is identical to that of Dijkstra's algorithm for sparse graphs). This iteration needs to be performed (σ_0 (0, 1) λ / ε) times for each grasp, sampling at points with distance ε/λ between them. Thus, σ_ε can be computed in $O(n^3p^2 + p^2/\varepsilon \log p)$ time.

5.2 Conclusion

We defined a new measure of grasp quality for two-point deform-closure grasps. This metric is based on balancing the potential energy needed to release the part against the potential energy that would result in plastic deformation. Given two jaw contacts at the perimeter nodes, we give a numerical algorithm to determine the optimal jaw separation based on this metric. For a part with *n* mesh nodes and *p* perimeter nodes, the algorithm computes an approximation to the optimal separation in time $O(n^3 p^2 + (p^2/\varepsilon) \log p)$.

We are currently studying how U_A varies with increasing mesh density, and the minimal mesh needed to define deform-closure. We are also investigating the optimal choice of initial contact nodes and the accounting for friction in our algorithm.

Acknowledgements

We thank Frank van der Stappen, Rama Koganti, Matt Zaluzec, Ron Alterovitz, Dezhen Song and Vladlen Koltun and the anonymous reviewers for their valuable comments.

References

[Bicchi00] A. Bicchi and Vijay Kumar, Robotic Grasping and Contact: A Review, Proceedings of IEEE International Conference on Robotics and Automation, pp348-353, 2000.

[Brost96] R.C.Brost and K.Y.Goldberg, "A complete algorithm for designing planar fixtures using modular components," IEEE Trans. On Robotics and Automation, vol. 12, no. 1, pp. 31 – 46, 1996.

[Cai96] Cai W., Hu S.J., Yuan J.X., Deformable sheet metal fixturing: principles, algorithms, and simulations. Transactions of the ASME. Journal of Manufacturing Science and Engineering, vol.118, (no.3), ASME, Aug. 1996. p.318-24.

[Cheong02] Cheong, J.-S., Goldberg K., Overmars M.H. and van der Stappen, A.F., Fixturing Hinged Polygons, Proceedings, IEEE International Conference on Robotics and Automation, pp. 876-881, 2002.

[Cheong03] Jae-Sook Cheong, Herman J. Haverkort, and A. Frank van der Stappen, On Computing All Immobilizing Grasps of a Simple Polygon with Few Contacts accepted for ISAAC (International Society for Analysis, its Applications and Computation), 2003.

[Corman90] T.H. Corman., C.E. Leiserson. and R.L. Rivest, Introduction to Algorithm, MIT Press, 1990.

[Doulgeri04] Doulgeri, Z. and Peltekis, J., Modeling and Dual Arm Manipulation of a Flexible Object, IEEE International Conference on Robotics and Automation, April 2004.

[Gopalakrishnan02] K. “Gopal” Gopalakrishnan, Ken Goldberg, Gripping parts at concave vertices, IEEE International Conference on Robotics and Automation, 2002, Page(s): 1590 - 1596, Volume: 2 , 2002.

[Gopalakrishnan03] K. “Gopal” Gopalakrishnan, Matthew Zaluzec, Rama Koganti, Patricia Deneszczuk and Ken Goldberg, “Unilateral” Fixturing of Sheet Metal Parts Using Modular Jaws with Plane-Cone Contacts, Proceedings, IEEE International Conference on Robotics and Automation, pp 3953-3958, 2003.

[Gopalakrishnan04] K. "Gopal" Gopalakrishnan and Ken Goldberg, D-Space and Deform Closure: A Framework for Holding Deformable Parts, IEEE International Conference on Robotics and Automation (ICRA), May 2004.

[Guibas99]Guibas, L.J.; Holleman, C.; Kavraki, L.E.; A probabilistic roadmap planner for flexible objects with a workspace medial-axis-based sampling approach, Intelligent Robots and Systems, 1999. IROS '99. Proceedings. 1999 IEEE/RSJ International Conference on , Volume: 1 , 17-21 Oct. 1999, Page(s): 254 -259 vol.1.

[Heinrich00]Heinrich D. and Worn H., editors, Robot manipulation of deformable objects, Springer Verlag, May 2000.

[Hirai01] Hirai S., Tsuboi T., Wada T., Robust Grasping Manipulation of Deformable Objects, Proc. IEEE Symposium on Assembly and Task Planning, pp.411-416, Fukuoka, May, 2001

[Holleman98] Holleman, C.; Kavraki, L.E.; Warren, J., Planning paths for a flexible surface patch, Robotics and Automation, 1998. Proceedings. 1998 IEEE International Conference on , Volume: 1 , 16-20 May 1998, Page(s): 21 -26 vol.1.

[Howard97] Howard, A.M. and Bekey, G.A., Recursive Learning for Deformable Object Manipulation,8th International Conference on Advanced Robotics, pp. 939-944, Monterey, CA, July 7-9, 1997.

[Lamiraux99] Lamiraux, F.; Kavraki, L.E.; Path planning for elastic plates under manipulation constraints, Robotics and Automation, 1999. Proceedings. 1999 IEEE International Conference on , Volume: 1 , 10-15 May 1999, Page(s): 151 -156 vol.1.

[Li02] Li B., Shiu B.W., Lau K.J., Fixture configuration design for sheet metal assembly with laser welding: a case study. International Journal of Advanced Manufacturing Technology, vol.19, (no.7), Springer-Verlag, 2002. p.501-9.

[Li02b] Li H. F., Ceglarek D., Shi J., A Dexterous Part-Holding Model for Handling Compliant Sheet Metal Parts, ASME Transactions, Journal of Manufacturing Science and Engineering. Vol. 124, No. 1, pp. 109-118, 2002.

[Markenscoff90] X. Markenscoff, L. Ni and C. H. Papadimitriou, The Geometry of Grasping, International Journal of Robotics Research, Vol. 9, No. 1, pp 61-74, 1990.

[Mason01]Mason M.T., Mechanics of Robotic Manipulation, The MIT Press, 2001.

[Menassa91] Menassa R., De Vries W., Optimization Methods Applied to Selecting Support Positions in Fixture Design, ASME Journal of Engineering for Industry, vol 113, pp. 412-418, 1991.

[Milgram02] Milgram, R.J. and Trinkle, J.C., "The Geometry of Configuration Spaces for Closed Chains in Two and Three Dimensions," Homology Homotopy and Applications, 2002.

[Mishra87]B. Mishra, J. Schwarz, and M. Sharir, On the existence and Synthesis of Multifinger Positive Grips, Algorithmica 2, 1987.

[Moll04] Moll, M. and Kavraki L., Path Planning for Minimal Energy Curves of Constant Length, Proceedings, IEEE International Conference on Robotics and Automation, 2004.

[Ramamurti98] Ramamurti V., Computer Aided Mechanical Design and Analysis, McGraw Hill Book Company, July 1998.

[Reuleaux1876] F. Reuleaux, The Kinematics of Machinery. New York: Macmillan 1876, republished by New York: Dover, 1963.

[Rimon96]Rimon E. and Burdick J., On force and form closure for multiple finger grasps, Proceedings of IEEE International Conference on Robotics and Automation, 1996, pp. 1795 -1800 vol.2.

[Rimon98]Elon Rimon and Joel Burdick, Mobility of bodies in contact - I, IEEE transactions on Robotics and Automation, 14(5): 696-708, 1998.

[Rimon99] Rimon, E. and Blake, A., Caging planar bodies by one-parameter two fingered gripping systems, International Journal of Robotics Research, v18, n3, March 1999, pp. 299-318.

[Rong99] Rong, Y. and Zhu, Y., "Computer-aided Fixture Design", Marcel Dekker, ISBN: 0-8247-9961-5, New York, 1999.

[Selig00] J. M. Selig. Geometrical Foundations of Robotics. World Scientific, 2000.

[Shiu97] Shiu, B.W., Ceglarek, D., Shi, J., 1997 "Flexible Beam-Based Modeling of Sheet Metal Assembly for Dimensional Control," Trans. of NAMRI, Vol. XXV, pp. 49-54.

[Shiu03] Shiu, B.W., Apley, D., Ceglarek, D., Shi, J., 2003 "Tolerance Allocation for Sheet Metal Assembly using Beam-Based Model", Trans. of IIE, Design and Manufacturing,, Vol. 35, No. 4, pp. 329-342. (April 2003). The paper was also summarized and featured in the March 2003 edition of Industrial Engineer (formerly IIE Solutions).

[Somoff1900] P. Somoff, Uber gebiete von schraubengeschwindigkeiten eines starren korpers bieverschiedener zahl von stuz achen, Zeitschrift fur Mathematic and Physik, vol. 45, pp. 245-306, 1900.

[vanderStappen99] Van der Stappen A.F., Wentink C., Overmars M.H., Computing form-closure configurations, Proceedings of IEEE International Conference on Robotics and Automation, vol.3, pp. 1837 -1842, 1999.

[Wakamatsu96] Wakamatsu, H., Hirai, S., and Iwata, K., Static Analysis of Deformable Object Grasping Based on Bounded Force Closure, Proc. IEEE Int. Conf. on Robotics and Automation, Vol.4, pp.3324-3329, Minneapolis, April, 1996.

[Zhu03] Zhu X., Ding H., Wang J., Grasp analysis and synthesis based on a new quantitative measure, IEEE Transactions on Robotics and Automation, vol. 19, no. 6, pp. 942-953, Dec. 2003.

Coordinating Multiple Droplets in Planar Array Digital Microfluidics Systems

Eric Griffith and Srinivas Akella

Rensselaer Polytechnic Institute, Troy, New York 12180, USA

Abstract. This paper presents an approach to coordinate the motions of droplets in digital microfluidic systems used for biochemical analysis. A digital microfluidic system typically consists of a planar array of cells with electrodes that control the droplets. The primary challenge in using droplet based systems is that they require the simultaneous coordination of a potentially large number of droplets on the array as the droplets move, mix, and split. This paper describes a general-purpose system that uses simple algorithms and yet is versatile. First, a semi-automated approach to generate the array layout in terms of components is explained. Next, simple algorithms to select destination components for the droplets and a decentralized scheme for components to route the droplets on the array are discussed. These are then combined into a reconfigurable system that has been simulated in software to perform DNA polymerase chain reaction and other analyses. The algorithms have been able to successfully coordinate hundreds of droplets simultaneously and perform one or more chemical analyses in parallel. Since it is challenging to analytically characterize the behavior of such systems, methods to detect potential instabilities are proposed.

1 Introduction

The field of microfluidics has been revolutionized recently by the creation of miniature biochemical analysis systems using microfabrication technology. These systems are often termed *micro total analysis systems* or "lab on a chip" systems. These systems offer a number of advantages, including size reduction, power reduction, and increased reliability. However, current systems are typically tailored to a specific task. Therefore an important goal is to create reconfigurable and reprogrammable systems capable of handling a variety of analysis tasks.

Digital microfluidics systems (DMFS) that use techniques such as electrowetting and dielectrophoresis are promising candidates for reconfigurable systems ([21], [14],[19], [7]). We focus on microfluidic systems that manipulate discrete droplets by electrowetting, where the interfacial tension of the droplets is modulated with a voltage [19]. Droplets are microliters in volume, and have been moved at 12–25 cm/sec on planar arrays of 0.15 cm wide electrodes. (See [10] and [7] for details.) The ability to control individual droplets on a planar array enables complex analysis operations to be performed in biochemical "lab-on-a-chip" systems (Figure 1). For example, they can be used to perform polymerase chain reactions for DNA sequence analysis. For many

M. Erdmann et al. (Eds.): Algorithmic Foundations of Robotics VI, STAR 17, pp. 219–234, 2005.

biochemical analysis operations, no special purpose devices are required aside from the array itself. These systems have the potential to process hundreds of samples quickly. The primary challenge of using droplet based systems is developing algorithms for the simultaneous coordination of a potentially large number of droplets. Planning optimal paths through the array for each droplet would be computationally intractable for a large number of droplets.

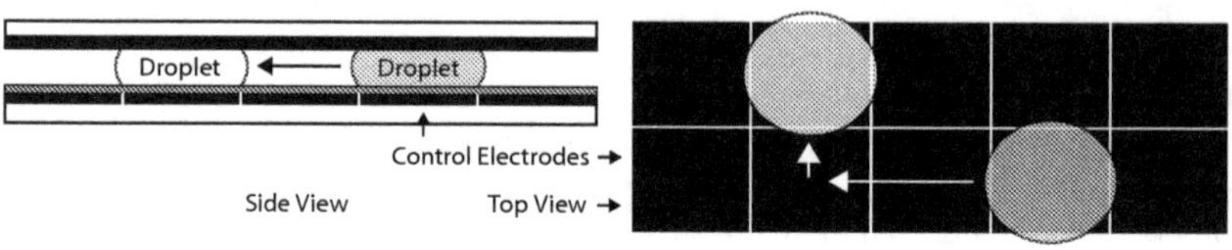

Fig. 1. Droplets on an electrowetting array (side and top views). Droplet motion is controlled by turning the control electrodes on and off. Based on [19].

This paper describes an approach to creating a general-purpose DMFS. First, a semi-automated approach to design the array layout in terms of modular components is explained, along with the motivating design choices. Next, simple algorithms to select destination components for the droplets and a decentralized scheme for components to route the droplets are discussed. These are then combined into a reconfigurable system that has been simulated in software to perform DNA polymerase chain reaction and other analyses. The algorithms have been able to successfully coordinate hundreds of droplets simultaneously and perform one or more chemical analyses in parallel. Since it is challenging to analytically characterize the behavior of such systems, methods to detect potential instabilities due to congestion are proposed.

2 Related Work

Digital Microfluidic Systems: Pollack, Fair, and Shenderov [21] demonstrated rapid manipulation of discrete microdroplets by electrowetting-based actuation. Ding, Chakrabarty, and Fair [8] described an architectural design and optimization methodology for scheduling biochemical reactions using electrowetting arrays. They identified a basic set of droplet operations and used an integer linear programming formulation to minimize completion time. Droplet paths and areas on the array for storage, mixing and splitting operations are predefined by a human. Fair et al. [10] describe experiments on injection, dispensing, dilution, and mixing of samples in an electrowetting DMFS. Cho, Moon, and Kim [7] demonstrated creating, merging, splitting, and move operations using electrodes covered with dielectrics, and identified conditions under which these operations can be performed in an air environment. Fan, Hashi, and Kim [11] developed a cross-reference grid of single layer electrodes to manipulate droplets with limited row-column addressing.

Paik, Pamula, and Fair [19] studied the effects of droplet aspect ratios and mixing strategies on the rate of droplet mixing. Jones et al. [14] demonstrated dielectrophoresis based liquid actuation and nanodroplet formation.

Multiple robot coordination: The coordination of droplets in a DMFS is closely related to multiple robot motion coordination. However very few motion planning algorithms can guaranteeably coordinate more than 10 or 20 robots. Each droplet in a DMFS can be viewed as a simple robot that moves on a 4-connected array (Bohringer [4,5]). Bohringer outlines an approach for moving droplets from start to goal locations, subject to droplet separation constraints, obstacles, and control circuitry limitations. He uses an A* search algorithm to generate optimal plans for droplets. To overcome the exponential complexity of this approach, he plans the droplet motions in prioritized order ([9]). However note that in a DMFS, the droplets must additionally sometimes combine for different durations to mix, and then be split.

Hopcroft, Schwartz, and Sharir [13] showed that even a simplified two-dimensional case of motion planning for multiple robots is PSPACE-hard. Reif and Sharir [22] and Kant and Zucker [15] developed techniques to plan motions of a single robot among moving obstacles. This can be generalized to obtain a heuristic solution for planning the motions of multiple robots. Erdmann and Lozano-Perez [9] order robots by assigned priority and sequentially search for collision-free paths. Recent efforts have focused on probabilistic approaches. Svestka and Overmars [26] developed a PRM planner for path coordination of multiple car-like robots. Sanchez and Latombe [23] use lazy PRM variants for coordinated path planning of multiple robot arms.

When the paths of the robots are specified, a path coordination problem, first studied by O'Donnell and Lozano-Perez [18] for two robots, arises. LaValle and Hutchinson addressed a similar problem in [16] where each robot was constrained to a C-space roadmap during its motion. Simeon, Leroy, and Laumond [25] coordinate over 100 car-like robots, where robots with intersecting paths are partitioned into smaller sets. Akella and Hutchinson [1] developed a mixed integer linear programming (MILP) formulation for the trajectory coordination of 10–20 robots by changing robot start times. Peng and Akella [20] developed an MILP formulation to coordinate many robots with simple double integrator dynamics along specified paths. Conflict resolution among multiple aircraft in a shared airspace (Tomlin, Pappas, and Sastry [28], Bicchi and Pallottino [3], Schouwenaars et al. [24]) is also closely related to multiple robot coordination.

3 Components for Array Layout

Many common operations for biochemical analyses can be performed on an array without additional special purpose hardware. These operations include: dispensing droplets onto the array, collecting droplets from the array, transporting droplets around the array, mixing droplets together, and splitting droplets apart. The array layout design presented here uses a system of "virtual" components. Each component type is responsible for performing one or more types of operation. Multiple instances of the same type of component can be present, and each component instance is allotted an area of the array to perform its operations in. By linking components that can perform all the operations in an analysis, a DMFS can be created to perform that analysis.

In our system, we have defined six component types. The *street*, *connector*, and *intersection* components transport droplets around the array. The *source* component adds droplets to the array, and the *sink* component removes droplets from the array. The *work area* component manages mixing and splitting of droplets. New component types can be defined and integrated into the system for operations that do not require special purpose hardware.

3.1 The Components

This section provides an overview of the functionality of each component type. Each component is responsible for managing all of the droplets that are in its portion of the array. At each clock cycle, components attempt to move all droplets in their section of the array. Each component maintains connections with its neighboring components, which are used to pass control of droplets when they move between components. The connections are entrance/exit pairs where the exit from one component is the entrance into another. Components may have to wait to move their droplets until the components they are connected to have moved theirs. Figure 2 shows an example system with each of the component types present. The details of the layout itself are presented in Section 5. Figure 3 depicts the individual components.

The Street Component: The street component is the general-purpose droplet transportation component. It moves droplets in one direction through at least two array cells. Streets are one-way to prevent two droplets from moving in opposite directions through the component. A street attempts to advance all droplets within it in synchrony at each clock cycle. If moving any droplet would cause it to be adjacent to another, it is not allowed to move.

The Connector Component: The connector component is similar to a street component, except that a droplet only moves through a single cell in it. The distinction is made because droplets in the connector are adjacent to two components simultaneously. Thus, when a droplet attempts to enter a connector, the connector must ensure that there is no other droplet adjacent to the connector.

The Intersection Component: The intersection component is vital to the droplet routing system. Droplets enter an intersection and then move into its middle cell. Once there, the intersection routes the droplet to the appropriate exit based on the algorithm described in Section 4.3.

The Work Area Component: The work area component is where mixing and splitting take place. Each work area has a transit area and multiple *work units*. Each work unit may function as a *mixer* and/or as a *splitter*. Mixers

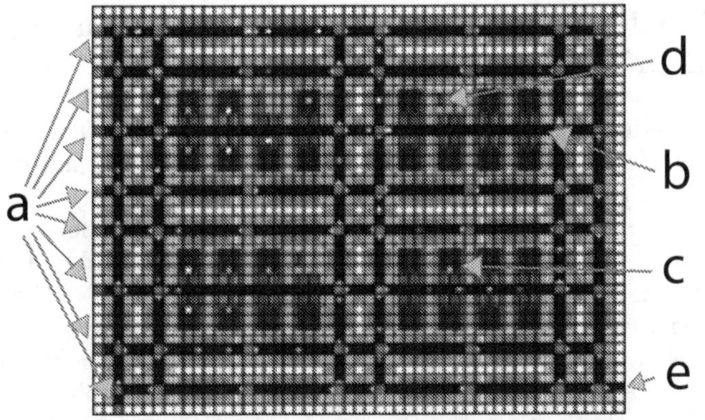

Fig. 2. Array layout for the PCR reaction described in Section 7. On the left side of the array are **(a)** eight sources, which supply the input sample droplets to the system. There are **(b)** four work areas on the array, in which droplets are **(c)** mixed together and **(d)** split apart. In the lower right corner of the array is a **(e)** sink, which collects the droplets of the final products and moves them off the array.

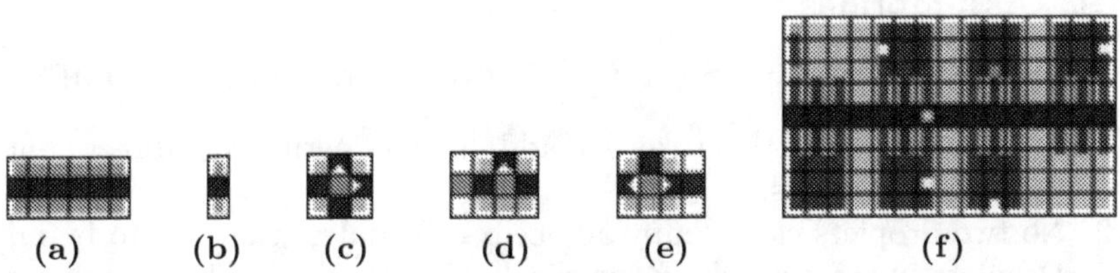

Fig. 3. The components. **(a)** A street component. Droplets move through the dark blue squares. **(b)** A connector component. **(c)** An intersection component with two entrances and two exits. The arrowheads indicate the exits. **(d)** A source component connected to an intersection component. **(e)** An intersection component connected to a sink component. **(f)** An active work area, showing several work units with droplets. Each work unit can function as a mixer and as a splitter.

merge two droplets into a new droplet. The mixer moves the droplet around to speed up mixing, as in [19]. Splitters split the mixed droplet into two droplets. A work unit may be used as a mixer until a mixing operation is completed, and then part of that same area may be used as a splitter. A work area can mix and split multiple droplets at the same time. When a droplet gets to a work area, the work area sends it to a specific work unit. The work units manage the droplets assigned to them. Once a mix and split operation is complete, the resulting droplets are sent out of the work area.

The Source Component: The source component represents droplet entry points into the array. Each source introduces specified droplet types at specified intervals. Droplets entering the array are assigned a goal operation.

The Sink Component: The sink component represents droplet exit points from the array. Each sink removes specified droplet types from the array.

4 Droplet Destination Selection and Routing Algorithms

To be practical, a DMFS should be able to handle a large number of droplets simultaneously. Ideally, it should do this in a way that optimizes some quantity, such as throughput or completion time. We significantly reduce the computational cost of planning the droplet motions at the expense of optimality. By dividing the array into components, restrictions are placed on where operations can take place on the array. The interconnection of the components can be viewed as a network with the intersections as the routing devices and the streets and connectors as the "wires." This reduces the motion planning problem to a network routing problem on a directed graph.

Selecting destinations for droplets is a two-fold process. First, the droplet must be assigned an operation to perform. Once the droplet has an assigned operation, a component for that operation may be selected. The methods for these processes, as well as the droplet routing, are discussed in this section.

4.1 Assumptions

We assume the following when modeling the operations of the DMFS:

1. Individual cells of the array are addressable, permitting direct control of individual droplets.
2. No two droplets can occupy adjacent cells unless they are to be mixed.
3. Droplets in the array have identical volumes, except during mixing.
4. Every mix operation is followed by a split operation.
5. Mixing operations take longer to perform than transport operations.
6. All the operations can be performed on the array without additional special purpose hardware.

4.2 Droplet Destination Selection

The system is supplied with parameters, described in Section 5, which it uses to maintain a list of components available for certain operations. Work areas can perform a mixing operation with any droplet type, and sinks remove specific types of droplets. Each work area and sink adds itself to a sorted list of components accepting droplets for operations. There is also a sorted list of higher priority containing requests from components for specific droplet types required to complete an operation. Currently, only work areas needing one of two droplet types for a mixing operation place requests in this list.

When a new droplet enters the system, or is created through a mixing operation, the corresponding source or work area assigns it an operation to perform. When the droplet enters an intersection, the intersection tries to find a component to send the droplet to. First, it checks the higher priority list for existing requests for the droplet's type and operation to perform. If any exist, then the droplet is assigned to the first requesting component and that request is removed from the list. Failing that, it is assigned to the first component that can accept droplets of its type, and that component is sent to the end of the list (if it is a sink or is a work area with work units available). If no components are available, then the next intersection the droplet enters attempts to assign it a destination. In the worst case scenario, this operation takes $O(w+s)$ time for one droplet at one intersection, where w is the number of work areas and s is the number of sinks in the system.

4.3 Droplet Routing

After a droplet is assigned a destination component, the droplet must be routed to the component. The routing method we use can be viewed as a deflection routing variant of the Open Shortest Path First (OSPF) network protocol ([6], [27]). The routing method relies on each intersection maintaining shortest path information to the other components, computed from the component graph. The *component graph* is a directed graph where each node is a component and each edge is a connection between adjacent components. The directed edge points from the component with the exit to the component with the entrance. The distance along an edge is, generically, taken to be the length of the component containing the exit for the edge.

When the system is initialized, each intersection computes and stores its routing table, which maps the shortest path to each component to a corresponding exit to take from the intersection. The specified exit is the first leg of the shortest path to the component in question. Each intersection constructs its routing table by running Dijkstra's algorithm on the component graph to compute shortest paths from the intersection. The one restriction is that a shortest path should not travel through work areas, sources, or sinks, unless those components are the destination. Dijkstra's algorithm on a sparse graph using a standard implementation has a runtime of $O(n \log n)$ where n is the number of nodes in the graph. Since Dijkstra's algorithm must be run from each of the intersections, the routing system has an initial overhead of $O(i \cdot n \log n)$ where i is the number of intersections in the array.

At each clock cycle, the intersections are processed in a fixed (random) order to select their droplet routing moves. Subsequently, synchronous motion of droplets is executed. The droplet routing is straightforward once the routing table has been constructed. If a droplet entering the intersection has no destination, then the intersection attempts to assign it one. If that fails, then the droplet is sent to a random, valid exit. Otherwise, the intersection finds the destination component in its routing table and selects the exit that is the best choice for the droplet. If the droplet is able to move toward that exit, it does so. Otherwise, the intersection randomly chooses a valid exit for the droplet. If no viable exit is available, then the droplet waits. The amount of time required for selecting an exit for a droplet is $O(1)$ because the intersection must check at most three possible exits and the routing table has constant access time.

5 A General-Purpose Digital Microfluidics System

A general-purpose DMFS can be created by combining the component based layout design approach (Section 3) with the accompanying algorithms (Section 4). A general-purpose layout must be capable of handling arbitrary analyses that require the movement, mixing, and splitting of different types of droplets. Further, the layout should contain a sufficient number of work areas and be able to efficiently transport droplets around the array.

Efficient droplet transportation is crucial because the usage of the system is not known in advance. Therefore, all parts of the array should be accessible. We group street components in pairs to simulate two-way streets, allowing droplets to closely follow the same path between two locations in both directions. To allow for intersections between these two-way streets, we group four intersections together in a rotary-like arrangement. See Figure 4.

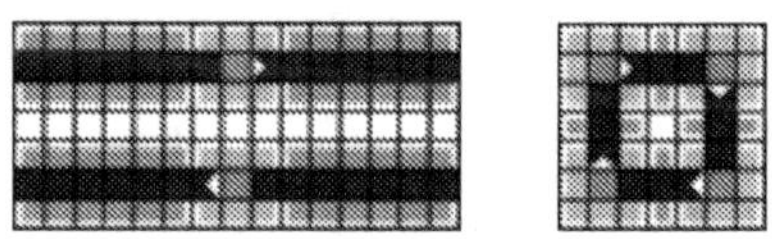

(a) (b)

Fig. 4. Simulating two-way transportation: **(a)** Two-way street **(b)** Rotary.

We group these two-way streets and rotaries with a work area to form a pattern of components. This pattern, shown in Figure 5, is designed to provide a balance between number of work areas and ease of access. The array layout itself is a periodic tiling of this pattern, completed with an alternating sequence of rotaries and streets along its upper and right edges. The extra intersection components in the two-way streets are for connecting sources and sinks around the perimeter of the layout. The intersections in the vertical streets are connected to provide shorter paths to and from work areas.

To generate the layout, the user must specify a set of parameters dependent on the hardware. These are the physical size of the array and the locations of sources and sinks. To fully define the system, the user specifies parameters based on the chemical analyses to be performed, including the types of droplets introduced at sources, when and how often they are produced, the types of droplets to send to the sinks, and information about the various intermediate operations to perform with the droplets on the array.

The size of the array is limited to be of width $9 + 22i$ and height $9 + 16j$, where the layout is i tiles wide and j tiles tall. The pattern tile is 22 cells

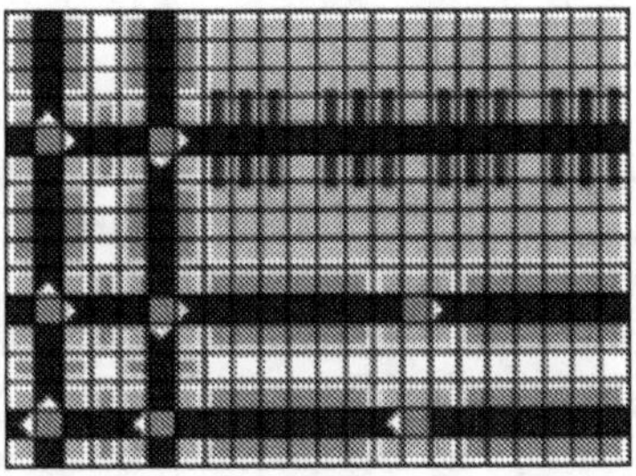

Fig. 5. The layout pattern tile that is a modular building block for the array.

wide and and 16 cells tall. The sequence of two-way streets and rotaries that completes the layout adds an additional 7 units. The sources and sinks must be placed around the perimeter of the array such that they can be connected to an intersection. The remaining 2 units come from setting aside one square around all sides of the perimeter of the pattern for sources and sinks. A complete example layout with eight sources and one sink can be seen in Figure 2.

5.1 DMFS Operation

The DMFS can be operated in two modes:

1. **Batch mode:** Here all the droplets necessary for a reaction are input at the source components in one batch. The droplets are coordinated to complete the reaction, and then the next batch of droplets is processed. The droplets are input in a synchronized manner based on when they are required for the reactions. After each mix and split operation, one of the two resulting droplets is sent to a waste output. A reaction performed in batch mode will typically require a smaller number of tiles in the array since the number of droplets in the system is small.
   One advantage of batch mode is that the droplet routing is almost entirely deterministic, and we can easily analyze the system behavior. Only when no work units are available for a droplet will an intersection component route the droplet randomly.
2. **Continuous mode:** Here the source components input the droplets at a fixed rate. (The rate for each droplet type is specified by the human designer.) One advantage of this mode is that it produces a larger volume of product droplets than the batch mode in the same amount of time, especially when no droplets are discarded as waste droplets. Another advantage is that multiple reactions yielding different products can be processed in parallel to better utilize the array. A potential disadvantage is that system behavior is harder to analyze. We explore this issue, and in particular, the likelihood of system instability, in Section 6.

6 Stability Analysis

The behavior of a general-purpose system changes with the chemical analysis it performs. A system operating in continuous mode may or may not be stable depending on its parameters. In an unstable system, droplets enter the system faster than the system is able to process them, and a steady-state flow cannot be guaranteed ([12]). If a system is not stable, in time it will become heavily congested and may finally become deadlocked. Hence instability is to be avoided.

We have identified a set of conditions for a system to be classified as unstable. At least one of the following two conditions must hold for a system to become congested due to instability. The first is for some droplets to be unable to follow the shortest paths to their destinations. The second condition is for droplets to be unable to be assigned a destination. When either condition is met, it is an indication that the system is not processing droplets fast enough. A system is deadlocked when droplets are unable to reach their destination. Deadlock is a sufficient condition for a system to be unstable.

We treat congestion as an indicator of instability and try to identify conditions that lead to or result from congestion. By selecting system parameters such as input droplet rates and source and sink locations to avoid these conditions, the resulting system will likely be stable. We use two methods to identify these conditions. At the operation level, we analyze a graph of the operations to be performed based on the system parameters. At the component level, we model droplet flow in the system using the component graph.

6.1 Analysis Graph

The (biochemical) *analysis graph* provides an organized representation of the behavior of the system. It is a directed graph, with an input node for each droplet type entering the system, an output node for each droplet type leaving the system, and a mix node for each mixing operation performed in the system. The nodes are connected based on the droplet types they require and produce, and the edges represent transport operations. Each node stores the duration of its operation, and the analysis graph is augmented with additional information (Figure 6).

The first augmentation is the rate at which droplets will enter and leave each node. Droplet rate can be most intuitively described in terms of a source. If a source introduces a particular droplet type once every k cycles, then the rate would be $\frac{1}{k}$ droplets per cycle. Also, there would be $k-1$ empty cells between each introduced droplet. This rate is propagated through the graph. If these droplets perform a mixing operation, then, on average and assuming no unusual delays, one of these droplets would begin its mixing operation every k cycles. Similarly, one droplet would complete the mixing operation every k cycles.

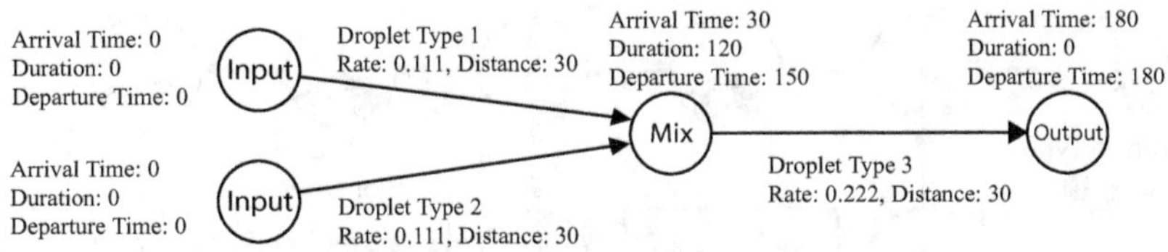

Fig. 6. A simple analysis graph with four nodes: two input nodes, one mix node, and one output node. Droplet types, rates, and expected distances traveled are indicated along the edges. Expected arrival time, duration, and departure time (in units of clock cycles) are indicated at the nodes.

The second augmentation is the best case expected distance to travel between performing operations. Each node's operation can be performed at one or more components on the array; specific components are chosen during destination selection for the droplets. The expected distance is the average of the shortest path between each possible originating component and each possible destination component. In a system with a source component introducing droplets for mixing into a system with two work areas, the distance from the input node to the mix node would be the average of the lengths of the shortest paths from the source to each of the two work areas.

The third augmentation is the best case expected time at which the first droplets will enter each node. There is a separate arrival time from each parent node of a given node, and it is computed as the expected departure time from the parent node plus the best case expected travel time from the parent node to the current node. The expected departure time from a given node is the average of the arrival times, plus the duration of the operation.

The graph is then checked to ensure that the following properties hold:

1. Every path from every source node must lead to a sink node.
2. Droplets must enter a node at the same rate from each parent node.
3. Arrival times to a node should be about the same from each parent node.

If any of these properties is violated, it is likely to result in inefficient processing of droplets. In the event that the properties are violated, the system parameters can be adjusted until they result in a graph without violations.

6.2 Droplet Flow Analysis

The analysis graph provides a reasonable estimate of the overall system stability. However a more detailed component level analysis of system stability may be required. There may be certain bottleneck components that become congested from too many droplets moving through them, and so slow down droplet flow. These bottlenecks can result in an unstable system when they prevent droplets from reaching their destinations. Some simple experiments have demonstrated that this can arise in larger arrays (for example, arrays

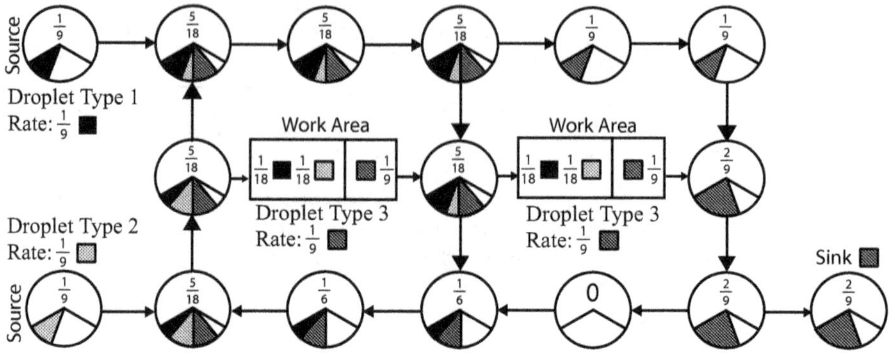

Fig. 7. A component graph for a system with the analysis graph in Figure 6. For clarity, the component graph uses a simplified layout. Each circular node contains a pie chart for flow through the component. The actual flow is the shaded portion of the maximum flow rate of $\frac{1}{3}$ in each node. In this example, inflow is equal to outflow. The total flow through each component is depicted by the fraction in the node. The work areas receive flow from both sources and produce flow destined for the sink. The rate of flow and droplet type (color) produced by each source and work area is indicated by the node.

of size 295×297), where there is a lot of droplet traffic around the perimeter near the sources and sinks but relatively light traffic in the interior. Modeling the droplet flow through the system is an effective tool to identify unstable systems, especially those that are unstable due to these bottleneck, congested components.

The droplet flow modeling attempts to predict the expected flow through the component graph (described in Section 4.3) using the rate information from the analysis graph. (See Figure 7.) The idea is to determine the rate at which droplets enter and leave each component on the array, which is the flow rate through that component, when the system is in a steady state. To approximate the expected flow rate through components, we developed an iterative, network-style analysis. For this analysis, we make the assumption that work areas are uniformly utilized. That is, a droplet being assigned to a work area is equally likely to be assigned to any work area on the array.

Each component is initially assigned inflow and outflow rates of 0. Sources generate a certain amount of flow, dictated by the analysis graph, destined for each work area. For example, if a source produces droplets at a rate of $\frac{1}{4}$ in a system with 2 work areas, it generates a flow of $\frac{1}{8}$ to each work area. Similarly, work areas generate a certain amount of flow destined for each work area and to each sink. At each iteration, the output of each node in the graph is defined as a function of the input to the node. The input of a node is the sum of the outputs, from the previous iteration, of its parent nodes plus any flow it generates at that iteration. For intersections, the input flow is divided amongst the possible exits based on where the flow is destined. If a particular node becomes congested, nodes sending flow to it try to redirect excess flow away.

Nodes in the graph are assigned a maximum inflow capacity. For all nodes except work areas, this is set as $\frac{1}{3}$, which is the maximum rate that droplets may make a turn through an intersection without overlapping. The maximum inflow rate to a work area is computed based on the duration of its operations in the analysis graph. For example, consider a system where mixing operations take 100 cycles to complete. Each mixing operation requires two droplets and each work area can support 8 simultaneous mixing operations. If the droplets entered a work area at a rate of 1 droplet every 7 cycles, the first mixing operation should complete shortly after, or even before, the eighth operation begins. Droplets would likely not be able to enter the work area at a faster rate than this because there would not be any free work units. To respect the maximum inflow rates, parent nodes must adjust their outflows, which may result in some of the inflow not translating into outflow. A congested node is defined as a node with inflow greater than outflow.

Once the computed flow through the system converges, the component graph is analyzed. If the system is stable, the inflow will equal the outflow at every street, connector, work area, and intersection node. Otherwise, the system is likely unstable. A simplified example system with steady state flow information is depicted in Figure 7.

7 Example Systems

We have simulated example systems operating in batch mode and in continuous mode. We now describe an example system based on the polymerase chain reaction operations outlined in [8]. The reaction involves eight input droplet types and seven mixing operations. See Figure 8 for an analysis graph of the system. Immediately following each mixing operation, the resulting droplet is split into two droplets. The array is set up with four work areas, eight sources, each introducing an input droplet type, and one sink to collect the final product (Figure 2). The size of the array in this example is 53×41. The system parameters were set with the aid of the stability analysis described above. In continuous mode, the system has an average of 66 droplets on the array. The routing computations are performed at a rate of 73000 cycles a second, enabling rapid simulation of the system to verify stability. Animations of the PCR reaction, in both batch mode and continuous mode, as well as other systems, including those running multiple reactions, are available at `www.cs.rpi.edu/~sakella/microfluidics/`.

8 Conclusions and Future Work

We described a new approach to creating a general-purpose DMFS by dividing the planar array into a collection of components and coordinating

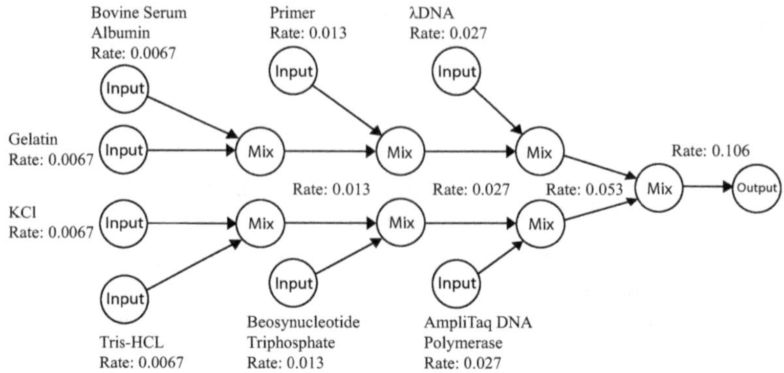

Fig. 8. PCR analysis graph. Input nodes are labeled with the samples they introduce and the rate at which they introduce them. Edges out of mix nodes are labeled with the droplet rate resulting from the operation.

the motions of droplets by implementing a decentralized routing algorithm. We explored techniques to identify potentially unstable systems, and applied them to successfully demonstrate a stable DNA polymerase chain reaction in simulation.

The system described here can semi-automatically generate a layout given a set of system parameters, and then perform real-time droplet manipulation. It has successfully coordinated hundreds of droplets, and is a proof of concept that decentralized network-like motion planning can work for a digital microfluidics system. Further, the software can be easily modified to act as a controller for a physical array. The same array can perform a variety of chemical analyses, and has been demonstrated to even perform multiple analyses in parallel.

The current system can be enhanced in a number of ways for greater flexibility and efficiency. The overall design of the components and the system allows for the introduction of new component types. For example, sensing components that permit sensor monitoring of the reactions, and storage components, where droplets can be temporarily stored, can be incorporated. The system can also support simpler hardware that permits only limited row-column addressing of electrodes ([11]). Automatically sequencing operations to achieve desired droplet concentrations would be a useful extension. Automatically generating the array for a given reaction requires optimizing the number of tiles and their layout, as well as the locations of the sinks and sources on the array. Optimizing tile design to optimize droplet flow rates, and optimizing array layout for a given tile pattern are interesting problems.

Techniques for stability and performance analysis, drawing upon methods from queueing theory and networking ([17], [6], [2]), can be applied to this system to provide insight. Detailed analysis of the system can help identify components where congestion or deadlock may occur, and automatically set droplet input rates to avoid such problems. More sophisticated dynamic routing techniques can be explored to more effectively load balance the transport sections of the array and reduce bottlenecks at components. The design and control of fully reconfigurable arrays, where any part of the array may be used for any desired operation, pose particularly interesting challenges.

Acknowledgments

Many thanks to Karl Bohringer for introducing us to this problem and providing encouragement and advice along the way. We benefited from discussions on network models with Bulent Yener, Biplab Sikdar, and Jayasri Akella. This work was supported in part by NSF under CAREER Award No. IIS-0093233.

References

1. S. Akella and S. Hutchinson. Coordinating the motions of multiple robots with specified trajectories. In *IEEE International Conference on Robotics and Automation*, pages 624–631, Washington, DC, May 2002.
2. D. P. Bertsekas and R. G. Gallagher. *Data Networks*. Prentice-Hall, Englewood Cliffs, N.J., second edition, 1992.
3. A. Bicchi and L. Pallottino. On optimal cooperative conflict resolution for air traffic management systems. *IEEE Transactions on Intelligent Transportation Systems*, 1(4):221–231, Dec. 2000.
4. K.-F. Böhringer. Optimal strategies for moving droplets in digital microfluidic systems. In *Seventh International Conference on Miniaturized Chemical and Biochemical Analysis Systems (MicroTAS '03)*, Squaw Valley, CA, Oct. 2003.
5. K. F. Böhringer. Towards optimal strategies for moving droplets in digital microfluidic systems. In *IEEE International Conference on Robotics and Automation*, New Orleans, LA, Apr. 2004.
6. J. Brassil and R. Cruz. Nonuniform traffic in the Manhattan street network. In *IEEE International Conference on Communications (ICC '91)*, pages 1647–1651, June 1991.
7. S. K. Cho, H. Moon, and C.-J. Kim. Creating, transporting, cutting, and merging liquid droplets by electrowetting-based actuation for digital microfluidic cells. *Journal of Microelectromechanical Systems*, 12(1):70–80, Feb. 2003.
8. J. Ding, K. Chakrabarty, and R. B. Fair. Scheduling of microfluidic operations for reconfigurable two-dimensional electrowetting arrays. *IEEE Transactions on Computer-Aided Design of Integrated Circuits and Systems*, 20(12):1463–1468, Dec. 2001.
9. M. Erdmann and T. Lozano-Perez. On multiple moving objects. *Algorithmica*, 2(4):477–521, 1987.
10. R. B. Fair, V. Srinivasan, H. Ren, P. Paik, V. Pamula, and M. G. Pollack. Electrowetting-based on-chip sample processing for integrated microfluidics. In *IEEE International Electron Devices Meeting (IEDM)*, 2003.
11. S.-K. Fan, C. Hashi, and C.-J. Kim. Manipulation of multiple droplets on NxM grid by cross-reference EWOD driving scheme and pressure-contact packaging. In *IEEE Conference on MEMS*, pages 694–697, Kyoto, Japan, Jan. 2003.

12. D. Gross and C. M. Harris. *Fundamentals of Queueing Theory.* Wiley, New York, third edition, 1998.
13. J. E. Hopcroft, J. T. Schwartz, and M. Sharir. On the complexity of motion planning for multiple independent objects: PSPACE-hardness of the "warehouseman's problem". *International Journal of Robotics Research*, 3(4):76–88, 1984.
14. T. B. Jones, M. Gunji, M. Washizu, and M. J. Feldman. Dielectrophoretic liquid actuation and nanodroplet formation. *Journal of Applied Physics*, 89:1441–1448, Jan. 2001.
15. K. Kant and S. W. Zucker. Toward efficient trajectory planning: The path-velocity decomposition. *International Journal of Robotics Research*, 5(3):72–89, Fall 1986.
16. S. M. LaValle and S. A. Hutchinson. Optimal motion planning for multiple robots having independent goals. *IEEE Transactions on Robotics and Automation*, 14(6):912–925, Dec. 1998.
17. N. F. Maxemchuk. Routing in the Manhattan street network. *IEEE Transactions on Communications*, 35(5):503–512, May 1987.
18. P. A. O'Donnell and T. Lozano-Perez. Deadlock-free and collision-free coordination of two robot manipulators. In *IEEE International Conference on Robotics and Automation*, pages 484–489, Scottsdale, AZ, May 1989.
19. P. Paik, V. K. Pamula, and R. B. Fair. Rapid droplet mixers for digital microfluidic systems. *Lab on a Chip*, 3:253–259, 2003.
20. J. Peng and S. Akella. Coordinating multiple robots with kinodynamic constraints along specified paths. In J.-D. Boissonnat, J. Burdick, K. Goldberg, and S. Hutchinson, editors, *Algorithmic Foundations of Robotics V*, pages 221–237. Springer-Verlag, Heidelberg, Germany, 2003.
21. M. G. Pollack, R. B. Fair, and A. D. Shenderov. Electrowetting-based actuation of liquid droplets for microfluidic applications. *Applied Physics Letters*, 77:1725–1726, 2000.
22. J. Reif and M. Sharir. Motion planning in the presence of moving obstacles. In *Proceedings of the 26th Annual Symposium on the Foundations of Computer Science*, pages 144–154, Portland, Oregon, Oct. 1985.
23. G. Sanchez and J. Latombe. On delaying collision checking in PRM planning — application to multi-robot coordination. *International Journal of Robotics Research*, 21(1):5–26, Jan. 2002.
24. T. Schouwenaars, B. De Moor, E. Feron, and J. How. Mixed integer programming for multi-vehicle path planning. In *European Control Conference 2001*, Porto, Portugal, 2001.
25. T. Simeon, S. Leroy, and J.-P. Laumond. Path coordination for multiple mobile robots: A resolution-complete algorithm. *IEEE Transactions on Robotics and Automation*, 18(1):42–49, Feb. 2002.
26. P. Švestka and M. Overmars. Coordinated path planning for multiple robots. *Robotics and Autonomous Systems*, 23(3):125–152, Apr. 1998.
27. A. S. Tanenbaum. *Computer Networks.* Prentice Hall, Upper Saddle River, NJ, third edition, 1996.
28. C. Tomlin, G. J. Pappas, and S. Sastry. Conflict resolution for air traffic management: A study in multi-agent hybrid systems. *IEEE Transactions on Automatic Control*, 43(4):509–521, Apr. 1998.

Topological Mapping with Sensing-Limited Robots

Wesley H. Huang and Kristopher R. Beevers

Rensselaer Polytechnic Institute
110 8th Street, Troy, New York 12180, U.S.A.
{whuang,beevek}@cs.rpi.edu

Abstract. Most mobile robot mapping and exploration research makes use of long-range, powerful sensors such as laser rangefinders to create maps. In this paper we take a different approach, creating maps using robots with limited sensing capabilities, most notably in their sensor range. Our prototype robots use only five infrared range sensors with a maximum range of 80 cm; in general, these robots cannot "see" both sides of a hallway. We present an algorithm for such a robot to build a topological map in the presence of sensing and odometry error. In doing so, we develop a paradigm to extend topological mapping to open spaces, long considered a deficiency of topological mapping; we also introduce an evidential approach to the problem of "closing the loop."

1 Introduction

Imagine you are navigating through your house in the dark — you've woken up in the middle of the night, hungry for a snack. You get out of bed, but leave the lights off so you don't wake everyone else. You can't see anything, so you run one hand along the wall as you walk down the hallway. There are small discontinuities in the wall, like the closet door, but you keep walking straight. You feel the wall "end," so you turn the corner and keep going. Perhaps at some point, you leave the wall and walk across a (known obstacle-free) room, holding your hand out in front to detect when you reach the opposite wall. You eventually reach the kitchen without much trouble.

This blind navigation is an apt analogy for how a sensing-limited robot "sees" the world. By "sensing-limited," we mean that the robot has sparse and very limited-range information about the world. These limited sensing capabilities result in a restricted view of the world — features of the environment can only be detected by observing a time series of sensor readings.

Small robots with limited sensing capabilities are inexpensive, and are therefore ideal in applications where many potentially disposable robots are required to map and explore a building in parallel, e.g. in search and rescue or urban reconnaissance scenarios. In this paper however, we focus solely on the problems of exploration, mapping, and navigation for a single robot.

Our problem is for a sensing-limited robot to explore an unknown environment, creating a topological map that it can later use to navigate between

M. Erdmann et al. (Eds.): Algorithmic Foundations of Robotics VI, STAR 17, pp. 235–250, 2005.

any two previously-visited locations. The resulting paths should be efficient to the extent that safety, limited-range sensing, and errors in sensing, robot motion and odometry allow.

In this paper, we demonstrate that such a robot can complete this task, though there are some limitations based on the size and geometric complexity of the world. We should not expect the same performance from a sensing-limited robot as we would from a robot with richer sensing modalities (e.g. a laser scanner). While we permit lesser performance in terms of time and scope, we still expect reasonable accuracy.

This paper presents a mapping strategy that constructs a topological map in which vertices are places where a robot behavior terminates, and edges represent a sequence of behaviors that move the robot from one place to another. We develop a new approach to the problem of "closing the loop" — recognizing that the robot has returned to a place in the world that it has already visited. Since sensing-limited robots cannot uniquely recognize a place based on instantaneous sensor readings, we create a hypothesis that the robot has closed the loop based on its geometric position estimate, and attempt to verify this hypothesis using an evidence-based approach that compares characteristics of subsequent edge traversals.

Our behavior-based map representation also allows us to introduce techniques that enable our maps to transcend the typical weakness of topological maps in representing open spaces. First, we introduce the concept of *portals*, which link open spaces (mapped by wall-following) to narrow corridors (mapped by hall-following). Second, we introduce the idea of *forays*, which cross open spaces to produce more direct links between vertices in the map.

1.1 Assumptions

We consider a robot that has a small number of short-range sensors. These sensors return the distance to the nearest obstacle along a straight line; they are "short-range" in that their maximum range is small compared to the dimensions of the environment. There must be enough sensors to robustly execute the behaviors described in Sec. 3.1. As an example, our robot has five infrared range sensors: one forward sensor and two sensors on each side.

There is known error in the robot's movement, odometry, and sensing. We assume these errors are random and zero-mean, and that we can merge and compound measurements and compute confidence bounds for a given confidence level.

We assume an enclosed, static, rectilinear environment. Rectilinearity is a strong assumption — it removes uncertainty in the robot's orientation — but it allows us to focus on the fundamental mapping problems. However, the environment may be "non-smooth:" there may be small discontinuities in the walls of the environment, such as rectilinear protrusions and recesses in a hallway due to door frames, structural columns, etc.

2 Related work

There are two traditional paradigms for robotic mapping: metric maps and topological maps. In metric maps, the geometry of the world is explicitly represented, either through exact or approximate representations. In topological maps [8], "places" in the world (typically hallway junctions) are represented by vertices in a graph, and paths between places are represented by edges. Metric maps provide a detailed world representation but require more storage and are sensitive to measurement errors. Topological maps offer a concise representation but (as traditionally implemented) cannot represent open spaces.

Many researchers have combined these approaches by using metric maps at nodes in topological maps. For example, Tomatis *et al.* [17] use a topological map to represent a network of hallways and use metric maps to represent rooms. In contrast, our work extends the topological mapping paradigm to represent open spaces. Our use of a wall-following behavior to generate an initial map of an open area is related to the idea of "coastal navigation" [13] which recognizes that areas near walls and obstacles have "high information content" due to the features they produce. Once the robot has built a map of the world boundaries, it can safely explore the interior.

A fundamental problem in topological mapping is recognizing an already-visited place ("closing the loop"). With sufficiently rich sensing, this problem is easily solved by recording a unique "sensing signature" [9] for each place. Without such sensing capabilities, the robot must continue exploring to deduce whether a loop has been closed. Kuipers' "rehearsal procedure" [10] encapsulates the general idea of using the map topology to make this decision. Choset and Nagatani [3] describe an approach where structural characteristics of the map (e.g., the degree of vertices and the order of incident edges) are the primary criteria for verification. Tomatis *et al.* [17] embed this comparison in a POMDP that should show a single peak upon loop-closing.

Our approach to this problem is based on accumulating evidence to verify or reject loop-closing hypotheses. The evidence is based on the odometry error model and is combined using Dempster-Shafer theory [14]. This approach is related to that of Cox and Leonard [4], who maintain multiple hypotheses about the state of a dynamic world. The probability of each hypothesis is assigned and updated using a Bayesian framework. The main advantage of a Dempster-Shafer based method is the ability to represent "ignorance."

Our work is related to the simultaneous localization and mapping (SLAM) problem. However, SLAM is typically solved using rich sensing (a scanning laser range-finder) and produces metric maps. See Thrun's recent survey [16] for an overview of work on SLAM.

Others have addressed mapping and exploration with limited sensing. Butler *et al.* [2] describe coverage (equivalent to metric mapping) using robots with only contact sensors but with near-perfect odometry. Doty and Seed [5] have shown preliminary results in creating a "landmark map" using a robot with four short-range infrared sensors and one long-range SONAR sensor.

3 Basic mapping algorithm

The "basic map" created by our robots is a topological map created by circumnavigating open spaces with a wall-following behavior and traversing narrow corridors with a hall-following behavior. In this paper, we focus primarily on mapping of open spaces. First, we describe the required robot behaviors and then detail construction of the map.

3.1 Behaviors

In order to create the basic map, we require that our robot be capable of executing the following simple behaviors:

- The "turn" behavior causes the robot to rotate a specified angle.
- The "move-to-wall" behavior causes the robot to move forward until one of its sensors detects a wall. The robot then moves so that its "wall-following sensors" (side sensors) are aligned with this wall.
- The "wall-following" behavior (Fig. 1) causes the robot to move forward, maintaining a range r_0 to the wall with its wall-following sensors; any small discontinuities in the wall cause the robot to adjust and continue wall-following. The behavior terminates when the robot detects a *well-defined* corner. A well-defined *exterior* corner occurs when the wall "falls away" beyond a range of r_{max}; likewise, a well-defined *interior* corner occurs when the wall juts inward within r_{min}. Any discontinuity that remains within the range $[r_{\text{min}}, r_{\text{max}}]$ of the robot's wall-following sensors is an *ill-defined* corner (currently unused in our maps).

  In the case of an exterior corner, wall-following terminates at a distance of r_0 past the discontinuity; for an interior corner, wall-following stops at a distance of r_0 *before* the discontinuity. Note that exterior corners will typically be detected by the wall-following sensors, whereas interior corners will be detected by front, side or bump sensors. The wall-following behavior must be capable of terminating in a reasonably repeatable location, regardless of the direction of travel. Because of these conditions, no two well-defined corners of the same type can be within a distance

$$r_* = \sqrt{r_0^2 + r_{\text{min}}^2} \tag{1}$$

- The "hall-following" behavior is similar to the wall-following behavior, except that it is only used when the robot is traversing a "corridor" of width $\leq 2r_0$. In this case, both of the robot's side sensors can "see" walls, and the robot moves through the center of the corridor.

Discussion The wall-following and hall-following behaviors are difficult to design because they are responsible for properly navigating through the environment. One requirement is that they must terminate in the same location,

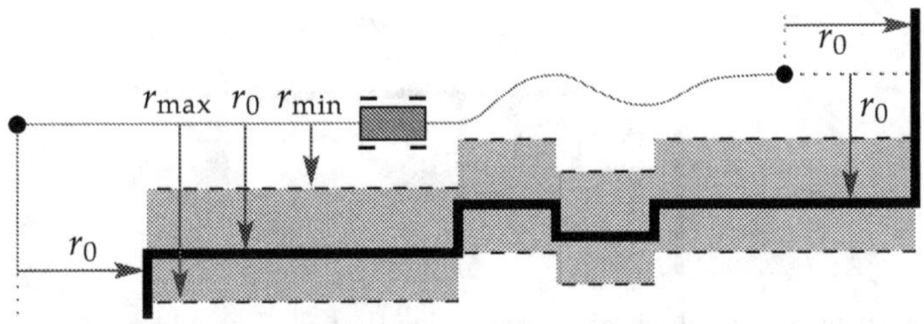

Fig. 1. Wall-following and feature detection. The "safe" wall-following range $[r_{min}, r_{max}]$ is indicated by the highlighted area. Vertices are placed at a distance of r_0 from both incident walls of a well-defined corner. The leftmost corner is exterior; the rightmost is interior. The rest of the discontinuities are ill-defined.

regardless of the direction of travel. This is done by maintaining the proper r_0 offset during wall-following and by staying centered while hall-following.

One way for the robot to make lateral adjustments while wall- or hall-following is to make only rectilinear motions, mimicking every discontinuity of the wall. This would be time consuming and unnecessary: our primary interest is in capturing the overall structure of the environment. Our current implementation instead makes gradual corrections while continuing forward. The tradeoff for this approach is that distance estimates may be inaccurate, and there are situations where small spacing between discontinuities may cause feature misidentification since the robot is not at offset r_0 and not parallel to the wall. Adjusting the responsiveness of the behaviors trades off the risk of feature misidentification for accuracy in the distance measurements.

The "perceptual aliasing" problem — the inability to detect small obstacles due to low sensor density — is another issue when creating maps with sensing-limited robots: the robot may encounter an obstacle without first sensing it. For example, the robot may bump into a chair leg that is not seen by its sparsely-placed range sensors. For such situations, bump sensors are necessary and can be thought of as extremely short-range sensors to be incorporated into the mapping behaviors accordingly.

3.2 Mapping process

We first consider an environment consisting only of open spaces. (We remove this assumption in Sec. 3.4.) Our basic map-making process is as follows:

- From an arbitrary starting location, the robot initiates the move-to-wall behavior, which aligns the robot with some wall in the environment.
- The wall-following behavior is activated, and its termination point becomes the start vertex v_0 — the first vertex in our topological map.
- After finding v_0, the robot turns as appropriate to follow the next wall (incident to v_0). Vertices are added for each well-defined feature that

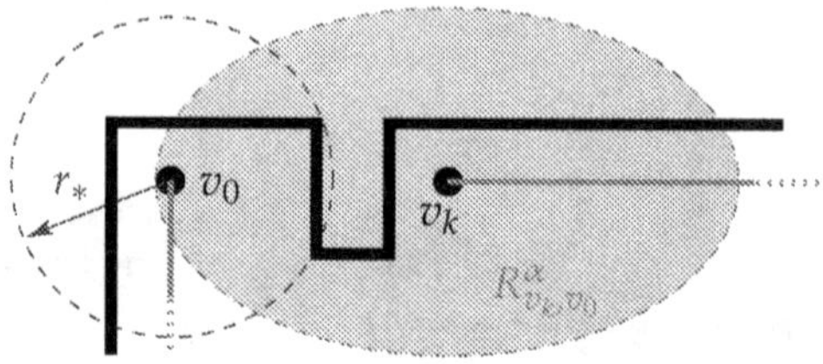

Fig. 2. A situation in which we hypothesize that $v_k = v_0$. The region $R^{\alpha}_{v_k,v_0}$ represents the positional uncertainty in the location of v_k with respect to v_0. Since this region contains v_0, it is possible that the two vertices are the same.

terminates the wall-following behavior. Each new vertex is connected, with an undirected edge, to the previously discovered feature (i.e. $v_0 \leftrightarrow v_1$, $v_1 \leftrightarrow v_2$, and so on).

- This process repeats until the robot returns to v_0. This is guaranteed to happen, since the environment is enclosed. Recognizing when it occurs is the problem of "closing the loop."

3.3 Closing the loop

For each vertex v_i in the map, we use the model of the robot's odometry error to maintain a probability density function U_{v_i,v_0} describing the robot's location with respect to v_0. The uncertainty in this distribution grows monotonically as the map expands. For a confidence level α specified to our mapping algorithm, let $R^{\alpha}_{v_i,v_0}$ represent confidence bounds of U_{v_i,v_0}.

If the robot reaches a vertex v_k that is the same type (interior or exterior corner) and $R^{\alpha}_{v_k,v_0}$ contains v_0, then we create a hypothesis that vertex v_k closes the loop. See Fig. 2 for an example of such a situation. Note that if the entire region $R^{\alpha}_{v_k,v_0}$ falls completely within r_* of v_0 (generally not the case), we can immediately declare the hypothesis correct with at least the specified confidence since no other vertex can be within that radius of v_0.

Once a loop-closing hypothesis is made, we attempt to verify it. One way to do this is to *backtrack* — travel back and forth between v_k and v_0, taking additional measurements of all the edges. Under the assumption of random zero-mean errors, this will reduce the size of $R^{\alpha}_{v_k,v_0}$ until it either lies entirely within r_* of v_0 or until v_0 does not lie in $R^{\alpha}_{v_k,v_0}$. This is the only certain way to verify or reject a hypothesis, but it is impractical as many trips back and forth between v_k and v_0 are required.

Instead, we have the robot continue traversing edges in the "forward" direction. If it comes upon a structural difference (e.g. vertices of different types or incident edges at different orientations), the hypothesis can be rejected. In order to verify the hypothesis, however, we compare lengths of edges that should match under the hypothesis. The degree to which the measurements match (or don't match) provides evidence in support of (or against) the correctness of the hypothesis. The robot continues traversing edges until sufficient evidence accumulates to accept or reject the hypothesis.

Dempster-Shafer approach Dempster-Shafer theory gives us a natural way to assign "belief" to a set of possibilities — in this case, that the hypothesis is correct, which we denote as C, or incorrect, denoted I. It also provides a way to represent "ignorance" about these possibilities. Ignorance can be thought of as uncertainty about which possibility is supported by evidence, and is denoted as $\{C, I\}$.

Suppose that we have made a hypothesis H_0 that two vertices in our map are the same. Continuing our forward traversal of the walls of the environment, we take a series of new wall length measurements. Under H_0, each revisited edge in the map has an associated set of measurements $L = \{\ell_1, \ell_2, \ldots, \ell_n\}$, with $n \geq 2$ which should all be of similar length. Dempster-Shafer theory is used to merge the evidence provided by multiple sets of such measurements (each corresponding to a different edge in the map).

We compare edge measurements directly instead of comparing vertex locations because errors in each edge measurement are independent. Using vertex locations for evidence would reuse early edge measurements when comparing subsequent vertices, giving them undue weight in providing evidence.

Belief function Given a set L of measurements, and their associated uncertainty distributions, we must determine the degree of support they provide for our hypothesis. To do this, we define a *belief function*, denoted m_L, where $m_L(C)$, $m_L(I)$ and $m_L(\{C, I\})$ are our belief, given L, that H_0 is correct, our belief that H_0 is incorrect, and our "ignorance" (the degree to which L provides evidence neither for nor against H_0), respectively. These are basic probabilities meeting the requirements of Dempster-Shafer theory such that:

$$m_L(C) + m_L(I) + m_L(\{C, I\}) = 1 \tag{2}$$

Our particular belief function draws on methods of statistical inference. It essentially computes probabilities that the measurements in L were taken from the same wall, given the robot's error model. The "probability information content" [15] of these probabilities is used to measure our degree of ignorance. Specifically, the belief function we use is:

$$m_L(C) = \Phi(L)\eta_L \tag{3}$$

$$m_L(I) = (1 - \Phi(L))\,\eta_L \tag{4}$$

Here, $\Phi\colon L \to [0, 1]$ returns the probability of obtaining L, assuming L contains measurements from the same wall. The particular Φ-function we use is a goodness-of-fit test based on the *z-score* of the measurements in L. (The z-score represents the deviation of a sample from the expected mean, expressed in units of the standard deviation of the underlying distribution.) We use the square of the z-score, computed using a distribution with standard deviation $\hat{\sigma}$ generated according to the robot's error model for the current "best estimate" $\hat{\ell}$ (based on L) of the length of the wall:

$$z^2 = \sum_{\ell_i \in L} (\ell_i - \hat{\ell})^2 / \hat{\sigma}^2 \tag{5}$$

The value of z^2 follows a χ^2 distribution with $n = |L|$ degrees of freedom. So, the degree of support provided by L for H_0 is computed as:

$$\Phi(L) = \int_{z^2}^{\infty} \frac{y^{n/2-1} e^{-y/2}}{2^{n/2} \Gamma(n/2)} \, dy \tag{6}$$

The value $\eta_L \in [0,1]$ in Eqns. 3 and 4 represents the "probability information content" [15] of $\Phi(L)$, computed as follows:

$$\eta_L = 1 + \Phi(L) \log_2(\Phi(L)) + (1 - \Phi(L)) \log_2(1 - \Phi(L)) \tag{7}$$

The probability information content of $\Phi(L)$ essentially measures the "distance" of $\Phi(L)$ from the uniform distribution. Thus, given a $\Phi(L)$ with high entropy (less information), we profess more ignorance, and for a $\Phi(L)$ with low entropy (more information), we profess less ignorance.

Evidence accumulation Upon first making a hypothesis, we must initialize the overall belief in the hypothesis. Recall that geometrically, no two features of the same type and orientation can be within a radius r_* (see Sec. 3.1). This suggests the metric

$$m(C) = \iint_{B_{v_0}^{r_*}} U_{v_0, v_h}(x, y) \, d\mathbf{A} \tag{8}$$

where v_0 is the start vertex, $B_{v_0}^{r_*}$ is the ball of radius r_* about v_0, and $H_0 \equiv v_0 = v_h$. In other words, the initial value of $m(C)$ is the probability, according to our error model, that $|v_0 - v_h| \leq r_*$ (in which case they must represent the same feature). We initialize $m(I)$ to zero: at the moment we make H_0, we have no basis for assigning belief mass to the possibility that H_0 is incorrect.

As we take new measurements, we must combine the evidence they provide with previous evidence we've acquired, in order to update our global belief in the correctness of the hypothesis. To combine evidence, we use Dempster's rule of combination. In our particular application, the combination rules are as follows: given global belief m and belief m_L attributed to a specific set of measurements L,

$$m \oplus m_L(C) = \frac{m(C)m_L(C) + m(C)m_L(\{C,I\}) + m(\{C,I\})m_L(C)}{1 - m(C)m_L(I) - m(I)m_L(C)} \tag{9}$$

$$m \oplus m_L(I) = \frac{m(I)m_L(I) + m(I)m_L(\{C,I\}) + m(\{C,I\})m_L(I)}{1 - m(C)m_L(I) - m(I)m_L(C)} \tag{10}$$

We continue building evidence about a hypothesis until our overall belief in its correctness (or incorrectness) surpasses some threshold β. In general, we choose $\beta = \alpha$, the confidence bound used in generating the hypothesis. In [1], we provide additional discussion about making loop-closing decisions, and present extensions to the evidential loop-closing method for dealing with more complicated mapping scenarios.

Discussion The worlds in which it is most difficult to close the loop are those with structural self-similarity. Two ways in which a world might be self-similar occur when the world: (1) consists of walls that spiral inward or outward; or (2) consists entirely or partially of repeating similar sequences of walls. Our evidential approach fares well in "spiral" worlds (by their nature, these worlds yield comparisons between different-length edges). Worlds of "repeating similar sequences" are generally more difficult, though those consisting only partially of such sequences can often be properly mapped despite the presence of error, given a high enough evidence threshold.

Continuing the forward traversal of the environment after making a loop-closing hypothesis presents several issues. The key concern is that one can never be completely certain that a hypothesis is correct — a deficiency of nearly all hypothesis-based methods for closing loops. The backtracking method resolves this issue, but only after indefinitely many measurements. We believe that the idea of building evidence in support of and against the hypothesis is both intuitive and reasonable for use in practical situations.

A danger of the hypothesis-based approach is that incorrect hypotheses may be accepted as correct (or vice versa). One way to deal with this is to keep multiple hypotheses and never commit to any one of them (i.e. retain the hypothesis that the loop has not been closed). If several hypotheses seem correct, the robot can direct its exploration toward differentiating features. When a single hypothesis has emerged as correct, the map associated with it can be used for navigation, etc. If it is later deemed incorrect, additions to the map depending on that hypothesis can be discarded.

We have made a few other observations about the loop-closing problem:

- When the robot has actually returned to v_0, all inconsistent hypotheses will have been disproved by structural mismatches.
- The robot's orientation must pass through a full 360° (net) before it can return to the start vertex.
- False positive or false negative loop-closings can be disproved or proved later, but it is preferable to find the right hypothesis during basic mapping. As such, evidence thresholds should be reasonably high given no prior knowledge of the environment.
- Prior knowledge of some aspects of the world (e.g. feature density, distribution of potential wall lengths, etc.) might lead to more efficient loop-closing by allowing us to design more informed belief functions and letting us be more aggressive in accepting or rejecting hypotheses.

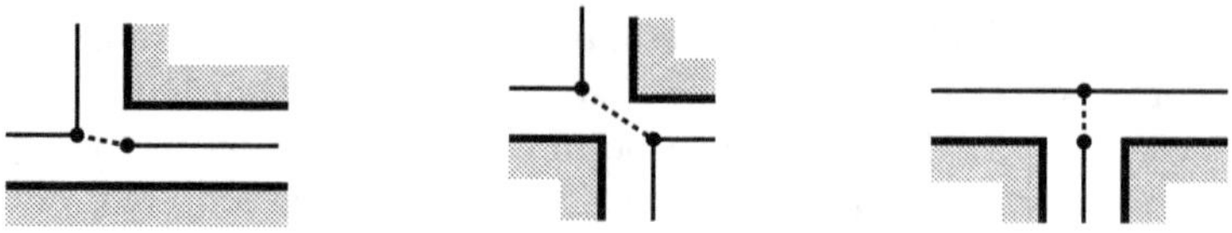

(a) (b) (c)

Fig. 3. Portals occur under special wall-following or hall-following terminating conditions. In each figure, the robot moves from left to right. In Fig. (a), the interior-corner portal is detected while following a wall. In Fig. (b), the portal, detected while turning an interior corner, leads directly to another open space. In Fig. (c), the portal is detected while turning an exterior corner.

3.4 Portals and narrow corridors

We have so far addressed only the case where open space lies to one side of the robot. When the area around the robot is narrow enough that the robot can see walls on both sides, the robot should switch from wall-following to hall-following. We define *portals* (Fig. 3) to be special undirected edges in the map that make the connection between these two different mapping modes.

In our basic mapping, we can use portals to skip over narrow corridors that may lead to other open areas, leaving them to be explored later. Often, this has the advantage of making the loops in the basic map smaller and therefore easier to close. Although doorways are prime candidates for portals, there is nothing special about them except that they constitute a narrow corridor.

Due to limited space, we omit a complete discussion of detecting portals and exploring narrow corridors. Though recognizing a portal condition is generally simple — side sensors on the non-wall-following side suddenly detect an object — a number of circumstances exist in which portals might occur, and each needs to be handled specifically. Our wall-following and hall-following behaviors use a "probing" strategy to find portals: upon encountering what appears to be a well-defined feature, they examine the area around the feature, attempting to determine if it constitutes a boundary between an open space and a narrow corridor. This method is also used when hall-following to differentiate between different types of hallway junctions.

For complete specifications of the wall-following and hall-following behaviors and discussion of portal detection, see [1].

4 Refinements

After we have completed the basic mapping process, navigation between any two points in the map is straightforward: we search the map (graph) for the shortest sequence of behaviors (based on time estimates) between the points and then execute this sequence. However, navigation through the basic map is inefficient in many cases: the robot may need to circumnavigate half of a building to reach a point across a wide hallway.

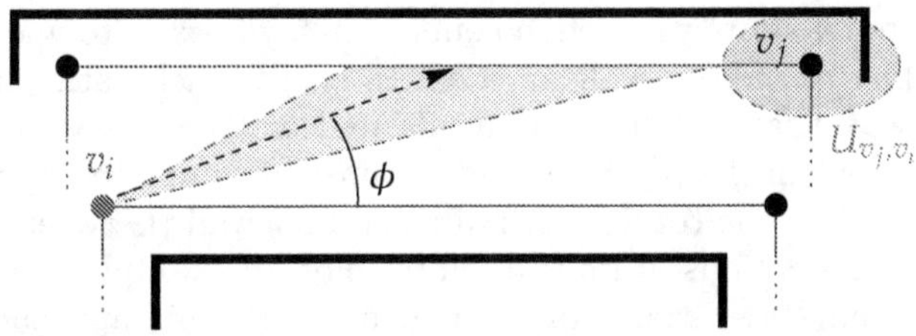

Fig. 4. A refinement from vertex v_i to vertex v_j. The robot turns to angle ϕ and "forays" toward the opposite wall. The robot's angular uncertainty (highlighted) is small enough that it is confident it will land on the target wall.

The second phase of our algorithm introduces *refinements* into the map which improve its usefulness for navigating through open spaces (and for discovering "islands" in these spaces when they exist). Refinements to our basic map are special directed edges that connect disjoint places in the environment with paths that pass through open (featureless) regions. Refinements offer a tool for extending topological maps to such spaces in a natural way.

Our refinements are based on the geometry of the basic map. Because of measurement error during basic mapping, the map is not geometrically consistent. Prior to generating candidate refinements, we must embed the vertices of the map in a metric space. This problem has been addressed by several researchers, including Duckett *et al.* [6], Lu and Milios [12], and Golfarelli *et al.* [7]. Any of these methods suffice; the reference frame for the vertices can be placed arbitrarily.

4.1 Forays

Suppose we wish to add a path through free space to connect vertex v_i in our basic map with vertex v_j. If the robot experienced no movement error, we could do this by simply turning the robot towards v_j and driving straight. However, to allow for error in the basic map and in the robot's motion, we direct the robot towards a wall adjacent to v_j. This *foray* across open space, then, consists of the sequence of behaviors: (1) turn toward the target wall (incident to the target vertex); (2) move-to-wall; (3) wall-follow to the target vertex.

We already have a means for estimating the relative uncertainty of vertex locations, using uncertainty distributions based on the robot's error model. Confidence limits specify a region within which we believe (with the desired confidence) the vertex lies.

We assume that the uncertainty in the robot's location estimate as it forays from v_i, due to the robot's inability to drive in a straight line, can be bounded by an angular range originating at v_i. The robot's positional uncertainty grows with the length of the foray.

There are two potential refinements from a vertex v_i to another vertex v_j, corresponding to the two walls incident to v_j. For each path, the robot forays in such a way that it lands on the desired wall. So, a valid refinement is possible only when the positional uncertainty accumulated by the robot as it forays is less than the total length of the target wall (between the confidence bounds in the locations of its endpoints). For an example foray, see Fig. 4.

We can make two types of refinements with this approach: *passive refinements*, which can be computed directly from the basic map, and *active refinements*, which require further exploration of the world.

Passive refinements Passive refinements use knowledge of the area swept out by the robot's sensors as it performed its initial exploration. The union of all the area swept out by the robot's sensors is the "known space." A simple approximation can be obtained by storing the shortest range detected by the sensors opposite the wall-following sensors when traversing each wall.

Using the geometry of our basic map, we can immediately determine which candidate refinements pass only through the known space, and add these refinements to our map. Fig. 4 shows a simple example of such a refinement; though the robot could not see both sides of the hallway while building the basic map, the areas swept out by its non-wall-following sensors overlap, so we may infer the entire hallway is obstacle-free.

Active refinements & exploration Potential refinements passing partially or entirely through unknown space must be actively explored to verify that they can be traversed, since there may be obstacles in the unknown space.

Suppose that, when foraying, the robot encounters an obstacle. In this case the robot *must* be able to safely return to a known place in its basic map — otherwise, it is lost and will have to rebuild its basic map, or at least relocalize itself in the map. Prior to making forays, we discard those that are so long that we are unable to ensure the robot will be able to return to a known location in its original map. The identification of these "unsafe" forays is based on the robot's odometry error model and the estimated distance between the origin and destination of each foray.

For the remaining exploration targets, we compute a sequence in which to explore them. One method for constructing this sequence might be to use a traveling-salesman like approach; unexpected obstacles might require the sequence to be re-planned. After planning a sequence, each foray is executed in turn.

If we encounter no obstacles until we reach the target wall, we keep the refinement; otherwise we discard it. If an obstacle is encountered, the robot begins the basic mapping process to map the obstacle. If its positional uncertainty accumulates too much before it encounters a well-defined feature, it gives up and returns to its original map. Otherwise, it creates a basic map of the obstacle, connecting the obstacle to its original map with refinements.

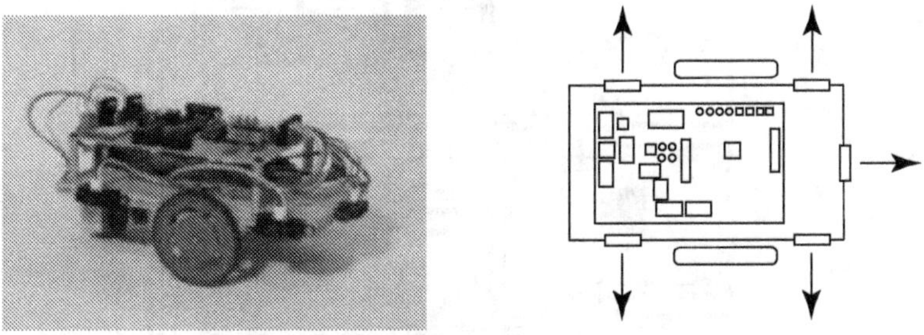

Fig. 5. Our prototype robot is a differential drive mobile robot approximately 20 cm long. It has five Sharp GP2D12 infrared range sensors, 256 CPR encoders on each wheel, and an Atmel ATMEGA64 microcontroller as the main processor.

4.2 Discussion

The refinements described here are reminiscent of the LMT [11] approach to compliant motion planning in which successive preimages of the goal are computed, taking into account uncertainty in the robot's motion. Note that refinements serve to both explore unknown parts of the space as well as to create more efficient pathways through the world. The actions associated with a foray are a simple sequence of behaviors. More complex sequences could be planned that involve conditional branches, loops, or longer sequences of behaviors. For example, the robot could drive directly toward an interior corner vertex; it might land on either of the two supporting walls and must wall-follow in the appropriate direction to reach the goal vertex. In the extreme, we could form a completely connected graph, creating a "program" of behaviors to take the robot between any pair of vertices.

The computation required would be considerable and mostly unnecessary — most of these links would never be used. The refinements described earlier can serve as the building blocks for more sophisticated behavior sequences.

5 Experimental Results

We have implemented our mapping algorithm both in simulation and on a mobile robot (Fig. 5), with promising results. Simulated experiments were performed with a variety of simple worlds, and with several more complicated environments. Real-world experiments were performed in both specially-tailored and unmodified building environments.

Our simulations, which include the exploration of refinements, have been effective in mapping large worlds (Fig. 6), even under harsh error conditions. In over 200 trials in small worlds, with varying degrees of odometry error, 97% of the maps were structurally correct after initial loop-closing.

Our hardware experiments have focused primarily on verifying the evidential loop-closing method. Despite the robot's sparse sensing and the often

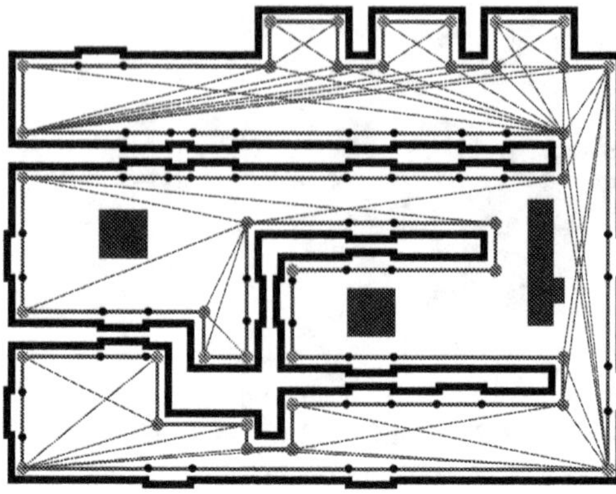

Fig. 6. Map created in simulation. Refinements to the basic map are included here; islands have not been mapped.

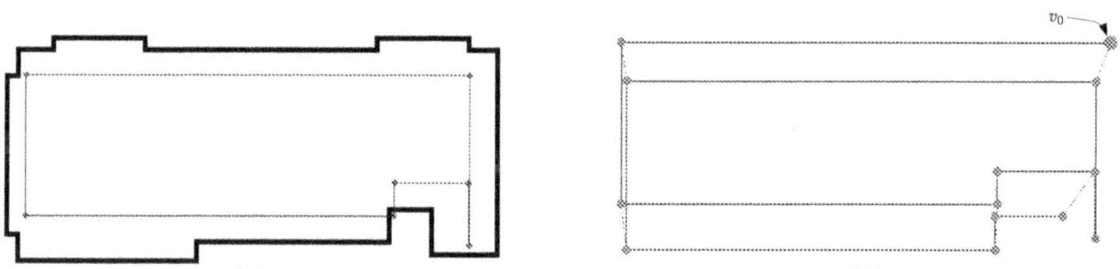

(a) (b)

Fig. 7. Basic map created using the hardware from Fig. 5. In Fig. (a), the final map (after loop-closing and embedding) is overlaid on the floor plan. The environment is approximately 4.5 m × 2 m. In Fig. (b), the raw data for the map is displayed. Dotted lines are drawn between vertices that match under the correct hypothesis; the start vertex is the large dot at upper right. Here, the robot needed five measurements to accumulate 0.99 belief in the correct loop-closing hypothesis.

large errors it experienced, the algorithm closed loops correctly in most circumstances. Real-world maps were made of both small areas (Fig. 7) and large areas, including the 12 m × 30 m first floor of an academic building.

Our mapping algorithm — especially the loop-closing method — relies on zero-mean error, particularly in the robot's odometry. In reality, the error experienced by a robot is rarely zero-mean. For example, suppose the robot travels down a long hallway with a tile floor, and then returns down a heavily carpeted hallway. The difference in floor surface will yield biased (non-zero-mean) odometric error. So far in our experiments however, the negative effects of the zero-mean assumption have been minimal, even in large environments.

Overall, our simulated and real-world experiments indicate that the mapping algorithm closes loops correctly well over 95% of the time for non-self-similar worlds. For additional experiments and further discussion, see [1].

6 Conclusions and Future Work

In this paper, we have presented a topological mapping algorithm for robots capable of only sparse, limited-range sensing. We have introduced an evidential approach to the problem of "closing the loop" — recognizing when the robot has returned to a place it has already visited — based on the Dempster-Shafer theory of evidence. Finally, we have presented paradigms for incorporating open spaces into topological maps: "portals" that naturally connect narrow regions and open spaces and refinements that make "forays" through open spaces. Our experiments, both in simulation and with a real robot, verify the efficacy of our approach.

Our mapping approach is not without limitations. In particular, our robots are limited in the worlds they can map. These limitations are primarily dependent on the robot's sensing deficiencies; for example, we should not expect a robot with large odometry error to be capable of efficiently making complete maps of vast open spaces. While creating precise maps of large spaces with a sensing-limited robot may be possible, our approach favors mapping and navigation efficiency over extensiveness.

Our approach is also limited by the capabilities of the behaviors it depends upon. Wall- and hall-following behaviors must be capable of terminating in repeatable locations and handling small spacing between discontinuities. While this is not strictly possible in a real-world implementation, our experiments have shown that reasonably capable behaviors can be developed.

6.1 Future work

Our approach to topological mapping should transfer directly to general polygonal worlds with a suitable wall-following behavior. We are also interested in more deeply exploring some of the fundamental questions brought forth by the topological mapping problem:

- How are the quality of the map and a robot's mapping ability in general affected by sensing limitations? By the extent of sensor error? By the complexity of the world?
- What benefits are provided by specific assumptions about the world, such as rectilinearity or near-rectilinearity?
- How can we best use geometry to make inferences when loop-closing?
- What can we do in a curved world, with no "well-defined" features?

Acknowledgment

Thanks to Jonathan Fink and Alden Roberts for their help with the robot hardware. This work was supported by the NSF through award IIS–9983642.

References

1. K. R. Beevers. Topological mapping and map merging with sensing-limited robots. Master's thesis, Rensselaer Polytechnic Institute, Troy, NY, April 2004.
2. Z. J. Butler, A. A. Rizzi, and R. L. Hollis. Complete distributed coverage of rectilinear environments. In B. R. Donald, K. M. Lynch, and D. Rus, editors, *Algorithmic and Computational Robotics: New Directions (WAFR 2000)*, pages 51–61. A K Peters, 2001.
3. H. Choset and K. Nagatani. Topological simultaneous localization and mapping (SLAM): Toward exact localization without explicit localization. *IEEE Transactions on Robotics & Automation*, 17(2), April 2001.
4. I. Cox and J. Leonard. Modeling a dynamic environment using a Bayesian multiple hypothesis approach. *Artificial Intelligence*, 66:311–344, 1994.
5. K. Doty and S. Seed. Autonomous agent map construction in unknown enclosed environments. In *Proc. MLC-COLT Workshop on Robot Learning*, pages 47–55, New Brunswick, N.J., July 1994.
6. T. Duckett, S. Marsland, and J. Shapiro. Learning globally consistent maps by relaxation. In *Proc. 2000 IEEE Intl. Conf. on Robotics & Automation*, volume 4, pages 3841–3846, 2000.
7. M. Golfarelli, D. Maio, and S. Rizzi. Elastic correction of dead-reckoning errors in map building. In *Proc. 1998 IEEE/RSJ Intl. Conf. on Intelligent Robots & Systems*, pages 905–911, B.C., Canada, 1998.
8. B. Kuipers. Modeling spatial knowledge. *Cognitive Science*, 2:129–153, 1978.
9. B. Kuipers and P. Beeson. Bootstrap learning for place recognition. In *Proc. 18th Natl. Conf. on Artificial Intelligence*, Edmonton, Canada, 2002.
10. B. Kuipers and Y.-T. Byun. A robot exploration and mapping strategy based on a semantic hierarchy of spatial representations. *Journal of Robotics and Autonomous Systems*, 8:47–63, 1991.
11. T. Lozano-Pérez, M. Mason, and R. Taylor. Automatic synthesis of fine-motion strategies for robots. *International Journal of Robotics Research*, 3(1):3–24, 1984.
12. F. Lu and E. Milios. Globally consistent range scan alignment for environment mapping. *Autonomous Robots*, 4(4):333–349, October 1997.
13. N. Roy and S. Thrun. Coastal navigation with mobile robots. *Advances in Neural Processing Systems*, 12:1043–1049, 1999.
14. G. Shafer. *A Mathematical Theory of Evidence*. Princeton University Press, 1976.
15. J. Sudano. Pignistic probability transforms for mixes of low- and high-probability events. In *Proc. 4th International Conference on Information Fusion*, pages 23–27, Montreal, August 2001.
16. S. Thrun. Robotic mapping: A survey. Technical Report CMU-CS-02-111, Carnegie Mellon University, Pittsburgh, PA, February 2002.
17. N. Tomatis, I. Nourbakhsh, and R. Siegwart. Hybrid simultaneous localization and map building: closing the loop with multi-hypothesis tracking. In *Proc. 2002 IEEE Intl. Conf. on Robotics & Automation*, May 2002.

Locating and Capturing an Evader in a Polygonal Environment

Volkan Isler[1⋆], Sampath Kannan[1⋆⋆], and Sanjeev Khanna[1⋆⋆⋆]

University of Pennsylvania, Department of Computer and Information Science
Philadelphia, PA, USA
{isleri,kannan,sanjeev}@cis.upenn.edu

Abstract. This paper contains two main results: First, we revisit the well-known *visibility based pursuit-evasion* problem and show that, in contrast to deterministic strategies, a single pursuer can locate an unpredictable evader in any simply-connected polygonal environment using a randomized strategy. The evader can be arbitrarily faster than the pursuer and it may know the position of the pursuer at all times but it does not have prior knowledge of the random decisions made by the pursuer. Second, using the randomized algorithm together with the solution of a known lion and man problem [12] as subroutines, we present a strategy for two pursuers (one of which is at least as fast as the evader) to quickly capture an evader in a simply-connected polygonal environment.

1 Introduction

Pursuit-evasion games are among the fundamental problems studied by Robotics researchers. In a pursuit-evasion game, one or more pursuers try to capture an evader who, in turn, tries to avoid capture indefinitely. A typical example is the homicidal chauffeur game where a driver wants to collide with a pedestrian and the goal is to determine conditions under which he can (not) do so. Among the numerous applications of this game are collision avoidance and air traffic control.

Recently, there has been increasing interest in capturing intelligent evaders with certain sensing capabilities who are contaminating a complex environment. Aside from its obvious applications in search-and-rescue and surveillance, this problem provides a natural test-bed for many mobile robot algorithms.

Perhaps the most well-understood game in this context is the *visibility-based pursuit-evasion* where one or more pursuers try to locate an evader in a polygonal environment [13,10,2]. In this game, the evader is very powerful: it has unbounded speed and global visibility, meaning that it knows the location

⋆ Supported in part by NSF-IIS-0083209, NSF-IIS-0121293 and MURI DAAH-19-02-1-03-83.

⋆⋆ Supported in part by NSF Grants CCR0105337 and CCR9820885.

⋆⋆⋆ Supported in part by an Alfred P. Sloan Research Fellowship and an NSF Career Award CCR-0093117.

M. Erdmann et al. (Eds.): Algorithmic Foundations of Robotics VI, STAR 17, pp. 251–266, 2005.

of the pursuers at all times. In [15,4] the authors study a similar game in a probabilistic framework and propose a greedy algorithm.

The main ingredient of a pursuit-evasion game is the presence of an adversarial evader who actively avoids capture. Due the this aspect, the solutions to pursuit-evasion problems are game theoretic and distinguish themselves from their target finding counterparts (e.g. [11,14]) where the evader's motion is independent from the pursuer's.

In our present work, we propose randomized pursuer strategies for the visibility based pursuit-evasion problem. Randomization is a powerful technique which allows us to solve many problems that are not solvable by deterministic algorithms and has found wide-spread applications in many areas ranging from computational geometry to cryptography.

As we show in the following sections, it turns out that randomization provides a drastic increase in the power of the pursuers. For example, it is known that there are simply-connected environments where $\Theta(\log n)$ pursuers are required [2] in order to locate the evader with deterministic strategies. In contrast, we show that a single pursuer can locate the evader in any simply-connected environment with high probability, even if the evader knows the pursuer's location at all times and has unbounded speed (Theorem 1). The power of randomized strategies comes from the fact that the evader has no prior knowledge of the random decisions inherent in such strategies. It is worth noting that the randomized strategies work against any evader strategy and require no prior information about the strategy of the evader.

We also address the harder task of capturing the evader. For this problem we present a strategy for two pursuers, one of which is at least as fast as the evader. The strategy is based on the randomized strategy to locate the evader and the solution of a known lion and man problem [12] which is reviewed in Section 3.1. The same strategy can be used to capture the evader while protecting a door. This problem was introduced in [7] to model scenarios where the goal is to locate the evader which may leave the polygonal area through a door and win the game.

The two-pursuer strategy can be modified so that a single pursuer can also capture the evader. However, the expected time to capture in this case, though finite, may be significantly longer than the expected time to capture with two pursuers.

1.1 Randomized strategies

The power of randomization in the context of pursuit-evasion games is nicely illustrated by the example in Figure 1. A similar example can be found in [2].

In this example, a single pursuer can never locate the evader using a deterministic strategy: Let us distinguish four points A, B, C and D as shown in the figure. We will slightly abuse the notation and use these points to refer to the maximal convex region inside the polygon that contains these points as well. Now suppose the pursuer has a deterministic strategy of visiting those

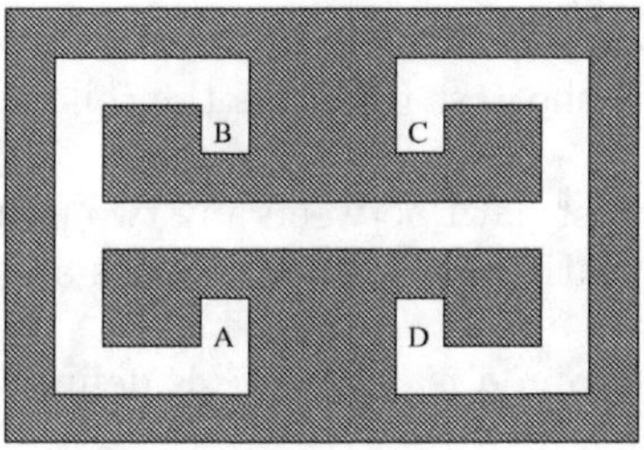

Fig. 1. A single pursuer can not capture an evader using deterministic strategies.

points in the order A, B, C, D. In this case, the evader can first hide at B and escape to D while the pursuer is visiting A. Afterwards, it can repeat the same strategy and escape to B when the pursuer is at C. If the pursuer visits the points in a different order, it is easy to see that the evader can find a similar strategy and avoid the pursuer. Therefore, in this polygon one pursuer can never locate the evader.

Now consider the following randomized strategy: Instead of committing to a deterministic strategy, the pursuer selects one of the regions $\{A, B, C, D\}$ uniformly at random and visits that point. It is easy to see that if the pursuer guesses the region where the evader is located, then the evader can not escape and the probability of this desired event is $\frac{1}{4}$. The crucial observation is that since the evader does not know which region the pursuer will visit, it can not choose a strategy based on the order of points visited by the pursuer.

One might suspect that the desired event of locating the evader is not guaranteed since the pursuer can keep guessing a wrong corridor forever. However, this probability can be made arbitrarily small by repeating the same strategy a few times. This is because, if k is the number of trials, the probability of missing in all k trials is only $(\frac{3}{4})^k$ in this example and this probability decreases exponentially in k. In general, if the probability of capture is p, the expected number of rounds to capture is $\frac{1}{p}$. Note that each round is independent. We can obtain the expected time to locate the evader as follows: Since the length of a round is bounded by the time to travel between two furthest points in the polygon (say T), the expected time to capture is $\frac{T}{p}$. We can convert the expected time to a high probability argument by repeating the experiment roughly $\frac{1}{p} \log \frac{1}{p}$ times using the Chernoff bound. For details of this technique the reader is referred to [8].

1.2 Preliminaries

Let P be the input polygon including its interior and V be the set of vertices of P. Unless stated otherwise, n denotes the number of vertices of the polygon. Two points $u, v \in P$ can *see* each other if the line segment uv lies entirely in P.

We use $d(u,v)$ to denote the length of the shortest path from u to v that remains inside P. The shortest path has the following property.

Property 1. The shortest path between any two points u and v inside a polygon P is a polygonal path whose inner vertices are vertices of P.

The *shortest path tree from a point x in P* is defined as $\cup_{v \in V(P)} d(x,v)$, where $V(P)$ is the set of vertices of P.

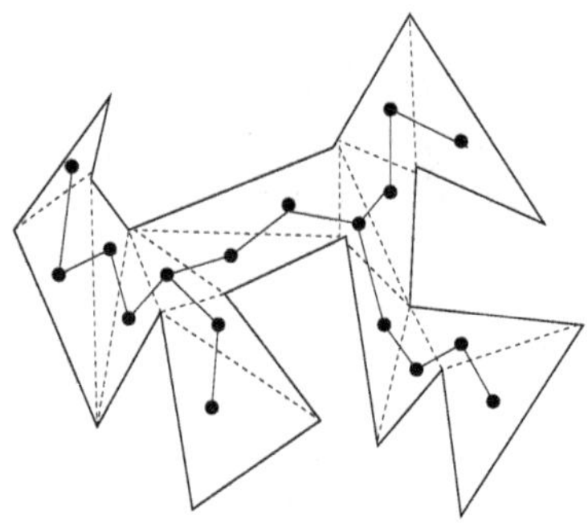

Fig. 2. Triangulation of a polygon and its dual tree.

The *diameter* of the polygon P, denoted $diam(P)$, is defined as $\max_{u,v \in P} d(u,v)$. Without loss of generality, we assume that the units are normalized so that the time it takes to travel the diameter is also $diam(P)$. A polygon is *simply-connected* if any simple closed curve inside the polygon can be shrunk to a point. All the polygons considered in this paper are simply-connected.

The *triangulation* of a polygon is a decomposition of the polygon into triangles by a maximal set of non-intersecting diagonals (see Figure 2). The dual of a triangulation is a graph whose vertices correspond to the triangles. There is an edge between two vertices if the corresponding triangles share a side. It is well known that the triangulation of a simply-connected polygon has exactly $n-2$ triangles where n is the number of vertices of the polygon. In addition, the dual of the triangulation is a tree [9].

2 Locating the evader

In this section, we present a randomized algorithm for the visibility-based pursuit-evasion problem [2]:

An evader whose location is unknown to the pursuer is hiding inside a polygon P. The evader has global visibility (i.e. it knows the location of the pursuer at all times) and can be arbitrarily faster than the pursuer. Moreover, the evader is unpredictable – that is, no prior information about its strategy is available. The pursuer's goal is to locate the evader by establishing line-of-sight visibility: He wins the game if he ever sees the evader. The evader wins the the game if it can avoid being located forever.

Next, we show that for any simple polygon P, the pursuer can locate the evader in $O(n \cdot diam(P))$ expected time.

2.1 The pursuer strategy

Given polygon P, the pursuer first triangulates the polygon. The pursuer's strategy is divided into rounds of length at most $diam(P)$. Let T be the triangulation tree (see Figure 2) rooted at the triangle that contains the pursuer's initial location at the beginning of a round. For any triangle t let $t_1, .., t_k$, $k \leq 3$ be the children of t. We use the notation $T(t)$ to denote the subtree of T rooted at the triangle t. Figure 3 is provided for quick reference to the notation used in this section.

The pursuer's strategy relies on the following observation: Suppose the pursuer is inside triangle t and the evader is located inside a triangle contained in $T(t_j)$ for some j. Then, while the pursuer is located at t, the evader can not enter any triangle contained in $T(t_i)$, $i \neq j$ without being seen by the pursuer. This is because the triangle t is a separator for the subtrees $T(t_i)$. Moreover, this property is preserved if the pursuer moves to the triangle t_j.

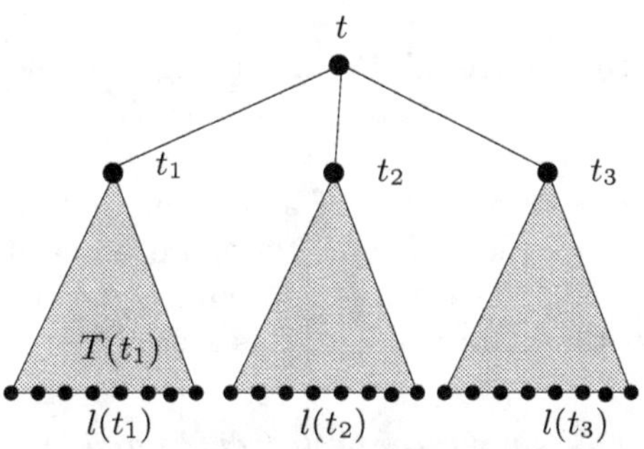

Fig. 3. Notation used for Lemma 1. Each vertex of the tree corresponds to a triangle in the triangulation tree.

Therefore, had the pursuer known the subtree that contains the evader, he could gradually move towards it while preventing the evader to move from one subtree to another. This process guarantees that the pursuer can enter the triangle which contains the evader and this clearly implies that the evader would be located. Of course, the pursuer does not know where the evader is. This is where we will utilize randomization.

The pursuer will guess the subtree that contains the evader according to the following rule:

Let $l(t)$ denote the number of leaves of the subtree $T(t)$. Suppose the pursuer is located in triangle t and let $t_1, .., t_k$ be the children of t (see Figure 3).

Let $L = \sum_{i=1}^{k} l(t_i)$. With probability $\frac{l(t_i)}{L}$, the pursuer picks the child t_i and moves there. The round is over whenever the pursuer arrives at a leaf of T.

Next, we show that using this guessing strategy, the pursuer efficiently locates the evader.

Lemma 1. *Let T be the triangulation tree rooted at the triangle that contains the pursuer's initial location in the beginning of the round. At each round, if the pursuer follows the guessing strategy described above, he can locate the evader with probability at least $\frac{1}{L}$.*

Proof. The lemma is proven by induction on the height of T. The basis, where the height of t is 0, corresponds to the case where the input polygon is a triangle. The pursuer trivially locates the evader with probability 1 in this case.

Let $p(t)$ be the probability that the evader is located within a round, starting from triangle t and inductively assume that the lemma is true for all trees of height less than or equal to j.

Given a triangulation tree of height $j+1$, the probability of success starting from the root t is:

$$p(t) \geq \min\left\{\frac{l(t_1)}{L}p(t_1), \ldots, \frac{l(t_k)}{L}p(t_k)\right\} \tag{1}$$

Note that all the subtrees $T(t_i)$ have height at most j, therefore by the inductive hypothesis we have $p(t_i) \geq \frac{1}{l(t_i)}$ for all i and the lemma follows.

Clearly, the number of leaves of any triangulation tree is less than the number of vertices of the polygon, therefore at each round the evader is located with probability at least $\frac{1}{n}$. Moreover, since the length of a round is $diam(P)$, we have the main result of this section:

Theorem 1. *In any simply connected environment P, against any evader strategy, the expected time to locate the evader with a single pursuer is at most $n \cdot diam(P)$ where n is the number of vertices and $diam(P)$ is the diameter of the polygon.*

Remark 1. Any simply-connected polygon can be partitioned into a minimum number of disjoint convex polygons in polynomial time [5,1]. It is easy to see that the dual of such a partition will be a tree. Therefore, instead of using a triangulation dual, the pursuer can execute the strategy described above using the dual of the convex partition. However, in general this does not improve the expected capture time. For example, for the polygon shown in Figure 5, the number of leaves of the triangulation dual is equal to the number of leaves of a minimum convex partition.

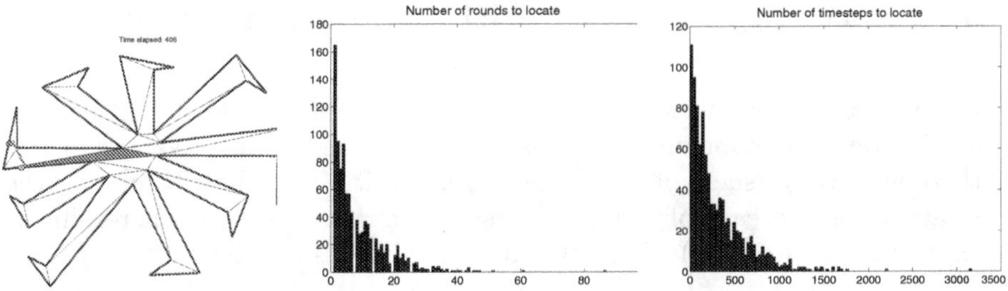

Fig. 4. **Left:** An instance of the simulator showing the triangulation of the environment as well as the hiding location of the evader. **Middle:** The histogram of the number of rounds required to locate evader in 1000 simulations. **Right:** The histogram of the time-steps required to locate evader in 1000 simulations.

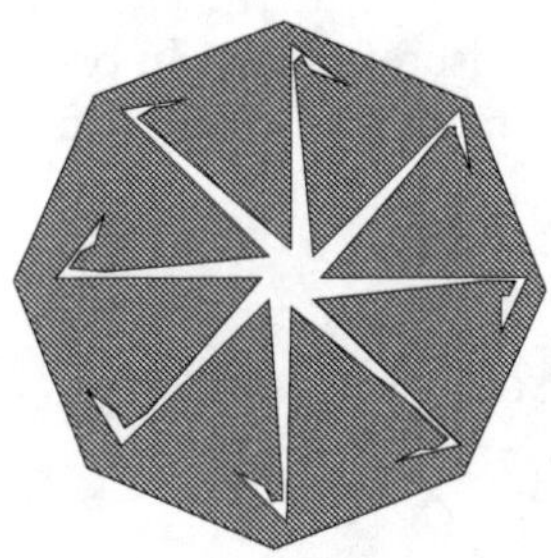

Fig. 5. For any randomized pursuer strategy, the expected time to capture the evader in this star with hooks is $O(n \cdot diam(P))$.

2.2 Lower bounds

One might suspect that the expected time to locate an evader can be improved using a more sophisticated strategy. Unfortunately, this is not possible: The polygon in Figure 5 is a $k-$star with hooks attached at the end of each spike (in the figure $k = 8$). The evader's strategy is to choose a hook at random and hide there until the end of the game. In order to locate the evader, the expected number of spikes searched by the pursuer is $\frac{k}{2}$ and it takes $diam(P)$ steps to travel from one spike to another. Since the number of vertices is a constant multiple of k, the time it takes to locate the evader is $\Omega(n \cdot diam(P))$. In fact, using the well-known technique due to Yao, this argument can be extended to show that the expected time to capture the evader for *any* randomized pursuer strategy is $\Omega(n \cdot diam(P))$ (see [8] for details).

We present the results of a simulation of the pursuer and the evader strategy for this environment in Figure 4.

3 Capturing the evader with two pursuers

In this section, we move on to the more challenging task of capturing the evader, defined as moving to the same point as the evader. We assume that there are two pursuers in the game. Later, we will show how to modify their strategy for the case of a single pursuer (at the expense of increasing the expected capture time). We make the following assumptions:

- The pursuers can communicate with each other at all times. [1]
- The evader has global visibility. It is unpredictable, i.e., no prior information about its strategy is available.
- Both pursuers have only line of sight visibility. One pursuer is at least as fast as the evader.
- The game is played in discrete time. In order to simplify presentation, we assume that the players move in turns: first the evader moves, followed by the pursuers.

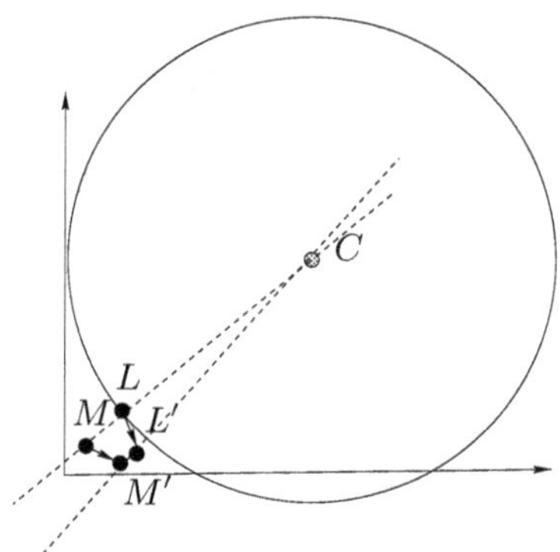

Fig. 6. Lion's strategy

The strategy of one of the pursuers is based on the solution to a problem known as the *lion and man* problem [12]. We present an extension of this strategy in the case of a (possibly non-convex) polygonal environment. One of the major difficulties for our pursuers is that the evader may not be visible at all times, in which case the lion's strategy is not well-defined. The second pursuer will use the strategy presented in the previous section to tackle this difficulty.

We start with a review of the lion's strategy.

[1] This assumption can be relaxed to line-of-sight communication, i.e. the pursuers can communicate only if they see each other.

3.1 The lion and man problem

The *lion and man* problem with discrete time in the nonnegative quadrant of the plane is attributed to David Gale [3]. Let the initial positions of the lion and man be $L_0 = (x_0, y_0)$ and $M_0 = (x'_0, y'_0)$, respectively. In each round, first the man moves to any point in the quadrant at distance at most 1 from his current position, and then the lion does the same. The lion wins if he moves to the current position of the man. The man wins if he can keep escaping for infinitely many rounds. In [12], Sgall proves that, when both $x'_0 < x_0$ and $y'_0 < y_0$, the lion always catches the man in a finite number of rounds. The number of moves required is bounded by a quadratic function in x_0, y_0 and the slope (or its inverse) of the line segment L_0M_0.

3.2 Lion's strategy

Let the initial positions of the lion and man be $L_0 = (x_0, y_0)$ and $M_0 = (x'_0, y'_0)$, respectively. In the beginning of the game, the lion finds a point C on the line M_0L_0 such that L_0 is inside the segment M_0C and the circle with center C, radius $|CL_0|$ and passing through L_0 intersects both axes. Among all possible such circles, it chooses the one whose center is closest to the origin. C remains fixed throughout the game.

Let L and M denote the current positions of the lion and the man respectively (see Figure 6). Let M' denote the point the man moves to, $|MM'| \leq 1$. If $|LM'| \leq 1$, the lion catches the man. Otherwise, it moves to a point L' on the line $M'C$ such that $|L'L| = 1$. There are two such points, it chooses the one closer to the man.

Definition 1. We will refer to this move as the *lion's move from L with respect to C and M'.*

The lion's move maintains the following:

Lemma 2 ([12]). *If the lion does not catch the man in the current move then*

(i) M' has both coordinates strictly smaller than C,
(ii) L' is inside the segment $M'C$, and
(iii) $|L'C|^2 \geq 1 + |LC|^2$.

Proof. See [12].

3.3 The strategy to capture the evader

Let $p_1(t), p_2(t)$ and $e(t)$ denote the locations of the pursuers and the evader, respectively, at time t. In the beginning of the game the two pursuers move together and search for the evader using the strategy described in the previous

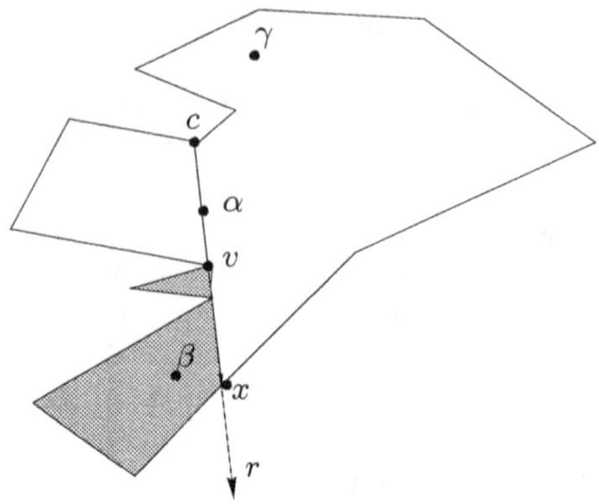

Fig. 7. Pocket with respect to c and v

section. Without loss of generality, we assume that the game starts at $t = 0$ where $p_1(0) = p_2(0) = o$ and $e(0)$ is visible from o. We will sometimes refer to point o as the *origin*. The origin will be fixed until the evader is captured. Let $d_1(t) = d(p_1(t), o)$, $d_2(t) = d(p_2(t), o)$, and $d_e(t) = d(e(t), o)$.

Definition 2. Suppose $e(t)$ is visible from $p_1(t)$ but $e(t+1)$ is not visible from $p_1(t)$. This means that the shortest path P from $p_1(t)$ to $e(t+1)$ is composed of at least two line segments (Property 1). The first vertex on the path from $p_1(t)$ to $e(t+1)$ is called a *pseudo-blocking vertex.*

Let r be the ray starting from a vertex c and passing from another vertex v that is not adjacent to c. In the sequel, c will be the center of the circle for the lion's move and v will be the pseudo-blocking vertex. Consider the first time the ray r leaves the polygon P after it passes through v and let x be the point on $r \cap P$ just before this happens (see Figure 7). The line segment vx splits the boundary of the polygon into two chains. The chain which does not contain the point c, together with the line segment vx defines a polygon. We will refer to this polygon as *the pocket with respect to c and v*. The line segment vx is referred to as the *entrance of the pocket.*

We will utilize the following properties of pockets:

Property 2. Let α be a point on the line segment cv and β be a point in the pocket with respect to c and v. The line segment αv is contained in the shortest path from α to β (Figure 7).

Property 3. Let R be a pocket with respect to c and v inside a polygon P. Any path from $\beta \in R$ to $\gamma \in P - R$ crosses the entrance of the pocket (Figure 7).

Looking ahead, let us describe how we will utilize these properties: Suppose pursuer p_1 is moving towards the evader and the evader disappears. Let v be the current pseudo-blocking vertex. If p_1 moves towards v, Property 2

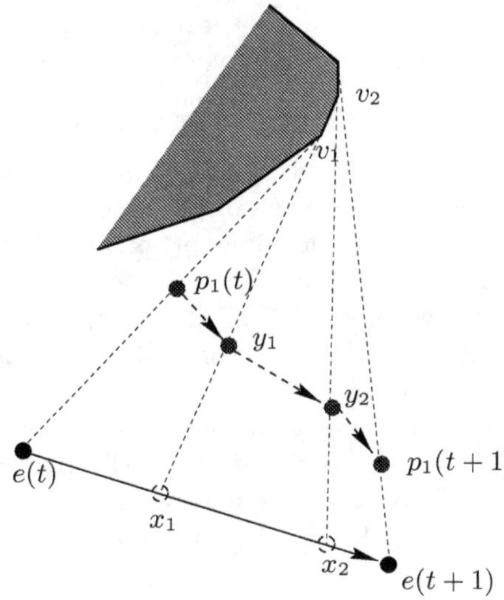

Fig. 8. The extended lion's move

implies that it is still moving on the shortest path from the evader to the origin. If the evader becomes visible before p_1 reaches v, Property 3 implies that it must cross the entrance of the pocket and p_1 can continue its strategy (described in the next section) as if the evader has not disappeared.

If the evader is not visible when p_1 arrives at v, then v becomes a *blocking vertex*. At this point, the second pursuer will enter the game.

Next, we present the details of the strategies of p_1 and p_2.

3.4 Strategy of Pursuer p_1

As stated earlier, we assume that pursuer p_1 is at least as fast as the evader. At time step t, p_1 moves according to the following strategy:

If the evader is visible, he performs a *extended lion's move* which is defined as follows: Let τ be the shortest path from $e(t)$ to $e(t+1)$. Without loss of generality, p_1 will pretend that the evader followed τ. As a point x moves from $e(t)$ to $e(t+1)$ along τ, the vertices on the shortest path from x to the origin o may change. However, the number of changes is at most n: The first vertex on the shortest path from x to o must be one of the vertices of the polygon. Since τ is the shortest path from $e(t)$ to $e(t+1)$, each vertex of the polygon can be this first vertex for at most one contiguous sub-path in τ. Let $x_1, \ldots, x_{k-1}$ correspond to the points on τ where such changes occur, we define $x_0 = e_t$ and $x_k = e(t+1)$. Let v_i be the first vertex on the shortest path from x_i to o. The extended lion's move consists of $k-1$ phases. During Phase i, $i = 1, \ldots, k$, pursuer p_1 performs the lion's move with respect to v_i and x_i (see Figure 8). Note that the time spent by the pursuer in Phase i is equal to the time spent by the evader in traveling from x_{i-1} from x_i.

If the evader was visible in the previous time step, but is not visible any more, let v be the pseudo-blocking vertex. Pursuer p_1 moves towards v until he reaches it. If the evader becomes visible before p_1 arrives at v, he continues with the lion's move. Otherwise, v becomes a blocking vertex.

If the evader is still not visible after p_1 reaches the blocking vertex, he waits for p_2 to report the location of the evader. Let R be the current pocket defined with respect to the blocking vertex and the current center. There are two possibilities.

1) The evader reveals itself to p_1. Then, by Property 3, this must happen before the evader crosses the entrance of R. In this case p_1 continues the game with the lion's move.

2) Pursuer p_2 finds the evader located at e. Let v' be the first vertex on the shortest path from v to e and R' be the pocket with respect to v and v'. In this case, v' becomes a pseudo-blocking vertex, R' becomes the new pocket and p_1 continues his strategy by moving towards v'.

3.5 Strategy of Pursuer p_2

The task of pursuer p_2 is to search for the evader when it is not visible to p_1. When the evader disappears from the sight of p_1, pursuer p_2 waits until p_1 reaches the blocking vertex. Afterwards, p_2 locates the evader using the strategy described in the previous section and reports the location of the evader to p_1.

3.6 Properties of Pursuer p_1's strategy

Lemma 3. *For all times t, pursuer p_1 maintains the following invariants until the evader is caught:*

(I1) $p_1(t)$ is on the shortest path from o to $e(t)$.

(I2) $d_1(t+1)^2 \geq d_1(t)^2 + \frac{1}{n}$ if $p_1(t) \neq p_1(t+1)$.

Proof of Invariant I1: We prove the invariant by induction. Assume that it holds at time t.

First consider the case where p_1 can see e at time t. Let the first vertex on the shortest path from $e(t)$ to o be u. It follows that $p(t)$ is in the line segment joining u to $e(t)$, since if $p(t)$ is between o and u on the shortest path, he would not be able to see $e(t)$.

Let x denote the evader's position at an arbitrary time in the time interval $[t, t+1)$. Suppose when the evader is at x, the first vertex on the shortest path from x to o changes from u to v. Note that p_1 can see the evader until this point. Then, the shortest path from x to o passing through u and the shortest path from x to o passing through v have the same length. This implies that u, v, and x have to be collinear. For otherwise, a shorter path from x to o can be found in the interior of the polygon formed by these two presumed shortest paths from x to o, which is a contradiction.

This implies that either u is an ancestor or a descendant of v in the shortest path tree rooted at o. If u is an ancestor, at the point x where the switch occurs, p_1 could either be on the segment vx in which case it can continue the lion's move in the next phase or p_1 is on the segment uv, in which case e will become invisible to p_1 after x. In this case, p_1 must be either moving towards a pseudo-blocking vertex or waiting at a blocking vertex. In both cases, the invariant is maintained by Property 2. If u is a descendant of v, then p_1 is already on the segment ux and hence on the segment vx. Hence it can continue the lion's move in the next phase. The invariant is therefore maintained as a corollary of Lemma 2.

Otherwise, if p_1 does not see the evader at time t, he must be either waiting at a blocking vertex or moving towards a pseudo-blocking vertex. In both cases, the invariant is maintained by Property 2.

Proof of Invariant I2:

If p_1 is moving towards a pseudo-blocking vertex, his distance to the origin is increasing by 1 and the invariant is maintained.

Next, we show that the extended lion's move maintains the invariant: Suppose the lion's move has $k \leq n$ phases and consider phase i of the extended lion's move where the evader moves from the point x_{i-1} to x_i. Suppose, during this phase the pursuer p_1 moved from point y_{i-1} to y_i (see Figure 8) and let v_i be the center of the circle for the lion's move during this phase.

Let $\alpha_i = d(o, y_i) - d(o, y_{i-1})$.

As a corollary of Lemma 2 we have

$d(y_i, o)^2 \geq d(y_{i-1}, o)^2 + \alpha_i^2$.

Summing up over all phases we get the total progress to be $\sum_{i=1}^{k} \alpha_i^2$.

This expression when subject to $\sum_{i=1}^{k} \alpha_i = 1$ is minimized when all $\alpha_1 = \ldots = \alpha_k = \frac{1}{k}$. Therefore we have $d_1(t+1)^2 \geq d_1(t)^2 + \frac{1}{k}$ which implies the invariant I2. □

The combined strategy of the two pursuers can be viewed as follows: Pursuer p_1 moves only when it knows the shortest path from the evader to the origin o. Performing the lion's move is equivalent to growing a disk inside the polygon whose center is at the origin o and passes through the current location of p_1. By invariant I1, the evader can never enter the disk. Further, the disk is still protected if p_1 does not move. Invariant I2 implies that, whenever p_1 moves, the disk monotonically grows and the evader is eventually squeezed between p_1 and the polygon boundary.

Pursuer p_2 moves only when p_1 does not know the evader's path to the origin. It locates the evader using the randomized strategy given in the previous section and reports its location to p_1 so that p_1, in turn, can keep growing the disk and eventually capture the evader.

3.7 Expected time to capture

Let T_1 be the time it takes the faster pursuer (who performs the lion's move) to travel the diameter of the polygon. By Invariant I2 (Lemma 3), this pursuer will capture the evader in nT_1^2 steps. However, in the meantime, the other pursuer may have to search for the evader. Each such search ends in time $T_2 \cdot n \cdot \log n$ with high probability, where T_2 is the time for pursuer p_2 to travel the diameter of the polygon. Moreover, once a vertex becomes a blocking vertex, it will never become a blocking vertex again (it will be included in the ball defined by the origin and p_1), therefore at most n such searches will be necessary. In conclusion, the expected time to capture the evader is $O(nT_1^2 + T_2 \cdot (n^2 \log n))$ with high probability.

3.8 Capturing the evader with a single pursuer

Suppose we have only pursuer p_1. In this case, instead of waiting for p_2 to find the evader, p_1 can guess the first vertex on the shortest path from the evader to his current location and move there.

Consider the shortest path tree T from the origin o to the vertices of the polygon. For each vertex v, let $l(v)$ be the number of leaves of the subtree $T(v)$ of T rooted at the vertex v. Then the probability that the pursuer's guess will be successful if he is located at v is at least $\frac{1}{l(v)}$. If the guess is correct and the evader is visible, the pursuer continues with the lion's move. However, in case of a wrong guess the evader may end up in an advantageous location and move towards the origin o, in which case the pursuer must restart the game. Further, if all the guesses are correct, no vertex can be a blocking vertex more than once. Continuing this way we can obtain a worst-case lowerbound on the probability of success. Unfortunately, this bound can be possibly exponentially small in the number of reflex vertices in the environment. However, the expected time to capture the evader is still finite for any simply connected environment and this strategy may still be practical for simple settings.

One might suspect that an analysis similar to the one in Section 2 can be applied to prove that the expected time to capture is polynomial. The reason such an analysis does not apply directly is that even if the pursuer and the evader are co-located in a leaf triangle, the capture game still continues and the evader can move to another triangle in the tree. Therefore the number of guesses may exceed the depth of the tree, resulting in a possibly exponential capture time. This poses an interesting trade-off between the pursuer's visibility and the capture time. If the pursuer can somehow track the evader at all times (perhaps using a satellite), then Lemma 3 implies that he could capture the evader in time $O(n \cdot diam(P)^2)$. If this is not possible though, he can either use a second pursuer for locating the evader and still capture it in polynomial time or simultaneously search and capture which results in a much longer capture time.

3.9 Polygonal rooms with a door

In [7], Lee et al. studied the following variant of the pursuit-evasion problem: The input is a pair (P, d) where P is the polygonal room the game is played in and d is a *door*, a point marked on the boundary of P. The goal is to devise a strategy for the pursuer to eventually see the evader, in such a way that the evader can not escape through the door. The authors presented a characterization of polygons where a single pursuer with very narrow visibility (represented by a single ray) can locate the evader before it reaches the door.

In a similar scenario, the two pursuer algorithm presented in Section 3 can be used to *capture* an evader before it exits through the door. The only modification necessary is the following: Initially, pursuer p_1 is located at the door d and waits until pursuer p_2 locates the evader. Afterwards, he continues with the lion's move with respect to d. This ensures that the evader can never enter the disk whose origin is d and passes through the current location of p_1. Therefore, the door is always protected until the evader is captured.

4 Conclusion and Future Work

In this paper, we studied the visibility-based pursuit evasion game and showed that using a randomized strategy a single pursuer can locate an unpredictable evader in any simply-connected polygonal environment. The evader may be arbitrarily faster than the pursuer and it may know the location of the pursuer at all times.

The randomized strategy has some desirable properties: First, as shown in [2], there are polygonal environments which require an arbitrary number of pursuers if they are restricted to deterministic strategies. Therefore on such environments, a randomized strategy is mandatory for locating the evader with a single pursuer. Moreover, even if the polygon is deterministically searchable by a single pursuer, it is known that some of these polygons require revisiting parts of the polygon $\Omega(n)$ times [2]. Hence, the expected time to capture with a randomized strategy is comparable to the time to capture with a deterministic strategy. However, the randomized strategies may be preferable to the deterministic strategies, as they do not require complicated data structures and costly preprocessing.

Second, the randomized strategy to locate the evader does not require an exact map of the environment: It is based on the dual graph of the triangulation, therefore it is insensitive to errors in the map of the environment. An interesting research direction is to incorporate the navigation strategies in [14] which require a minimal representation of the environment.

Another interesting extension is the case of non-polygonal environments. The randomized strategy can be used to locate the evader in non-polygonal, simply-connected environments. For example, this could be done by replacing the triangulation tree (Lemma 1) with the decomposition studied in [6].

We have also studied the more challenging problem of capturing the evader. For this problem, we presented a strategy for two pursuers (one of which is as fast as the evader) to capture the evader in expected time polynomial in the number of vertices and the diameter of the environment. The strategy can be modified for a single pursuer, however it is not clear whether the expected time to capture remains a polynomial in the number of vertices. We leave this as a future research direction.

References

1. B. Chazelle and D. Dobkin. Computational geometry. *Optimal convex decompositions, G. Toussaint (editor)*, pages 63–133, 1985.
2. L. J. Guibas, J.-C. Latombe, S. M. LaValle, D. Lin, and R. Motwani. A visibility-based pursuit-evasion problem. *International Journal of Computational Geometry and Applications*, 9(4/5):471–, 1999.
3. R. Guy. Unsolved problems in combinatorial games. In *Proc. Symp. Applied Mathematics*, pages 183–189, 1991.
4. J. P. Hespanha, G. J. Pappas, and M. Prandini. Greedy control for hybrid pursuit-evasion games. In *Proceedings of the European Control Conference*, pages 2621–2626, Porto, Portugal, September 2001.
5. M. Keil. Decomposing a polygon into simpler components. *SIAM J. Computing*, 14:799–817, 1985.
6. S. M. LaValle and J. Hinrichsen. Visibility-based pursuit-evasion: The case of curved environments. *IEEE Transactions on Robotics and Automation*, 17(2):196–201, Apr. 2001.
7. J.-H. Lee, S.-M. Park, and K.-Y. Chwa. Searching a polygonal room with one door by a 1-searcher. *International Journal of Computational Geometry and Applications*, 10(2):201–220, 2000.
8. R. Motwani and P. Raghavan. *Randomized Algorithms.* Cambridge University Press, 1995.
9. J. O'Rourke. *Computational Geometry in C.* Cambridge University Press, 1998.
10. S.-M. Park, J.-H. Lee, and K.-Y. Chwa. Visibility-based pursuit-evasion in a polygonal region by a searcher. *Proceedings of the International Colloquium on Automata, Languages and Programming (ICALP)*, 2076, 2001.
11. A. Sarmiento, R. Murrieta-Cid, and S. Hutchinson. An efficient strategy for rapidly finding an object in a polygonal world. In *Proc. IEEE/RSJ International Conference on Intelligent Robots and Optimal Systems (IROS)*, 2003.
12. J. Sgall. Solution of David Gale's lion and man problem. *Theoret. Comput. Sci.*, 259(1-2):663–670, 2001.
13. I. Suzuki and M. Yamashita. Searching for a mobile intruder in a polygonal region. *SIAM Journal on Computing*, 21(5):863–888, 1992.
14. B. Tovar, S. M. LaValle, and R. Murrieta. Optimal navigation and object finding without geometric maps or localization. In *Proc. IEEE International Conference on Robotics and Automation*, 2003.
15. R. Vidal, O. Shakernia, J. Kim, D. Shim, and S. Sastry. Probabilistic pursuit-evasion games: Theory, implementation and experimental evaluation. *IEEE Transactions on Robotics and Automation*, 18:662–669, 2002.

Semi-Differential Invariants for Recognition of Algebraic Curves

Yan-Bin Jia and Rinat Ibrayev

Iowa State University, Ames, IA 50011, USA

Abstract. This paper studies the recognition of low-degree polynomial curves based on minimal tactile data. Differential and semi-differential invariants have been derived for quadratic curves and special cubic curves that are found in applications. Such an invariant, independent of translation and rotation, is computed from the local geometry at up to three points on a curve. Recognition of the curve reduces to invariant verification with its canonical parametric form determined along the way. In addition, the contact locations are found on the curve, thereby localizing it relative to the touch sensor. Simulation results support the method in the presence of small noise. Preliminary experiments have also been carried out. The presented work distinguishes itself from traditional model-based recognition in its ability to simultaneously recognize and localize a shape from one of several classes, each consisting of a continuum of shapes, by the use of local data.

1 Introduction

Human can feel a shape through touch. Essentially, the action is performed to detect some geometric features on the shape which are then subconsciously synthesized in the brain. Typical geometric features include, for instance, smoothness, saliences, concavities, etc.

Supported by touch sensing, the robot can also obtain shape information without the help of a vision system. Such shape inference is important, for example, when camera occlusion becomes inevitable or when motion is involved. Since tactile data are local (and one-dimensional for point contact), seemingly they convey only a limited amount of geometric information. But how much shape knowledge can the robot really acquire then?

Fig. 1 illustrates a hand with two tactile fingers touching an object. Suppose through local movements the fingers are able to estimate information such as the curvatures at several points of contact. And suppose the shape is known to be from a finite family of parametric curves. Then we would like to *recognize the shape as well as determine the finger placement*.

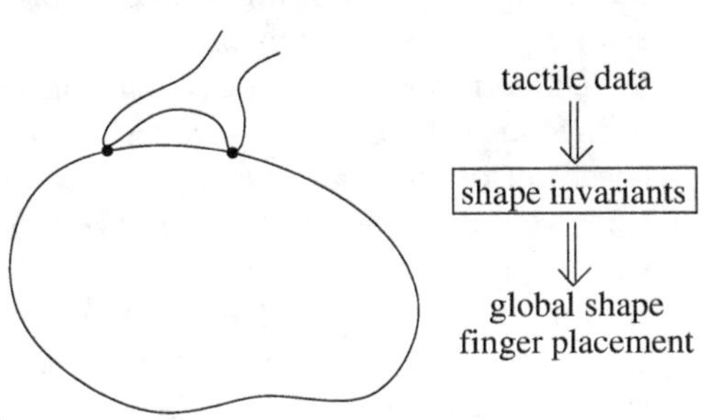

Fig. 1. A robotic hand touching an object to recognize its shape.

M. Erdmann et al. (Eds.): Algorithmic Foundations of Robotics VI, STAR 17, pp. 267–282, 2005.

This problem draws several distinctions from traditional model-based recognition. First, every model here is not a real shape but rather a *continuum* of shapes parametrized in the same form. Second, we would like to keep the sensor data to the minimum. This is because a touch sensor, unlike a vision system, does not generate global shape data. Third, we hope to determine where the tactile data were obtained on the shape.

The characteristics of our problem naturally suggest an approach based on differential and semi-differential invariants. Such invariants of a shape are independent of its position and orientation, the computation of which is often a burden. In this paper, we are interested in invariants that are also *independent of point locations on a shape at which they are evaluated.* Given the local nature of touch sensing, such shape descriptors should be computable from measurements at just a few points. Our investigation will be focused on quadratic and cubic spline curves.

1.1 Related Work

There are two primary recognition strategies in model-based vision. The first one hinges on the recovery of viewing parameters (thus the pose). Kriegman and Ponce [12] constructed implicit shape equations from image contours and then solved for viewing parameters through data fitting. The second approach is to develop descriptors that are invariant to Euclidean, affine, or projective transformation, or to camera-dependent parameters [15,19].

Algebraic invariants are expressions of the coefficients of polynomial equations describing curved shapes. The foundation was due to Cayley, Sylvester, Young, and among others, Hilbert [8], who offered a procedure that constructs all independent algebraic invariants for a given curve or surface. In real applications, polynomials are fit to image data and their coefficients are extracted for invariant evaluation. Keren [10] and Forsyth *et al.* [6] presented efficient methods for finding algebraic invariants and demonstrated on recognition of real objects. Civi *et al.* [4] also conducted object recognition experiments with algebraic invariants of Euclidean, affine, and projective groups.

One drawback of algebraic invariants is the requirement of global shape data. This is almost impossible to provide by a touch sensor, or by a vision system in case of occlusion.

Differential invariants depend on local data and deal with situations like occlusion well. Up till now, vision- and invariant-based recognition has focused on differential invariants that are independent of various transformation groups but not of point locations on a shape. Calabi *et al.* [3] introduced the "signature curve" that is invariant to Euclidean or affine transformation. Rivlin and Weiss [18] derived differential invariants for a shape by applying to its quartic fit the same transformation that turns an osculating curve (a cubic) into the canonical form.

Semi-differential invariants combine global constraints and local information to ease the correspondence issue faced by non-invariant-based methods and also relieve the burden on estimating higher order derivatives for differential invariants. The theoretical foundation for this type of invariants was presented by Moons *et al.* [14]. Pajdla and Van Gool [16] used simple semi-differential invariants to match curves extracted from range data in the presence of partial occlusion.

In touch sensing, shape recognition has long built on the notion of "interpretation tree", which represents all possible correspondences between geometric features of an object and tactile data. Grimson and Lozano-Pérez [7] identified and localized a 3-D polyhedron from a set of models using tactile measurements of positions and surface normals. Fearing [5] described how a cylindrical tactile fingertip could recover the pose of a generalized convex cone from a small amount of data.

Allen and Michelman [1] fit a superquadric surface to sparse data obtained by a Utah-MIT hand around an object as its reconstructed shape. Moll and Erdmann [13] showed how to simultaneously estimate the shape and motion of an unknown convex object from tactile readings on multiple manipulating palms under frictionless contact.

A method based on the interpretation tree or least-squares fitting needs to recover the pose. This may be costly and often unnecessary. Not until very recently did differential invariants start to find applications with touch sensing. For spheres, cylinders, cones, and tori, Keren *et al.* [11] constructed descriptors using curvatures and torsions and their higher order derivatives and computed them from a large amount of tactile measurements.

In this work, we have extended our recent result [9] and derived semi-differential invariants that not only recognize the classes but also recover the algebraic descriptions of quadratic and certain cubic curves from curvature and derivative measurements.

2 Curve Invariants

The touch sensor in contact with a 2D object can "feel" its local geometry, which is described by the curvature. At the contact point denote by ϕ the tangential angle formed by the tangent of the boundary curve $\alpha(t) = (x(t), y(t))$ with the x-axis. The *curvature* κ is the rate of change of ϕ with respect to arc length s, that is,

$$\kappa = \frac{d\phi}{ds} = \frac{x'y'' - x''y'}{(x'^2 + y'^2)^{3/2}}. \tag{1}$$

Curvature is independent of parametrization, rotation, and translation. We are also interested in the derivative of curvature with respect to arc length:

$$\kappa_s = \frac{d\kappa}{dt}\frac{dt}{ds} = \frac{\kappa'(t)}{(x'^2 + y'^2)^{1/2}}. \tag{2}$$

Section 7 will look at how curvature and derivative can be reliably estimated from real data. Until then we just assume that *these two quantities are measurable.*

2.1 Signature Curve

A *differential invariant* of a curve is a real-valued function that depends on the curve and its derivatives but not on a specified transformation group or parametrization. The transformation group considered in this paper is the plane Euclidean group $SE(2)$. More intuitively, the value of a differential invariant depends on the point location on the curve but not on the curve's rotation and orientation.

Theorem 1. *Every differential invariant of a plane curve is a function of the curvature κ and its derivatives with respect to arc length.*

The *Euclidean signature curve* of a curve $\alpha(t)$ is the set of all points $(\kappa(t), \kappa_s(t))$ evaluated along the curve. An example is shown in Fig. 2. The

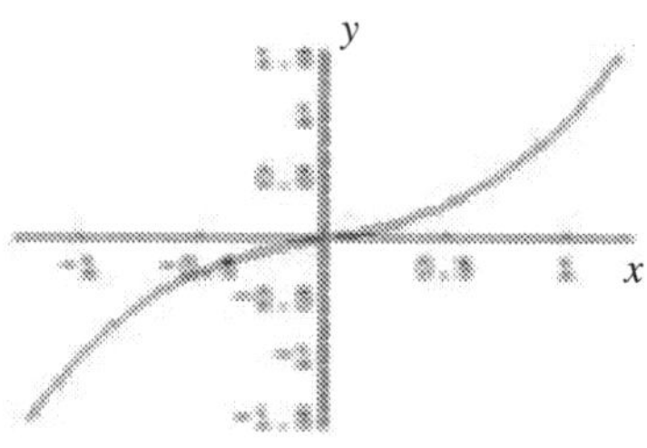

(a)

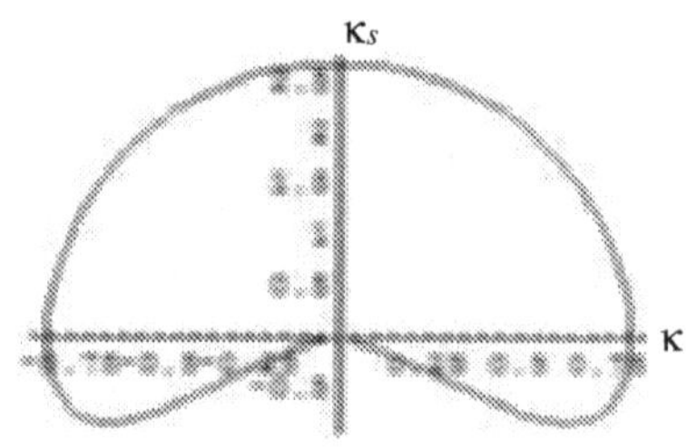

(b)

Fig. 2. (a) A cubical parabola $y = 0.6x^3 + 0.4x$; (b) its signature curve.

following result is well known [3]:

Theorem 2. *Two smooth curves are equivalent up to an Euclidean transformation if and only if their signature curves are identical.*

The above result has led to the development of shape recognition methods [18,17,3] based on matching signature curves. Construction of the signature curve, nevertheless, requires global shape information, which the touch sensor does not provide. So our aim is to make use of the local geometry at a small number of points to perform the recognition task.

2.2 Semi-Differential Invariants as Curve Descriptors

Suppose the curve $\alpha(t) = (x(t), y(t))$ is known to be from a family. Often we can derive a *canonical parametric form* of the family through proper rotation, translation, and reparametrization. This canonical form should have minimum number of indeterminates (other than t) to parametrize the family. These independent indeterminates are referred to as *shape parameters* and denoted by $a_1, \ldots, a_n$. For instance, the class of all ellipses are parametrized with the semimajor axis a and the semiminor axis b. The canonical form expresses the curve in a coordinate system determined by its geometry.

Since the parameter t, which specifies the location of contact (with the touch sensor), is not measurable, we try to eliminate it from the expressions (1) and (2). This can always be done through computing the resultant, yielding an equation for the signature curve:

$$f[a_1, \ldots, a_n](\kappa, \kappa_s) = 0. \tag{3}$$

The simplest case is when the function f can be split into two parts and rewritten as

$$I(\kappa, \kappa_s) = g(a_1, \ldots, a_n).$$

Then I is an expression whose value depends on the shape of the curve not on any specific point at which it is evaluated. It is thus an invariant for the curve, or a *curve invariant.* That the expression assumes the same value at different points is a necessary condition for an unknown curve to be from the family. The family of parabolas will be given as an example in Sect. 3.1.

When a curve family has n shape parameters, we need n independent differential invariants to uniquely identify a curve from the family. If only one point on the curve is considered, this requires up to the nth derivative of the curvature. Numerical computation of high order derivatives is very unreliable. The solution is to trade the order of derivative for extra points. So we consider the curvatures and derivatives at n points and derive *semi-differential invariants.* They are functions of the $2n$ curvatures and derivatives but assume values depending on $a_1, \ldots, a_n$ only. For some curves, such as ellipses and hyperbolas (Sects. 3.2–3.3), semi-differential invariants can be found via algebraic manipulation.

However, derivation of semi-differential curve invariants appears to be very difficult, if not impossible, for many curves. It is more likely that we have to solve for the shape parameters $a_1, \ldots, a_n$ using curvature and derivative estimates at $m \geq n$ points. Viewed differently, the semi-differential invariants now have values equal to the shape parameters but their evaluation can only be done through solution or minimization.

From now on we focus on curves parameterized by polynomials. From (1) and (2) we derive two polynomial equations in t:

$$\kappa^2 \left(x'^2 + y'^2\right)^3 - (x'y'' - x''y')^2 = 0, \tag{4}$$

$$\frac{\kappa_s}{\kappa^2} \left(x'y'' - x''y'\right)^2 - \left(x'y''' - x'''y'\right)\left(x'^2 + y'^2\right) + 3(x'x'' + y'y'')(x'y'' - x''y') = 0. \tag{5}$$

Note that κ and κ_s are measurable and thus treated as constants. Suppose $x(t)$ and $y(t)$ have degrees d_x and d_y, respectively. The two polynomials on the left hand sides of (4) and (5) may have a resultant with degree as high as $d_x + d_y + 8\max\{d_x, d_y\} - 12$ in $a_1, \ldots, a_n$. This is not computationally tractable once d_x or d_y exceeds 3.

2.3 Semi-Signature Curve

We can lower the degree of the resultant polynomial in $a_1, \ldots, a_n$ by considering the slope $\lambda = y'/x'$. This can be rewritten as a polynomial equation:

$$\lambda x' = y'. \tag{6}$$

Assume that $x' \neq 0$ at every point measured by the touch sensor. Equations (1) and (5) are rewritten as

$$\kappa x'^2 (1+\lambda^2)^{3/2} - (y'' - x''\lambda) = 0, \tag{7}$$

$$\frac{\kappa_s/\kappa^2 + 3\lambda}{1+\lambda^2}(y'' - x''\lambda)^2 + 3x''(y'' - x''\lambda) - (y''' - x'''\lambda)x' = 0. \tag{8}$$

With the slope λ treated as a separate variable, we compute the resultants of (6) with (7) and (8), respectively, to eliminate t and obtain

$$f_1[a_1, \ldots, a_n](\lambda, \kappa, \kappa_s) = 0 \quad \text{and} \quad f_2[a_1, \ldots, a_n](\lambda, \kappa, \kappa_s) = 0.$$

The two functions f_1 and f_2 have degrees not exceeding $3\max\{d_x, d_y\} - 3$ and $3\max\{d_x, d_y\} - 5$, respectively, in $a_1, \ldots, a_n$.

What is the reason for using the slope λ determined by the canonical form? Though not able to measure the slope, the touch sensor can measure the tangent rotation from one point to another. The slope λ_1 at the first point and the slope λ_i at the ith point are related as

$$\lambda_i = \frac{\lambda_1 + \tan \Delta\theta_{1i}}{1 - \lambda_1 \tan \Delta\theta_{1i}}, \qquad \theta_{1i} \text{ tangent rotation.} \tag{9}$$

So only one new variable λ_1 has been introduced.

We refer to the point set $\{(\lambda, \kappa, \kappa_s)\}$ as the *semi-signature curve* of $\alpha(t)$. An example is shown in Fig. 3. Its projection onto the κ-κ_s plane is the signature curve. The semi-signature curve is dependent on the chosen canonical parametrization, more specifically, on the orientation of the original curve under the parametrization. Now we may employ similar methods to derive expressions in terms of $\lambda, \kappa, \kappa_s$ but whose values depend on the shape parameters only. These "pseudo-invariants" together with (9) are solved for the slopes first, and then the shape parameters. The details will be described in Sect. 4 on cubical and semicubical parabolas and cubic spline curves.

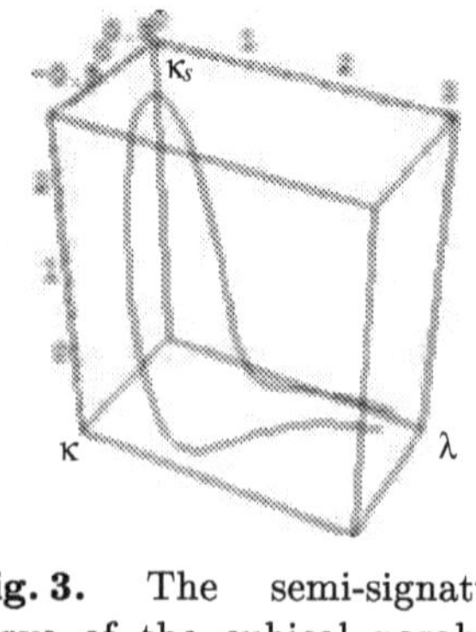

Fig. 3. The semi-signature curve of the cubical parabola in Fig. 2.

3 Quadratics

It is well known that every quadratic curve is one of three types: the ellipse, the hyperbola, and the parabola. Together the three types of curves are referred to as the *conics*. We start with deriving some invariants for these curves.

3.1 Parabola

Parabolas are identified with all the curves parametrized by quadratic polynomials: $x = a_2t^2 + a_1t + a_0$ and $y = b_2t^2 + b_1t + b_0$, where $a_2b_1 - a_1b_2 \neq 0$. Every parabola has a canonical parametrization up to rotation and translation:

$$x = at^2 \quad \text{and} \quad y = 2at, \qquad a > 0.$$

We obtain the curvature and its derivative with respect to arc length:

$$\kappa = -\frac{1}{2a(t^2+1)^{3/2}} \quad \text{and} \quad \kappa_s = \frac{\kappa'}{v(t)} = \frac{3t}{4a^2(t^2+1)^3}.$$

Eliminating t from the above equations leads to an equation that describes the signature curve of the parabola:

$$\frac{1}{(2a)^{2/3}} = \kappa^{2/3}\left(\frac{\kappa_s^2}{9\kappa^4}+1\right) \equiv I_p(\kappa, \kappa_s). \tag{10}$$

The expression $I_p(\kappa, \kappa_s)$ has value independent of t. It is an invariant which has a one-to-one correspondence to the shape of the parabola. Fig. 4 illustrates three parabolas distinguished by I_p. Since κ and κ_s are measurable,

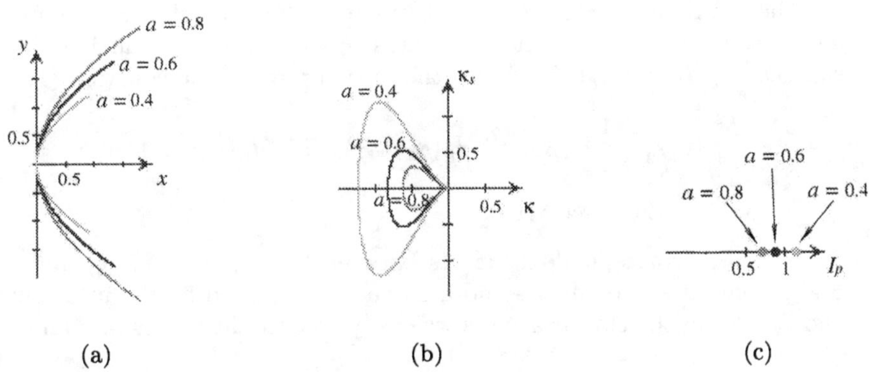

Fig. 4. (a) Three parabolas in the form $y^2 = 4ax$; (b) their signature curves $\{(\kappa, \kappa_s)\}$; (c) corresponding values of the invariant I_p. The invariant is evaluated using any point on a signature curve.

from (10) we can determine the shape parameter a.

3.2 Ellipse

Let us start with the canonical parametrization:

$$x = a\cos(t) \quad \text{and} \quad y = b\sin(t), \qquad a, b > 0.$$

The curvature and its derivative with respect to arc length are

$$\kappa = \frac{ab}{\left(a^2\sin^2(t) + b^2\cos^2(t)\right)^{3/2}} \quad \text{and} \quad \kappa_s = \frac{-3ab\left(a^2 - b^2\right)\sin(t)\cos(t)}{\left(a^2\sin^2(t) + b^2\cos^2(t)\right)^3}.$$

Elimination of t from the above two equations results in an equation describing the signature curve (see Fig. 5(b)):

$$\frac{a^2 + b^2}{(ab)^{4/3}} - \frac{1}{(ab\kappa)^{2/3}} - I_{\mathrm{p}}(\kappa, \kappa_s) = 0, \tag{11}$$

where I_{p} is an expression of κ and κ_s defined in (10). Since there are two unknown quantities a and b, at least two points on the ellipse are required.

Let κ_i and κ_{si}, $i = 1, 2$, be the curvature and its derivative at the ith point. Then we end up with two equations in the form of (11). Subtracting one of them from the other yields the following (assuming $\kappa_1 \neq \kappa_2$):

$$\begin{aligned} \frac{1}{(ab)^{2/3}} &= \frac{(\kappa_1\kappa_2)^{2/3}}{\kappa_1^{2/3} - \kappa_2^{2/3}}\Big(I_{\mathrm{p}}(\kappa_1, \kappa_{s1}) - I_{\mathrm{p}}(\kappa_2, \kappa_{s2})\Big) \\ &\equiv I_{\mathrm{c}1}(\kappa_1, \kappa_2, \kappa_{s1}, \kappa_{s2}). \end{aligned} \tag{12}$$

The expression $I_{\mathrm{c}1}$ is a semi-differential invariant, since it involves the geometry at more than one points. Its value $1/(ab)^{2/3}$ is independent of the two points that are used.

The invariant $I_{\mathrm{c}1}$ alone cannot distinguish ellipses with the same product ab, or equivalently, with the same area. So we find a second invariant by substituting $I_{\mathrm{c}1}$ for $1/(ab)^{2/3}$ into the second term of the equation (11):

$$\begin{aligned} \frac{a^2 + b^2}{(ab)^{4/3}} &= \frac{1}{\kappa_1^{2/3} - \kappa_2^{2/3}}\Big(\kappa_1^{2/3} I_{\mathrm{p}}(\kappa_1, \kappa_{s1}) - \kappa_2^{2/3} I_{\mathrm{p}}(\kappa_2, \kappa_{s2})\Big) \\ &\equiv I_{\mathrm{c}2}(\kappa_1, \kappa_2, \kappa_{s1}, \kappa_{s2}). \end{aligned} \tag{13}$$

A one-to-one correspondence exists between the tuples $(I_{\mathrm{c}1}, I_{\mathrm{c}2})$ and (a, b). Fig. 5 compares two ellipses and a circle distinguished by the invariants $I_{\mathrm{c}1}$ and $I_{\mathrm{c}2}$. From the two invariants we easily recover the values of a and b.

3.3 Hyperbola

A hyperbola has the canonical parametric form

$$x = a\cosh(t) = a\,\frac{e^t + e^{-t}}{2} \quad \text{and} \quad y = b\sinh(t) = b\,\frac{e^t - e^{-t}}{2}, \qquad a, b > 0.$$

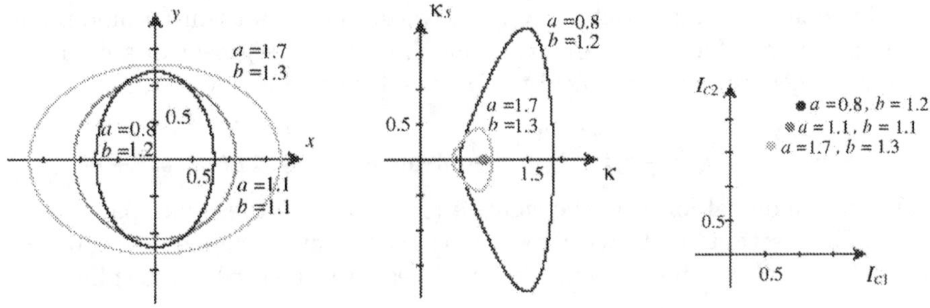

(a) (b) (c)

Fig. 5. (a) Three ellipses in the form $x^2/a^2 + y^2/b^2 = 1$; (b) their signature curves (the one for the circle with $a = b = 1.1$ degenerates into a point $(1/1.1, 0)$); (c) corresponding values of the invariant pair (I_{c1}, I_{c2}). The invariants are evaluated using any two points on the same signature curve.

Similar to the case of an ellipse, we are able to eliminate t from the equations $\kappa = \kappa(t)$ and $\kappa_s = \kappa_s(t)$ and obtain the following:

$$\frac{a^2 - b^2}{(ab)^{4/3}} + \frac{1}{(ab\kappa)^{2/3}} - I_{\mathrm{p}}(\kappa, \kappa_s) = 0, \tag{14}$$

where I_{p} is again defined in (10). Taking the curvatures and derivatives at two points on the hyperbola, from the two copies of equation (14) we derive

$$I_{c1}(\kappa_1, \kappa_2, \kappa_{s1}, \kappa_{s2}) = -\frac{1}{(ab)^{2/3}} \quad \text{and} \quad I_{c2}(\kappa_1, \kappa_2, \kappa_{s1}, \kappa_{s2}) = \frac{a^2 - b^2}{(ab)^{4/3}}.$$

These two invariants are in the same forms as for an ellipse but their values are in different expressions of a and b. In particular, I_{c1} is always negative for the hyperbola.

The invariants I_{c1} and I_{c2} completely determine the hyperbola. Computation of a and b from them is very straightforward.

3.4 Invariants for Conics

Both I_{c1} and I_{c2} are also semi-differential invariants for a parabola, assuming values 0 and $1/(2a)^{2/3}$, respectively. With curvature and derivative information at any two different points, the sign of I_{c1} tells the type of a conic. When the invariant is positive the curve is an ellipse, when it is negative the curve is a hyperbola, and when it is zero the curve is a parabola. The invariants I_{c1} and I_{c2} thus describe the correlations between any two points on a conic.

4 Cubics

There is no classification of all cubic curves. So it seems very difficult to construct invariants that recognize all of them. However, we are interested in cubic splines, whose continuity in curvature enables them to model curved shapes in graphics and geometric modeling. Every segment of a cubic spline has the general and canonical parametric forms as follows:

$$\begin{aligned} x &= a_3t^3 + a_2t^2 + a_1t + a_0, & & x = t^2, \\ y &= b_3t^3 + b_2t^2 + b_1t + b_0; & \text{equivalently,} \quad & y = at^3 + bt^2 + ct. \end{aligned} \tag{15}$$

In the canonical form on the right, a (> 0), b, c are the shape parameters.

This section starts with two subclasses of cubic splines — cubical and semi-cubical parabolas — and then moves on to general cubic splines.

4.1 Cubical Parabola

This class of curves has the canonical parametric form:

$$x = t \quad \text{and} \quad y = at^3 + ct, \qquad a > 0.$$

Figs. 2 and 3 plot an example and its signature and semi-signature curves. Unfortunately, it is not obvious how to eliminate the parameter t from the expressions of the curvature κ and its derivative κ_s. So we employ the method in Sect. 2.3 and utilize the slope $\lambda = \frac{y'}{x'} = 3at^2 + c$. First, we obtain

$$\kappa^2 = \frac{12a(\lambda - c)}{(1+\lambda^2)^3} \qquad \text{and} \qquad \kappa_s = \frac{6a(1+\lambda^2) - 36a\lambda(\lambda - c)}{(1+\lambda^2)^3}.$$

Note that we can obtain a and c using λ, κ, and κ_s:

$$a = \frac{\left(\kappa_s + 3\lambda\kappa^2\right)\left(1+\lambda^2\right)^2}{6} \quad \equiv \quad I_{\mathrm{cp}1}(\lambda, \kappa, \kappa_s), \tag{16}$$

$$c = \lambda - \frac{\kappa^2\left(1+\lambda^2\right)}{2\left(\kappa_s + 3\lambda\kappa^2\right)} \quad \equiv \quad I_{\mathrm{cp}2}(\lambda, \kappa, \kappa_s). \tag{17}$$

The expressions $I_{\mathrm{cp}1}$ and $I_{\mathrm{cp}2}$ map any point on the semi-signature curve to the shape parameters a and c, respectively. They are invariants of the curve provided that the slope λ can be determined.

Measure the tangent rotation $\Delta\theta_{12}$ from point 1 to point 2. Write $\delta_{12} = \tan \Delta\theta_{12}$. Since the value of c should be same, we have

$$I_{\mathrm{cp}2}(\lambda_1, \kappa_1, \kappa_{s1}) = I_{\mathrm{cp}2}(\lambda_2, \kappa_2, \kappa_{s2}). \tag{18}$$

Substitution of (9) with $i = 2$ into (18) results in a quartic polynomial:

$$d_4\lambda_1^4 + d_3\lambda_1^3 + d_2\lambda_1^2 + d_1\lambda_1 + d_0 = 0, \tag{19}$$

which can be solved for λ_1 (and hence λ_2). These coefficients are

$$\begin{aligned} d_0 &= \kappa_{s1}\left(\kappa_2^2\left(5\delta_{12}^2 - 1\right) + 2\kappa_{s2}\delta_{12}\right) + \kappa_1^2\left(3\kappa_2^2\delta_{12} + \kappa_{s2}\right), \\ d_1 &= 2\delta_{12}\left(\kappa_{s1}\left(3\kappa_2^2 - \kappa_{s2}\delta_{12}\right) + 2\kappa_1^2\left(3\kappa_2^2\delta_{12} + \kappa_{s2}\right)\right), \\ d_2 &= \kappa_{s1}\left(\kappa_2^2\left(5\delta_{12}^2 - 1\right) + 2\kappa_{s2}\delta_{12}\right) + \kappa_1^2\left(18\kappa_2^2\delta_{12} - \kappa_{s2}\left(5\delta_{12}^2 - 1\right)\right), \\ d_3 &= 2\delta_{12}\left(\kappa_{s1}\left(3\kappa_2^2 - \kappa_{s2}\delta_{12}\right) + 2\kappa_1^2\left(3\kappa_2^2\delta_{12} + \kappa_{s2}\right)\right), \\ d_4 &= 5\kappa_1^2\delta_{12}\left(3\kappa_2^2 - \kappa_{s2}\delta_{12}\right). \end{aligned}$$

4.2 Semi-Cubical Parabola

The canonical parametrization for this class of curves takes the form

$$x = t^2 \quad \text{and} \quad y = at^3 + bt^2, \qquad a > 0.$$

Reparametrize the curve with the slope is $\lambda = y'/x' = 3at/2 + b$ and obtain

$$a = \sqrt{-\frac{8\kappa^3(1+\lambda^2)^{5/2}}{9(\kappa_s + 3\lambda\kappa^2)}} \quad \equiv \quad I_{\text{scp1}}(\lambda, \kappa, \kappa_s), \tag{20}$$

$$b = \lambda + \frac{\kappa^2(1+\lambda^2)}{\kappa_s + 3\lambda\kappa^2} \quad \equiv \quad I_{\text{scp2}}(\lambda, \kappa, \kappa_s). \tag{21}$$

Using two points, we can set up an equation:

$$I_{\text{scp2}}(\lambda_1, \kappa_1, \kappa_{s1}) = I_{\text{scp2}}(\lambda_2, \kappa_2, \kappa_{s2}).$$

This equation together with (9) again yield a quartic polynomial in λ_1. The invariants for this class of curves are I_{scp1} and I_{scp2}.

4.3 Cubic Spline

It is time now to turn to general cubic splines described by (15). Different from its two subclasses, we cannot replace t with the slope λ in the expressions of the curvature and its derivative. So we resort to solving equations (6)–(8), which are simplified to the following:

$$3at^2 + 2(b-\lambda)t + c = 0, \tag{22}$$

$$Lt^2 - 3at - (b-\lambda) = 0, \tag{23}$$

$$6at + M\Big((b-\lambda)^2 - 3ac\Big) + 3(b-\lambda) = 0, \tag{24}$$

where $L = 2(1+\lambda^2)^{3/2}\kappa$ and $M = \frac{\kappa_s + 3\kappa^2\lambda}{(1+\lambda^2)\kappa^2}$. We substitute c in (24) with (22):

$$9a^2Mt^2 + 6a\Big(1 + M(b-\lambda)\Big)t + M(b-\lambda)^2 + 3(b-\lambda) = 0. \tag{25}$$

Next, the resultant of equations (23) and (25) is computed to eliminate t:

$$81Ma^4 + 18L\Big(1 + 3M(b-\lambda)\Big)a^2 + L^2(b-\lambda)\Big(M(b-\lambda) + 3\Big)^2 = 0. \tag{26}$$

Since M can get very large when κ is small, we divide the left hand side of (26) by $81Ma^4$ and denote the resulting expression as the function $g(a, b, \lambda)$.

With curvatures and derivatives estimated at $l \geq 3$ points, the shape parameters a, b, and the slope λ_1 at the first point can be estimated through a least-squares optimization:

$$\min_{a,b,\lambda_1} \sum_{i=1}^{l} g(a, b, \lambda_i)^2,$$

where λ_is depend on λ_1 according to (9). To determine the third parameter c, we just need to eliminate the t^2 and t terms from (22)–(24).

5 Locating Contact

The parameter value t determines the contact location on the curved shape with the touch sensor. Since the tangent at the contact is measurable, t also determines the relative pose of the shape to the sensor. We have the following:

$$t = \begin{cases} \frac{\kappa_s}{3\kappa^2}, & \text{if parabola;} \\ \sin^{-1}\left(\sqrt{\frac{\left(\frac{ab}{\kappa}\right)^{2/3} - b^2}{a^2 - b^2}}\right), & \text{if ellipse;} \\ \sinh^{-1}\left(\sqrt{\frac{\left(\frac{ab}{\kappa}\right)^{2/3} - b^2}{a^2 + b^2}}\right), & \text{if hyperbola;} \\ \pm\sqrt{\frac{\lambda - b}{3a}}, & \text{if cubical parabola;} \\ \frac{2(\lambda - b)}{3a}, & \text{if semi-cubical parabola;} \\ -\frac{M\left((b-\lambda)^2 - 3ac\right) + 3(b-\lambda)}{6a}, & \text{if cubic spline.} \end{cases}$$

6 Simulation

In simulation, we approximate the curvature and its derivative by finite difference quotients using arc length s and tangential angle ϕ:

$$\kappa \approx \frac{\phi(s + \Delta s) - \phi(s - \Delta s)}{2\Delta s} \quad \text{and} \quad \kappa_s \approx \frac{\phi(s + \Delta s) - 2\phi(s) + \phi(s - \Delta s)}{(\Delta s)^2},$$

The arc length between two points on the curve, close to each other, is approximated by their Euclidean distance. The rotation of the tangent from one point to another uses the exact value since it can be measured quite accurately in practice. Given the errors of finite differences, it is not very meaningful to introduce simulated noise, which may either reduce or magnify such errors.[1]

6.1 Verification of Invariants

The first group of simulations were conducted to verify the invariants of the three conics, cubical parabolas, and semi-cubical parabolas. The results are summarized in Table 1, where estimation errors of κ and κ_s accounted for the discrepancies between the actual values of the invariants and their estimates.

In the second group of simulations shown in Table 2, we demonstrate that *an invariant for one of the curve classes varies for another.*

Finally, Table 3 reveals how much the recovered shape parameters $\bar{a}, \bar{b}, \bar{c}$ differ from the real ones a, b, c. From the table we see that on the average the relative errors are around 1% except for cubic splines.

[1] A method introduced in [3] approximates the osculating circle with one that passes through three local points. This curvature estimation scheme was extended in [2]. These methods are able to generate slightly better estimates than finite differencing but the simulation outcomes would not have been altered.

Table 1. Invariant verification on five specific curves. Each invariant, labeled with its defining equation, is evaluated 100 times using points randomly selected from the corresponding curve. The shape parameters of each curve are easily recoverable from the definitions of its invariants.

inv.	I_p	I_{c1} (12)		I_{c2} (13)		I_{cp1}	I_{cp2}	I_{scp1}	I_{scp2}
	(10)	ellipse	hyperbola	ellipse	hyperbola	(16)	(17)	(20)	(21)
real	0.2198	0.1857	−0.2678	1.2055	0.3222	6.9963	2.6127	1.3730	6.5107
min	0.2168	0.1801	−0.2729	1.1749	0.2937	6.7687	1.7312	1.4111	6.3945
max	0.2230	0.1863	−0.2655	1.2083	0.3615	7.0289	3.1684	1.4447	6.5834
mean	0.2198	0.1852	−0.2675	1.2035	0.3210	6.9355	2.5022	1.4220	6.5154

Table 2. Evaluating three invariants on data obtained from curves of different classes. Each cell displays the summary over 100 evaluations on a curve.

\ data inv.\	conic (ellipse)	cubical parabola	semi-cub. parabola	cubic spline
I_{c1}	———	−6.38(min) −0.04(max) −0.73(mean) 1.22(stdev)	−22.84 28.37 3.37 6.76	−45.24 −4.94 −16.50 10.86
I_{cp2}	−11.97 15.46 −0.04 2.53	———	8.54 19.03 13.76 3.07	11.66 1721.04 55.52 217.34
I_{scp2}	−265.80 5.83 −3.22 26.75	7.80 65.22 29.17 17.19	———	−150.68 1715.73 38.97 182.24

Table 3. Relative errors $\sqrt{((a-\bar{a})/a)^2 + \left((b-\bar{b})/b\right)^2 + ((c-\bar{c})/c)^2}$ on estimating shape parameters. Summary over 100 curves from each class (only 25 curves from the cubic spline class) randomly generated under uniform distributions of its shape parameters within prescribed ranges.

	ellip.	hyper.	par.	cub. par.	semi-cub. par.	cubic spline
min	0.02%	0.10%	0.01%	0.02%	0.04%	0.61%
max	7.99%	9.71%	3.35%	7.49%	8.09%	29.23%
mean	0.40%	1.15%	0.36%	0.83%	1.23%	11.27%

6.2 Curve Recognition

We now use invariants to recognize a curve out of the six classes of quadratic and cubic curves. For example, consider the ellipse in Figure 6(a). The values of κ and κ_s are estimated at $t_1 = 0.36$, $t_2 = 1.86$, and $t_3 = 4.23$. Invariant I_p has values 0.8971 and 0.4030 at the first two points, so the curve is not a parabola. Invariant I_{c1} yields values 0.3447, 0.3446, and 0.3449 at the three resulting pair of points, from which we infer that the curve is an ellipse. The recovered shape parameters (from I_{c1} and I_{c2}) are $a \approx 2.8609$ and $b \approx 1.7275$.

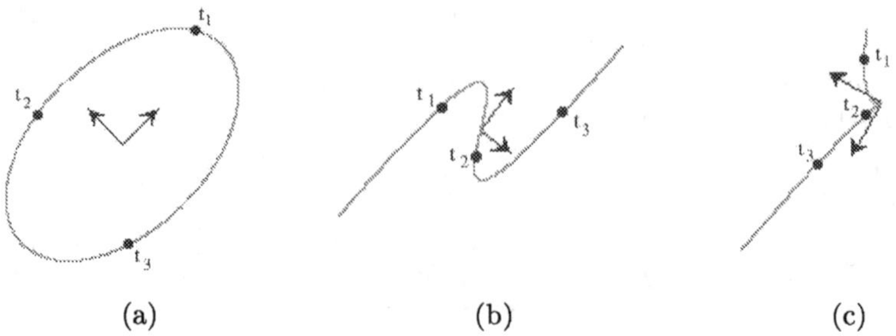

Fig. 6. Recognition of three shapes based on local geometry at three points. (a) An ellipse with $a = 2.8605$ and $b = 1.7263$; (b) a cubical parabola with $a = 3.2543$ and $b = -2.3215$; and (c) a semi-cubical parabola with $a = 2.5683$ and $b = 1.4102$.

For the cubical parabola in Figure 6(b), a test on the invariant I_{c1} has failed. So we know that the curve is not quadratic. Hypothesizing cubical parabola, we solve for λ_1 from (19). Subsequent tests on invariant I_{cp1} (for shape parameter a) yield values 3.2244, 3.2536, and 3.1872, and on invariant I_{cp2} (for b) yield values −2.3237, −2.3237, and −2.3972. The recognition of the semi-cubical parabola in Fig. 6(c) is similar.

7 Preliminary Experiments

The key for the applicability of our invariant-based approach lies in obtaining reliable estimates of curvature κ and its derivative κ_s from real data. The tactile data used in our experiments were generated by a joystick sensor mounted on an Adept Cobra 600 robot. Despite the Adept's high precision, we found that difference quotients and other estimation methods [3,2] based on the Taylor expansion were still too sensitive to small measurement errors.

To reliably estimate κ at a point, say, p_0, the sensor measures n points (including p_0) in its neighborhood on the shape. Then a quadratic curve is fit over the point sequence. Differentiating this curve at p_0 gives us the curvature.

Similarly, we estimate curvature $\hat{\kappa}_1$ at a second point p_1 following p_0. This is done by sensing l more points along the shape while removing the first l points from the sequence (l is very small so p_0 remains in the new sequence). The arc length $\hat{s}_1$ between p_0 and p_1 is estimated by numerically integrating the new curve fit. Continuing this process, we generate a sequence of pairs $(0, \hat{\kappa}_0), (\hat{s}_1, \hat{\kappa}_1), \ldots, (\hat{s}_{m-1}, \hat{\kappa}_{m-1})$. Now, fit a quadratic curve over this sequence and differentiate it to obtain curvature derivative estimates at the m points. Fig. 7 shows the results for an ellipse and a cubic spline shape.

8 Discussion

We have introduced an invariant-based method that aims at unifying shape recognition, recovery, and pose estimation with tactile information. Differen-

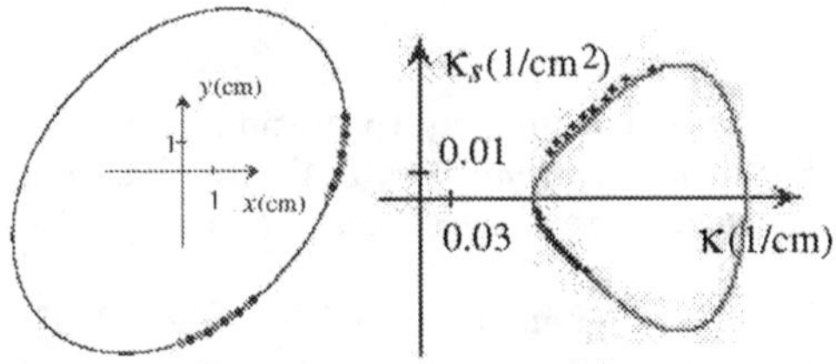

(a) (b)

	I_{c1}	a	b
real	0.373836	2.5	1.75
min	0.350559	2.38636	1.62636
max	0.404903	2.67234	1.83549
mean	0.377728	2.511269	1.719586

(c)

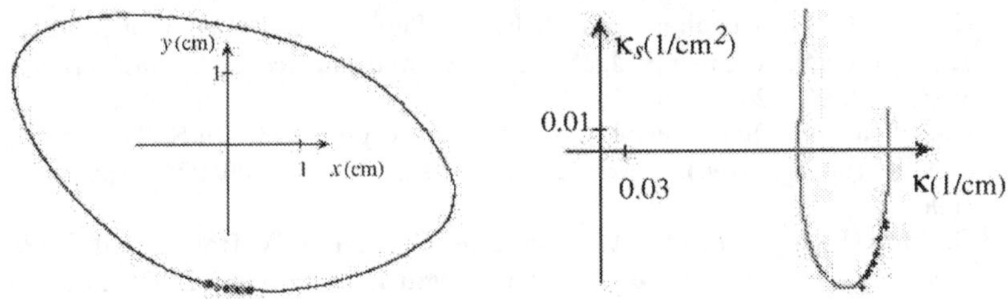

(d) (e)

Fig. 7. Estimating κ and derivative κ_s from tactile data generated by a joystick sensor: (a) an ellipse and 20 sample points; (b) estimates $(\hat{\kappa}, \hat{\kappa}_s)$ at these points plotted against the signature curve; (c) the invariant I_{c1} evaluated at 80 pairs of the same points; (d) a closed cubic spline with 8 sample points from one of its segment; (e) estimates (κ, κ_s) and the signature curve of the segment.

tial and semi-differential invariants have been developed for several classes of low-degree algebraic curves. Evaluated over a few points, these invariants capture intrinsic information about a curve. They allow us to recover the shape parameters of the curve as well as obtain the locations of contact where the data are supposed to be taken.

Although only quadratic curves and special cubic curves are treated, it is straightforward to extend the results to objects bounded by segments of these types. The invariant-based scheme also extends to a curve in implicit form, which can be used along with curvature and its derivative to eliminate the two coordinates. Analytical feasibility of the scheme lies in deriving an equation that describes the signature (or semi-signature) curve. For a curve in 3D, such invariants will have to involve torsion (and possibly its derivatives).

Future work includes improvement on the robustness of curvature and derivative estimation, invariant design for more general plane curves, and extension to 3D surfaces.

Acknowledgment Support for this research has been provided in part by Iowa State University, and in part by the National Science Foundation through a CAREER award IIS-0133681. The authors would like to thank David Kriegman for pointing to differential invariants for shape recognition, and Liangchuan Mi for providing the tactile data used in the experiments. Also thanks to the reviewers for their helpful comments.

References

1. P. K. Allen and P. Michelman. Acquisition and interpretation of 3-D sensor data from touch. *IEEE Trans. Robot. and Automation*, 6(4):397–404, 1990.
2. M. Boutin. Numerically invariant signature curves. *Intl. J. Comp. Vision*, 40(3):235–248, 2000.
3. E. Calabi, P. J. Olver, C. Shakiban, A. Tannenbaum, and S. Haker. Differential and numerically invariant signature curves applied to object recognition. *Intl. J. Comp. Vision*, 26(2):107–135, 1998.
4. H. Civi, C. Christopher, and A. Ercil. The classical theory of invariants and object recognition using algebraic curve and surfaces. *J. Math. Imaging and Vision*, 19:237–253, 2003.
5. R. S. Fearing. Tactile sensing for shape interpretation. In S. T. Venkataraman and T. Iberall (eds.), *Dextrous Robot Hands*, pp. 209–238. Springer-Verlag, 1990.
6. D. Forsyth, J. L. Mundy, A. Zisserman, C. Coelho, A. Heller, and C. Rothwell. Invariant descriptors for 3-D object recognition and pose. *IEEE Trans. Pattern Analysis and Machine Intell.*, 13(10), 1991.
7. W. E. L. Grimson and T. Lozano-Pérez. Model-based recognition and localization from sparse range or tactile data. *Intl. J. Robot. Res.*, 3(3):3–35, 1984.
8. D. Hilbert. *Theory of Algebraic Invariants.* Cambridge University Press, 1993.
9. R. Ibrayev and Y.-B. Jia. Tactile recognition of algebraic shapes using differential invariants. In *Proc. IEEE Intl. Conf. Robot. and Automation*, pp. 1548-1553, 2004.
10. D. Keren. Using symbolic computation to find algebraic invariants. *IEEE Trans. Pattern Analysis and Machine Intell.*, 16(11):1143–1149, 1994.
11. D. Keren, E. Rivlin, I. Shimshoni, and I. Weiss. Recognizing 3D objects using tactile sensing and curve invariants. *J. Math. Imaging and Vision*, 12(1):5–23, 2000.
12. D. J. Kriegman and J. Ponce. On recognizing and positioning curved 3-D objects from image contours. *IEEE Trans. Pattern Analysis and Machine Intell.*, 12(12):1127–1137, 1990.
13. M. Moll and M. A. Erdmann. Reconstructing the shape and motion of unknown objects with active tactile sensors. In J.-D. Boissonnat *et al.* (eds.), *Algorithmic Foundations of Robotics V*, pp. 293–309. Springer-Verlag, 2004.
14. T. Moons, E. J. Pauwels, L. J. van Gool, and A. Oosterlinck. Foundations of semi-differential invariants. *Intl. J. Comp. Vision*, 14(1):25–47, 1995.
15. J. L. Mundy and A. Zisserman. Introduction — towards a new framework for vision. In J. L. Mundy and A. Zisserman (eds.), *Geometric Invariance in Computer Vision*, pp. 1–39. The MIT Press, 1992.
16. T. Pajdla and L. Van Gool. Matching of 3-D curves using semi-differential invariants. In *Proc. IEEE Intl. Conf. Comp. Vision*, pp. 390–395, 1995.
17. E. J. Pauwels, T. Moons, L. J. van Gool, P. Kempenaers, and A. Oosterlinck. Recognition of planar shapes under affine distortion. *Intl. J. Comp. Vision*, 14(1):49–65, 1995.
18. E. Rivlin and I. Weiss. Local invariants for recognition. *IEEE Trans. Pattern Analysis and Machine Intell.*, 17(3):226–238, 1995.
19. I. Weiss. Geometric invariants and object recognition. *Intl. J. Comp. Vision*, 10(3):207–231, 1993.

Modeling Macromolecular Machines Using Rigid-Cluster Networks

Moon K. Kim[1] and Gregory S. Chirikjian[1]

Department of Mechanical Engineering, The Johns Hopkins University, Baltimore MD 21218, USA

Abstract. Many macromolecules consist of domains which act more or less as rigid bodies during large conformational changes. These collective motions are thought to be strongly related with the functions of a system. This fact encourages us to model a biomolecular structure as a set of rigid clusters which are interconnected with distance constraints. In this way the intermediate conformations in an anharmonic pathway can be determined by the translational and rotational displacements of large clusters in such a way that steric constraints are observed. In this paper, we present a geometry-based interpolation technique called rigid-cluster interpolation. This is an extension of the coarse-grained elastic network interpolation (ENI) method that we previously developed to generate physically realistic pathways between two different conformations of a macromolecule. We present the derivation of the rigid-cluster model and apply it to various "macromolecular machines". Simulation results show that this method generates sterically feasible pathways of large systems in a very short time. The computational cost of the interpolation does not scale heavily with the size of structures, but rather depends strongly on the number of rigid-clusters.

1 Introduction

Motions and conformational transitions of macromolecules are involved in all of the functions of the cell. Hence, comprehending such biological mechanisms is central in understanding the phenomena of life. As a large amount of structural information about macromolecules continues to be obtained experimentally and deposited in the Protein Data Bank [1], research areas such as computational biology, bioinformatics, and protein dynamics continue to grow rapidly.

Molecular dynamics (MD) simulation is a popular tool for the dynamic analysis of macromolecules. It is based on full-atom empirical potential functions describing chemical properties of macromolecules. However, this methodology becomes computationally inefficient or impractical as the size of the systems of interest continues to increase.

Robot motion planning techniques including probabilistic roadmap methods (PRMs) have been utilized for studying the kinetics of protein folding and ligand-protein binding problems [2,3]. These methods enable one to generate multiple pathways faster than the conventional molecular dynamics techniques by sampling a molecular conformation space at random. Zang

M. Erdmann et al. (Eds.): Algorithmic Foundations of Robotics VI, STAR 17, pp. 283–296, 2005.

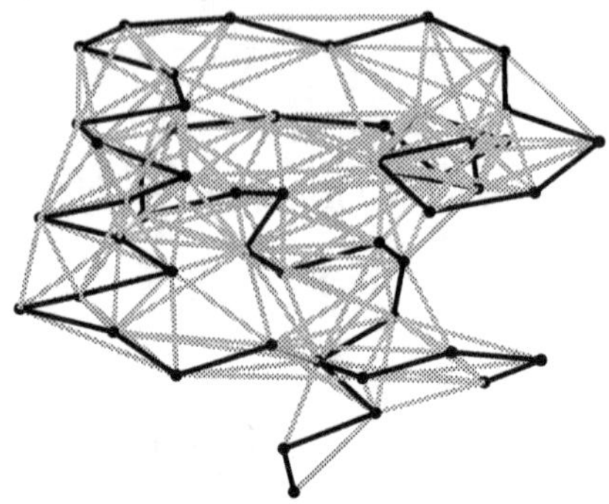

Fig. 1. Representation of a coarse-grained elastic network model. The grey lines represent the spring connections between representative point masses (e.g., alpha carbons) with distance constraints. The backbone trace of a protein is shaded in black.

and Kavraki [4] adopted a robotics-based parameterization (called the atomgroup local frames method) in which all atoms are grouped into atomgroups like rigid-bodies and a conformational change is described by only torsion angles between distinct atomgroups. It is especially useful in both energy minimization and conformational search problems which usually require huge computational loads.

Atilgan et al. [5] developed a coarse-grained elastic network model in which a biomolecular structure is analogous to a mechanical system composed of a network of springs connecting representative point masses. For example, only alpha carbons in a protein are treated as point masses and spatially proximal points are assumed to be linked with linear springs as shown in Fig. 1. Only structural (i.e., geometric) information is used to define a simple harmonic potential function. Normal mode analysis (NMA) based on this coarse-grained elastic network model has been performed to study the dynamics of a macromolecule around an equilibrium state [6]. This is computationally more efficient than conventional approaches such as MD or even NMA using full-atom empirical potentials, and captures very well the low-frequency vibrational modes of the structures.

Furthermore, Kim et al. [7,8] proposed the elastic network interpolation (coarse-grained ENI) method to generate anharmonic and large conformational transitions in macromolecules. The key idea is to uniformly interpolate the distances in two different conformations within the context of the coarse-grained elastic network model. Coarse-grained ENI generates a feasible reaction pathway between two different conformations. It is suitable to describe the global motions of large systems within reasonable time on a personal computer.

In this paper, a new interpolation method (called rigid-cluster ENI) is addressed. It enables us to dramatically save the computational cost for describing conformational changes of a macromolecule when compared with coarse-grained ENI based on point mass systems. The complete derivation of rigid-cluster ENI is presented here and then several macromolecular systems which act like macroscopic machines are used as examples of rigid-cluster systems in order to illustrate and validate this methodology. Since most conformational changes in macromolecules can be classified into hinge and shear motions associated with the collective behavior of atoms [9], a biomolecular structure often can be simply represented as a set of rigid-clusters.

2 Method

The basis of the rigid-cluster network model is to interconnect a set of rigid clusters with distance constraints. These distance constraints are greater in number than the degrees-of-freedom (DOFs) in the model. This overconstraint is resolved with a quadratic penalty function, which in essence treats each distance constraint as a linear spring. This model is equivalent to the replacement of collections of point masses with rigid clusters in the conventional coarse-grained elastic network model and the derivations of these two models are similar to each other [10]. However, the displacement in the rigid-cluster network model is represented as a 6×1 vector such that

$$\boldsymbol{\delta}_i(t) = [\boldsymbol{v}_i^T(t)\ \boldsymbol{\omega}_i^T(t)]^T \in \mathbb{R}^6, \tag{1}$$

where $\boldsymbol{v}_i(t)$ is the translational displacement and $\boldsymbol{\omega}_i(t)$ is the orientational displacement of cluster i.

In the C_α coarse-grained model, the mass (or inertia) matrix M is simply constructed as an array of sub-diagonal blocks such that

$$M_i = m\mathbb{I}_3 \in \mathbb{R}^3, \tag{2}$$

where i is the index of each residue, m is the mass of the C_α atom, and $\mathbb{I}_3$ is the 3×3 identity matrix. The sub-diagonal block of the mass matrix corresponding to cluster i in the rigid-cluster model is of the form

$$M_i = \begin{pmatrix} m_i\mathbb{I}_3 & \mathbb{O}_3 \\ \mathbb{O}_3 & I_i \end{pmatrix} \in \mathbb{R}^6, \tag{3}$$

where m_i is the total mass of cluster i, $\mathbb{O}_3$ is the 3×3 zero matrix, and I_i is the moment of inertia matrix of cluster i [10].

These mass matrices together with stiffness matrices play an important role in NMA, whereas ENI only utilizes a geometric penalty function (which is somewhat different from the simplified potential energy function) to produce an anharmonic pathway between the two end conformations of the same system denoted by sets of Cartesian coordinates $\{\boldsymbol{x}_i\}$ and $\{\boldsymbol{y}_i\}$, respectively. One can first build a rigid-cluster elastic network model for each of them as shown in Fig. 2.

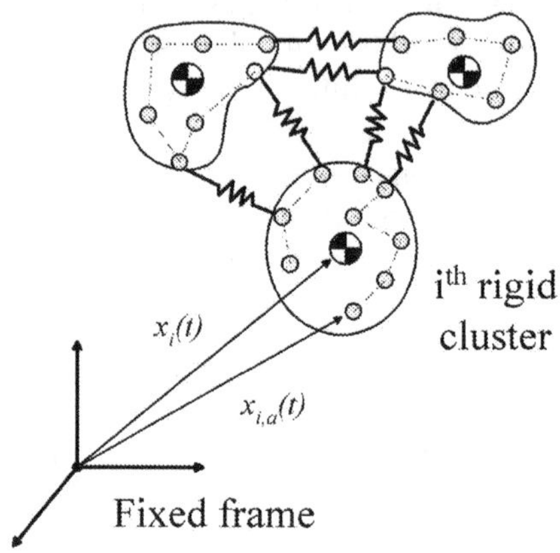

Fig. 2. Schematic of a rigid-cluster model. Each rigid-cluster consists of representative point masses. Interconnection between rigid-clusters is constructed by the spring connection of a pair of point masses which are located in two distinct clusters, but spatially close to each other. The backbone trace of a set of point masses is represented by a dashed line.

Assume that each conformation can be treated as an assembly of N rigid-clusters. The center of mass of the ith cluster is

$$\boldsymbol{x}_i(t) = \frac{1}{m_i} \sum_{a=1}^{n(i)} m_{i,a} \boldsymbol{x}_{i,a}(t), \tag{4}$$

where $n(i)$ is the number of point masses in cluster i, $m_{i,a}$ is the mass of residue a of cluster i, and $\boldsymbol{x}_{i,a}(t)$ is the vector of Cartesian coordinates of residue a of cluster i at time t. Hence,

$$m_i = \sum_{a=1}^{n(i)} m_{i,a}. \tag{5}$$

Fig. 3 shows rigid-body motions of the ith cluster with respect to time t. The translational displacement of cluster i is defined as

$$\boldsymbol{v}_i(t) = \boldsymbol{x}_i(t) - \boldsymbol{x}_i(0), \tag{6}$$

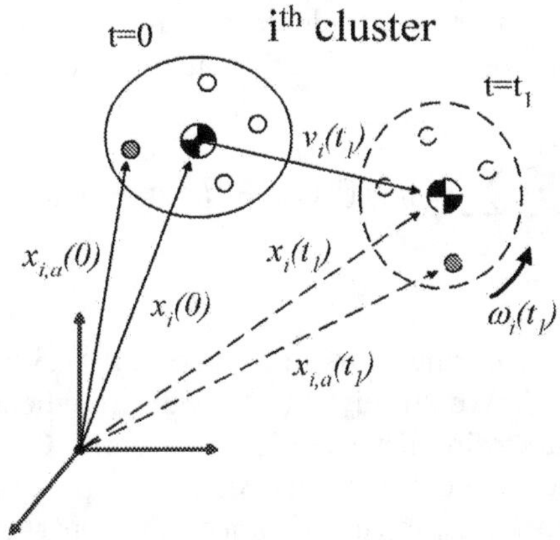

Fig. 3. Schematic of the pose of the ith rigid-cluster with respect to time. The initial pose of cluster i at $t = 0$ changes with the translational and orientational displacement. Small rigid-body motions at each timestep are assumed in this context but exaggerated here for illustration.

where $\boldsymbol{x}_i(0)$ is calculated from the initial coordinates of the given system. Note that $\boldsymbol{v}_i(t)$ is not a velocity, but a displacement. The orientational displacement of cluster i, $\boldsymbol{\omega}_i(t)$, is defined as the rotation matrix

$$R(\boldsymbol{\omega}_i(t)) \triangleq \exp[mat(\boldsymbol{\omega}_i(t))], \tag{7}$$

where $\boldsymbol{\omega}_i(t)$ is parallel to the axis of rotation, $\|\boldsymbol{\omega}_i(t)\|$ is the angle of rotation about that axis, and "mat" is the operator which converts a 3×1 vector into a skew-symmetric matrix such that

$$mat(\boldsymbol{x}) = \begin{pmatrix} 0 & -x_3 & x_2 \\ x_3 & 0 & -x_1 \\ -x_2 & x_1 & 0 \end{pmatrix}. \tag{8}$$

Note that $\boldsymbol{\omega}_i(t)$ is not angular velocity, but rather a small orientational displacement. Therefore, the position of residue a of cluster i at time t is written as

$$\boldsymbol{x}_{i,a}(t) = R(\boldsymbol{\omega}_i(t))(\boldsymbol{x}_{i,a}(0) - \boldsymbol{x}_i(0)) + \boldsymbol{x}_i(0) + \boldsymbol{v}_i(t). \tag{9}$$

Assuming infinitesimal rigid-body displacements from a baseline equilibrium conformation, the rotation matrix $R(\boldsymbol{\omega}_i(t))$ is approximated such that

$$R(\boldsymbol{\omega}_i(t)) \approx \mathbb{I}_3 + mat(\boldsymbol{\omega}_i(t)). \tag{10}$$

Substitution into Eq. 9 yields

$$\boldsymbol{x}_{i,a}(t) \triangleq \boldsymbol{x}_{i,a}(0) + H_{i,a}\boldsymbol{\delta}_i(t), \tag{11}$$

where $H_{i,a} = [\mathbb{I}_3 \;\; - mat(\boldsymbol{x}_{i,a}(0) - \boldsymbol{x}_i(0))] \in \mathbb{R}^{3\times 6}$.

We propose a cost function to calculate incremental conformational changes of the system which consists of N rigid-clusters such that

$$C = \frac{1}{2}\sum_{i=1}^{N-1}\sum_{j=i+1}^{N}\left\{\sum_{a=1}^{n(i)}\sum_{b=1}^{n(j)} k_{i,a,j,b}(\|\boldsymbol{x}_{i,a} + H_{i,a}\boldsymbol{\delta}_i - \boldsymbol{x}_{j,b} - H_{j,b}\boldsymbol{\delta}_j\| - l_{i,a,j,b})^2\right\}, \tag{12}$$

where $k_{i,a,j,b}$ is the "spring constant" between residue a of cluster i and residue b of cluster j. We simply set the value 1 whenever two residues are judged to be in contact in either conformation and 0, otherwise. In this way there is a penalty for the distance between these points to deviate after each incremental motion. $l_{i,a,j,b}$ is the distance between these two residues in an intermediate conformation. In order to artificially force the conformational transition, this can be chosen as

$$l_{i,a,j,b} = (1-\alpha)\|\boldsymbol{x}_{i,a} - \boldsymbol{x}_{j,b}\| + \alpha\|\boldsymbol{y}_{i,a} - \boldsymbol{y}_{j,b}\|, \tag{13}$$

where α is the coefficient specifying how far a given state is along the transition from $\{\boldsymbol{x}_i\}$ to $\{\boldsymbol{y}_i\}$ (for review see [7,8]). We only consider inter-connections among rigid-clusters from the conventional elastic network model because intra-connections within a cluster are preserved under rigid-body motions.

The squared term of Eq. 12 can be approximated to second order as

$$\begin{aligned}
&(\|\underbrace{[\boldsymbol{x}_{i,a} - \boldsymbol{x}_{j,b}]}_{X} + \underbrace{[H_{i,a}\boldsymbol{\delta}_i - H_{j,b}\boldsymbol{\delta}_j]}_{Y}\| - \underbrace{l_{i,a,j,b}}_{l})^2 \\
&\approx \|Y\|^2 - \frac{l}{\|X\|}Y^T(\mathbb{I}_3 - \frac{XX^T}{\|X\|^2})Y \xleftarrow{2^{nd}\ order\ terms} \\
&\quad + 2(1 - \frac{l}{\|X\|^2})X^T Y) \xleftarrow{1^{st}\ order\ terms} \\
&\quad + \|X\|^2 - 2l\|X\| + l^2 \xleftarrow{constants} .
\end{aligned} \tag{14}$$

The above 2nd order terms can be written in the quadratic form

$$\underbrace{\boldsymbol{\Delta}_{i,j}^T[H_{i,a}\ \ -H_{j,b}]^T\{\mathbb{I}_3 - \frac{l}{\|X\|}(\mathbb{I}_3 - \frac{XX^T}{\|X\|^2})\}[H_{i,a}\ \ -H_{j,b}]\boldsymbol{\Delta}_{i,j}}_{G_{i,a,j,b}}, \tag{15}$$

where $\boldsymbol{\Delta}_{i,j} = [\boldsymbol{\delta}_i^T\ \boldsymbol{\delta}_j^T]^T \in \mathbb{R}^{12}$ and $Y = [H_{i,a}\ \ -H_{j,b}]\boldsymbol{\Delta}_{i,j}$. $G_{i,j}$ can be defined as

$$G_{i,j} = \sum_{a=1}^{n(i)}\sum_{b=1}^{n(j)} k_{i,a,j,b}G_{i,a,j,b}. \tag{16}$$

Likewise, the 1st order terms can be expressed in a simple way such that

$$\underbrace{2(1-\frac{l}{\|X\|^2})X^T[H_{i,a} \ - H_{j,b}]}_{g_{i,a,j,b}}\boldsymbol{\Delta}_{i,j},$$
$$g_{i,j} = \sum_{a=1}^{n(i)}\sum_{b=1}^{n(j)} k_{i,a,j,b} g_{i,a,j,b}. \tag{17}$$

Eq. 12 consequently becomes a simplified form with respect to $\boldsymbol{\Delta}_{i,j}$

$$C = \frac{1}{2}\sum_{i=1}^{N-1}\sum_{j=i+1}^{N}\left\{\boldsymbol{\Delta}_{i,j}^T G_{i,j}\boldsymbol{\Delta}_{i,j} + g_{i,j}\boldsymbol{\Delta}_{i,j} + B_{i,j}\right\}, \tag{18}$$

where $G_{i,j}$ is a 12×12 matrix, $g_{i,j}$ is a 1×12 low vector, and $B_{i,j}$ is the sum of constant terms derived in Eq. 14 such that

$$B_{i,j} = \sum_{a=1}^{n(i)}\sum_{b=1}^{n(j)} k_{i,a,j,b}(\|X\|^2 - 2l\|X\| + l^2). \tag{19}$$

If we denote $G_{i,j}$ and $g_{i,j}$ such that

$$G_{i,j} = \begin{pmatrix} P_{i,j} & Q_{i,j} \\ Q_{j,i} & S_{i,j} \end{pmatrix}, \quad \boldsymbol{g}_{i,j} = [\boldsymbol{u}_{i,j}\ \boldsymbol{v}_{i,j}], \tag{20}$$

where P, Q, S are 6×6 matrices, $Q_{i,j} = Q_{j,i}^T$, and $\boldsymbol{u}, \boldsymbol{v}$ are 1×6 row vectors, then Eq. 18 can be expressed in terms of $\boldsymbol{\delta}(t) = [\boldsymbol{\delta}_1^T, \cdots, \boldsymbol{\delta}_N^T]^T \in \mathbb{R}^{6N}$

$$C(\boldsymbol{\delta}) = \frac{1}{2}\boldsymbol{\delta}^T\Gamma\boldsymbol{\delta} + \frac{1}{2}\gamma\boldsymbol{\delta} + B, \tag{21}$$

where Γ is a $6N \times 6N$ matrix, γ is a $1 \times 6N$ row vector, and B is a constant. They can be constructed as follows:

$$\Gamma_{i,j} = \begin{cases} \sum_{a=1}^{i-1} S_{a,i} + \sum_{b=i+1}^{N} P_{i,b} & if \quad i = j \\ Q_{i,j} & if \quad i \neq j \end{cases}, \quad for \quad 1 \leqq i,j \leqq N,$$
$$\gamma_i = \sum_{a=1}^{i-1} \boldsymbol{v}_{a,i} + \sum_{b=i+1}^{N} \boldsymbol{u}_{i,b}, \tag{22}$$
$$B = \frac{1}{2}\sum_{i=1}^{N-1}\sum_{j=i+1}^{N} B_{i,j},$$

where $\Gamma_{i,j}$ is the (i,j) submatrix of Γ and $\gamma = [\gamma_1, \cdots, \gamma_N]$.

We finally obtain $\boldsymbol{\delta}$ which minimizes the cost of Eq. 22 by setting its derivative to be zero such that

$$\frac{\partial C(\boldsymbol{\delta})}{\partial \boldsymbol{\delta}} = \Gamma\boldsymbol{\delta} + \frac{1}{2}\gamma^T = \mathbf{0}. \tag{23}$$

Note that here Γ always has three zero eigenvalues corresponding to rigid-body translation of the whole system. In other words, there is no unique solution $\boldsymbol{\delta}$ satisfying Eq. 23 algebraically and the system is physically free to float around in space. To anchor the system in space, one can either set $\boldsymbol{v}_i = \mathbf{0}$ for any single value of i or add the constraint of linear momentum conservation such that

$$\sum_{i=1}^{N} m_i \boldsymbol{v}_i = \mathbf{0}. \tag{24}$$

Now we get a new conformation by plugging this uniquely determined $\boldsymbol{\delta}$ into Eq. 9. This generates one small incremental change. Intermediate conformations are iteratively generated in this context by altering the coefficient α from 0 to 1 in Eq. 13. Typically we iterate 100 times and index each intermediate conformation. Here index k indicates the intermediate conformation which is $k\%$ along the path from $\{\boldsymbol{x}_i\}$ to $\{\boldsymbol{y}_i\}$. For example, index 0 represents the initial conformation whereas index 100 stands for the final conformation.

3 Results

In this section, rigid-cluster ENI is tested for two example protein structures which essentially consist of several rigid domains. Lactoferrin is divided into three rigid domains during its transition. ATP synthase consists of three α, three β, and one γ domains. It is modeled here as 10 rigid-clusters because each β domain can be divided into two rigid clusters. Although the loop regions are typically more flexible than other secondary structures such as α-helices and β-sheets, they are assumed to be parts of the rigid clusters in this context. Clustering has been performed using a windowed root-mean-square-deviation (RMSD) computation between corresponding segments in both end conformations.

3.1 Lactoferrin

Roughly speaking, the lactoferrin structure can be divided into three rigid-domains (i.e., one head and two side lobes as shown in Fig. 4). Thr90 and Val250 act like hinges in the middle of two lobes so that they can open and close the ion binding site. To apply the rigid-cluster ENI method, lactoferrin is represented by three rigid-clusters defined in Table 1. RMSD of each cluster is approximately 1Å. This value is relatively small compared with the large conformational change between the closed (PDB code: 1LFG) and open (PDB code: 1LFH) forms [11], which is 7Å in RMSD.

Fig. 4 shows the pathway computed from rigid-cluster ENI. It appears that two side lobes move apart from each other resulting in the exposure of the ion binding site. Next, we measure RMSD of all intermediates and virtual

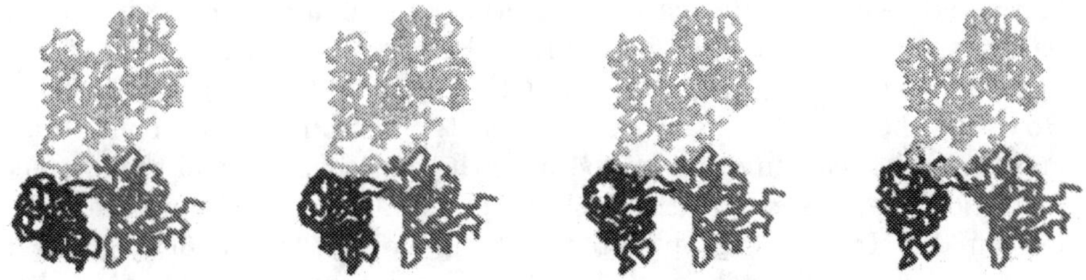

Fig. 4. The conformational change of lactoferrin generated by rigid-cluster ENI. Lactoferrin is assumed to have three rigid-clusters: head (yellow), left (blue) and right (red) lobes. Two lobes are opened and closed by the hinge motion around Thr90 and Val250. The closed (left) and open (right) conformations are displayed with two intermediate conformations (middle) calculated from rigid-cluster ENI.

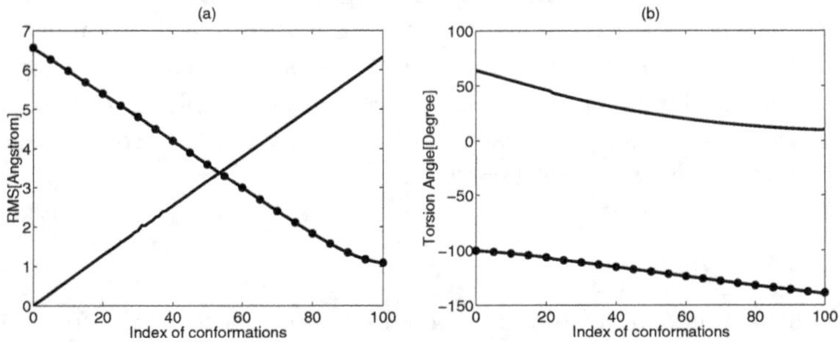

Fig. 5. Characteristics of the simulated conformational transition in lactoferrin. (a) The solid and solid-dot lines indicate RMSD of all intermediates with respect to the closed and open conformations, respectively. (b) Virtual torsion angles for Thr90 (solid line) and Val250 (solid-dot line) monotonically vary during the transition. They act like hinges that open the two side lobes. These results show that the rigid-cluster ENI is able to generate a feasible pathway observing distance constraints.

Table 1. Three rigid-clusters of lactoferrin

Cluster	Residue number	RMSD[a]
Head	Gly321 $\sim$ Lys691	0.99
Left lobe	His91 $\sim$ Val250	0.53
Right lobe	Gly1 $\sim$ Thr90 , Pro251 $\sim$ Leu320	1.15

[a] RMSD between two corresponding rigid-clusters that come from the closed and open conformations of lactoferrin, respectively. The unit is Å.

torsion angle changes during the transition to evaluate characteristics of the pathway generated by rigid-cluster ENI. Fig. 5a shows that RMSD of all intermediates monotonically increases with respect to the closed conformation. However, there is 1Å difference between the computed open conformation and the real open conformation because the domain shapes in the open conformation are not exactly the same as those in the closed conformation as measured in Table 1. Fig. 5b plots the changes of virtual torsion angles at the two hinge points Thr90 and Val250. These monotonic angle changes indicate the computed conformational transition seems to be naturally favorable and energetically feasible.

In addition, the minimum distance among all pairs of alpha carbons is measured for all intermediate conformations. The minimum distance is 2.87Å between Cys627 and Pro628. This value does not change during the transition. That is, there exists no inter-connection with distance shorter than that of the intra-connection between Cys627 and Pro628. This fact shows that rigid-cluster ENI does not produce any steric clashes among the three rigid clusters in lactoferrin. The computational performance in rigid-cluster ENI is significantly better than that of coarse-grained ENI. In this particular case, the DOF in the rigid-cluster model is only $6 \times 3 = 18$, whereas that of coarse-grained ENI is $3 \times 691 = 2073$.

3.2 ATP synthase

ATP synthase is a molecular motor which synthesizes ATP from ADP and phosphate using the energy induced by transmembrane proton absorption, as well as a molecular pump in which ATP is hydrolyzed and a proton is pumped out of the proton channel [12]. Both synthetic and hydrolytic cycles are performed by a rotary motion of the center γ domain [13]. The general structure of ATP synthase is shown in Fig. 6. It consists of three α, three β, and a single γ domains. Functionally, the α and β domains behave as a "stator" and the γ domain acts as a "rotor" in a molecular motor system.

The schematic in Fig. 7 shows the chemical reactivity of a catalytic site. The reactivity of each site is related to the orientation of the γ domain (it is thought to rotate 120° every synthetic and hydrolytic cycle). These chemical changes presumably induce large conformational changes in ATP synthase. Fig. 8 shows the hinge-bending motion between the upper and lower portions of the β domain. This is associated with the change of state of the catalytic site from the empty state, denoted as E, to the ATP bound state denoted as T [13]. From this large and collective motion, each β domain can be modeled as two rigid-clusters. Three residues Gly130, Gly179, and Leu329 are chosen as the hinge points. Consequently, the upper portion consists of Gln130∼Gly178 and Leu329∼Ala474 and the lower portion is composed of Thr9∼Glu129 and Gly179∼His328.

Now, ATP synthase is represented by 10 clusters (i.e., $3\alpha+2(3\beta)+\gamma = 10$). Table 2 shows RMSD between clusters that belong to the same group. For

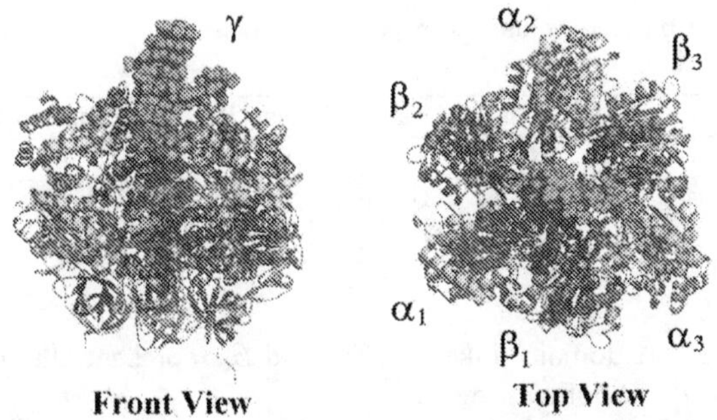

Fig. 6. Structure of ATP synthase. Its X-ray crystal coordinates are obtained from the Protein Data Bank (PDB code: 1BMF [12]). The functional behavior of this protein is like a molecular motor which consists of three α domains (yellow), three β domains (pink), and one γ domain (green). Three catalytic sites are located at the $\alpha\beta$ interface. The γ domain points out one of the catalytic sites in the top view.

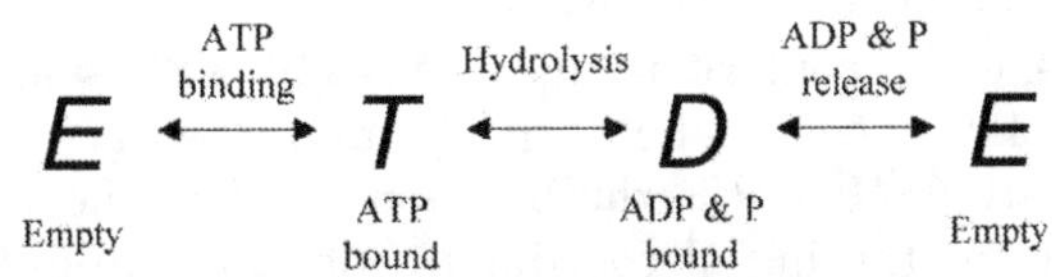

Fig. 7. Schematic of the reactivity cycle of a catalytic site in ATP synthase. Each catalytic site sequentially takes three different chemical states corresponding to the orientation of the γ domain. This is modified from [13].

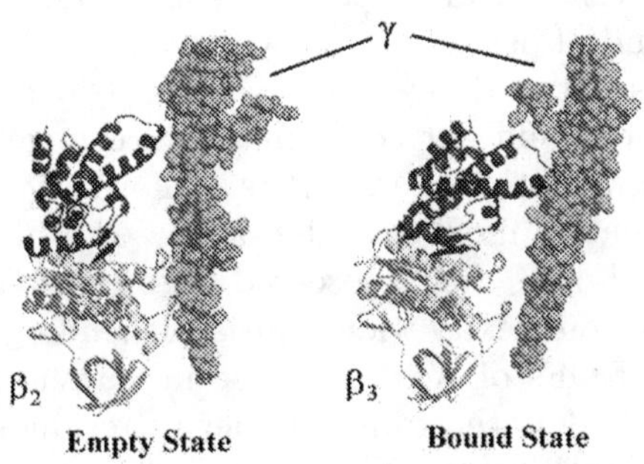

Fig. 8. Conformational changes in the β domain. These cross-sectional views show the conformational changes in the β-domain in the empty and ATP-bound states, respectively. The β domain is divided into two rigid-clusters (blue and yellow). The hinge motion of the β-domain rotates the upper cluster by about 30° during hydrolysis and synthesis.

Table 2. RMSD between clusters in ATP synthase

Domain[a]	Cluster	RMSD[b]	Cluster	RMSD	Cluster	RMSD
α	$\alpha_1\alpha_2$	1.32	$\alpha_1\alpha_3$	1.18	$\alpha_2\alpha_3$	1.34
β^u	$\beta_1^u\beta_2^u$	1.16	$\beta_1^u\beta_3^u$	0.60	$\beta_2^u\beta_3^u$	1.29
β^l	$\beta_1^l\beta_2^l$	1.11	$\beta_1^l\beta_3^l$	0.34	$\beta_2^l\beta_3^l$	1.06

[a] α_i is the ith α-domain. Likewise, β_i^u and β_i^l represent the upper and lower segments of ith β-domain, respectively.
[b] The unit of RMSD is Å.

example, three α domains, denoted as $\alpha_1, \alpha_2, \alpha_3$, look similar to one another but they are not identical. However, as an idealization they are assumed to be the same as one another to construct a simple mathematical model describing the rotary motion of ATP synthase.

To determine the conformational change of ATP synthase, new coordinates at $\theta = 120°$ need to be derived from those at $\theta = 0°$ by using a rigid-body rotation. Here θ is the rotation angle of the γ domain. That is, $\theta = 0°$ stands for the initial coordinates obtained from the X-ray crystal structure, while $\theta = 120°$ indicates the new conformational asymmetry of ATP synthase for the next cycle. Rigid-cluster ENI is then applied to calculate intermediate conformations between the two end conformations (i.e., $\theta = 0°$ and $\theta = 120°$). Fig. 9 illustrates the simulated conformational transition of ATP synthase. During a 120° counterclockwise rotation of the γ domain from its initial position as shown in Fig. 6, the upper portion of the β_1 domain rotates outward ($D \rightarrow E$) but the β_2 domain rotates inward ($E \rightarrow T$). There is no significant conformational change in the β_3 domain ($T \rightarrow D$). RMSD between the intermediate conformations computed here and the target conformation monotonically decreases from 6Å to zero. No steric clashes among rigid-clusters are observed during the transition.

Computational complexity widely varies depending on the choice of modeling parameters. DOFs of ATP synthase in various models are calculated in Table 3. Here 2981 is the total number of residues in ATP synthase. It is observed that rigid-cluster ENI tremendously reduces the computational cost.

4 Conclusion

In this paper, the ENI method developed previously by the authors is extended to rigid-cluster systems. The conformational change of a rigid-cluster system can be represented by rigid-body motions such as hinge and shear motions. Hence, for many macromolecules it is not necessary to model rigid

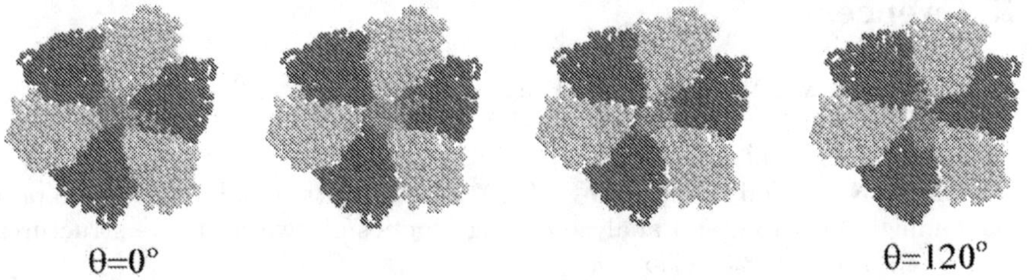

Fig. 9. Conformational transitions in ATP synthase generated by rigid-cluster ENI. The major conformational change is the rotary motion of the γ domain by a $120°$ rotation per catalytic cycle.

Table 3. DOF and computation time on a PC of ATP synthase in various parameterizations.

Method[a]	Residue	DOF[b]	Total DOF[c]	Fraction[d]	Time[e]
[FA]	22,722	3	68,166	100%	Impractical
[CG]	2,981	3	8,943	13.1%	57min
[RC]	10 clusters	6	60	0.1%	<1min

[a] [FA], [CG], and [RC] are the abbreviation of full-atom, coarse-grained, and rigid-cluster model, respectively.
[b] The number of DOF per residue or cluster.
[c] The total DOF of ATP synthase in this particular modeling.
[d] The relative fraction of the DOF of each method with respect to that of [FA].
[e] The elapsed time for calculating a single intermediate conformation on a 1.5GHz Pentium with 512MB memory.

domains of the structure at atomic detail. Only six parameters are required to capture the translational and rotational motions of each rigid-cluster. This strategy has a big impact on computational cost because it only depends on the number of rigid-clusters modeled, regardless of the size of the molecules. Rigid-cluster ENI is applied to lactoferrin and ATP synthase. Even though the systems are modeled by far fewer parameters than in coarse-grained ENI, the results in both cases are in good agreement with those of coarse-grained ENI. Rigid-cluster ENI may serve as a very powerful and practical tool for the study of conformational changes in many other large macromolecules where the computational cost of molecular dynamics is prohibitive.

References

1. Berman H.M., Westbrook J., Feng Z., Gilliland G., Bhat T.N., Weissig H., Shindyalov I.N., Bourne P.E. (2000) The Protein Data Bank. Nucleic Acids Res. **28**, 235–242
2. Amato N.M., Dill K.A., Song G. (2003) Using motion planning to map protein folding landscapes and analyze folding kinetics of kwown native structures. J. Comp. Biol. **10**, 239–255
3. Apaydin M.S., Brutlag D.L., Guestrin C., Hsu D., Latombe J.C., Varma C. (2003) Stochastic roadmap simulation: An efficient representation and algorithm for analyzing molecular motion. J. Comp. Biol. **10**, 257–281
4. Zang M., Kavraki L.E. (2002) A new method for fast and accurate derivation of molecular conformations. J. Chem. Inf. Comp. Sci. **42**, 64–70
5. Atilgan A.R., Durell S.R., Jernigan R.L., Demirel M.C., Keskin O., Bahar I. (2001) Anisotropy of fluctuation dynamics of proteins with an elastic network model. Biophys. J. **80**, 505–515
6. Kim M.K., Jernigan R.L., Chirikjian G.S. (2003) An elastic network model of HK97 capsid maturation. J. Struct. Biol. **143**, 107–117
7. Kim M.K., Chirikjian G.S., Jernigan R.L. (2002) Elastic models of conformational transitions in macromolecules. J. Mol. Graph. Model. **21**, 151–160
8. Kim M.K., Jernigan R.L., Chirikjian G.S. (2002) Efficient generation of feasible pathways for protein conformational transitions. Biophys. J. **83**, 1620–1630
9. Gerstein M., Krebs W. (1998) A database of macromolecular motions. Nucleic Acids Res. **26**, 4280–4290
10. Schuyler A.D., Chirikjian G.S. (2004) Normal mode analysis of proteins: a comparison of rigid cluster modes with C-alpha coarse graining. J. Mol. Graph. Model. **22**, 183–193
11. Norris G.E., Anderson B.F., Baker E.N. (1991) Molecular replacement solution of the structure of apolactoferrin, a protein displaying large-scale conformational change. Acta. Cryst. B **47**, 998–1004
12. Abrahams J.P., Leslie A.G.W., Lutter R., Walker J.E. (1994) Structure at 2.8Å resolution of F_1-ATPase from bovine heart mitochodria. Nature **370**, 621–628
13. Wang H., Oster G. (1998) Energy transduction in the F_1 motor of ATP synthase. Nature **396**, 279–282

Fast Tree-Based Exploration of State Space for Robots with Dynamics

Andrew M. Ladd and Lydia E. Kavraki

Rice University, Houston TX 77005, USA

Abstract. This paper presents a new motion planning algorithm which we call the Path-Directed Subdivision Tree exploration planner (PDST-EXPLORE). It is a sampling-based method which builds a tree and takes a substantially different approach from other exploration planners such as RRT [18] and EST [12]. PDST-EXPLORE is a general purpose planner but is designed to overcome difficulties inherent in planning for robots with non-trivial dynamics. Specifically, our planner represents samples as path segments rather than individual states and uses non-uniform subdivisions of the state space to estimate coverage. This change avoids many of the problems that previous sampling-based planners have had with milestone placement, metrics and coverage estimation. We use a deterministic update schedule together with randomized path generation to adaptively strike a balance between greedy exploration and methodical search. We have obtained a proof of probabilistic completeness for the planner which assumes very little about the specific robot system that the planner operates on. Finally, we have implemented the planner for planar kinodynamic point robots, differential drive robots and blimp-like robots. The experimental results demonstrate the efficiency of the planner's implementation as well as its robustness in covering the entire reachable free space.

1 Introduction

This paper treats the problem of developing efficient planning techniques for robot systems with non-trivial dynamics operating in the presence of obstacles [18]. There are many interesting examples of dynamic motion planning problems such as car-like robots [16,3], tractor-trailer robots [16], snakeboard [19], helicopter robots [10], and spaceship robots [7]. We present a general purpose planner that is designed to address the challenges posed by robot systems with non-trivial second-order dynamics. In these cases, planning is complicated by constrained motion, drift and the importance of minimizing a path cost such as time or fuel usage. This paper relates the design and implementation of the Path-Directed Subdivision Tree exploration planner (PDST-EXPLORE). It is a tree-based planner which departs from previous frameworks, such as Rapidly-Exploring Random Trees RRT [18] and Expansive Spaces Trees EST [12], in order to avoid well-known weaknesses of these approaches [4,1].

There are various techniques applicable to dynamics planning problems in the literature. We group these into several families: exact methods, path

M. Erdmann et al. (Eds.): Algorithmic Foundations of Robotics VI, STAR 17, pp. 297–312, 2005.

conversion methods and sampling-based methods. Exact methods which produce optimal paths are known but are **PSPACE** [9]. Polynomial time algorithms that produce approximately optimal paths are known for acceleration bounded point robots and for curvature constrained planning [20]. These exact algorithms produce optimal or approximately optimal solutions, however the main barrier to usage is that they are not practical algorithms and are only applicable to specific instances. Path conversion techniques make use of structure [2] to lift solutions obtained for simplified systems to feasible solutions for the given problem [22,21]. Although path conversion methods often offer a computationally efficient way of generating dynamic motions, the resulting paths tend to be of low quality, having many redundant or stopping motions. In the area of sampling-based planners, we divide existing planners into two categories: roadmap methods, such as the Probabilistic Roadmap (PRM) planner, and tree-based planners. The roadmap family of planning techniques [13] use global sampling and local planners to build roadmaps. In the dynamics case, implementing the local planner primitive, as it used in PRM, requires solving the steering problem [6,15], which is typically computationally expensive. The PRM approach offers a probabilistically complete solution [14] but the large number of calls to the steering primitive is prohibitive. Finally, tree-based planners, such as the (RRT) planner [18,5] and the (EST) planner [12], make use of forward integration of controls (propagation) to incrementally construct trees in the state space (exploration planning). Typically, steering primitives are then used to connect pairs of trees (connection planning). The tree-based methods enjoy the computational advantage of the relatively cheap propagation primitive and are thus the starting point for the research developed in this paper.

We choose to focus on improving the basic instance of a tree-based planner: the exploration planner. The purpose of an exploration planner is to capture the structure of the reachability set of the root state by incrementally constructing a tree of paths beginning from the given root. Exploration planners are usually employed as primitives in larger schemes, such as in bidirectional planning [18] or for the Probabilistic Roadmap of Trees planner [1], and dominate the run-time [18]. Our objective is to improve the state-of-the-art of exploration planning for dynamic planning problems. Exploration planners for such systems, namely RRT and EST, incrementally construct a tree in reachability space by alternating between two actions, selection and propagation, at each iteration of the planner. We refer to this design as the select-propagate architecture. In the selection phase, a state from the sample set is chosen using an algorithmic procedure guided by a data structure which is constructed and maintained on-line. This data structure is designed to estimate the current coverage of the sample set and to bias the sampling toward areas of the space which have few or no samples. Propagation generates new branches in the tree extending from the state chosen by the selection phase.

PDST-EXPLORE is a general purpose exploration planner but focuses on the case of motion planning with second-order dynamics as it is particularly challenging for exploration planners [4]. It is these challenges that inspired the PDST-EXPLORE planner and motivate its design. The efficiency of exploration planners is strongly tied to the implementation of the selection phase during expansion. The RRT approach uses Voronoi bias computed with proximity queries to guide expansion [18] and the EST approach maintains local density estimates to achieve the same [12]. Both approaches determine a probability distribution over the set of all current samples. The randomized (or quasi-random) selection approach is the foundation of probabilistic (or resolution) completeness for both methods. During propagation it is usually desirable to create multiple new samples along the propagated path (checkpointing) [4,12]. It is now well-known that the performance of these methods when applied to dynamic planning is very sensitive to the metric that determines proximity or density [4]. In particular, the drift present in second-order systems invalidates inherent assumptions in the way the metric is used [4]. Although specific remedies to this difficulty have been proposed [4], the metric sensitivity issue still impacts the efficiency of the planner. It has been empirically observed that RRT and EST enjoy rapid initial convergence but global convergence is much slower [1]. We posit that this is due to the randomized (or quasi-random) selection schedule leading to redundant growth as coverage increases.

PDST-EXPLORE preserves the features of exploration planning that have been successful: the use of cheap propagation primitives and the select-propagate architecture for growing the tree. Our contribution is a new selection algorithm and corresponding introspection data structure which are radically different from the ones typically employed in sampling-based planning. Namely, we propose the following design changes:

1. The samples are path segments and not states.
2. Metrics are not employed during selection and their properties are not used to prove completeness.
3. We use a deterministic, greedy selection schedule.

Outline This paper is organized as follows: in Section 2 we define the motion model that we use and state the problem that we are addressing. In Section 3, we describe the PDST algorithm and state a theorem about probabilistic completeness. We have tested the efficiency of our planner on the planar kinodynamic point robot, the differential drive robot and a blimp robot in 3-D. The planar workspaces we used are maze-like environments with features of various scale and character. Such workspaces are very challenging for existing planners [18,12,4,5]. Our experiments and results are detailed in Section 4, together with a novel application of Maneuver Automata theory to trajectory generation [11]. We conclude in Section 5 with a brief discussion.

2 Problem Statement

The state space of the robot is denoted by Q and is a smooth m-dimensional manifold. Let U be the set of controls for the robot. The motion of the robot system is governed by a differential equation. This equation is of the form $\dot{q} = f(q, u)$ for $q \in Q$ and $u \in U$. The function f is smooth in q. In addition to the constraint imposed by f, the motion of the robot may be further constrained by inequality constraints of the form $g(q, \dot{q}) \leq 0$ where g is smooth. Any path through Q that satisfies these constraints is *feasible*. In addition to the dynamic constraints, there are constraints determined by obstacles in Q. The set of states which are free is an m-dimensional submanifold of Q which is denoted Q_{free}. A state $q \in Q_{\text{free}}$ is a free state and state $q \notin Q_{\text{free}}$ is in collision. A feasible path which lies entirely in Q_{free} is said to be *collision-free*.

We give a formal definition of an exploration planner which is convenient for exposition and to state our theorem of probabilistic completeness. We define the exploration planning problem as follows: given $q_0 \in Q_{\text{free}}$ and $A \subset Q_{\text{free}}$, the exploration planner must compute a collision-free path from q_0 to any $q_f \in A$. This formulation captures how RRT and EST can be used a uni-directional planners and allows to analyze coverage of the reachability set of q_0.

3 The PDST-EXPLORE Planner

Like RRT and EST, PDST-EXPLORE is a sampling-based planner which uses select-propagate architecture to incrementally construct a tree. However, we propose radical changes to the way samples are represented and to the implementation of the selection algorithm. We will begin this section by giving a brief overview of and introduction to the operation of our planner.

In exploration planners to date, the sample set is represented as a set of states connected by paths defining a tree. In the PDST-EXPLORE planner, we represent the samples as a set of paths which are connected at branch states. By switching to this representation, the selection algorithm adaptively places branch states as needed. This seemingly superficial change avoids the need for check-pointing and drastically reduces the number of stored sample objects.

Previous exploration planners construct and maintain a biased probability distribution over the samples and then select the state to propagate from by randomly selecting from this distribution. The distribution bias is determined by proximity [18] or density [12]. In PDST-EXPLORE, priorities are assigned to each sample. The priority of a given sample is initialized to the iteration number the sample was created on and doubles each time that sample is selected. The selection algorithm for PDST-EXPLORE selects the sample with the lowest weighted priority at each iteration. The weighting for the priority are determined by a partition of the state space into cells. At the end of iteration, one cell is subdivided and samples contained in that cell are subdivided as well. The cell subdivision rule, priority scheme and sample growth are designed to satisfy probabilistic completeness, to cover the reachability set greedily and to support efficient implementation.

3.1 Space and Sample Representation in PDST-EXPLORE

In this subsection we describe how the PDST-EXPLORE planner represents the state space and the samples. The space representation is a non-uniform cell partition of state space which is used to estimate coverage and to guide expansion. The samples are collision-free path segments which define a tree.

Definition of Subdivision Scheme PDST-EXPLORE maintains a complete subdivision of the state space into a set of cells. The subdivision is refined after each iteration. Each cell has a volume. Refining a cell creates two new cells, the union of which is the original cell, each with non-zero volume. This final property is necessary for our proof of correctness. A straightforward implementation of the cell partition can be obtained with a Binary Space Partition Tree [8].

Formally, a cell C is a subset of Q. The measure or volume of a cell is given by $\mu(C)$. The function μ is a measure in the formal sense and is normalized so that $\mu(Q) = 1$. A subdivision, S, is a finite partition of Q into cells of non-zero volume, $S = \{C_1, ..., C_K\}$. Cell subdivision is deterministic and consists of splitting a cell C into two parts, $\texttt{left}(C)$ and $\texttt{right}(C)$, which form a disjoint partition of C and each have non-zero measure. A subdivision S is refined by subdividing a single cell. The measure μ and refinement rules are called a subdivision scheme. It turns out that the subdivision scheme need not satisfy any additional properties beyond the ones we have detailed in order to permit the proof of probabilistic completeness. For this reason, the choice of the subdivision scheme can be made to optimize the efficiency of the planner.

Definition of Sample (Mass) A change proposed in this paper is to use paths rather than states as samples. We introduce the term *mass* to refer to a sample consisting of a non-empty path segment together with some additional structure. Masses also have a non-negative scalar *weight* and can be subdivided. It is possible to generalize mass to mean arbitrary subsets of state space but we eschew this in favor of clarity of exposition.

A mass is a collision-free path π with given duration T. Since the masses define a tree structure we need to define the root of the tree and the path from the root to each state contained in a mass. Specifically, every mass M has a root state $\texttt{root}(M) \in Q$ and associated with a path from $\texttt{root}(M)$ to each state along π. If q is state along the path π, we write $\texttt{path}(M, q)$ for the collision-free sub-path from $\texttt{root}(M)$ to q.

Each mass M has a weight $\nu(M) \geq 0$. The function ν is a measure in the formal sense. In our implementations, we have used $\nu(M) = |\pi| = T$, the duration of the path. Other notions of length or measure could be used. If π' is a sub-path of the path π, then the mass $M' = \pi'$ is a sub-mass of M if $\texttt{root}(M) = \texttt{root}(M')$ for any state q along both π and π', $\texttt{path}(M, q) = \texttt{path}(M', q)$.

Propagation The `propagate` operation creates a new mass by randomly extending from an existing mass. In our experiments, we have implemented `propagate` using two operations: `choose` and `generate`. The `choose` operation chooses a random state from the mass. For mass $M = \pi$ with duration T, $\texttt{choose}(M)$ generates a random parameter t uniformly from $[0, T]$ and returns $\pi(t)$. The operation `generate` creates a new mass by integrating a randomly chosen control function starting from the given state. Let M be a mass and let q be the result of $\texttt{choose}(M)$, the call to $\texttt{generate}(M, q)$ chooses $\kappa : [0, \infty] \to U$ from a distribution of control functions and integrates it starting from q to obtain a new path. The design of such distributions of control functions has been described in the literature [18,12,10,5]. The new path is collision detected and a collision-free sub-path, π', is obtained. The new mass M' is created from π' and the canonical path to $\texttt{root}(M)$ is constructed from the path from the root to q, $\texttt{path}(M, q)$.

3.2 PDST-EXPLORE Description

In this subsection, we describe the operation of the PDST-EXPLORE algorithm given as Algorithm 1. We split the pseudo-code of the algorithm into three parts: initialization (lines 1-4), select and propagate (lines 6-9) and the update code (lines 10-30). As we describe the operation of Algorithm 1, we will detail the calculation of various quantities and the data invariant maintained during the execution. Certain aspects of the interaction between the priority scheme, selection and space partitioning might seem somewhat arbitrary. These design decisions were made for two reasons: observed experimental efficiency and to permit the proof of Theorem 1.

Initialization and Invariants We now describe the initialization phase of PDST-EXPLORE (lines 1-4). An input of q_0, A and N is provided. The state $q_0 \in Q_{\text{free}}$ will be the root state for all masses created by the run. The subset $A \subset Q_{\text{free}}$ is the goal set for the run and planner will halt if a generated mass has non-empty intersection with A. The integer N determines the number of iterations that the algorithm runs for. Before we continue, some key invariants need to be discussed to shed light on the design of PDST-EXPLORE.

The algorithm maintains a few structures during the run. Fundamentally, these are a subdivision of Q called S, a finite set of masses written $\mathbf{M}$ and

Algorithm 1 PDST-EXPLORE(q_0, A, N)

```
Set the subdivision to S = {Q}.
Create a singleton mass M_0 = {q_0}.
Set the mass set M = {M_0}.
Set priority(M_0) = 1.
for Each iteration = 1, ..., N do
  Choose C_s ∈ S, the cell with the lowest priority.
  Choose M_s ∈ masses(C), the mass in C_s with the lowest priority.
  Set q to choose(M).
  Set M_c to generate(M_s, q).
  if M_c = ∅ then
    priority(M_s) := 2(priority(M_s) + iteration).
  else
    if M_c ∩ A ≠ ∅ then
      Take q_f ∈ M_c such that q_f ∈ M_c.
      return path(M_c, q_f).
    end if
  else
    priority(M_s) := 2(priority(M_s) + 1).
    while M_c ≠ ∅ do
      C_next := stab(S, M_c)
      M_r := C_next ∩ M_c.
      M_c := M_c − M_r.
      if not the first time through and if density(C_next) > avgdensity(S)
      then
        break out of the while loop.
      else
        M := M ∪ {M_r}.
        priority(M_r) := iteration.
      end if
    end while
    Subdivide cell C_s.
  end if
end for
return FAILED.
```

a positive integer priority for each $M \in \mathbf{M}$ denoted priority(M). We discuss low-level implementations for these structures later in this section. The subdivision S and $\mathbf{M}$ always satisfy the invariant property that for every $M \in \mathbf{M}$, there is a unique $C \in S$ such that $M \subset C$. In other words, each mass element lies uniquely in a some cell of the subdivision. When a cell is subdivided, the masses contained in that cell are also subdivided.

During initialization, the structures are created and assigned trivial values. The subdivision is initialized to the trivial subdivision $S = \{Q\}$. A special singleton mass M_0 is created. We have $M_0 = \{q_0\}$ and is associated with the trivial path at q_0. The mass set is initialized by $\mathbf{M} = \{M_0\}$. The initial priority of M_0 is set to 1.

Select and Propagate Selection and propagation occur on lines 6-9 and execute at the beginning of each iteration. The selection step chooses a mass M_s from the current mass set $\mathbf{M}$. The mass M_s has the lowest priority among all masses contained in the cell $C_s \in S$ with the lowest priority. The priority of cell C, $\mathtt{priority}(C)$, is determined as a function of priorities of the masses contained in C and the volume of C. The intuition is that small cells are less important than large cells.

$$\mathtt{priority}(C) = \frac{\min\{\mathtt{priority}(M) : M \subset C \text{ and } M \in \mathbf{M}\}}{\mu(C)}.$$

If there are no $M \in \mathbf{M}$ such that $M \subset C$ then $\mathtt{priority}(C) = \infty$.

After selecting the mass M_s, a new mass is propagated from the old mass. The new mass, which we call M_c, is created by calling a randomized operator $\mathtt{propagate}(M_s)$. It is possible that $M_c = \emptyset$ in which case we say that propagation failed.

Update After selection and propagation occur, the mass set, priorities and subdivision are updated. The update takes place on lines 10-30. There are three cases: if $M_c = \emptyset$, if $M_c \cap A \neq \emptyset$ and the final case where M_c is incrementally inserted into $\mathbf{M}$ according to S.

If the call to $\mathtt{propagate}$ fails ($M_c = \emptyset$) then the priority of M_s is adjusted to $2(\mathtt{priority}(M_s) + \mathtt{iteration})$. The addition of the iteration counter is to penalize the rate of selection for that mass beyond the normal doubling scheme since propagation failed. If $M_c \cap A \neq \emptyset$ then there is $q_f \in M_c$ such that $q_f \in A$. The exploration planner has succeeded and can return $\mathtt{path}(M_c, q_f)$.

If the propagation succeeded but M_c did not intersect A, then M_c is partitioned according to S and incrementally inserted into $\mathbf{M}$. The insertion proceeds until it reaches a cell of above average density. The density thresholding technique helps limit the creation of redundant samples. At least one insertion occurs regardless of density. The insertion loop is on lines 10-30. After the insertion is complete, the selected cell C_s is subdivided. Additionally, the priority of M_s is updated to $2(\mathtt{priority}(M_s) + 1)$ to penalize its selection.

The density of a cell is determined by the weight of the masses in the cell and the volume of the cell. The density of a cell C is determined by the sum of the masses contained in a given cell divided by the volume of the cell

$$\mathtt{density}(C) = \sum_{M \in \mathbf{M} : M \subset C} \frac{\nu(M)}{\mu(C)}.$$

The average density of the subdivision S, $\mathtt{avgdensity}(S)$, is determined by cells in the subdivision which are non-empty $S_p = \{C \in S : \text{ there is } M \in$

$\mathbf{M}$ such that $M \subset C\}$. Specifically, the average density is the sum of the densities of the non-empty cells divided by the number of non-empty cells,

$$\texttt{avgdensity}(S) = \sum_{C \in S_p} \frac{\texttt{density}(C)}{|S_p|}.$$

The insertion loop proceeds by determining a cell C_{next} such that $M_c \cap C_{\text{next}}$ is non-empty and then inserting $M_i = M_c \cap C_{\text{next}}$ into $\mathbf{M}$. After the insertion, M_c is set to the remainder $M_c - M_i$ and the loop proceeds. After the first insertion, the insertion loop ends if $M_c = \emptyset$ or if $\texttt{density}(C_{\text{next}}) > \texttt{density}(S)$. The choice of C_{next} is determined by operation called **stab** such that $C_{\text{next}} = \texttt{stab}(M_c, S)$. The stabbing operation finds the cell containing the first state in M_c, i.e. if $M_c = \pi$ then $\texttt{stab}(M_c, S)$ returns the cell $C \in S$ such that $\pi(0) \in C$. This operation is well-defined since the cells in S are a partition of the state space. After insertion has completed, the cell C_{next} is subdivided. The cell C_{next} is removed from the subdivision is removed from S and the cells $C_l = \texttt{left}(C_{\text{next}})$ and $C_r = \texttt{right}(C_{\text{next}})$ are added. Then each $M \in \mathbf{M}$ that was contained in C_{next} is split into $M_l \subset C_l$ and $M_r \subset C_r$. The mass M is discarded and M_l and M_r are added to $\mathbf{M}$. Empty masses are discarded.

Data Structures Obtaining a general implementation of PDST-EXPLORE is fairly straightforward. The mass representation, subdivision scheme and their associated operations can be made abstract. A generic binary space partition tree is then the primary data structure and the masses are referenced by the leaf nodes (cells). A hash table is used to store priorities and the cell priorities and densities are stored at each cell. All non-empty cells are placed in a priority queue and sorted by increasing priority. We use binary heap with hash table back-pointers to implement the priority queue and its operations. The stabbing operations are implemented top-down. The algorithmic overhead in PDST-EXPLORE at the mth iteration is proportional to $O(DS + \log(m))$, where D is the maximum depth of the tree, S is the number of stab operations and $\log(m)$ is incurred by the binary heap. In practice, we have observed that the bulk of the run-time is spent in path generation and collision detection.

Probabilistic Completeness The general formulation of PDST-EXPLORE that we provide supports an abstract proof of probabilistic completeness similar to one made for PRM [14]. We begin by defining a simple random walk exploration planner which we call RANDOM-WALK-EXPLORE. Given input (q_0, A), the planner begin by setting M to singleton mass $\{q_0\}$. At each iteration: if M intersects A then return success and report the path, otherwise set $M := \texttt{propagate}(M)$. If $M = \emptyset$ then return failure otherwise loop again. We can now state the Theorem below.

Theorem 1. *For a given robot system and given a formulation for the mass and the subdivision scheme, the exploration planner* RANDOM-WALK-EXPLORE *succeeds with strictly positive probability on an input* (q_0, A) *if and only if the exploration planner* PDST-EXPLORE *eventually succeeds on input* (q_0, A) *with probability* 1.

Although we do not have space to present the proof of this Theorem as it is fairly involved, we can briefly describe the argument. For the input (q_0, A), both exploration planners can only succeed with strictly positive probability if there a sequence of calls to `propagate` which creates mass in A with strictly positive probability. Using measure-theoretic techniques similar to those applied to our analysis of PRM [14], we prove the existence of a finite stationary sampling sequence. It remains to show that PDST-EXPLORE eventually makes progress, thus decreasing the length of stationary sequence required to generate mass in A by one. Then, applying induction, we can conclude that the planner succeeds. The technical difficulties in the proof arise as a result of the deterministic selection schedule and the subdivision of the masses and cells. Since the effect of random results to `propagate` on the sequence of mass selections is difficult to judge directly, a non-deterministic model of `propagate` can be taken and it can be shown that some sub-mass of any generated mass is selected infinitely often over the course of an infinite run. From this observation, the rest of proof can be constructed after formalizing the effect of subdivision on the masses.

4 Experimental Results

In this section, we begin by describing a novel extension to Maneuver Automata theory [11] which uses PRM sampling to generate Maneuver Automata. The generated Maneuver Automata are then used for trajectory generation. We continue by outlining the specific robot systems we have implemented PDST for. Finally, we describe our experiments and present our results.

4.1 Maneuver Automata

We propose using the Maneuver Automata [11] to implement the `generate` function used in our exploration planner. This is novel extension to the Maneuver Automata literature and can applied to any planner that uses propagation for a robot that satisfies a certain symmetry property. The advantage we gain is that we can restrict ourselves to nice family of motions and eliminate numerical integration from the call to `generate`. These techniques only apply when special structure exists in the motion of the robot, however the class of applicable robot systems is an important one. A sufficient condition occurs when the state space Q is a direct product of a Lie Group G and a shape manifold Z. Furthermore, the Lie Group G operating on Q must preserve path feasibility [11]. If this is the case, we say that G is a symmetry group for Q and Maneuver Automata theory can be applied.

For our purposes a Maneuver Automata is a finite directed multi-graph $MA = (V, E)$. The vertex set V is a finite subset of Z. An element of E is (z_1, λ, z_2) is a directed edge between vertices z_1 and z_2 together with a control function $\lambda : [0, T] \to U$ of duration T. The control function λ must satisfy the property that if $q_1 = (e, z_1)$ then result of integrating the control function λ starting at state q_1 produces a state $q_2 = (g, z_2)$ where g can be arbitrary. In other words, λ gives a control schedule for transitioning from any state with shape z_1 to some state with shape z_2. Such a transition defines a set of paths equivalent under G-symmetries which is called a *maneuver* motion. A second kind of motion can be effected from a state $q = (g, z)$ where $z \in V$ by executing the zero control $0 \in U$ for any amount of time. These motions can be called *trim* motions and can be represented by Lie group exponentiation. Precisely, if $q = (g, z)$ for $z \in V$ and for any time $t \geq 0$ there exists $g_z \in G$ such that integrating the constant control $0 \in U$ starting at q produces a motion $\alpha_z(t)$ which can be written as $\alpha_z(t) = (g \exp(g_z, t), z)$. The Maneuver Automata can be used to generate motions by alternating between the fixed duration maneuver motions and the anytime trim motions. If an automata MA satisfies certain properties then the set $Q_{MA} = \{q = (g, z) : g \in G \text{ and } z \in V\}$ is strongly connected by these generated motions.

The graph $MA = (V, E)$ is a kind of roadmap in shape space. Continuing this analogy, we can sample MA using PRM techniques. So, given $z_1, z_2 \in Z$, we need a local planning primitive which can compute a control function which defines a maneuver motion to connect z_1 to z_2 . The local planner and the sampling distribution over Z can be used to implement PRM. It seems likely that Quasi-Random-Lattice methods would be particularly effective for this task [17].

4.2 Robot Systems

In this subsection, we describe the robot systems for which we implemented PDST. In each case, we specify the state space, the dynamics, inequality constraints, the symmetries, primitive trajectories and cell subdivision scheme.

2-D Kinodynamic Robot The state space for the 2-D kinodynamic robot is denoted by $Q = \mathbb{R}^2 \times \mathbb{R}^2$. A state $q = (x, y, \dot{x}, \dot{y})$, where (x, y) is the robot's position in the plane and $(\dot{x}, \dot{y})$ is the robot's velocity. The symmetry group we use for this robot is $G = \mathbb{R}^2$, the group of 2-D translations. The shape manifold for the robot is $Z = \mathbb{R}^2$ and represents positionless velocities. Every $z = (\dot{x}, \dot{y}) \in Z$ defines a trim primitive α_z and corresponds to the straight line at that fixed velocity. Maneuvers between trim primitives consist of the robot acceleration toward the different velocity state at maximum acceleration.

We impose a constraint that velocity is bounded, i.e. $||(\dot{x}, \dot{y})|| \leq v_{\max}$. The subdivision scheme builds a *kd*-tree in the state space by equal splits on the first and second dimensions.

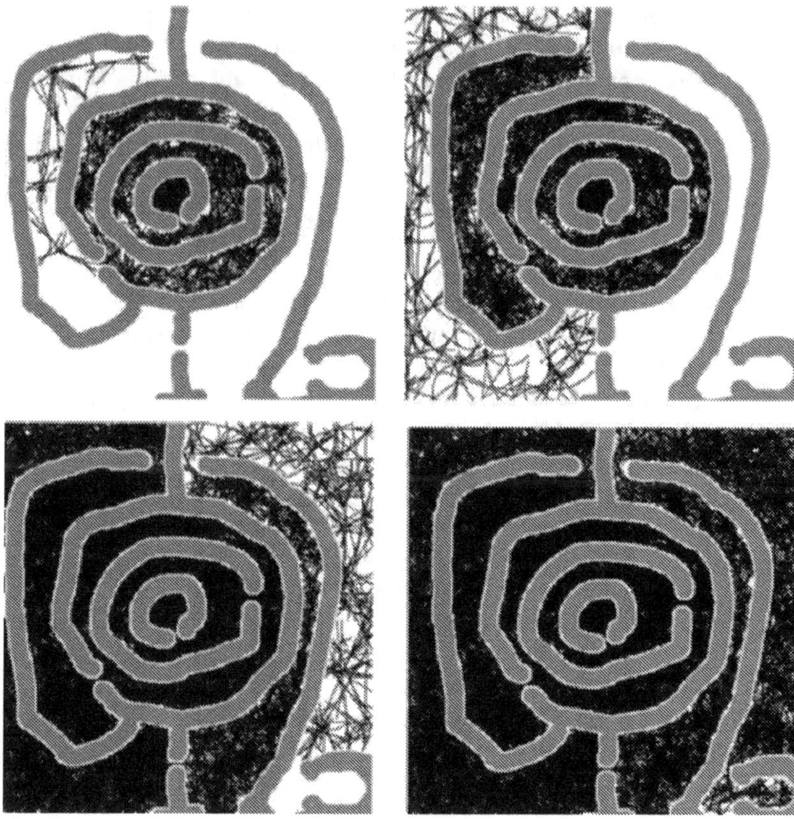

Fig. 1. Execution snapshots of PDST-EXPLORE for a differential drive robot

Differential Drive Robot The state space for this robot is $Q = \mathbb{R}^2 \times S \times \mathbb{R}^2$. A state is given by $q = (x, y, \theta, v_l, v_r)$. The vector (x, y, θ) is the robot's position and orientation and (v_l, v_r) are the robot's wheel velocities. The symmetry group we use for this robot is $G = \mathbb{R}^2 \times S = SE(2)$, the group of 2-D rigid motions. The shape manifold for the robot is $Z = \mathbb{R}^2$ and represents positionless wheel velocities. Every $z = (\dot{x}, \dot{y}) \in Z$ defines a trim primitive α_z. The canonical path for the trim primitive defined by $z = (v_l, v_r)$ is

$$\alpha_z(t) = \begin{cases} (v_f t, 0, 0, v_l, v_r) & \omega = 0 \\ (\frac{v_f}{\omega}\sin(\omega t), \frac{v_f}{\omega}(1 - \cos(\omega t)), \omega t, v_l, v_r) & \omega \neq 0, \end{cases}$$

where $\omega = \frac{v_r - v_l}{L}$, $v_f = \frac{v_l + v_r}{2}$ and L is the length of the wheel base of the robot. Let $z^1 = (v_l^1, v_r^1)$ and $z^2 = (v_l^2, v_r^2)$. The maneuver primitive that brings z^1 to z^2 is $[\pi_{z^1z^2}]$. It is determined by $a_{\max}$, the maximum wheel acceleration for the robot. Let

$$T = \max\left\{\left|\frac{v_l^2 - v_l^1}{a_{\max}}\right|, \left|\frac{v_r^2 - v_r^1}{a_{\max}}\right|\right\}$$

be the duration of $[\pi_{z^1z^2}]$ which has canonical representative

$$\pi_{z^1z^2}(t) = \left(v_l^1 + (v_l^2 - v_l^1)\frac{t}{T}, v_r^1 + (v_r^2 - v_r^1)\frac{t}{T}\right).$$

We impose a constraint which bounds the maximum wheel velocity, $|v_l|, |v_r| \leq v_{\max}$. The subdivision scheme we use for this space builds a *kd*-tree in the state space and uses equal splits on the first, second and third dimensions.

Fig. 2. Workspaces (from left to right) spiral-1, varied-1, varied-2 and slot

Blimp Robot The the state space for this robot is $Q = \mathbb{R}^3 \times S \times \mathbb{R}^3 \times S$. A state $q = (x, y, z, \theta, \dot{x}, \dot{y}, \dot{z}, \dot{\theta})$ is the robot's position, orientation and velocity. The symmetry group that we use for this robot is $G = \mathbb{R}^3$, the group of translations in 3-D. The shape manifold for this robot is $\mathbb{R}^3 \times S$. Every $z = (\dot{x}, \dot{y}, \dot{z}, \dot{\theta}) \in Z$ represents the velocities of the robot. Each $z = (\dot{x}, \dot{y}, \dot{z}, 0)$ defines a trim primitive α_z. The canonical path for the trim primitive defined by such a z is $\alpha(t) = (\dot{x}t, \dot{y}t, \dot{z}t, 0, \dot{x}, \dot{y}, \dot{z}, 0)$. The controls for this robot are a_f, a_z and a_θ. The robot is subject to the following constraints: $\ddot{x} = \cos(\theta)a_f$, $\ddot{y} = \sin(\theta) \cdot a_f$, $\ddot{z} = a_z$ and $\ddot{\theta} = a_\theta$. Furthermore, $a_f \in [0, a_f^{\max}]$, $a_z \in [-a_{\max}^z, a_{\max}^z]$ and $a_\theta \in [-a_{\max}^\theta, a_{\max}^\theta]$. In particular, since a_f must be positive the robot's motion is highly constrained. The calculation of the maneuver primitives are accomplished using a controller which tries to minimize the amount of time taken to switch between two shapes. During the switch, the z-dimension is controlled independently and the controller attempts to minimize the change in z by keeping $|\dot{z}| = 0$ for as long as possible. The controller that we use is expensive to compute but is effective at minimizing the time used. The trajectories taken through shape space to connect two shapes are not reversible and can differ greatly in the total time used. The time step used to integrate the motion of the robot needs to be very small and once the path is computed we re-sample using a variable sized time step which approximates the motion in the state space. The subdivision scheme we use for this robot builds a *kd*-tree in the state space and uses equal splits on the first, second, third and fourth dimensions.

4.3 PDST Experiments

In Figure 2, we depict some of the workspaces we used for the experiments. In each experiment, a maneuver automata was built offline which took less than two seconds for the kinodynamic robot and differential drive robot and between 50 and 70 seconds for the blimp example. During the experiment, the maneuver automata was loaded off the disk and the PDST-EXPLORE planner was run until the measured dispersion [17] in the free space became very small. Dispersion was measured on a high-resolution cell grid. Cells containing an obstacle were not considered in the dispersion measure. The threshold we used ensured that over 0.999 of the space was covered. The number of iterations was then reported. In every example, 384 trials were carried out. The number of iterations required to solve the problem tended to be very similar to the mean number of iterations with the occasional outlier requiring between two and six times more iterations. In Figure 1, we show snapshots of the execution of the exploration of the free space for a differential drive robot in the workspace chambers-1. The time costs and collision detect calls were very consistent across multiple runs. The raw data is presented in Figure 3. Cost in time per iteration is roughly $O(n \log n)$ experimentally, which is expected because of the binary heap.

problem	robot	avg. # iterations	avg. # time	avg. # collision detects
spiral-1	kino	54205	0.76 s	51274
chambers-1	kino	95963	1.88 s	94112
varied-1	kino	76549	1.28 s	77974
varied-2	kino	431736	5.8 s	290904
spiral-1	dd	86000	4.6 s	71808
chambers-1	dd	282708	13.9 s	160350
varied-1	dd	288067	22.9 s	297662
varied-2	dd	1069687	66.7 s	717530
six	blimp	10000	3.4 s	391737
slot	blimp	65000	22.0 s	2515246

Fig. 3. Average running times to obtain full coverage.

5 Discussion

In this paper, we have presented the PDST-EXPLORE algorithm, stated its probabilistic completeness theorem, discussed motion generation for planning using Maneuver Automata and then implemented our ideas for kinodynamic point robots in 2-D, second-order differential drive robots and a second-order blimp-like robots. In our experiments we have demonstrated that PDST-EXPLORE produces full coverage of the space efficiently. Many of our examples were for complicated and varied maze-like environments. In previous studies for planning with second-order dynamics that have reported run-times, times on the scale on a hour have been reported for finding a path in simple environments using robot systems similar to those implemented for this paper [18,5].

There is a great of flexibility in how the PDST-EXPLORE planner can be applied that we have not evaluated in this paper. Time or other cost variables could be incorporated into the space to encourage cost optimization. Our preliminary experiments in this direction have been promising. The cell subdivision scheme, space measure and mass measure can be varied significantly without taking away completeness since they can be chosen to reduce running time by improving convergence speed and by reducing overhead of the geometric computations. For example, the use of non-uniform measures or cell subdivision in a projective space such as task space might be useful for some applications.

The PDST-EXPLORE is only one component of an efficient planning framework for robots with dynamics. We envision using the Probabilistic Roadmap of Trees (PRT) [1] to continue developing this framework toward general purpose single and multiple-query planners for dynamic systems. In pursuit of this goal, we plan on studying path optimality issues, connection planning and adapting PDST-EXPLORE to higher dimensional systems. Most importantly, we plan to apply our techniques to more complex dynamical robot systems to determine the limitations of our approach.

Acknowledgements

Work on this paper has been partially supported by NSF 0308237, NSF 0205671, NSF EIA 0216467, an FCAR fellowship to A. Ladd and a Sloan Fellowship to L. Kavraki. We also thank AMD for donating hardware. A. Ladd thanks S. LaValle, E. Frazzoli, P. Cheng, O. Brock and J. Phillips for valuable input on planning and dynamics. Finally, we thank E. Plaku, K. Bekris and D. Ruths for their feedback on the manuscript.

References

1. M. Akinc, K. E. Bekris, B. Y. Chen, A. M. Ladd, E. Plaku, and L. Kavraki. Multiple query probabilistic roadmap planning using single query primitives. In *Int. Symp. Robotics Research*, 2003.
2. F. Bullo, A. D. Lewis, and K. Lynch. Controllable kinematic reductions for mechanical systems: concepts, computational tools and examples. In *Mathematical Theory of Networks and Systems*, Notre Dame, IN, August 2002.
3. L. G. Bushnell, D. M. Tilbury, and S. S. Sastry. Steering three-input nonholonomic systems: the fire truck example. *IJRR*, 14(4):366–381, 1995.
4. P. Cheng and S. LaValle. Reducing metric sensitivity in randomized trajectory design. In *IROS*, 2001.
5. P. Cheng and S. LaValle. Resolution-complete rapidly-exploring random trees. In *ICRA*, 2002.
6. P. Cheng and S. LaValle. Exploiting group symmetries to improve precision in kinodynamic and nonholonomic planning. In *IROS*, 2003.

7. P. Cheng, Z. Shen, and S. M. LaValle. RRT-based trajectory design for autonomous automobiles and spacecraft. *Control Sciences*, 11(3-4):51–78, 2001.
8. M. de Berg, M. van Kreveld, and M. Overmars. *Computational Geometry: Algorithms and Applications*. Springer, Berlin, 1997.
9. B. Donald, P. Xavier, J. Canny, and J. Reif. Kinodynamic motion planning. *J. of the ACM*, 40:1048–1066, 1993.
10. E. Feron, E. Frazzoli, and M. Dahleh. Real-time motion planning for agile autonomous vehicles. In *AIAA Conf. on Guidance, Navigation and Control*, 2000.
11. E. Frazzoli. Maneuver-based motion planning for non-linear systems with symmetries. Submitted to Transactions on Robotics and Automation.
12. D. Hsu, R. Kindel, J. Latombe, and S. Rock. Randomized kinodynamic motion planning with moving obstacles. *IJRR*, 2001.
13. L. E. Kavraki, P. Svestka, J.-C. Latombe, and M. H. Overmars. Probabilistic roadmaps for path planning in high-dimensional configuration spaces. *TRA*, 12(4):566–580, June 1996.
14. A. Ladd and L. Kavraki. Measure theoretic analysis of PRM. *TRA*, pages 222–229, 2004.
15. F. Lamiraux, E. Ferre, and E. Vallee. Kinodynamic motion planning: Connecting exploration trees using trajectory optimization methods. In *ICRA*, 2004.
16. J. P. Laumond. Feasible trajectories for mobile robots with kinematic and environment constraints. In *Preprints of the International Conference on Intelligent Autonomous Systems*, pages 346–354, Amsterdam, The Netherlands, 1986. Elsevier Science Publishers, B.V.
17. S. LaValle, M. Branicky, and S. Lindemann. On the relationship between classical grid search and probabilistic roadmaps. *International Journal of Robotics Research*, 2003.
18. S. LaValle and J. Kuffner. Rapidly exploring random trees: Progress and prospects. In B. Donald, K. Lynch, and D.Rus, editors, *WAFR*, pages 293–308. A.K. Peters, 2001.
19. A. Lewis, J. P. Ostrowski, R. M. Murray, and J.W.Burdick. Nonholonomic mechanics and locomotion: the snakeboard example. In *ICRA*, pages 2391–2400, May 1994.
20. J. Reif and H. Wang. Non-uniform discretization approximations for kinodynamic motion planning. In J.-P. Laumond and M. Overmars, editors, *Algorithms for Robotic Motion and Manipulation*, pages 97–112. A K Peters, Wellesley, MA, 1997.
21. S. Sekhavat, P. Svestka, J.-P. Laumond, and M. H. Overmars. Multilevel path planning for nonholonomic robots using semiholonomic subsystems. *IJRR*, 17:840–857, 1998.
22. P. Švestka and M. Overmars. Motion planning for car-like robots using a probabilistic learning approach. *Int. J. of Robotics Research*, 16(2):119–143, 1997.

Incremental Grid Sampling Strategies in Robotics

Stephen R. Lindemann, Anna Yershova, and Steven M. LaValle

Dept. of Computer Science
University of Illinois
Urbana, IL 61801 USA
{slindema, yershova, lavalle}@uiuc.edu

Summary. We present algorithms for generating deterministic sample sequences using incremental grid-based sampling. Our algorithms are designed to generate dense sample sequences over spaces common in robotics, such as the unit cube, $SO(3)$, and $SE(3)$. Our sampling techniques provide the advantageous properties of uniformity, lattice structure, and incremental quality. In addition, the inherent structure of grid-based sequences not only enables them to be used in the place of other sampling techniques in existing algorithms, but also permits the development of new algorithms aimed at exploiting this structure.

1 Introduction

Numerous algorithms in robotics rely heavily on the generation of samples over a continuous state space. Motion planning algorithms, for example, use sampling along with efficient collision detection to find collision-free paths in the configuration space. Hence, it is of great importance that the underlying sample sequences be as good as possible. Previous work has argued that randomization is not the key to the success of modern motion planning algorithms; rather, there are deterministic sampling schemes which can be expected to perform as good or better than sampling uniformly at random [9, 10]. Furthermore, it was speculated that integrating sampling methods more closely into motion planning algorithms could be advantageous compared to viewing sampling as a "black box". One way to do this is by using a sample sequence with lattice structure, and developing algorithms which exploit that structure. In order to make this possible, it is of first importance to develop sample sequences which have lattice structure as well as high incremental quality, which is of great importance in many situations. Incremental quality is the property that if the sequence is stopped after any number of samples, the samples taken up to that point uniformly cover the space. Previous work addressed this by the introduction of incremental grid-based sequences which optimized discrepancy [11]. These sequences have high incremental quality; however, the

M. Erdmann et al. (Eds.): Algorithmic Foundations of Robotics VI, STAR 17, pp. 313–328, 2005.

methods described require exponential space to store the optimal ordering. This renders these methods unsuitable for high-dimensional problems. In this paper, we introduce new grid-based sequences which optimize different uniformity measures. The orderings we find are simple and compact to represent, making them suitable for high-dimensional problems. In addition to this, we show how to use these sample sequences to sample other topological spaces common in robotics, such as $SO(3)$ and $SE(3)$. The methods we present use Platonic solids as a means to apply sampling techniques originally developed for the unit cube to these spaces, and were originally formulated in [23].

We begin by briefly overviewing important concepts in the area of uniformity measures and uniform sampling techniques.

2 Uniformity Measures and Uniform Sampling Techniques

Uniform sampling criteria and techniques have been developed by numerous mathematicians over the past century. Excellent overviews of the subject include [12, 13]. Here we briefly introduce only the concepts needed for this paper. Let $X = [0,1]^d \subset \mathbb{R}^d$ define a space over which to generate samples. Define a *range space*, $\mathcal{R}$, as a collection of subsets of X. Let $R \in \mathcal{R}$ denote one such subset. Reasonable choices for $\mathcal{R}$ include the set of all axis-aligned rectangles, the set of all balls, or the set of all convex subsets.

Let $\mu(R)$ denote the Lebesgue measure (or volume) of subset R. If the samples in P are uniform in some ideal sense, then it seems reasonable that the fraction of these samples that lie in any subset R should be roughly $\mu(R)$ (divided by $\mu(X)$, which is simply one). We define the *discrepancy* [22] to measure how far from ideal the point set P is:

$$D(P, \mathcal{R}) = \sup_{R \in \mathcal{R}} \left| \frac{|P \cap R|}{N} - \mu(R) \right| \tag{1}$$

in which $|\cdot|$ applied to a finite set denotes its cardinality.

Whereas discrepancy is based on measure, a metric-based criterion, called *dispersion*, can be introduced:

$$\delta(P, \rho) = \sup_{q \in X} \min_{p \in P} \rho(q, p). \tag{2}$$

Above ρ denotes any metric, such as Euclidean distance or ℓ^∞. Intuitively, this corresponds to the radius of the largest empty ball (assuming all ball centers lie in $[0,1]^d$).

Discrepancy can be thought of as enforcing two criteria: first, that no region of the space is left uncovered; and second, that no region is too full. Dispersion eliminates the second criterion, leaving only the first. It can be shown that low discrepancy implies low dispersion [13]. It is easy to conceive of a uniformity criteria that emphasizes the second criterion, rather than the first. A criterion such as this would require that points be spread as far away from each other as possible; we call this *mutual distance*, and introduce it in Section 3. As in the case of dispersion, a bound on discrepancy implies one on mutual distance.

Motivated by integration and optimization problems, sampling issues have been studied extensively in the applied mathematics community. Sample sets and sequences were developed to replace the random sequences traditionally used for these applications; they received the name *quasi-Monte Carlo* to denote this connection. Due to the fundamental importance of numerical integration, and the close connection between discrepancy and integration error, most of the quasi-Monte Carlo literature focuses on the discrepancy measure. Low-discrepancy sampling methods can be divided into three categories: Halton/Hammersley sampling [3, 4, 20], lattices [5, 7, 12, 18, 21], and (t, s)-sequences and (t, m, s)-nets [13, 14].

3 Optimal Grid Sampling Techniques

In contrast to the sampling techniques discussed above, we introduce methods based on incremental sampling of grid points. In [11], grid sample sequences which optimized discrepancy were introduced. In that work, it was shown how to construct discrepancy-optimal grid sampling sequences, and several interesting properties of such sequences were proven. In this section, we examine optimal grid sampling sequences with respect to different quality measures. The new sequences will retain the good properties of the discrepancy-optimal sequences and will add key new advantages.

Before proceeding to describe the sequences, several definitions will be useful. Consider a classical grid in the d-dimensional unit cube, $I^d = [0,1]^d \subset \mathbb{R}^d$; we define a multiresolution classical grid of resolution level l, P_l^n to be a grid with 2^{dl} points (i.e., 2^l points per axis). More formally, $P_l^n = \{(i_1/2^l, \cdots, i_n/2^l) : i \in \mathbb{Z}, 0 \leq i \leq 2^l - 1\}$. One may also define the *grid region* associated with point j at resolution level l: $G_{j,l} = [j_1, j_1 + 1/2^l) \times \cdots \times [j_n, j_n + 1/2^l)$.

Our previous work focused on optimizing discrepancy; however, discrepancy is only one of several interesting and useful uniformity criteria. In addition to this, discrepancy-optimal sequences currently suffer from the requirements of both exponential time to compute and exponential space to store. By optimizing different uniformity criteria, we will derive sequences which eliminate the exponential space requirement, leading to sequences which have very compact, efficient representations. Our techniques are capable of optimizing either dispersion or another criterion we call *mutual distance*; the mutual distance of a set S is

$$\rho_m(S) = \min_{x,y \in S} \rho(x, y)$$

Intuitively, mutual distance corresponds to forcing sequence points to be as far from each other as possible. Both dispersion and mutual distance are metric-based criteria, in contrast to discrepancy, which is measure-based. The techniques we develop in this section will exploit this fact to generate sequences with compact representations.

The rest of this section proceeds as follows. First, we will describe the representation that our ordering takes and how it is used to generate samples. Second, we prove the existence of optimal orderings admitting the representation given. Third, we show how we compute optimal orderings and approximations of the optimal orderings. Primary focus will be given to mutual distance; however, we will address dispersion-optimal orderings as well.

3.1 The digital construction method

The most widely-used quasi-Monte Carlo sampling techniques are (t, m, s)-nets and (t, s)-sequences. Typically, these sample sets and sequences are generated by a method known as the *digital construction method.* The digital construction of a (t, m, s)-net (in base 2) takes a set of generator matrices and computes sample locations based on multiplying these matrices by a vector which whose components are the bit representation of the sample index. Our approach for computing grid orderings which optimize dispersion or mutual distance uses this same idea.

Consider the integers $[0, 1, \ldots, 2^d - 1]$. Each integer can be written in the following form:

$$x = \sum_{i=0}^{d-1} a_i 2^i, \; a_i \in \mathbb{Z}_2.$$

Hence, there is a straightforward bijection between this set of integers and the group $\mathbb{Z}_2^d$. Now consider a $d \times d$ matrix, with each element $e_{ij} \in \mathbb{Z}_2$. If this matrix is full rank, it defines a linear transformation for the vector space $\mathbb{Z}_2^d$ over $\mathbb{Z}_2$ (using standard matrix/vector multiplication). Via this transformation, and the previously-mentioned bijection, the bit representations of the integers $[0, 1, \ldots, 2^d - 1]$ are mapped to d-dimensional vectors corresponding to grid points. Each transformation represents a unique ordering of the set of all grid points in the grid of resolution 1. See Figure 1 for an example. It is clear that the space of all grid orderings is much larger than the orderings that can be generated by matrices of this type. It is desirable that our orderings take this form, due to its compact representation and the ease of computation it affords. Fortunately, there exist orderings which both optimize either dispersion or mutual distance and which admit this representation; we proceed to show that this is the case.

3.2 Optimal orderings using the digital method

Our approach for constructing an optimal generator matrix is to build it one column at a time, rather than computing it as a unit. This is possible because

The following matrix defines an ordering in four dimensions.

$$\begin{pmatrix} 1\,0\,0\,0 \\ 1\,1\,0\,0 \\ 1\,0\,1\,0 \\ 1\,1\,1\,1 \end{pmatrix}$$

An integer $x = a_0 + a_1 2^1 + a_2 2^2 + a_3 2^3 \in [0, 15]$ is written in vector form as $(a_0\ a_1\ a_2\ a_3)'$. Multiplying this vector on the left by the given matrix yields another vector, whose elements are interpreted as the coordinates of a 4-dimensional grid. Applying the transformation to the sequence $\{0, 1, \ldots, 15\}$ yields an ordering which optimizes mutual distance for the grid of resolution level 1. Points in higher resolution levels are obtained by recursively applying this ordering.

Fig. 1. An example of the digital construction method in 4 dimensions.

only the first i columns affect the location of the first 2^i samples (in the bit representation of the first 2^i integers, the final $d-i$ elements are zero and hence the final $d-i$ columns of the matrix do not affect the product). Hence, we will select d generator vectors, one at a time, and use these as the columns of our generator matrix.

Before proceeding to show how to construct matrices to generate optimal orderings, we must first define our optimality criteria more precisely. A grid ordering which optimizes mutual distance is defined to be a sequence of points such that, given the first k elements of the sequence, the $(k+1)$-th element is chosen to maximize the mutual distance of the set $\{x_1, x_2, \ldots, x_{k+1}\}$. First, we will show how to construct generator matrices which optimize mutual distance; then, we will comment briefly on the analogous dispersion-optimal constructions. We begin with a useful lemma:

Lemma 1. *Let $G_i = \{g_1, g_2, \ldots, g_i\}$ be i linearly independent vectors in $\mathbb{Z}_2^d$, and let $S_i = Span(G_i) = \{x : x = \sum_j a_j g_j, a_j \in \mathbb{Z}_2\}$. Take some vector $u \in \mathbb{Z}_2^d \setminus S_i$, and construct the sets $G_{i+1} = G_i \cup u$ and S_{i+1}. If the distance from u to its nearest neighbor in S_i is d, then the distance from every element of $S_{i+1} \setminus S_i$ to its nearest neighbor in S_i is also d.*

Proof. The set $S_{i+1} \setminus S_i$ is simply the left coset of S_i generated by u. It is obvious that the difference between any element of the coset and any element of S_i is an element of the coset; its magnitude is the same as the distance between u and some element of S_i. Hence the distance from every element of $S_{i+1} \setminus S_i$ to its nearest neighbor in S_i is d.

While simple at first glance, this lemma is important for understanding grid orderings constructed using the digital method. After choosing i generator vectors and generating the points corresponding to them, the remaining $2^d - 2^i$ grid points are divided into equivalence classes (each of size 2^i), corresponding to different choices of the next generator vector g_{i+1}. The above lemma shows that all vectors in a given equivalence class have the same distance to the set of points generated by the first i generator vectors.

Proposition 1. *The first generator vector is* $(1\ 1\ \cdots\ 1)'$.

Proof. The first point in the sequence (for any choice of initial generating vector) is the zero vector. Hence, the first generator vector g_1 should be chosen to be maximally distant from the zero vector. That vector is clearly $(1\ 1\ \cdots\ 1)'$.

Proposition 2. *Let* $G_i = \{g_1, g_2, \ldots, g_i\}$ *be* i *linearly independent vectors in* $\mathbb{Z}_2^d$ *with* $g_1 = (1\ 1\ \cdots\ 1)'$, *and let* $S_i = Span(G_i) = \{x : x = \sum_j a_j g_j, a_j \in \mathbb{Z}_2\}$. *If the next generator vector* g_{i+1} *is chosen to be some vector* $u \in \mathbb{Z}_2^d$ *which is maximally distant to* S_i, *then mutual distance is maximized for all samples in the sequence with indices in the range* $(2^i, 2^{i+1}]$.

Proof. In order that the mutual distance be maximized for sample index 2^d+1, the vector chosen must be maximally distant from the set S_i, as assumed. Since selecting this vector as the next generating vector will result in the generation of the next 2^i elements of the sequence, we must show that making such a selection will maximize mutual distance for the next 2^i elements. Assume that the distance from u to S_i is r. Then, the mutual distance of the resulting set is $\rho_m(S_i \cup u) = \min(r, \rho_m(S_i))$. Taking $g_{i+1} = u$, we know by Lemma 1 that each element in the set $S_{i+1} \setminus S_i$ (that is, each of the next 2^i elements in the sequence) is also of distance r to S_i. Also, it is the case that $\rho_m(S_{i+1} \setminus S_i) = \rho_m(S_i)$. So, the addition of the next 2^i elements does not change the mutual distance of the set, which after the addition of the vector u was already $\min(r, \rho_m(S_i))$. Thus at each point of the sequence with index in the range $(2^i, 2^{i+1}]$, the mutual distance remains maximized.

Hence, by choosing the $(i+1)$-th generator vector to maximize the distance from the set generated by the previous i generator vectors, we maximize the mutual distance for the next 2^i elements of the sequence. By constructing our generator matrix from d such vectors, then, we obtain an ordering for the first 2^d points of the sequence which optimizes our criterion of mutual distance. To obtain the entire infinite sequence, then, it remains only to specify how to apply to generator matrix to samples whose index is greater than or equal to 2^d (note that the first sample has index 0, so the first 2^d elements have index up to $2^d - 1$). As in the case of the discrepancy-optimal sequence [11], we apply the ordering recursively to each of the 2^d grid regions corresponding to the first 2^d points. After the top-level grid region is selected by applying the generator matrix to the first d bits of the sample index, the position within that region is found by applying the generator matrix to the next d bits of the sample index. This process is repeated until the final sample location is found.

Theorem 1. *The sequence generated by the above procedure optimizes mutual distance at every point in the sequence.*

Proof. We show this by induction on the resolution level l. By Propositions 1 and 2, the sequence optimizes mutual distance for the first 2^d samples, which corresponds to $l = 1$. Now, assume that the sequence optimizes mutual distance for resolution level l, and examine the sequence during resolution level $l + 1$. It is clear that within a particular grid region, the points are added in an order which maximizes the mutual distance (this follows directly from the fact that the generator matrix maximizes mutual distance, and the matrix is applied recursively to each grid region). Now, it is also the case that if some point is added to a particular grid region (chosen recursively as described above), a corresponding point is added to every other grid region in the space before any another point is added to the initial grid region (this is because the grid region is determined by the lowest dl bits of the sample index, which cycle through all possibilities before a higher bit changes). This implies that if some point x of resolution level $l + 1$ occurs in the sequence and decreases the mutual distance, the next 2^{dl} points to be added will be points corresponding to x, but in different grid regions. This implies that they will not reduce the mutual distance, because they will have the same distance from their neighbors that x had from its neighbors. Note that points from neighboring grid regions do not interact in such a way as to decrease the mutual distance, because there will always be an analogous point in the same grid region with identical distance, and we know that all intra-grid region distances follow a mutual distance-maximizing ordering. Since each point x which reduces the mutual distance is added according to the optimal ordering given by the generator matrix, and all other points do not reduce the mutual distance, we see that each point added in resolution level $l + 1$ is added in a way that maximizes the mutual distance. Hence the inductive hypothesis is seen to be true and the theorem proved.

The construction of dispersion-optimal generator matrices proceeds similarly. The most significant difference is that the optimality criteria must be slightly modified. Lemma 1 implies that given some G_i and corresponding S_i, for *any* choice of g_{i+1}, the dispersion of the sequence will remain constant until all samples of set S_{i+1} have been taken. This means that the optimality criteria used for mutual distance is inadequate for dispersion; we must make a slight modification. Assume that the set G_i is given. Then the optimal choice for the next generator vector g_{i+1} is defined to be that which minimizes the dispersion of the point set S_{i+1}. Note that a sequence optimizing this criteria will also be optimal in the point-wise sense (as in the case of mutual distance), but the reverse is not necessarily true. For dispersion-optimal sequences, the first generator vector can also be seen to be $(1\ 1\ \cdots\ 1)'$. Each successive generator vector is chosen to optimize the above criterion. After the entire generator matrix has been constructed, we apply it recursively (as in the case of mutual distance). Under the appropriate definitions, it can be shown that an optimal sequence is obtained.

3.3 Useful Properties for Motion Planning

In [11], several properties of discrepancy-optimal grid sequences were shown. The same properties hold for sequences optimizing mutual distance or dispersion; we present the two main properties here. For each of these properties, dimension is considered to be constant. However, at various points we note dependence on dimension; in all cases, dimension has only a limited effect on practical performance.

A first consideration is the amount of time required to generate each sample. If it is computationally expensive to generate the sample sequence, this may offset time gained through the quality of the sequence. Hence, we give bounds on the time required to generate a particular sample.

Property 1. The position of the i-th sample in the d-dimensional sampling sequence S_d can be generated in $O(\log i)$ time.

Proof. The bit representation of the i-th sample contains $O(\log i)$ bits. Thus the vector representing the sample will have this length. There will be $O(\log i)$ recursions (the number of recursions is the number of bits divided by the dimension). Each recursion requires $O(d^2)$ operations, which is constant time since d is constant. Hence the total time is $O(\log i)$.

Property 2. Let the number of samples taken so far be N. Then, the existence of a neighbor (in the current resolution level) of any of these samples can be found in $O(\log N)$ time.

Proof. Given the index of sample i, one can find its position in $O(\log i)$ time. Then, depending on the total number of samples, N, taken, add the appropriate quantity to the calculated position, reflecting the desired neighbor. To calculate the index of this sample, one can easily use the inverse of the procedure used to generate the position (the inverse of the generator matrix is easily computed, since it is full rank). Then a hash table can be queried to see if a sample of that index exists, in constant time. The cost to compute the inverse index is at most $O(\log N)$, so the total cost is $O(\log N)$.

These properties indicate the potential of the sequences we have introduced. These samples can be generated very quickly (especially given the fact that all mathematical operations can be done using simple bit operations). In fact, samples such as these are likely to cost much less than "good" random points, such as those generated by sophisticated nonlinear congruential techniques. In addition to this, sampling from a set of grid points gives implicit structure to the sample set. This structure can be exploited efficiently, which could be of value when using the sample sequence for applications such as motion planning.

4 Deterministic Sample Sequences for Spheres and SO(3)

Thus far, we have considered sampling in the unit cube $I^d = [0,1]^d \subset \mathbb{R}^d$. Problems in robotics, however, typically result in much more complex topological spaces. Algorithms for tasks such as motion planning typically take samples from the unit cube and transform them to the appropriate space. However, it may be possible to design sample sequences explicitly with particular topological spaces in mind. In this section, we consider the case of spheres and $SO(3)$. We will then discuss how to sample $SE(3)$ using these methods. We begin with some brief definitions.

4.1 Definitions and Quality Measures for Points on Spheres and SO(3)

We consider generating samples over spheres and $SO(3)$. Let S^d represent a d-dimensional sphere, embedded in $\mathbb{R}^{d+1}$ as

$$S^d = \{x \in \mathbb{R}^{d+1} \mid \|x\| = 1\}.$$

The set of all rotations in $\mathbb{R}^3$ is denoted as $SO(3)$, which is defined as the set of all 3×3 orthonormal matrices. It will be helpful to sometimes represent $SO(3)$ as the set, H, of unit quaternions, each of which is expressed as $h = a + bi + cj + dk$, with the identification $h \sim -h$ [8]. Note that it appears that $H = S^3$, except that antipodal points on S^3 are identified in the definition of H. This leads to a close relationship between sampling on sphere and sampling on $SO(3)$.

Discrepancy and dispersion may be defined for these spaces, analogous to the case of $\mathbb{R}^d$ given in Section 2. The definition of discrepancy is identical; a rotation-invariant measure μ is used (the Haar measure over the set of all rotation matrices in $SO(3)$), and typical range spaces are the set of all spherical caps (intersections of the sphere with half spaces) or the set of all spherical slices (intersections of two half-spheres) [2, 15]. Likewise, the definition of dispersion is identical; the metric ρ is required to be rotation-invariant.

We now proceed to outline a general method for generating samples on spheres and $SO(3)$. For more detail, see [23].

4.2 Generating Samples on Spheres and SO(3)

Our general approach to sampling is based on Platonic solids. In $\mathbb{R}^3$, a Platonic solid or *regular polyhedron*, is a polyhedron for which every face is a copy of a regular polygon, fixed over all faces, and the degree of every vertex is fixed. Let (v, e, f) denote the numbers of vertices, edges, and faces of a regular polyhedron. Although there are an infinite number of regular polygons,

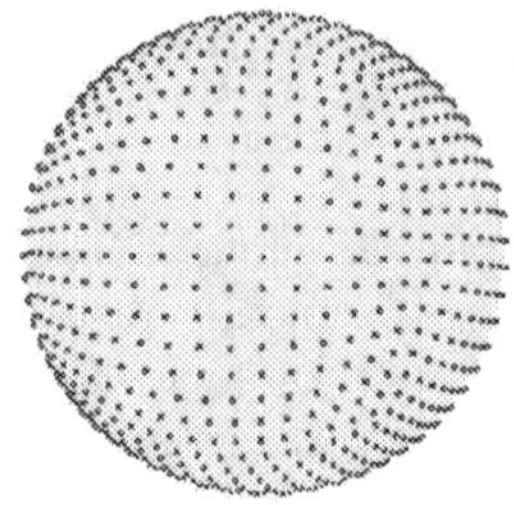

Fig. 2. Distribution of points on the sphere S^2 generated by a grid (Sukharev [19]) on each spherical face.

there are only five regular polyhedra: tetrahedron (4,6,4), cube (8,12,6), octahedron (6,12,8), icosahedron (12,30,20), and dodecahedron (20,30,12). The notion of regular polyhedron can be generalized to higher dimensions to obtain a *regular polytope*. In $\mathbb{R}^4$, it turns out that there are six regular polytopes: simplex (5,10,10,5), cube (16,32,24,8), cross polytope (8,24,32,16), 24 cell (24,96,96,24), 120 cell (600,1200,720,120), 600 cell (120,720,1200,600). The forth element in each sequence denotes the number of 3D cells (which are regular polyhedra). Finally, in $\mathbb{R}^d$ for any $d > 4$, there are only three regular polytopes: simplex, cube, and cross polytope.

We first address the problem of generating a uniformly distributed set of points over S^d. Consider inscribing any $(d+1)$-dimensional regular polytope inside of S^d, so that all of its n vertices lie in S^d. The set of vertices are beautifully arranged around S^d so that the points are evenly spaced. Furthermore, the edges of the polytope yield a regular lattice structure that is natural for building roadmaps in planning problems. For the case of sampling $SO(3)$, we simply use a set of vertices that lie in one hemisphere (making sure that no antipodal pairs of points appear in the set). The edges can be obtained directly from the polytope by making the appropriate identification of antipodal pairs.

Unfortunately, there are only a few combinations of n and d, for which these ideal samples may be constructed for S^d and $SO(3)$. This might be suitable for some applications, such as picking a set of candidate directions from S^d for gradient descent of a potential function; however, in general, we would like to a have a nice distribution of points for any value of n.

To the best of our knowledge, it is impossible to perfectly space n points around S^d, for any n and for $d > 1$. One simple idea that increases the number of samples is place one point in the center of each of the c d-cells of some regular polytope, and lift it to S^d. If we take the union of these points with the set of v polytope vertices, a nice point set of size $c + v$ may be obtained. If more points are placed; however, the problem becomes more complicated. Therefore, we are willing to tolerate some distortion in the distribution of points. It still seems useful, however, to borrow some of the properties of the regular polytopes to generate good samples. The general idea pursued in this paper is to sample uniformly on the surface of the regular polytope, and then transform generated distribution on the surface of the sphere. We next describe this general method and discuss the induced distortion.

Consider a $(d+1)$-dimensional regular polytope inscribed in the sphere S^d. Suppose there exists a good method of sampling the surface of this polytope. The faces (d-dimensional cells) of the polytope, if projected outward to the surface of the sphere, form a tiling of the surface with the d-dimensional *spherical polytopes.* Consider some particular face, F, and its corresponding spherical face, F'. Each point inside F can be described by the barycentric coordinate systems induced by vertices of F after its triangulation. Now imagine that a distribution of points is generated inside F. Each of the points in this distribution can be obtained through several steps of linear interpolation between the vertices of the barycentric coordinate systems. The distribution on F' can be obtained then through similar steps of interpolating between the vertices of F', except that the interpolation should be done on the surface of the sphere [16]. This idea is similar to the one proposed in [1] for stratified sampling of spherical triangles. As an example, consider a cube inscribed in the sphere S^2, and sample the surface of the cube by placing a Sukharev grid [10, 19] on each face of the cube. Using the proposed method we get a distribution of samples on S^2 as shown on Figure 2.

The distribution of points on the sphere S^d obtained by this method will introduce distortion since spherical arcs corresponding to the intervals inside F with the same length may have different lengths in F'. The amount of the distortion, and therefore bounds on the dispersion and discrepancy, can be obtained through the analysis of the maximal arc differences.

This idea can also be adapted to $SO(3)$ (and in general to the projective space of any dimension). Take a four-dimensional regular polytope inscribed in S^3 and use only half of the faces to generate the distribution on the surface. We pick the faces so that in the set of used faces, there must not exist a pair of antipodal points, one from each of two different faces. This way the obtained samples will cover exactly half of the sphere, which forms $SO(3)$ surface.

While this approach can incorporate any uniform sampling method, we use the grid sequences described above and in [11]. In what follows we introduce the concept of a *layered Sukharev grid sequence.*

4.3 Layered Sukharev Grid Sequences for Spherical Cubes

In the construction of our basic grid sequences, we considered only classical multi-resolution grids. We can use these grid sequences to construct Sukharev grids; a Sukharev grid of k points per axis is identical to a classical grid, with the exception that the samples are placed in the center of the k^d grid regions rather than in the bottommost corner. We will be using a layered version of this sequence; a layered Sukharev grid of resolution l is the union of all Sukharev grids with 2^i points per axis, for $0 \le i \le l$. A *layered Sukharev grid sequence* constructs the Sukharev grids one resolution at a time. Since each individual Sukharev grid is a shifted version of a classical grid, we can use the sequences previously described to generate the samples. In what follows we generalize layered Sukharev grid sequence to the sphere S^d. We first show how the points should be generated in each of the spherical cubes, and then how all these points can be combined into one sequence on the sphere.

Consider a face, F, of a $(d+1)$-cube inscribed in a sphere S^d. F is a d-dimensional cube, which in each of its corners has d edges. If we project all of these edges onto the surface of the sphere they form arcs, which delineate a spherical d-cube, F'. The lengths, α, of these arcs are equal for all edges of F. If we consider those equatorial angles that correspond to the edges coming from a common vertex of F, we can define an *angular coordinate system* for the spherical face F'. Indeed, the coordinates $(x_1, x_2, ... x_{n-1})$ with all possible values $x_i \in [0, \alpha]$ specify all possible points of F'.

The construction of the sequence, T, essentially follows the construction of the layered Sukharev grid sequence for the unit cube, except that instead of the Euclidean coordinate system we use the angular coordinate system defined above. We call this a *Sukharev spherical grid.* The dispersion of the resulting sequence can be calculated, as can the discrepancy, which is especially appropriate when using a discrepancy-optimal grid ordering to generate the Sukharev spherical grid. These results can be found in [23]. It should be fairly obvious that the time complexity of generating samples on a Sukharev spherical grid is the same as for generating ordinary grid samples. Consequently, Properties 1 and 2 from Section 3.3 hold.

4.4 Layered Sukharev Grid Sequences for S^d

Now that we have defined a sequence for each of the spherical cubes separately, we need to define an ordering by which these will be combined to form a sample sequence over the entire surface of the sphere. A straightforward way to do this is to place one point from each of the faces' sequences at a time. The order in which the faces should be considered can be explicitly computed using the same uniformity criteria used in computing the underlying grid sequence. This ordering, combined with the grid orderings of each individual spherical cube, yields the entire sphere sampling sequence.

5 Sample Sequences for Cross Products of $\mathbb{R}^n$ and S^d

We have defined the multiresolution grid sequences for the unit cube (Section 3), and the sphere S^d (Section 4). The spaces that arise in robotics are often the cross products of these. For example, the set of all rotations and translations of a 3d rigid body, denoted as $SE(3)$, can be represented by $\mathbb{R}^3 \times \mathbb{R}P^3$. The rotations and translations of m multiple rigid bodies is represented by $(\mathbb{R}^3 \times \mathbb{R}P^3)^m$.

When designing uniform sequences for such spaces the parameterization of the space together with the choice of measure and metric on the space should be defined carefully. For $SE(3)$ the Haar measure should be taken. However, since there is no natural metric on $SE(3)$, the weighted sum of the metrics on $\mathbb{R}^3$ and $SO(3)$ is usually used. While the weighted metric can be defined on general cross products of the spaces (which is assumed in the construction below), in some cases particular techniques for designing sequences might be advantageous.

In what follows we construct the multiresolution grid sequence for the space that is a cross product of multiple copies of $\mathbb{R}^n$ and S^d. We define it inductively, starting with any tuple of multiresolution grid sequences.

Let T_1 and T_2 be two multiresolution grid sequences. Let T_1 be defined over the space X_1, and T_2 be defined over the space X_2. Either of X_1 or X_2 may be $\mathbb{R}^n$ or S^d. Let $\dim(T_1) = d_1$ and $\dim(T_2) = d_2$ be the dimensions of these sequences. Let m_1 (m_2) be the number of points at the resolution level 0 of sequence T_1 (T_2) respectively. When weighted metric is defined on $X_1 \times X_2$ the values for m_1 and m_2 can be chosen so that they respect the appropriate weights of X_1 and X_2. Then the number of points at the resolution level l is $m_1 \cdot 2^{ld_1}$ and $m_2 \cdot 2^{ld_2}$ for sequence T_1 and T_2 respectively.

With each point $p = (p_1, ...p_d)$ at the resolution level l one may define the grid region associated with this point as $G_{p,l} = \left[p_1, p_1 + a/2^l\right) \times ... \times \left[p_d, p_d + a/2^l\right)$, in which $a = 1$ for the unit cube sequence and $a = \alpha$ (Section 4) for the sphere sequence.

Next consider the space $X = X_1 \times X_2$. The multiresolution grid sequence that we define for this space has $m_1 \cdot m_2 \cdot 2^{l(d_1+d_2)}$ points at the resolution level l. Each of these points can be expressed as $p = (p_1, p_2) = (p_{1,1}, ...p_{1,d_1}, p_{2,1}, ...p_{2,d_2})$, where $p_1 \in T_1$ and $p_2 \in T_2$. Each of these points has an associated grid region: $G_{p,l} = \left[p_{1,1}, p_{1,1} + a/2^l\right) \times ... \times \left[p_{1,d_1}, p_{1,d_1} + a/2^l\right) \times \left[p_{2,1}, p_{2,1} + b/2^l\right) \times ... \times \left[p_{2,d_2}, p_{2,d_2} + b/2^l\right)$.

The sequence for X is constructed one resolution level at a time. The order in which the points from each resolution level are placed in the sequence can be described as follows. The ordering, $L_X()$, of the first $m_1 \cdot m_2 \cdot 2^{(d_1+d_2)}$ points determine the order of the grid regions whithin X and should be precomputed in advance. Every successive $m_1 \cdot m_2 \cdot 2^{(d_1+d_2)}$ points in the sequence should be placed in these grid regions in the same order. Where exactly each point should be placed within each of the grid regions should be determined by the recursion procedure defined for $[0, 1]^{(d_1+d_2)}$.

6 Experiments

We have implemented our algorithm in C++ and applied to implementations of PRM-based planner [6] in the Motion Strategy Library. The experiments reported here were performed on a 2 Ghz Pentium IV running Linux and compiled under GNU C++.

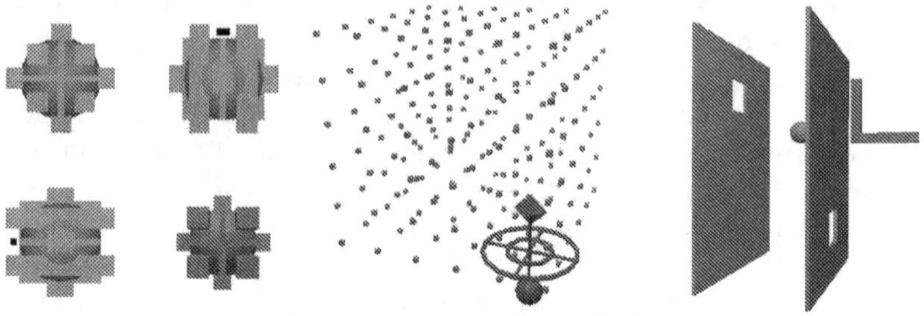

	Random Sequence	Layered Sukharev Grid Sequence
a)	1088	1067
b)	3460	3285
c)	3481	3202

Fig. 3. Problems involving: a) moving a robot (black) from the north pole to the south pole. Multiple views of the geometry of the problem are shown (obstacles are drawn in lighter shades); b) moving a robot from one corner of a 3d grid to the opposite corner; c) moving an L-shaped object through the holes in the obstacles. The comparisons of the number of nodes generated by different sampling strategies are shown in the table.

Performance results are shown in Figure 3. The robot in the model a) is allowed only to rotate; therefore, the configuration space is $\mathbb{R}P^3$. The robots in the models b),c) are allowed to translate and rotate; therefore the configuration space is $\mathbb{R}^3 \times \mathbb{R}P^3$. We compared the number of nodes generated by the basic PRM planner using a pseudo-random sequence (with quaternion components [17]), and the layered Sukharev grid sequences. The results for pseudo-random sequences were averaged over 50 trials. When we tested the deterministic sequences, we made sure that each particular problem does not have any advantage due to coincidental alignment with the grid directions of the sequence. Therefore, in each trial a fixed, random quaternion rotation was premultiplied to each sample, to displace the entire sequence. The results obtained were averaged over 50 trials (a different random rotation was used in each).

Based on our experiments we have observed that the performance of the deterministic sequences is equivalent to the performance of the random sequences for the PRM-based planner, which makes it an alternative approach to random sampling. It is important to note, however, that for some applications, such as verification problem, only deterministic guarantees are acceptable, making random sequences inappropriate.

7 Conclusions and Future Work

In conclusion, we have presented a family of new grid-based sample sequences. These sequences have the advantageous criteria of uniformity, lattice structure, and incremental quality. Our sample sequences, based on optimizing dispersion or mutual distance measures, have very compact representations and hence are applicable to high-dimensional problems, unlike the grid-based sequences of [11]. In addition to this, we have introduced methods for applying sampling methods designed for the unit cube to such topological spaces as $SO(3)$ and $SE(3)$. The sequences presented are widely applicable to algorithms in robotics and motion planning. Like traditional Monte Carlo or quasi-Monte Carlo sampling techniques, these sequences are uniform and have high incremental quality; hence, they may be easily substituted into existing algorithms in place of other sampling methods. In addition, these sequences have implicit lattice structure, which permits the development of new algorithms designed to exploit that structure.

There are several directions for future work. First, we plan to actually compute these orderings for very high dimensions. We believe that using approximation and heuristic search techniques, we will be able to find good (though not optimal) orderings for up to several thousand dimensions. Second, the sampling method presented for $SE(3)$ hints at a more general approach to forming products of multi-resolution sample sequences. We hope to study this more carefully and formally characterize how to find such sequence products. Finally, we plan to develop algorithms which specifically exploit the regularity implicit in grid sample sequences. By integrating the sampling method more tightly with planning rather than viewing it simply as a black box, it may be possible to see significant advantages not otherwise available.

Acknowledgments

This work was funded in part by NSF Awards 9875304, 0118146, and 0208891.

References

1. J. Arvo. Stratified sampling of spherical triangles. In *Computer Graphics (SIGGRAPH '95 Proceedings)*, pages 437–438, 1995.
2. M. Blümlinger. Slice discrepancy and irregularities of distribution on spheres. *Mathematika*, 38:105–116, 1991.
3. J. H. Halton. On the efficiency of certain quasi-random sequences of points in evaluating multi-dimensional integrals. *Numer. Math.*, 2:84–90, 1960.
4. J. M. Hammersley. Monte-Carlo methods for solving multivariable problems. *Ann. New York Acad. Sci.*, 86:844–874, 1960.
5. F. J. Hickernell, H. S. Hong, P. L'Ecuyer, and C. Lemieux. Extensible lattice sequences for quasi-Monte Carlo quadrature. *SIAM Journal on Scientific Computing*, 22:1117–1138, 2000.

6. L. E. Kavraki, P. Svestka, J.-C. Latombe, and M. H. Overmars. Probabilistic roadmaps for path planning in high-dimensional configuration spaces. *IEEE Trans. Robot. & Autom.*, 12(4):566–580, June 1996.
7. N.M. Korobov. The approximate computation of multiple integrals. *Dold. Akad. Nauk SSR*, 124:1207–1210, 1959.
8. S. M. LaValle. *Planning Algorithms.* [Online], 1999-2003. Available at http://msl.cs.uiuc.edu/planning/.
9. S. M. LaValle and M. S. Branicky. On the relationship between classical grid search and probabilistic roadmaps. In *Proc. Workshop on the Algorithmic Foundations of Robotics*, December 2002.
10. S. M. LaValle, M. S. Branicky, and S. R. Lindemann. On the relationship between classical grid search and probabilistic roadmaps. *International Journal of Robotics Research (to appear)*, 24, 2004.
11. S. R. Lindemann and S. M. LaValle. Incremental low-discrepancy lattice methods for motion planning. In *Proc. IEEE International Conference on Robotics and Automation*, pages 2920–2927, 2003.
12. J. Matousek. *Geometric Discrepancy.* Springer-Verlag, Berlin, 1999.
13. H. Niederreiter. *Random Number Generation and Quasi-Monte-Carlo Methods.* Society for Industrial and Applied Mathematics, Philadelphia, USA, 1992.
14. H. Niederreiter and C. P. Xing. Nets, (t,s)-sequences, and algebraic geometry. In P. Hellekalek and G. Larcher, editors, *Random and Quasi-Random Point Sets, Lecture Notes in Statistics, Vol. 138*, pages 267–302. Springer-Verlag, Berlin, 1998.
15. G. Rote and R. F. Tichy. Spherical dispersion with applications to polygonal approximation of the curves. *Anz. Österreich. Akad. Wiss. Math.-Natur, Kl. Abt. II*, 132:3–10, 1995.
16. Ken Shoemake. Animating rotation with quaternion curves. In *Proceedings of the 12th annual conference on Computer graphics and interactive techniques*, pages 245–254. ACM Press, 1985.
17. Ken Shoemake. Uniform random rotations. In D. Kirk, editor, *Graphics Gems III*, pages 124–132. Academic Press, 1992.
18. I. H. Sloan and S. Joe. *Lattice Methods for Multiple Integration.* Oxford Science Publications, Englewood Cliffs, NJ, 1990.
19. A. G. Sukharev. Optimal strategies of the search for an extremum. *U.S.S.R. Computational Mathematics and Mathematical Physics*, 11(4), 1971. Translated from Russian, *Zh. Vychisl. Mat. i Mat. Fiz.*, 11, 4, 910-924, 1971.
20. J. G. van der Corput. Verteilungsfunktionen I. *Akad. Wetensch.*, 38:813–821, 1935.
21. X. Wang and F. J. Hickernell. An historical overview of lattice point sets. In K.-T. Fang, F. J. Hickernell, and H. Niederreiter, editors, *Monte Carlo and Quasi-Monte Carlo Methods 2000*, pages 158–167. Springer-Verlag, Berlin, 2002.
22. H. Weyl. Über die Gleichverteilung von Zahlen mod Eins. *Math. Ann.*, 77:313–352, 1916.
23. A. Yershova and S. M. LaValle. Deterministic sampling methods for spheres and SO(3). In *Proc. IEEE International Conference on Robotics and Automation*, 2004.

Toward Complete Path Planning for Planar 3R-Manipulators Among Point Obstacles

Guanfeng Liu[1], J.C. Trinkle[1] and R.J. Milgram[2]

[1] Department of Computer Science, Rensselaer Polytechnic Institute, Troy, NY 12180, USA liugf,trink@cs.rpi.edu
[2] Department of Mathematics, Stanford University, Stanford, CA 94305, USA milgram@math.stanford.edu

1 Abstract

The problem of a planning collision-free motion of a planar 3R-manipulator among point obstacles is studied using techniques from topology and homology. By completely characterizing the set of singular configurations (the points in configuration space corresponding to an intersection of the chain with itself or a point obstacle), the complementary space, the free space, is also completely characterized. This characterization dictates an exact, complete, motion planning algorithm that builds on the authors' algorithm developed for 2R-manipulators [9]. Results obtained with a preliminary version of the algorithm are given.

2 Introduction

For almost two decades, robotics researchers have worked to develop theories and algorithms for important motion planning problems such as manufacturing automation, dexterous manipulation by multifingered hands, multi-vehicle coordination, robotic surgery, protein folding prediction, and space exploration. A broad overview of these problems and the corresponding planning methods can be found in the survey paper by Hwang and Ahuja [5] and in the text of Latombe [8]. The related history is briefly summarized below.

Lozano-Pérez [11] was among the first to propose the configuration space, C-space, setting for path planning. It is the space of all possible configurations of the robot, ignoring constraints such as joint limits and collisions. In C-space, a point defines a particular pose of the robot and a continuous curve represents a motion of the robot. C-space is normally partitioned into two mutually exclusive subsets:

1. **C-obstacle** - the set of configurations for which the robot's would overlap with one or more obstacles in the workspace.

M. Erdmann et al. (Eds.): Algorithmic Foundations of Robotics VI, STAR 17, pp. 329–344, 2005.

2. **C-free** - the complement in C-space of the closure of C-obstacle.

C-space transforms the motion planning problem into the problem of finding a continuous curve in C-free that connects given initial and final configurations of the robot.

Since the work of Lozano-Pérez, research in motion planning has branched in two primary directions: (i) the search for exact, complete[3] algorithms; (ii) the development of fast, approximate algorithms. Exact and complete algorithms are necessarily based on an exact representation of C-free. Efforts in the first direction began with Schwartz and Sharir [13, 12], who proposed to decompose C-free into a collection of non-overlapping cells and to represent cell connectivity using a graph. With this representation, the path existence question can be answered by graph search. The problem with this approach is that it relies on a cylindrical algebraic decomposition of C-free, and unfortunately, the complexity of the decomposition is doubly exponential in the dimension of C-space. Canny's roadmap algorithm [1] is the most efficient complete algorithm for general motion planning problems. This algorithm iteratively seeks a low-dimensional retraction of C-free by employing the techniques from differential topology. The complexity is only singly exponential in the dimension of C-space, and thus represents a major improvement. However, because of its intricacy, no full implementation of this algorithm exists.

To side-step the difficulties discussed above, researchers have developed exact algorithms for less general systems (e.g., [4]) and approximate hybrid algorithms combining the techniques of cell decomposition, retraction, and/or potential fields with various heuristics (*e.g,* [3, 2]). Most effort has been spent on the approximate methods, the most popular of which are variants of the probabilistic roadmap method (PRM) proposed by Kavraki *et al.* [7]. These algorithms use pseudo-random sampling and collision detection algorithms to construct a graph that approximates C-free well enough to answer planning queries quickly. These algorithms are *probabilistically complete*, which means that if one samples long enough (with truly random sampling), a path will be found if one exists. However, if a path does not exist, these algorithms cannot determine this fact definitively. In general, while PRMs are effective in many cases, they can be quite ineffective when the valid portion of C-space contains narrow passages. Further, the fact that, in the worst case, C-free contains a number of components exponential in the dimension of C-space, implies that, in the worst case, PRMs sampling requirements will grow exponentially with C-space dimension.

The above observations and the existence of new techniques from algebraic topology are the primary motivators of the work reported here and in the authors' previous papers [14, 9]. In [9], a solution to the path planning problem for planar 2R-manipulator among a finite number of point obstacles

[3] "Completeness" implies that if a path exists, one will be found, and if one does not exist, this fact will be reported. It also implies that the algorithm will terminate in finite time.

was given. Through the analysis of the set of singular configurations (those for which the chain intersects itself or a point obstacle), C-free was found to form parallel component sheaves locally. A global cell decomposition of C-free was then accomplished by identifying all bifurcations of local component sheaves. Under general-position assumptions, the exact number of components of C-free was determined (this number is an upper bound when the assumptions and constraints are relaxed). The local component sheaf representation of C-free allowed us to efficiently answer the existence question (*i.e.,* Are two given collision-free configurations in the same component of C-free?). Most importantly, the analysis also dictated the design of exact, complete, path generation algorithms.

The goal of this paper is to extend the results obtained for 2R-manipulators to the case of 3R-manipulators in the plane.

3 Basic notation and terminology

Let $\{\phi_1, \phi_2, \phi_3\}$ be the joint angles of a 3-link serial manipulator (or simply, 3-chain) with link lengths $\{l_1, l_2, l_3\}$ as shown in Figure 1. The C-space of a 3-chain is the three-dimensional torus, $T^3 = S^1 \times S^1 \times S^1$, where ϕ_i is the coordinate on the ith circle, for $i = 1, 2, 3$. Consequently, a three-dimensional cube $[-\pi, \pi]^3$ with the three pairs of parallel boundary surfaces identified can be used to represent T^3. Next, suppose that there is a finite set $\mathcal{O} = \{p_1, p_2, \cdots, p_n\}$ of point obstacles. The C-obstacle is an arrangement of two-dimensional surfaces in T^3 (as shown in Figure 8).

The path planning problem is to find a collision-free motion of the manipulator transferring it between given initial and final postures. In other words, given $\phi_{\text{init}} = (\phi_1, \phi_2, \phi_3)_{\text{init}} \in$ C-free and $\phi_{\text{final}} = (\phi_1, \phi_2, \phi_3)_{\text{final}} \in$ C-free determine a continuous map $\tau \in [0, 1] \mapsto (\phi_1(\tau), \phi_2(\tau), \phi_3(\tau)) \in$ C-free such that $\phi(0) = \phi_{\text{init}}$ and $\phi(1) = \phi_{\text{final}}$.

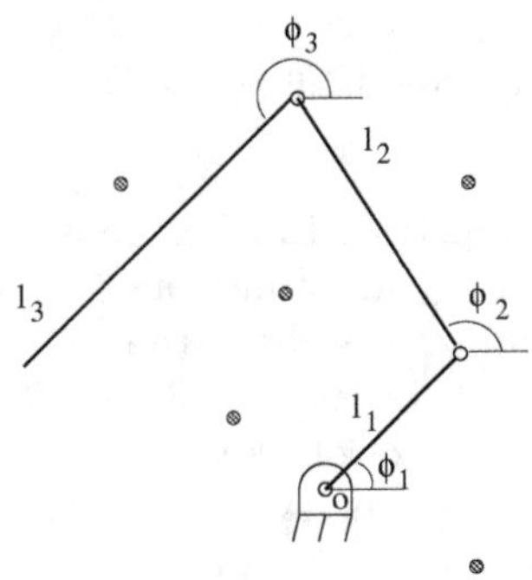

Fig. 1. A 3-chain among point obstacles (shown as small discs).

Our analysis technique is recursive in the sense that when we study an m-chain, we use the results obtained for the proximal (m-1)-chain formed by removing one link from the distal end of the m-chain, *i.e.*, the end not pinned to the ground. This pattern of analysis will be seen in the review of the 2-chain case in the next section and in the subsequent analysis of the 3-chain case.

4 Review of the 2-chain case

Consider the 2-chain manipulator shown in Figure 2. C-free of the proximal 1-chain is a set of open intervals on the circle S^1 formed by removing the points, $\theta_{1,i}$, where link 1 intersects the ith obstacle.[4] Fixing the first link and rotating the second link through a full circle gives a finite number of angles, $\theta_{2,i}(\phi_1)$, for which the second link intersects $\mathcal{O}$. These points are removed from the circle S^1 coordinatized by ϕ_2. To construct a representation of the C-free of the 2-chain, we start with the ϕ_1 circle with points removed. Then over each collision-free point, there is a finite number of disjoint open intervals of ϕ_2 that correspond to connected collision-free positions of link 2. The structure over each point is referred to as a *local component sheaf.*

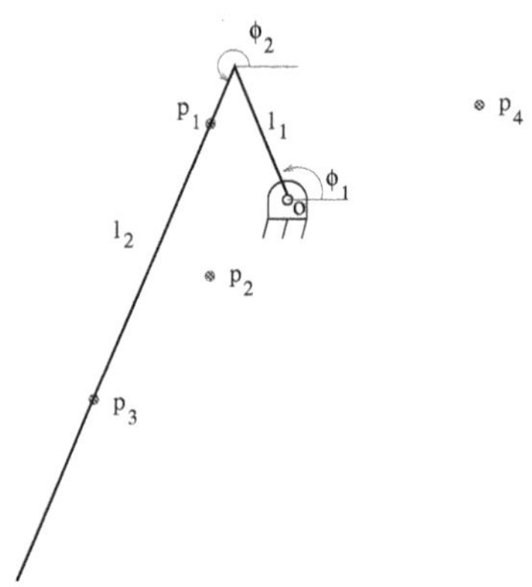

Fig. 2. A 2-chain that can reach all obstacles regardless of the angle of link 1.

This construction is illustrated in Figure 3 for the 2-chain shown in Figure 2. The angles $\theta_{1,1}$ and $\theta_{1,2}$ are values of ϕ_1 for which link 1 collides with $\mathcal{O}$. The "horizontal" *collision curves* above the ϕ_1-axis are curves corresponding to link 2 in contact with obstacles 1, 2, 3, and 4. For each point in the open interval, $(\theta_{1,1}, \theta_{1,2})$, the link 2 collision curves partition the ϕ_2 circle into a finite number of disjoint open intervals corresponding to collision-free configurations of link 2. Note that as ϕ_1 varies, the order of the collision curves above ϕ_1 changes at the points where they cross. These are the *critical points*

[4]This is at most n points, since n is the number of point obstacles.

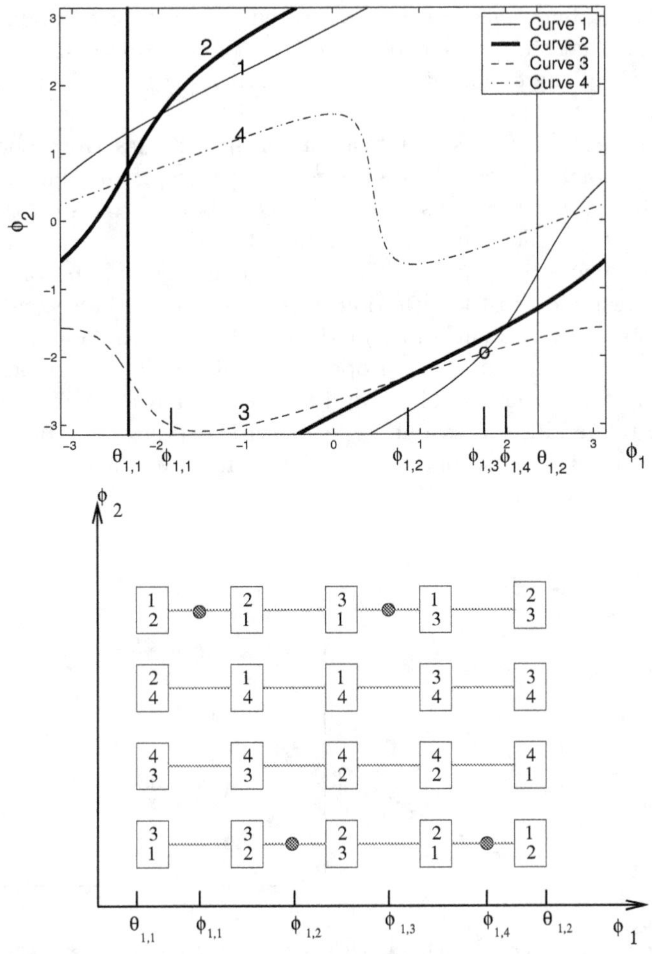

Fig. 3. Upper: C-space of 2-chain in Figure 2 showing all six collision curves (all are circles on the torus). Lower: Graph representing global structure of C-free over $(\theta_{1,1}, \theta_{1,2})$.

$\phi_{1,j}$ (where j ranges over an appropriate index set), when link 2 lies on a line containing two or more point obstacles.

For planning purposes, the free space over the open interval $(\theta_{1,1}, \theta_{1,2})$ can be represented as a graph, as shown in lower portion of Figure 3. A node (rectangle containing two curve indices) represents a portion of a cell of C-free bounded by the collision curves above and below and by the critical points of $\phi_{1,i}$ on the left and right. An arc connecting two nodes indicates that the nodes represent connected portions of a larger cell of C-free. A dot on the arc indicates that the corresponding free space chunks are not connected. Reversal of the label order corresponds to the crossing of two collision curves and so the corresponding cells are not connected. Inspection of Figure 3 reveals 8 cells (recall that the top and bottom edges of the square are identified).

If link 2 is long enough to reach all obstacles regardless of the value of ϕ_1, then the "horizontal" collision curves will be closed as they are in Figure 3, and in this case the critical points are all of the type described above. However, if link 2 is not long enough, then for some values of ϕ_1, the second link will not be able to reach all obstacles. This case is illustrated on the left side of Figure 4, from which one can infer that for a set of values of ϕ_1, p_3 or p_4 cannot be reached by link 2. Thus there is a second type of critical point, which corresponds to the situation where the tip of the second link touches a point obstacle. In a neighborhood of such a critical point $\phi_{1,j}$, for ϕ_1 on one side of $\phi_{1,j}$ there are $\gamma \leq n$ open intervals, while on the other side there are strictly fewer than γ. The graph representation used above can be easily modified to handle the second type of critical point as shown on the right side of Figure 4. Finally, one can see that C-free has the maximum number of

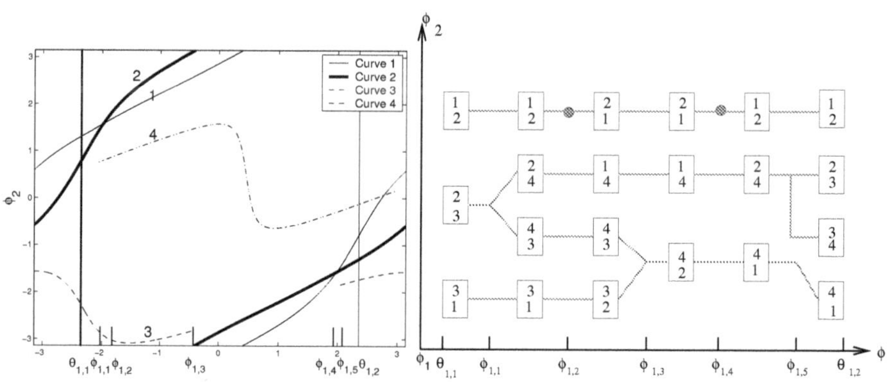

Fig. 4. Left: C-space of 2-chain with a shorter l_2 in Figure 2 showing collision curves, some of which are open. Right: Corresponding graph representing the global structure of C-free over $(\theta_{1,1}, \theta_{1,2})$.

components when all the curves are closed. In fact, this maximum number is the same for all generic arrangements of n obstacles, and is $2n^2 + n$ [9]. The number of components reduces monotonically as the length of link 2 reduces.

Since it is easy to determine cell membership of a configuration, graph search can be used to answer the path existence question. If a path is found to exist, it is easy to construct connected continuous curve segments through the cells to solve the path planning problem. All this is explained in detail in [9]. In addition, while we did not explicitly consider link-link collisions in the above discussion, one should note that the link-link collision curve is a single circle that is topologically equivalent to the link-obstacle collision curves, so the existence and planning problems can be solved without modifying the analysis or logic above. Solutions to sample planning problems can be found in [9].

5 Overview of the analysis for the three-link case

As indicated above, the analysis of planar three-chains can be handled by observing that if we have a configuration of a three-chain that avoids the obstacle set, $\mathcal{O}$, then the proximal two-chain also avoids $\mathcal{O}$. Consequently, there is a projection of C-free of the three-chain to C-free of the proximal two-chain. Once more, if we are only concerned with avoiding collisions with $\mathcal{O}$, then over each point of C-free of the two-chain, the set of possible angles of the third link will be a circle with a finite number of points deleted, so this map is *onto*. Note that the situation is more complex if we also choose to avoid self-intersections since the intersections of the third link with the first is generally a three-dimensional subset of T^3. In any case, as we will see, over each point of the C-free of the two-chain, there will be a finite number of open intervals that represent the points for which link 3 avoids the obstacles and the other links. To determine the structure of C-free of the three-chain, in direct analogy to the approach taken for planar 2-chains, we locate the points in C-free of the proximal two-chain *where the number of intervals changes*.

We initially restrict our attention to the case where self-intersections are not considered. In this case the changes will occur over critical curves in C-free of the two-chain, analogous to the critical points identified in C-free of the proximal one-chain in the previous section. These critical curves will be of two kinds.

1. Type 1: Curves where there are γ open intervals above all points, (ϕ_1, ϕ_2), in a small neighborhood of the curve except for points on the curve, above which there are $\gamma - 1$ (or fewer) intervals. These curves correspond to situations where link 3 can touch two (or more) point obstacles simultaneously.
2. Type 2: Curves where above all nearby points on one side of the curve there are γ open intervals, but only $\gamma - 1$ (or fewer) open intervals above nearby points on the other side. These curves arise from situations in which the tip of the three-chain touches a point obstacle.

Thus, to analyze the three-link case, one begins with the free space cells of the proximal two-chain and refines them by adding the critical curves. Recall that the critical points added to the free space cells (open intervals of the circle) of the proximal one-chain in the previous section did not change the topology of C-free of the one-chain. In Figure 3, ϕ_1 could still vary freely in the interval $(\theta_{1,i}, \theta_{1,i+1})$ across the critical points. Those points were only needed to understand the connectivity of free space "upstairs," *i.e.*, in the C-space of the two-chain. Similarly, in the three-link case, the critical curves running through the free space cells "downstairs" (in the C-space of the proximal two-chain) do not change the topology of those cells.

Figure 5 illustrates the refinement induced by a critical curve labelled "D." The graph representation below the plot and to the left is all that is needed to describe the free space of the proximal two-chain. However, in anticipation of adding the third link, we must include all critical curves such as D. These curves gives rise to additional "critical" points in the free space of the two-chain, and induces a refinement of the free space graph as shown below the plot and to the right. Note that nodes in the columns that contain D are connected by vertical arcs, because they do *not* disconnect the cells "downstairs."

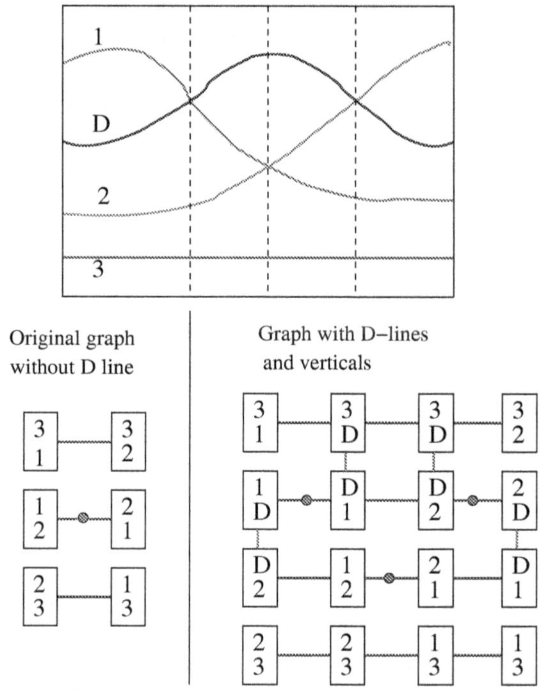

Fig. 5. Upper: Four cells of C-free generated by three point obstacles labelled "1," "2", and "3." Lower, left: The graph representation ignoring the critical curve D. Lower, right: The refined graph induced by the critical curve.

Figure 6 shows a rectangular free space cell "downstairs," crossed by a critical curve labelled "D". "Upstairs" there are the free space cells partitioned by four surfaces corresponding to the three-chain in contact with point obstacles 4, 5, 6, and 7. Two of the surfaces cross, at configurations where the three-chain simultaneously touches obstacles 4 and 5. The fact that this intersection separates the [4-5] cell from the [5-4] cell is the reason that the projection of the intersection curve is needed in the refinement of the free space of the proximal two-chain. This information is incorporated into the graph shown in the right portion of Figure 6 by growing the graph out of the

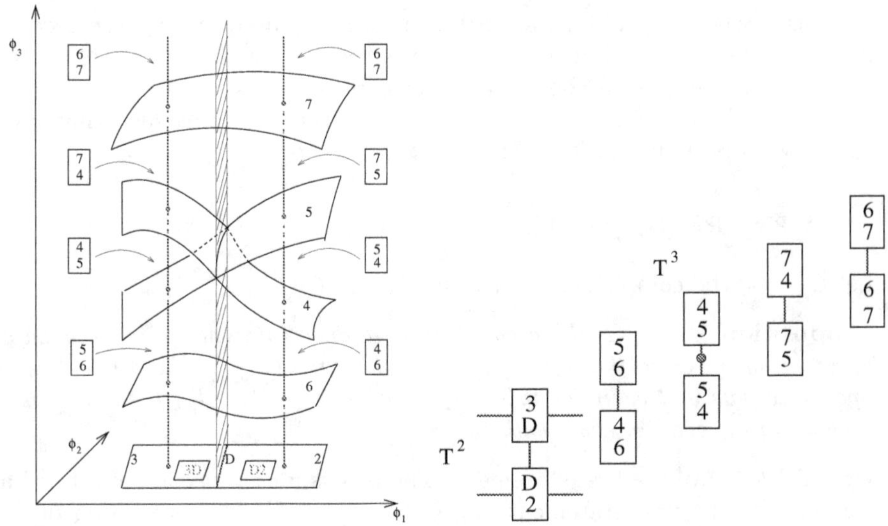

Fig. 6. Left: Component sheafs "above" two points in a cell of C-free of the proximal two-chain that are generated by adding a third link. Right: corresponding graph representing C-free for the 2-chain in T^2 ([3-D], [D-2]) and for the 3-chain in T^3 (the other 8 nodes corresponding to 5 components).

plane. The simple situation shown in the left portion of Figure 6, gives rise to four layers of the graph in the vertical direction - one for each surface. On the left is the free space cell [3-D,D-2] "downstairs." Above that are five free space cells "upstairs," [5-6,4-6], [4-5], [5-4], [7-4,7-5], and [6-7,6-7].

The discussion above shows the importance of understanding the subsets of the C-space of the three-chain where it intersects the obstruction set $\mathcal{O}$. We denote this intersection set Sing_2. We denote the subspace of configurations where the mechanism intersects itself Sing_1 and the union of these two spaces as

$$\text{Sing} = \text{Sing}_1 \cup \text{Sing}_2.$$

In the following section we analyze the set Sing_2. The analysis of Sing_1 will not be addressed here, interested readers can refer to our forthcoming technical report [10] for the details. However, we must emphasize here that Sing_1 for a three-chain has the same dimension as its C-space, while that for a two-chain has smaller dimension than its C-space.

6 Analysis of the set Sing_2

Let V_{p_i} be the collision surface which consists of all the configurations where the 3-chain intersects p_i. Under the assumptions that $\mathcal{O}$ is in general position and $l_2 > 2l_1$, $l_3 > 2(l_1 + l_2)$, V_{p_i} is the union of three surfaces, the first a 2-torus, where ϕ_2 and ϕ_3 are arbitrary and l_1 intersects p_i, the second a 2-torus, where ϕ_3 is arbitrary and l_2 intersects p_i, while the third is also a 2-torus consisting of all configurations where p_i intersects l_3. In fact, for all $(\phi_1, \phi_2) \in T^2$, there is a unique angle ϕ_3 for the last joint such that link 3 collide with p_i. C-free will be $T^3 - \cup_i V_{p_i}$.

6.1 A Key Property of V_{p_i}

Let $C_{V_{p_i}}$ be the complement of V_{p_i} in T^3, *i.e.*, $C_{V_{p_i}} = T^3 - V_{p_i}$.

Proposition 1. *$C_{V_{p_i}}$ is the product of three open intervals. If we weaken our assumption on $\mathcal{O}$ so that $l_1 < |p_i| < l_2 - l_1$, then $C_{V_{p_i}}$ is the product of two open intervals and a circle. If $l_1 + l_2 < |p_i| < l_3 - (l_1 + l_2)$ then $C_{V_{p_i}}$ is the product of an open interval with two circles.*

Proof: We prove the first statement. The proofs for the second and third are similar. For the first statement, note that $|p_i| < l_1$. The projection of $C_{V_{p_i}}$ to the ϕ_1 axis will be the interval $S^1 - \{\theta_1\}$, where θ_1 is the angle of the first joint for which the first link lies on the line joining p_i to the origin. For each $\phi_1 \in S^1 - \{\theta_1\}$, there is one and only one configuration (ϕ_1, θ_2) at which the second link of the manipulator will collide with p_i. Note also that for all $(\phi_1, \phi_2) \in S^1 - \{\theta_1\} \times S^1 - \{\theta_2(\phi_1)\}$, there is again only one configuration $(\phi_1, \phi_2, \theta_3)$ at which the third link will collide with p_i. Therefore, the inverse image of ϕ_1 will be the product of two open interval $S^1 - \{\theta_2\} \times S^1 - \{\theta_3\}$. This decomposition is locally a product so the total space of $C_{V_{p_i}}$ fibers over the interval with fiber an open rectangle. Since any fibration over the interval is a product of the fiber with the interval, the proposition follows.

Remark 1. This result generalizes directly to m-R manipulators as long as for every configuration of the first $j-1$ joints, there is an angle for the jth joint so that that link will pass through p_i, $j = 1, 2, \ldots$. The complement of $C_{V_{p_i}}$ on $T^m = S^1 \times \cdots \times S^1$ will be the direct product of m open intervals, and V_{p_i} will be the $m-1$-skeleton of T^m, the union of the m coordinate $m-1$ torii of T^m.

Once one has the V_{p_i}, in order to determine the entire structure of Sing_2 it is necessary to determine the intersections,

$$V_{p_{i_1}} \cap V_{p_{i_2}} \cap \ldots \cap V_{p_{i_r}}.$$

For reasonably general obstruction sets, $\mathcal{O}$, and lengths l_1, l_2, l_3, the intersections of 4 or more V_{p_i} will be empty, but it is easy to construct special sets $\mathcal{O}$ for any given 3R-manipulator where the intersections will be non-empty for much larger numbers of V_{p_i}.

Since any $\mathcal{O}$ and $\{l_1, l_2, l_3\}$ is arbitrarily close to a general $\mathcal{O}$, we make the assumption that four-fold intersections of the V_{p_i} are empty. In fact we can be more exact. We can assume, first,

- no three distinct points of $\mathcal{O}$ lie on a single line,
- no three lines, each through 2 distinct points of $\mathcal{O}$, intersect in a point,
- the first link does not collide with more than one point of $\mathcal{O}$ at a time, and
- by varying the lengths l_1, l_2, slightly, that the only configurations with three collision points will be isolated points in T^3 and will have the form $(1,1,1)$, $(1,0,2)$ $(0,2,1)$, $(0,1,2)$, where the first number is the number of collisions with l_1, the second represents the number of collisions with l_2, and the third represents the number of collisions with l_3.

Remark 2. It is worth noting that the assumptions above imply that there will only be five types of 2-point collisions, having the forms $(1,1,0)$, $(1,0,1)$, $(0,1,1)$, $(0,2,0)$, $(0,0,2)$.

Let us now look at the intersection of V_{p_i} and V_{p_j} in more detail. A configuration that collides with both p_i and p_j can do so in two distinct ways. In the first way p_i collides with l_r and p_j collides with l_s where $s \neq r$. In the second way, l_r collides with both p_i and p_j.

We consider the case where $p_i \in l_1$ and $p_j \in l_3$. Since the requirement that $p_i \in l_1$ determines ϕ_1, we see that the initial point of l_2 is fixed and this piece of $V_{p_i} \cap V_{p_j}$ consists of a closed segment of the ϕ_2 circle. This segment will be the entire circle if $l_3 > l_2 + d$ where d is the distance from the initial point of l_2 to p_j. It will be empty if $d > l_3 + l_2$.

If $p_i \in l_1$ while $p_j \in l_2$, then both ϕ_1 and ϕ_2 are fixed provided $|p_i| \neq l_1$ while ϕ_3 is free, so this piece of the intersection is a circle.

The final piece of this type is where $p_i \in l_2$ and $p_j \in l_3$. In this case l_2 is constrained to lie in a region where its initial point lies on the circle of radius l_1 about the origin, $p_i \in l_2$ and the end-point lies in the closed disk of radius l_3 about p_j. The curve given by the end-point of l_2 under the constraint that the initial point of l_1 is fixed, and l_2 contains p_i is a simple closed curve coordinatized by either the entire l_1 circle, or an arc on it.

Next we consider the double point arcs, $p_i, p_j \in l_s$. We can assume that l_1 never intersects two of the obstruction points. If two obstruction points lie on l_2, then the end of l_1 must lie on the line connecting p_i, p_j, and both p_i, p_j must be on the same side of this point. If this happens, then ϕ_1 takes either one or two distinct values, and ϕ_2 is also fixed, but ϕ_3 is free, so we get either one or two circles. Finally, if p_i and p_j lie on l_3, then the end-point of the l_1, l_2, configuration must lie on the line connecting p_i, p_j, and on one side of this line only. The l_1, l_2 workspace is an annulus and various possibilities occur.

7 Determination of the Components for a Key Example

Assumption. 1 *As in the previous section we assume first that $\mathcal{O}$ and $\{l_1, l_2, l_3\}$ are in general position. We also make the following assumptions: $l_3 > 2(l_2 + l_1)$, $l_2 > 2l_1 > 0$, and $|p_i| < l_1$ for each $p_i \in \mathcal{O}$.*

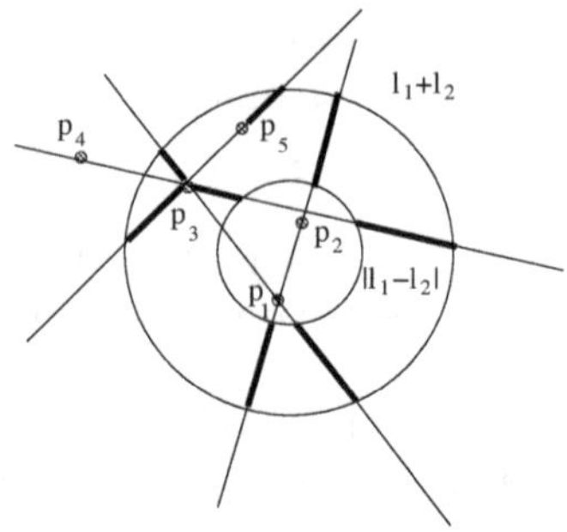

Fig. 7. The locations for the end-point of link 2 from which two obstacles can be touched by link 3 are drawn as bold line segments interior to the annulus.

As we will see, this results in a situation where there are no type-2 curves in the C-space of the first two links. Also, all collision surfaces will be closed, and all the generic intersections between collision surfaces will be present. When the assumptions are weakened, the arrangement of surfaces becomes simpler. Some surfaces open up, others disappear, and the intersections may also disappear.

From Proposition 1, we see that each V_{p_i} is a copy of the 2-dimensional skeleton of T^3. It is the union of three tori, T^2. The intersection between each two of these tori is a circle, and that between all three is a point, the point corresponding to the configuration where each link collides with p_i.

When we restrict to the first two links and the same $\mathcal{O}$ we get the assumptions studied in [9] so that $T^2 - \mathrm{Sing}_2$ is fully understood. Over each point of $T^2 - \mathrm{Sing}_2$, and for each $p_i \in \mathcal{O}$, there is an angle ϕ_3 so that l_3 collides with p_i due to the conditions in Assumption 1. Thus, for almost all the points of $T^2 - \mathrm{Sing}_2$, the inverse image in $T^3 - \mathrm{Sing}_2$ consists of n open intervals where n is the number of points in $\mathcal{O}$. The points where there are $\leq n-1$ intervals are those where the tip of the proximal 2-chain lies on one of the lines through two distinct points of $\mathcal{O}$.

Since each $p_i \in \mathcal{O}$ satisfies the condition $|p_i| < l_1 < |l_2 - l_1|$ it follows that any point in the intersection of the workspace of the two-link configuration - the annulus with radii $l_2 - l_1$ and $l_1 + l_2$ - with a line through two points of $\mathcal{O}$ lies on the same side of both points, and each such intersection consists of two intervals, each with endpoints on the $l_2 - l_1$ and $l_1 + l_2$ circles corresponding to the two different orders for p_i and p_j on l_3. *The inverse image of each such interval is a circle in T^2 that can be parameterized by ϕ_1.* These are the two point curves D discussed in Section 4.

What is not determined by the general discussion in Section 5 is the number and disposition of the intersections of the D. Such intersections correspond to the isolated points of $T^2 - \mathrm{Sing}_2$ where the inverse image consists of exactly $n-2$ open intervals, and correspond, on $T^3 - \mathrm{Sing}_2$, with the intersections of two of the D. Let the number of such intersections be N. Since the line through p_i and p_j intersects the line through p_i and p_k, $k \neq j$ at p_i, it follows that two such lines can intersect the proximal 2-chain only if p_i, p_j is disjoint from p_k, p_l, so we have

$$0 \leq N \leq \frac{n(n-1)(n-2)(n-3)}{8}.$$

However, since the sets of two-point curves at an intersection point are indexed by disjoint pairs of points, it follows that the four V_{p_t} surfaces upstairs that intersect along copies of these curves upstairs do not all intersect, but rather intersect two at a time. Consequently, these intersections downstairs do not contribute to the component count upstairs.

The final thing to note is that the regions of the downstairs torus that are cut out by these assumptions are cells. Moreover, the two point curves, D, needed in T^2 to take account of the points where the number of intervals over a point of $T^2 - \mathrm{Sing}_2$ is not n cuts each cell into further cells since each double point curve is parameterized by ϕ_1 under Assumption 1.

From the remarks above we know that each component of $T^3 - \mathrm{Sing}_2$ is the product of a cell in $T^2 - \mathrm{Sing}_2$ with an interval, and is consequently a cell. (This is not something that generalizes to other configurations $\mathcal{O}$ and 3-chains that do not satisfy Assumption 1.) There are $n(n-1)$ of the D curves, and each breaks into $2n-1$ components when restricted to the inverse image of the free part of the proximal T^2 - $n-1$ for intersections of the form $\langle 0,1,2\rangle$ and n more for intersections of the form $\langle 1,0,2\rangle$. Thus there are

$$n(n(2n-1)) + n(n-1)(2n-1) = n(2n-1)^2$$

components in all for 3-chains and obstruction sets $\mathcal{O}$ that satisfy the conditions of Assumption 1.

8 Preliminary results

While there are a small number of technical details of our approach that still require rigorous proof, we have begun implementation of the basic method. The planning algorithm uses the algorithm developed for planar 2R-manipulators presented in [9] to obtain the necessary free space graph downstairs, then computes a refinement of the graph based on critical curves, and finally "lifts" the graph into the 3-torus. The path existence problem is then solved by determining the cell membership of the initial and final configurations. If a path exists, then the details of the structure of the cell components containing the points are used to construct a collision-free path. In the algorithm description below, $\tilde{\phi} = (\phi_1, \phi_2)$ denotes a sub-configuration of the proximal two-chain, obtained by ignoring the third link.

Path planning algorithm for planar three-chain

- Input: n point obstacles p_i, $i = 1, \cdots, n$ which are in generic positions, a planar 3-R manipulator with arbitrary link lengths, and two arbitrary configurations $\phi(0)$ and $\phi(T)$ representing, respectively, the initial and final configurations
- Output: (i) "False" if no path exists, "True" otherwise; (ii) If a path exists, then return a path connecting the initial and final configurations.

1. Call the two-chain planning algorithm given in [9] to determine whether $\tilde{\phi}(0)$ and $\tilde{\phi}(T)$ are in the same component of free space downstairs. If not, return "False." Otherwise a return the cell E that contains both $\tilde{\phi}(0)$ and $\tilde{\phi}(T)$.
2. Calculate the singular curves and build the spatial graph above the planar graph.
3. Identify the nodes corresponding to the initial and final configurations.
4. Search the graph to determine path existence (most cells will be represented by more than one node). If a path does not exist, return "False." Otherwise continue.
5. Using the path through the graph, construct a path downstairs between $\tilde{\phi}(0)$ and $\tilde{\phi}(T)$.
6. Lift the path into T^3.

Figure 8 shows the results of path planning for a three-chain with four point obstacles in the workspace. To facilitate viewing this path, the projection of this path to (ϕ_1, ϕ_2) plane as well as other curves are drawn in Figure 9.

9 Conclusions and future work

We have developed most of the basic theory and developed an initial algorithm for planning collision-free motions of planar 3-chains with point obstacles in its workspace. There are three main avenues for further research that we plan to explore. First, we plan to exploit the methods of topology and homology to develop a planning algorithm with provably minimal complexity. Second, these methods also indicate that it will be possible to determine path existence without analyzing the entire free space. Such a method would use a candidate path connecting the initial and final configurations, but this path need not be valid. The third avenue of interest is the extension to allow two-dimensional obstacles.

Acknowledgements

The authors would like to thank Srinivas Akella for his careful reading of an earlier draft of this paper and for his insightful comments. This work was partially supported by NSF grant 0139701 and Rensselaer Polytechnic Institute.

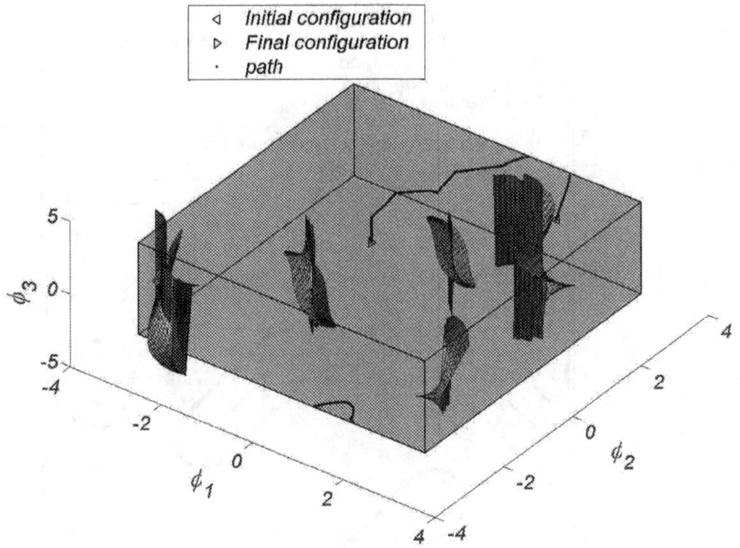

Fig. 8. C-space (T^3) with C-obstacle and a continuous path in C-free connecting the initial and final configurations. Note that due to the limitations of Matlab's "surf" function, a portion of the C-obstacle is plotted outside the cube.

References

1. J.F. Canny, *The complexity of robot motion planning.* Cambridge, MA: MIT Press, 1988.
2. P.C. Chen and Y.K. Hwang, *SANDROS: A dynamic graph search algorithm for motion planning.* IEEE Trans. on Robotics and Automation, **14**(3):390-403, 1998.
3. K.K. Gupta, *Fast collision avoidance for manipulator arms: A sequential search strategy.* IEEE Int'l. Conf. on Robotics and Automation, pp. 1724-1729, 1990.
4. D. Halperin, *On the complexity of a single cell in certain arrangements of surfaces related to motion planning.* Discrete and Comp. Geometry **11**:1-33, 1994.
5. Y. K. Hwang, N. Ahuja, *Gross motion planning - A survey.* ACM Computing Surveys **24**(3):219-291, 1992.
6. L.E. Kavraki, M.N. Kolountzakis, and J.C. Latombe, *Analysis of probabilistic roadmaps for path planning.* IEEE Trans. on Robotics and Automation, **14**(1):166-171, 1998.
7. L.E. Kavraki, P. Švestka, J.C. Latombe, and M.H. Overmars, *Probablistic roadmaps for path planning in high-dimensional configuration space.* IEEE Trans. on Robotics and Automation, **12**(4):566-580, 1996.
8. J.C. Latombe, *Robot motion panning.* Kluwer Academic Publishers, 1992.
9. G.F. Liu, J.C. Trinkle, R.J. Milgram, *Complete path planning for planar 2R-manipulators with point obstacles.* IEEE Int'l Conf. on Robotics and Automation, pp. 3263-3269, 2004.

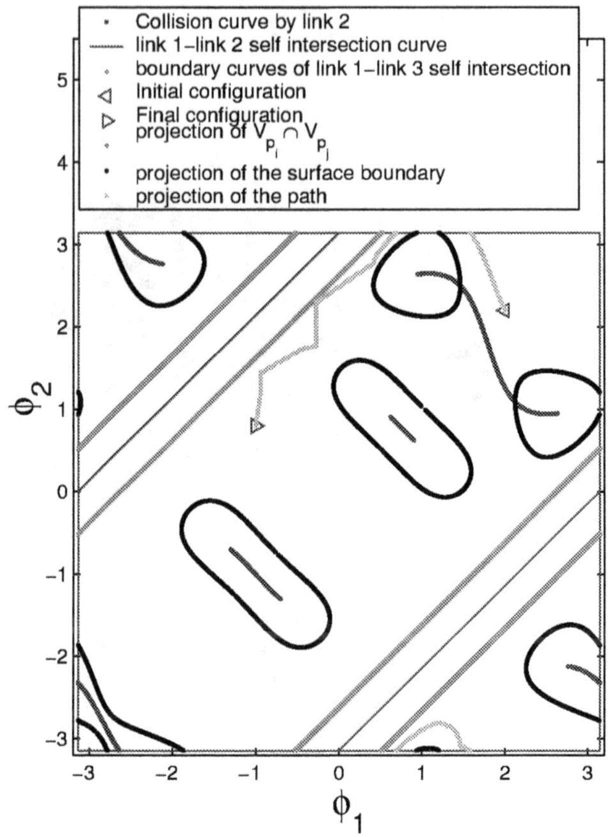

Fig. 9. The path and C-obstacle boundaries projected onto T^2.

10. G.F. Liu, J.C. Trinkle, R.J. Milgram, *Complete path planning for planar 3R-manipulators among point obstacles.* forthcoming tech. report, Dept. of Computer Science, Rensselaer Polytechnic Instiute, or through `www.cs.rpi.edu/~liugf/LTM-3R-planning.pdf`
11. T. Lozano-Pérez, *Spatial planning: A configuration space approach.* IEEE Transactions on Computers, **32**:108-120, 1983.
12. J. Schwartz, J. Hopcroft, and M. Sharir, *Planning, geometry, and complexity of robot motion.* Ablex, 1987.
13. J.T. Schwartz and M. Sharir, *On the piano movers II. General techniques for computing topological properties on real algebraic manifolds.* Adv. Appl. Math., **4**:298-351, 1983.
14. J.C. Trinkle and R.J. Milgram, *Complete path planning for closed kinematic chains with spherical joints.* Int'l Jrnl. of Robotics Research, **21**(9):773-789, 2002.
15. J.H. Yakey, S.M. LaValle, and L.E. Kavraki, *Randomized path planning for linkages with closed kinematic chains.* IEEE Trans. on Robotics and Automation, **17**(6):951-958, 2001.

Computing Protein Structures from Electron Density Maps: The Missing Fragment Problem

Itay Lotan[1], Henry van den Bedem[2], Ashley M. Deacon[2], and Jean-Claude Latombe[1]

[1] Computer Science Department, Stanford University, Stanford CA 94305.
E-mail: itayl@cs.stanford.edu, latombe@cs.stanford.edu

[2] Joint Center for Structural Genomics, Stanford Synchrotron Radiation Laboratory, Stanford Linear Accelerator Center, 2575 Sand Hill Road, Menlo Park CA 94025
E-mail: vdbedem@slac.stanford.edu, adeacon@slac.stanford.edu

Abstract. Rapid protein structure determination relies greatly on the availability of software that can automatically generate a protein model from an experimental electron density map. Tremendous advances in this area have been achieved recently. In favorable cases, available software can build over 90% of the final model. However, in less favorable circumstances, particularly at medium-low resolution, only about 2/3 completeness is attained. Manual completion of these partial models is usually feasible but time-consuming. The electron density in areas of missing fragments is often of poorer quality, especially for flexible loops, making manual interpretation particularly difficult. Except for the beginning and end of the protein chain, the end points of each missing fragment are known from the partial model. Due to the kinematic chain structure of the protein backbone, loop completion can be approached as an inverse kinematics problem. A fast, two-stage inverse kinematics algorithm is presented that fits a protein chain of known sequence to the electron density map between two anchor points. Our approach first samples a large set of candidates that meet the closure constraint and then refines the most promising candidates to improve the fit. The algorithm has been tested and used to aid protein model completion in areas of poor density, closing loops of up to 12 residues to within 0.3Å RMSD of the final refined structure. It has also been used to close missing loops of the same length in partial models built at medium-low resolution to within 0.6Å.

1 Introduction

A protein is a linear sequence of amino acids (residues) that form a polypeptide chain. The function of a protein is largely dictated by its folded, 3-D structure, which determines its ability to bind to other molecules, such as small ligands, other proteins, or DNA. However, in general, predicting the folded state of a protein of a given sequence is still beyond our ability.

X-ray crystallography and nuclear magnetic resonance (NMR) spectroscopy are two widely used experimental techniques to obtain atomic coordinates of macromolecular structures. NMR spectroscopy enables structural biologists

M. Erdmann et al. (Eds.): Algorithmic Foundations of Robotics VI, STAR 17, pp. 345–360, 2005.

to observe dynamic properties of proteins, but is limited in protein size. X-ray crystallography has no such limitation, but has the drawback that it requires crystallizing protein samples. Once a sequence of X-ray diffraction images is collected from a protein in crystalline form, a 3-D distribution of electronic charge in $\mathbb{R}^3$ of the atoms constituting the molecule can be calculated. Coordinates of each atom are obtained by interpreting this 3-D *electron density map* (EDM) [23]. This consists of placing the atoms in 3-D space to best match the EDM. The process from synthesizing the protein in the lab to depositing the coordinates of its folded structure in the Protein Data Bank (PDB) [5] may take several months, and sometimes years.

The Protein Structure Initiative (PSI), a National Institute of General Medical Sciences program in the US, aims to reduce the time and associated costs of determining a 3-D protein structure [43]. The long-range goal of the PSI is to produce structural coverage of a majority of sequenced genes. Tremendous advances in automated model building have been made over the past few years. Various software systems are now capable of generating a protein model from an EDM without human intervention [33,44,55,58].

The degree of completeness of these initial models, i.e., the fraction of residues correctly placed, varies with the quality of the experimental data, yet rarely reaches 100%. Mobility of some fragments of the molecule, temperature-dependent atomic vibration and other factors may lead to EDM of poor quality locally, making automatic interpretation difficult. Manually completing a partial protein model is often feasible, but time-consuming.

In practice, a large portion of the molecule is often resolved successfully. The endpoints of any gaps in the initial model are known, as well as the length and sequence of the missing fragment. Since a protein molecule can be modeled as a kinematic chain — with rigid groups of atoms as links and rotatable bonds as joints — fragment fitting in this case is similar to an inverse kinematics (IK) problem [48,49]. One needs to compute values for the degrees of freedom (DoFs) of the protein fragment that optimize its fit with the density while respecting the closure constraint — the fragment must bridge the gap between the two endpoints.

Exploiting this similarity, we have developed a fast, two-stage algorithm based on IK techniques to fit a protein fragment to the EDM between two anchor residues. The first stage samples a large number of closing conformations[1], guided by the EDM. These candidate fragments conformations are ranked according to density fit. Top-ranking conformations are then refined using an optimization procedure that locally searches the sub-space of closed conformation for the optimal fragment structure.

The algorithm has been successfully tested and used to complete initial models. At a resolution of 2.0Å, it closed gaps as long as 12 residues to within

[1] The chemical equivalent term for a configuration in robotics

0.3Å RMSD[2] of the manually resolved structure. In a partial model built at 2.8Å, it computed missing fragments of the same length to within 0.6Å all atom RMSD in areas of poor density.

2 Description of problem

Rapid protein structure determination depends critically on the availability of software that can automatically generate a protein model from an EDM. At high resolution, existing programs may build over 90% of the protein backbone [4]. At medium to low resolution levels (2.3Å $\leq d <$ 2.9Å), the model resulting from these programs is typically a gapped chain, and only about 2/3 completeness is attained. Programs targeting lower resolution levels, notably *RESOLVE* [58] and *MAID* [44], rely on pattern recognition techniques, unambiguous density and elementary stereochemical constraints. Thus, poorly defined areas of the EDM pose a considerable challenge.

In a crystallographic experiment, the phase angle of the diffracted beam is lost, and only the magnitudes are recorded [23]. The phase angle is recovered at a later stage, but may have a substantial error associated with it, leading to systematic errors in the EDM. The resolution at which diffraction data was collected also affects the interpretability of the map. High-resolution data allow the crystallographer to distinguish detail at the atomic level, whereas the EDM appears "blurred" at lower resolution. It is often difficult or even impossible to obtain high-resolution EDMs of some proteins. A protein crystal contains many replicas of the protein, which all contribute to the resulting EDM. If the structures of these replicas are not identical, localized disorder in the EDM may result. Temperature-dependent atomic vibrations and the existence of mobile regions in the protein are among the causes of local disorder.

Initial protein models, whether obtained manually or by automated procedures, are only approximately correct. Their coordinates typically serve as initial values for standard maximum likelihood (ML) algorithms, which improve the model coordinates using appropriate statistical techniques [53]. Iterating model building and refinement steps to improve the completeness and quality of atomic models has clear advantages [55,59]. Phase information from the updated model is combined with experimental phases to improve the EDM that is used to generate a new model. Missing fragments in a model hinder the ability of this procedure to converge to the correct phases and model, since significant parts of the EDM remain unaccounted for by the model. Thus, filling in the gaps with fragments whose coordinates are within the radius of convergence of ML algorithms may significantly improve the EDM at an early stage in the process.

---

[2] Root of the sum of the squared distances between corresponding atoms. It is computed after the fragments are optimally aligned in 3-D.

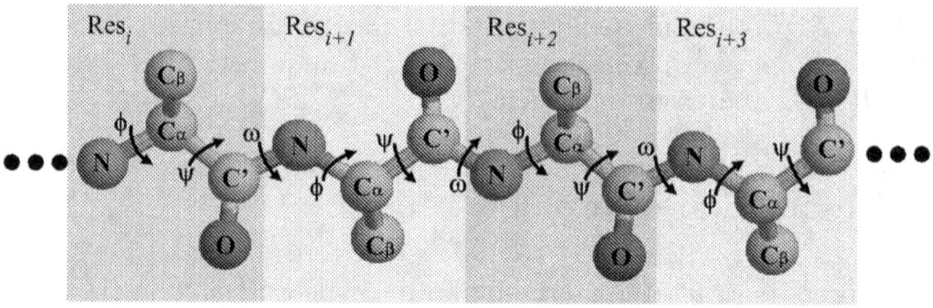

Fig. 1. The protein model used in this work

The input to our algorithm is the EDM, the partial structure resolved by the automatic model builder, and the two anchor residues that need to be bridged (denoted N-anchor and C-anchor). Since in the majority of partially-resolved structures, the amino acid sequence is correctly assigned, we assume the residue sequence of the missing fragment is known. Our goal is to propose candidate structures for the missing fragment that fall within the radius of convergence of existing refinement tools (1 - 1.5Å RMSD [53]).

The EDM is likely to have systematic errors due to erroneous phases. Consequently the parts that are solved — including the anchor residues of the missing fragment — may also be misplaced. Moreover, the density in the gap is somewhat disordered and thus not entirely reliable. Our protein model assumes *idealized geometry*, namely all bond lengths and bond angles are fixed to their idealized values [25]. These parameters vary very little in proteins [31], and their variance falls within the error margins of our input. Thus, the only DoFs in our model are the backbone torsional angles denoted the ϕ and ψ angles (one pair per residue). To simplify slightly our problem we will not have any side-chain DoFs and therefore no side-chain atoms in our protein model, with the exception of the Cβ atom (See Figure 1). Side-chains can be built onto the model once the backbone is fully in place. We will include all Cβ and O atoms because they help to orient the backbone though they do not add any DoFs.

When computing the missing fragment for a given gap, the two anchor residues will be incorporated at the ends of the fragment. We denote these copies of the anchors the *mobile* anchors. The anchors that are attached to the resolved structure are denoted the *stationary* anchors. Having mobile anchors provides us with a measure of closure. Closure is measured at both ends as the RMS distance of the three backbone atoms of the mobile anchor residue from their positions on the stationary anchor residue.

3 Related Work

3.1 Exact inverse kinematics solvers

The problem of fitting a protein backbone fragment between two anchor points is closely related to the IK problem in robotics [19]. It is known that for manipulators in a 3-D workspace there are a finite number of solutions to the IK problem when the number of DoFs does not exceed six. However, there is no analytical method that can compute these solutions for all types of manipulators. In the case of a 6-revolute-joint manipulator, which is the most relevant to this work, an analytic solution can always be obtained and the number of unique solutions is at most 16 [56]. An efficient algorithm was derived in [47] which was later applied to the manipulation and conformational analysis of small molecular chains [48,49]. For hyper-redundant robots, i.e. robots with a very large number of regularly distributed joints, methods based on curve approximations can be used [13]. Biologists have also developed their own specialized IK tools. A method for computing conformations of ring molecules, when rotation is possible around all backbone bonds, was introduced as early as 1970 [29]. An IK solution for three consecutive residues having ideal geometry was later devised [65]. Most recently a new formulation was developed that extends the domain to any three residues (not necessarily consecutive) with arbitrary geometry [17].

3.2 IK solutions by optimization

Optimization techniques have also been applied to the IK problem. A method for bounding the exact IK solutions within small intervals in the context of drug design was proposed in [68]. When the number of DoFs exceeds six, no analytical solution is known. Currently, only optimization-based solutions exist. The *random tweak* method [27,57] closes a loop by iteratively changing its DoFs until the desired distances between its two terminals are reached. It uses the Jacobian matrix of these distances (with respect to the torsional DoFs) to compute the DoF changes at each iteration. The *cyclic coordinate descent* (CCD) [64] method closes a kinematic chain by iteratively adjusting its DoFs until a desired closure tolerance is reached. This method has been applied to protein chains [9].

3.3 Motion planning for closed loops

Searching for the fragment that best matches an EDM has much in common with techniques used in roadmap-based motion planning for sampling robot configurations and connecting them. A number of works offer methods for planning the motion of closed kinematic loops. Common to many of them is the use of the probabilistic roadmaps (PRM) framework [37]. The work in [67] samples configurations by first ignoring the closure constraint and then enforcing the constraint through gradient descent. Nearby configurations are connected by the chaining of local steps that are generated in the null space of the Jacobian matrix. In [32] configurations are sampled by breaking the loop into an *active* part, for which forward kinematics sampling techniques are used, and a *passive* part, for which an exact IK solution is computed. Nearby samples are connected by using a local planner for the active part and letting the passive part follow the motion. An extension of this method [16] samples the active part of the chain one DoF at a time, making sure its endpoints are always reachable by the passive part. This method was recently applied to sampling conformations of mobile protein regions and to studying their flexibility [15]. Another extension to [32] is proposed in [66]. A roadmap for solving a slightly different problem, known as *point-to-point IK* for redundant robots, described in [3], also takes into account joint constraints and obstacle avoidance. A polynomial-time complete planner for the reconfiguration of closed loops with spherical joints and no obstacles is proposed in [60].

3.4 Methods for redundant manipulators

The refinement method we use exploits the redundancy of the protein chain to minimize a target function without breaking chain closure. It is closely related to the planning of instantaneous motion of redundant manipulators. Here the redundant DoFs are used for concurrent optimization of additional criteria beyond the completion of the main task (e.g., minimizing torque or kinetic energy) or the execution of additional lower priority tasks. Some examples include [8,12,38,39].

3.5 The loop closure problem in biology

Our problem is closely related to the *loop closure* problem in structural biology, where loops are predicted based on sequence information alone. The methods proposed for this problem can be roughly divided into two types: *ab initio* methods and *database search* methods. The ab initio methods search the loop conformation space, while database methods screen the PDB for known structures that closely meet the requirements of the desired loop.

Ab initio methods For loops that have six DoFs or less, exact methods that enumerate all possible solutions can be used, e.g. [17,29,47,49,65]. It is also possible to use the exact solver hierarchically as in [65] and thus handle longer loops. Examples of search methods include sampling biased by the database distribution of the ϕ/ψ angle pairs [52], uniform conformational search [7], sampling from a small set of ϕ/ψ pairs [20,21] or sampling from a small library of short representative fragments [41]. Other algorithms sample conformations randomly and then enforce the closure constraints, e.g. the *random tweak* method [27,57] or the CCD-based method [9]. Various methods exist for optimization of candidate loops, such as molecular dynamics [7,28,70] and Monte Carlo [1,14] simulations.

Database search methods Work on extracting loop candidates from the PDB started with [35]. Loops are chosen that have the right length and meet the required geometric constraints. The applicability of this approach was initially limited to loops of length 4 [26], but later work suggested an upper limit of 9 [62]. A recent study claims the limit should be raised to 15 [24].

Crystallography The crystallography community has developed its own semi-automated tools and techniques to assist in completing partial models, often combining database or ab initio methods with experimental data. The interactive graphics program *O* [34] selects fragments from a database to close loops. These fragments can then be refined against the EDM using torsion angle refinement based on grid summation [36]. A method combining random search of conformation space with grid- and gradient-based refinement techniques to close loops up to 8 residues in length is proposed in [54]. Both of these approaches share our two-stage approach, but do not use the EDM in the first stage. Even for medium-sized fragments, starting conformations are often outside the radius of convergence of the refinement step.

4 Our method

Our algorithm for computing missing fragments proceeds in two stages: candidate generation and refinement. In the first stage, candidate fragments are built using the CCD algorithm. Our implementation puts additional constraints on the DoFs to take EDM fit into account. Next, initial conformations are ranked according to density fit and top-ranking ones are passed on to the refinement procedure. Refinement is performed by minimizing a target function that quantifies the goodness of the fit of the fragment to the EDM. A minimization protocol searches for the global minimum of the target function while maintaining loop closure. Each candidate is optimized a number of times and the best scoring fragments are returned.

Our approach tries to compensate for the deficient density information by taking advantage of the closure constraint to guide the fragment to its correct positioning in space. In the first stage, the use of the closure constraint enables the generation of fragments that lie within 2Å RMSD of the PDB structure. The approximate enforcement of the closure constraint during refinement prevents the search from diverging and limits the searched space to motions that preserve closure.

Our implementation of the algorithm makes use of the following software packages: *Clipper* [18], the *CCP4 Coordinate Library* [42] and the exact IK solver of [17].

4.1 Stage 1: Generating fragment candidates

We start by constructing a protein fragment $\mathcal{C}$ of length L in a random conformation, where residue 0 is a copy of the N-anchor (the residue preceding the gap), and residue $L-1$ is a copy of the C-anchor (the residue after the gap). This fragment is attached to either the N-, or C-anchor, thus determining the *closing direction.*

Upon starting the procedure, the position of the mobile anchor will not coincide with the stationary anchor. The algorithm adjusts each backbone dihedral angle in turn such that the distances between the backbone atoms of the mobile anchor and the corresponding atoms of the stationary anchor are minimized. A total of 2000 iterations are allowed for closure up to a preset tolerance distance d_{closed}. Chains that do not close are discarded. We calculate an average density and conformational likelihood score for all conformations, and the 99-th percentile is submitted to stage two. Longer loops (9 or more residues) are split in the middle. Each half-chain is attached to its corresponding stationary anchor, and its terminal residue alternates between acting as stationary and mobile anchor in subsequent iterations.

Random Initial Conformations Five hundred random starting conformations are obtained from sampling $(\phi, \psi)_i, i = 0, \ldots, L-1$ angle pairs from PDB-derived distributions. A finite sum of 2-D Gaussians is thereto fitted to frequencies calculated from the Top500 database [46] of non-redundant protein structures, using the program EMMIX [50]. We obtained distributions for each of the 20 amino acids. The angles ϕ_0 and ψ_{L-1} remain fixed at their initial values.

Electron Density Constraints A change to the DoFs of a residue is calculated as follows. The CCD algorithm first proposes a distance minimizing dihedral angle ϕ_i for residue i. Based on ϕ_i, it proposes a minimizing ψ_i. (In our implementation, we change each DoF in turn, although this is not strictly necessary.) We denote these angles as a proposed angle pair $(\phi, \psi)_i^p$. To guide the fragment, a heuristic electron density constraint has been added to the CCD algorithm. For each pair i, we define a set of atoms $\mathcal{A}_i$ that is subject to change by angle pair i, and not affected by changes in angle pair $i+1$. Hence, $\mathcal{A}_i = \{C\beta_i, C_i, O_i, N_{i+1}, C\alpha_{i+1}\}$.

We evaluate electron density values corresponding to trial positions in a two dimensional, square neighborhood $U_{(\phi,\psi)^p}$ about $(\phi, \psi)^p$ in conformation space. A simple local scoring function is adopted; the sum of the electron density values at atom center positions of $\mathcal{A}_i$. We set $(\phi, \psi)_i$ to the trial position with maximum density score, i.e $(\phi, \psi)_i = \arg\max_{(\phi,\psi) \in U_{(\phi,\psi)_i^p}} s(\phi, \psi)$, where $s(\phi, \psi) = \sum_{A_j \in \mathcal{A}_i} \rho(A_j(c))$, and $A_j(c)$ denotes the center of atom A_j. The size of $U_{(\phi,\psi)^p}$ is reduced linearly in the number of CCD iterations to allow closure of the fragment.

4.2 Stage 2: Refining fragment candidates

Target function A candidate fragment is refined by minimizing the least squares residuals between the EDM ρ^o and the density calculated from the model ρ^c. This is a standard procedure used in crystallography for refinement of protein models [11,22]. The target function sums the squared differences between the observed density and the calculated density at each grid point in some volume V around the fragment:

$$T(q) = \sum_{g_i \in V} \left[S\rho^o(g_i) + k - \rho^c(g_i)\right]^2 . \tag{1}$$

The calculated density at each grid point is a sum of contributions of all atoms whose center lies within a cutoff distance from this point. The calculated density contribution of an atom is a sum of isotropic 3-D Gaussians [63]. The factors S and k scale ρ^o to ρ^c and are computed once at initialization using the partial model.

Optimization with closure constraints Our method is based on the approach described in [8,38] for optimizing an objective function while performing a task by taking advantage of robot manipulator redundancy. The redundant DoFs define a subspace of configuration space termed the *self-motion* manifold. Motions on this manifold do not influence the main task and thus can be used to move the manipulator configuration towards a minimum of the objective function. Since this manifold may be very complex these motions are in general difficult to compute. It is therefore common to locally approximate the self-motion manifold by its linear tangent space. This space is defined as the null-space of the linearized instantaneous kinematic relation between the linear and angular velocities of the end-effector (a frame attached to the end of the chain) and the joint velocities, also known as the chain's Jacobian matrix [19]. For an n-DoF chain in $\mathbb{R}^3$ at configuration q, the Jacobian $J(q)$ is a $6 \times n$ matrix satisfying the equation:

$$\dot{x} = J(q)\dot{q}. \tag{2}$$

Thus, $J(q) = df(q)/d(q)$ where $f(q)$ is the chain's forward kinematics function mapping joint coordinates to end-effector position and orientation. The rank of the Jacobian in $\mathbb{R}^3$ is at most 6 and thus the dimensionality of its null space is at least $n - 6$. In general, an instantaneous change in the configuration corresponding to a desired small change in end-effector position is computed by inverting Equation 2. We get:

$$\dot{q} = J^{\dagger}(q)\dot{x} + N(q)N^T(q)y, \tag{3}$$

where $J^{\dagger}$ is the pseudo-inverse of the Jacobian and $N(q)$ is an orthonormal basis for the null-space. The null space can now be used to optimize some objective function without affecting the motion of the end effector. When optimizing an objective function for a closed loop, the instantaneous change in position and orientation of the end, $\dot{x}$, is set to zero and y is taken to be the gradient vector of the objective function. Projecting y onto the null space of the Jacobian produces a motion that minimizes the objective function while not disturbing the chain closure. In general, the step size of the motions (the magnitude of the vector $\dot{q}$) needs to be small since the Jacobian approximation of the self-motion manifold is valid only in a small neighborhood about the current point in configuration space.

Implementation details

Computing the null space: We use the *atom group local frames* method [69] to represent the protein as a kinematic chain and the method prescribed in [10] to efficiently compute the Jacobian. The null space is computed by applying the SVD [30] to the Jacobian matrix giving the decomposition $J(q) = U\Sigma V^T$, where V is an orthonormal basis for the row-space of $J(q)$. The null-space basis $N(q)$ is simply the set of vectors of V that correspond to singular values (the diagonal elements of Σ) that are equal to zero.

Target function gradient: We derive an analytical expression for the gradient of the target function with respect to the torsional DoFs of the fragment. The gradient is computed using a recursive method [2], which is linear in the number of DoFs of the chain. The naïve approach in this case has quadratic complexity.

Minimization protocol: A gradient descent search for the minimum of the target function is prone to get stuck in local minima. To increase the radius of convergence, we use the Monte Carlo + Minimization (MCM) approach [45]. At each step, a large random move in conformation space is proposed, the new conformation is then minimized by gradient descent and the resulting local minimum is accepted or rejected using the Metropolis criterion [51]. The effect of minimization is to increase the acceptance probability of a trial move, enabling the search to make more progress. This comes at the cost of increasing the time of each simulation step. To further improve the ability to escape from local minima we envelop the MCM method with a simulated annealing (SA) protocol [40], which controls the probability of acceptance of uphill moves by varying the pseudo-temperature of the search.

Random moves: Two methods are used for generating large random closure-preserving moves for MCM. The first is to choose a random direction in the null-space of a randomly chosen sub-chain, and taking a step whose magnitude depends on the current temperature of the SA protocol. A similar method for motion on the self-motion manifold was described in [67]. Before performing minimization, we make sure the closure tolerance has not been exceeded. A second method for generating random steps is the use of an exact IK solver [17]. One of the solutions is chosen at random as the proposed move. The use of an exact solver allows jumping between unconnected parts of the self-motion manifold, which the Jacobian-based moves are unable to do.

Approximate closure The closure constraint is relaxed during this stage. A maximum RMSD of 0.5Å is allowed at both ends of the fragment. Since there are errors in the positions of the anchor residues, and because the resulting fragment will be subsequently refined together with the rest of the protein, this tolerance is acceptable. By allowing approximate closure, larger steps, can be taken in the null space of the Jacobian.

Refinement search protocol The refinement protocol is composed of three nested loops (See Figure 2). The inner loop performs MCM using the two methods described above for generating trial moves. The middle loop performs the SA protocol by gradually reducing the pseudo-temperature of the MCM. The outer loop enhances the SA protocol by restarting the search at decreasing pseudo-temperatures. Decreasing levels of smoothing are applied to the density after each restart. The magnitude of attempted null-space moves is reduced together with the pseudo-temperature of the simulation to increase the chance that these moves will be accepted.

Density smoothing The density map is smoothed by convolving it with an isotropic 3-D Gaussian kernel. The convolution is done using three 1-D kernels exploiting the separability of the Gaussian kernel. Since the convolution of a Gaussian with a Gaussian can be computed analytically by summing the means and variances of the two functions, computing ρ^c does not require any convolution. The variance of the summed Gaussians in ρ^c is simply augmented by the variance of the desired kernel.

5 Experimental results

The algorithm was developed and tested on missing fragments from various structures provided by the JCSG. Below, we provide results for two representative examples, protein structures TM0423 [6] (376 residues) and TM0813 (342 residues). Both are publicly available from the PDB under ID numbers 1KQ3 and 1J5X respectively. A complete account of the test results can be found in [61].

TM0423 is annotated as a glycerol dehydrogenase from the organism Thermotoga Maritima. Standard, publicly available crystallographic software was used to obtain a data set. We built an initial model using RESOLVE at 2.0Å. 88% of the residues in this protein were correctly built and assigned to the sequence.

```
for start_temp = high_start_TEMP downto low_start_TEMP {
    temp = start_temp;
    SmoothDensity(start_temp);
    for SA_steps = 1 to 8 {
        for MCM_steps = 1 to NUM_ITERS {
            M = ProposeRandomMove(temp);
            MinimizeMove(M);
            AcceptMove(M);
        }
        temp *= TEMP_dec_factor;
    }
}
```

Fig. 2. Pseudo-code for refinement search protocol

TM0813 is annotated as a conserved hypothetical protein from Thermotoga Maritima. High resolution data was excluded from the data set, and an initial model was built by RESOLVE at 2.8Å. The completeness of this model was found to be 61%.

We use the all-atom RMSD between top ranking fragments and their corresponding regions from the final, refined structure as a measure of performance. Note that since we use idealized geometry while the bond lengths and angles are allowed to vary in the final structure, a discrepancy of 0.2 - 0.3Å is expected even for an exactly correct fragment. In refinement, the protein structure may undergo rigid translations or rotations. More importantly, the stationary anchors can move independently. To exclude the effects of refinement steps, we tested the algorithm on an artificially created gap of length 7 in TM0423. We removed residues 19 to 25 from the initial model, and compared our top ranking candidates to this extracted fragment. We found a lowest RMSD of 0.35Å.

Next, we selected gaps from the initial models. TM0423 has a missing fragment of length 7, extending from residue 251 to residue 259, and a longer missing fragment of length 12, from residue 51 to 64. The minimum, all-atom RMSD achieved for both loops is 0.25. Figure 3(a) depicts the best scoring fragment together with the final, refined structure.

The low resolution model TM0813 is very incomplete, and we selected a fragment of length 12 extending from residue 83 to 96. The all-atom RMSD of candidate conformations submitted to the second stage of the algorithm did not get below 2.12Å. Nonetheless, the minimum, all atom RMSD achieved for these fragments upon completion of the algorithm is 0.56Å. The best scoring final conformation and its starting conformation for stage two are shown in Figure 3(b).

The average running time of the algorithm for fragments of length 4, 8, 12 and 15 is 40, 92, 134 and 178 minutes respectively on an EDM at 2.0Å resolution. Running times are faster at lower resolutions.

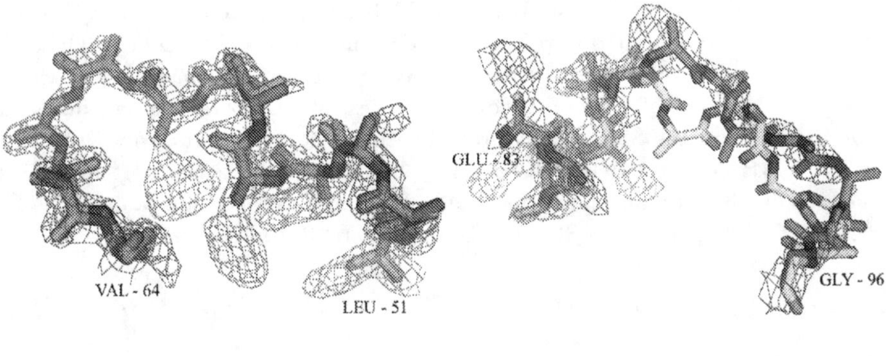

(a) (b)

Fig. 3. Fitting fragments into the density. (a) Highest scoring fragment for residues 51 to 64 from TM0423 in cyan, together with the PDB structure in magenta. The RMSD is 0.25Å. (b) Highest scoring fragment for residues 83 to 96 from TM0813 in magenta, together with its starting conformation in yellow (output of stage 1). The RMSD between the starting conformation and the PDB structure is 2.1Å. The refinement procedure reduces it to 0.6Å. The EDM is represented by an iso-surface shown in blue wire mesh.

6 Conclusion

Existing crystallographic software sometimes fails to model parts of a protein, resulting in an initial structure with gaps. In this paper we presented a novel method for the computation of these missing fragments. The use of robotics IK techniques enabled the inclusion of a closure constraint, thus augmenting reduced information available in areas of poor quality density. The experimental results demonstrated that our approach computed loops of up to 12 residues in good agreement with the final, refined structure. The method also performs well when using low resolution EDMs.

While facilitating interaction with other molecules, flexible regions in proteins are often poorly resolved in the electron density due to disorder in the crystal. Even if mobile fragments take on two clearly distinct conformations, it is often difficult for the crystallographer to identify both. Preliminary results on the protein TM0755 indicate that our method can identify and refine multiple conformations, even at sub-atomic resolution. For details, see [61]. Automated identification and modeling of ensembles of conformations is another interesting direction for future research.

We have shown that protein model-building procedures can greatly benefit from the use of robotics techniques and algorithms. Collaborative initiatives between the crystallography and robotics communities could therefore favorably impact the development of fully automated structure determination pipelines.

Acknowledgements: Test structures TM0423 and TM0813 used in this work were solved and deposited as part of the JCSG pipeline (www.jcsg.org). The JCSG is funded by the Protein Structure Initiative of the National Institutes of Health, National Institute of General Medical Sciences. SSRL operations are funded by DOE BES, and the SSRL Structural Molecular Biology program by DOE BER, NIH NCRR BTP and NIH NIGMS. Itay Lotan is supported in part by a Siebel Fellowship. Itay Lotan and Jean-Claude Latombe are also funded by NSF ITR grant CCR-0086013 and a Stanford BioX Research Initiative grant.

References

1. R. Abagyan and M. Totrov, *Biased probability monte carlo conformational searches and electrostatic calculations for peptides and proteins*, J. Mol. Biol. **235** (1994), 983–1002.
2. A. Abe, W. Braun, T. Noguti, and N. Gō, *Rapid calculation of first and second derivatives of conformational energy with respect to dihedral angles for proteins. general recurrent equations*, Comput. Chem. **8** (1984), no. 4, 239–247.
3. J.M. Ahuactzin and K.K. Gupta, *The kinematic roadmap: a motion planning based global approach for inverse kinematics of redundant robots*, IEEE Trans. Robot. Autom. **15** (1999), no. 4, 653–669.
4. J. Badger, *An evaluation of automated model-building procedures for protein crystallography*, Acta Cryst. **D59** (2003), 823–827.
5. H. Berman, J. Westbrook, Z. Feng, G. Gilliland, T. Bhat, H. Weissig, and et al, *The protein data bank*, Nucl. Acids Res. **28** (2000), 235–242.
6. L. S. Brinen, J. M. Canaves, X. P. Dai, A. M. Deacon, M. A. Elsliger, S. Eshaghi, R. Floyd, A. Godzik, C. Grittini, S. K. Grzechnik, and et. al., *Crystal structure of a zinc-containing glycerol dehydrogenase (tm0423) from thermotoga maritima at 1.5 angstrom resolution*, Proteins **50** (2003), no. 2, 371–374.
7. R. Bruccoleri and M. Karplus, *Prediction of the folding of short polypeptide segments by uniform conformational sampling*, Biopolymers **26** (1987), 137–168.
8. J.W. Burdick, *On the inverse kinematics of redundant manipulators: characterization of the self-motion manifolds*, IEEE Int. Conf. Robot. Autom. (ICRA), vol. 1, May 1989, pp. 264–270.
9. A.A. Canutescu and R.L. Dunbrack Jr, *Cyclic coordinate descent: A robotics algorithm for protein loop closure.*, Prot. Sci. **12** (2003), 963–972.
10. K.-S. Chang and O. Khatib, *Oprerational space dynamics: efficient algorithms for modeling and control of branching mechanisms*, IEEE Int. Conf. Robot. Autom. (ICRA) (San Francisco), April 2000, pp. 850–856.
11. M.S. Chapman, *Restrained real-space macromolecular atomic refinement using a new resolution-dependent electron-density function*, Acta Cryst. **A51** (1995), no. 1, 69–80.
12. I.D. Chen, Y.-C.; Walker, *A consistent null-space based approach to inverse kinematics of redundant robots*, IEEE Int. Conf. Robot. Autom, vol. 3, May 1993, pp. 374–381.
13. G.S. Chirikjian, *General methods for computing hyper-redundant manipulator inverse kinematics*, IEEE/RSJ Int. Conf. Intel. Robots and Sys. (Yokohama, Japan), July 1993.
14. V. Collura, J. Higo, and J. Garnier, *Modeling of protein loops by simulated annealing*, Prot. Sci. **2** (1993), 1502–1510.
15. J. Cortés, T. Siméon, M. Remaud-Siméon, and V. Tran, *Geometric algorithms for the conformational analysis of long protein loops*, J. Comp. Chem. **25** (2004), no. 7, 956–967.
16. J. Cortés, T. Simóon, and J.-P. Laumond, *A random loop generator for planning the motions of closed kinematic chains using prm methods*, IEEE Int. Conf. Robot. Autom. (ICRA), vol. 2, May 2002, pp. 2141–2146.
17. E.A. Coutsias, C Seok, M.P. Jacobson, and K.A. Dill, *A kinematic view of loop closure*, J. Comput. Chem. **25** (2004), 510–528.
18. K. D. Cowtan, *Clipper libraries*, 2004, http://www.ysbl.york.ac.uk/ cowtan/clipper/clipper.html.
19. J.J. Craig, *Introduction to robotics: manipulation nad control*, 2nd ed., Addison-Wesley, 1989.
20. C.M. Deane and T.L. Blundell, *A novel exhaustive search algorithm for predicting the conformation of polypeptide segments in proteins*, Proteins **40** (2000), no. 1, 135–144.
21. M.A. DePristo, P.I. de Bakker, S.C. Lovell, and T.L. Blundell, *Ab initio construction of polypeptide fragments: efficient generation of accurate, representative ensembles.*, Proteins **51(1)** (2003), 41–55.

22. R. Diamond, *A real-space refinement procedure for proteins*, Acta Cryst. **A27** (1971), no. 5, 436–452.
23. J. Drenth, *Principles of protein x-ray crystallography*, 2nd ed., Springer Verlag, New York, 1999.
24. P. Du, M Andrec, and R.M. Levy, *Have we seen all structures corresponding to short protein fragments in the protein data bank? an update*, Prot. Engin. **16** (2003), no. 6, 407–414.
25. R.A. Engh and R. Huber, *Accurate bond and angle parameters for x-ray protein structure refinement.*, Acta Cryst. **A47** (1991), 392–400.
26. K. Fidelis, P.S. Stern, D. Bacon, and J. Moult, *Comparison of systematic search and database methods for constructing segments of protein structure*, Prot. Engin. **7** (1994), no. 8, 953–960.
27. R.M. Fine, H. Wang, P.S. Shenkin, D.L. Yarmush, and C. Levinthal, *Predicting antibody hypervariable loop conformations. ii minimization and molecular dynamics studies of mcp603 from many randomly generated loop conformations*, Proteins **1** (1986), 342–362.
28. A. Fiser, R.K. Do, and A. Sali, *Modeling of loops in protein structures.*, Prot. Sci. **9(9)** (2000), 1753–73.
29. N. Gō and H.A. Scheraga, *Ring closure and local conformational deformations of chain molecules.*, Macromolecules **3** (1970), 178–186.
30. G. Golub and C. Van Loan, *Matrix computations*, 3rd ed., Johns Hopkins University Press, 1996.
31. A. Grosberg and A. Khokhlov, *Statistical physics of macromolecules*, AIP Press, New York, 1994.
32. L. Han and N.M. Amato, *A kinematics-based probabilistic roadmap method for closed chain systems*, Workshop Algo. Found. Robot. (WAFR) (B.R. Donald, K. Lynch, and D. Rus, eds.), March 2000, pp. 233–246.
33. T.R. Ioerger and J.C. Sacchettini, *The textal system: Artificial intelligence techniques for automated protein model building.*, Methods in Enzymology. (San Diego), vol. 374, Academic Press, 2003, pp. 244–270.
34. T.A. Jones and M. Kjeldgaard, *Electron-density map interpretation*, Methods in Enzymology (San Diego), vol. 277, Academic Press, 1997, pp. 173–230.
35. T.A. Jones and S. Thirup, *Using known substructures in protein model-building and crystallography*, EMBO J. **5** (1986), 819–822.
36. T.A. Jones, J.-Y. Zou, and S.W. Cowtan, *Ring closure and local conformational deformations of chain molecules.*, Acta Cryst. **A47** (1991), 110–119.
37. L.E. Kavraki, P. Svestka, J.-C. Latombe, and M. Overmars, *Probabilistic roadmaps for path planning in high-dimensional configuration spaces*, IEEE Trans. Robot. Autom. **12** (1996), no. 4, 566–580.
38. O. Khatib, *A unified approach for motion and force control of robot manipulators: The operational space formulation*, Int. J. Robot. Autom. **RA-3** (1987), no. 1, 43–53.
39. O. Khatib, L. Sentis, J.-H. Park, and J. Warren, *Whole body dynamic behavior and control of human-like robots*, International Journal of Humanoid Robotics **1** (2004), no. 1, 29–43.
40. S. Kirkpatrick, C.D. Gelatt, and M.P. Vecchi, *Optimization by simulated annealing*, Science **220** (1983), no. 4598, 671–680.
41. R. Kolodny, L. Guibas, M. Levitt, and P. Koehl, *Inverse kinematics in biology: the protein loop closure problem*, Submitted to IJRR, 2004.
42. E. Krissinel, *Ccp4 coordinate library project*, 2004, http://www.ebi.ac.uk/ keb/cldoc/.
43. S. A. Lesley, P. Kuhn, A. Godzik, A. M. Deacon, I. Mathews, A. Kreusch, G. Spraggon, H. E. Klock, D. McMullan, T. Shin, and et. al., *Structural genomics of the thermotoga maritima proteome implemented in a high-throughput structure determination pipeline*, Proc. Nat. Acad. Sci. **99** (2002), no. 18, 11664–11669.
44. D.G. Levitt, *A new software routine that automates the fitting of protein x-ray crystallographic electron-density maps.*, Acta Cryst. **D57** (2001), 1013–1019.
45. Z. Li and H. Scheraga, *Monte Carlo-minimization approach to the multiple-minima problem in protein folding*, Proc. Natl. Acad. Sci. **84** (1987), no. 19, 6611–6615.
46. S.C Lovell, I.W. Davis, W.B. Arendall III, P.I.W de Bakker, J.M. Word, M.G. Prisant, J.S. Richardson, and D.C. Richardson, *Structure validation by Cα geometry: ϕ, ψ and Cβ deviation*, Proteins **50** (2003), no. 3, 437–450.
47. D. Manocha and J. canny, *Efficient inverse kinematics for general 6R manipulator*, IEEE Trans. Robot. Autom. **10** (1994), no. 5, 648–657.
48. D. Manocha and Y. Zhu, *Kinematic manipulation of molecular chains subject to rigid constraints.*, Proc. Int. Conf. Intell. Syst. Mol. Biol. **2** (1994), 285–293.
49. D. Manocha, Y. Zhu, and W. Wright, *Conformational analysis of molecular chains using nano-kinematics*, Comput. Appl. Biosci. **11** (1995), no. 1, 71–86.
50. G.J. McLachlan, D. Peel, K.E. Basford, and P. Adams, *The emmix software for the fitting of mixtures of normal and t-components*, J. Stat. Software **4** (1999), no. 2.
51. N. Metropolis, A. Rosenbluth, M Rosenbluth, A. Teller, and E. Teller, *Equation of state calculations by fast computing machines*, J. Chem. Phys. **21** (1953), 1087–1092.

52. J. Moult and M.N.G. James, *An algorithm for determining the conformation of polypeptide segments in protein by systematic search*, Proteins **1** (1986), 146–163.
53. G. N. Murshudov, A. A. Vagin, and E. J. Dodson, *Refinement of macromolecular structures by the maximum-likelihood method*, Acta Cryst. **D53** (1997), 240–255.
54. T.J. Oldfield, *A number of real-space torsion-angle refinement techniques for proteins, nucleic acids, ligands and solvent.*, Acta Cryst. **D57** (2001), 82–94.
55. A. Perrakis, T.K. Sixma, K.S. Wilson, and V.S. Lamzin, *wARP: Improvement and extension of crystallographic phases by weighted averaging of multiple-refined dummy atomic models.*, Acta Cryst. **D53** (1997), 448–455.
56. M. Raghavan and B. Roth, *Kinematic analysis of the 6r manipulator of general geometry*, Int. Symp. Robot. Res. (Tokyo), 1989, pp. 314–320.
57. P.S. Shenkin, D.L. Yarmush, R.M. Fine, H.J. Wang, and C. Levinthal, *Predicting antibody hypervariable loop conformation. 1. ensembles of random conformations for ring-like structure*, Biopolymers **26** (1987), 20532085.
58. T.C. Terwilliger, *Automated main-chain model-building by template-matching and iterative fragment extension.*, Acta Cryst. **D59** (2002), 34–44.
59. ———, *Improving macromolecular atomic models at moderate resolution by automated iterative model building, statistical density modification and refinement.*, Acta Cryst. **D59** (2003), 1174–1182.
60. J.C. Trinkle and R.J. Milgram, *Complete path planning for closed kinematic chains with spherical joints*, Int. J. Robot. Res. **21** (2002), no. 9, 773–789.
61. H. van den Bedem, I. Lotan, J.-C. Latombe, and A. M. Deacon, *Automated protein model completion: an inverse kinematics approach*, Acta. Cryst. D, Submitted.
62. H.W.T. van Vlijmen and M. Karplus, *PDB-based protein loop prediction: parameters for selection and methods for optimization*, J. Mol. Biol. **267** (1997), no. 4, 975–1001.
63. D. Waasmaier and A. Kirfel, *New analytical scattering-factor functions for free atoms and ions*, Acta Cryst. **A51** (1995), no. 3, 416–431.
64. L.T. Wang and C.C. Chen, *A combined optimization method for solving the inverse kinematics problem of mechanical manipulators.*, IEEE Trans. Robot. Autom. **7** (1991), 489–499.
65. W.J. Wedemeyer and H.A. Scheraga, *Exact analytical loop closure in proteins using polynomial equations.*, J. Comput. Chem. **20** (1999), 819–844.
66. D. Xie and N.M. Amato, *Kinematics-based probabilistic roadmap method for high dof closed chain systems*, IEEE Int. Conf. Robot. Autom. (ICRA), 2004, To appear.
67. J. Yakey, LaValle S.M., and L.E. Kavraki, *Randomized path planning for linkages with closed kinematics chains*, IEEE Trans. Robot. Autom. **17** (2001), no. 6, 951–959.
68. M. Zhang and L.E. Kavraki, *Finding solutions of the inverse kinematics problems in computer-aided drug design*, Tech. Report TR02-385, Rice University, 2002.
69. ———, *A new method for fast and accurate derivation of molecular conformations*, J. Chem. Info. Comp. Sci. **42** (2002), no. 1, 64–70.
70. Q. Zheng, R. Rosenfeld, S. Vajda, and C. DeLisi, *Loop closure via bond closure and relaxation*, J. Comput. Chem **14** (1992), 556–565.

A Machine Learning Approach for Feature-Sensitive Motion Planning

Marco Morales, Lydia Tapia, Roger Pearce, Samuel Rodriguez, and Nancy M. Amato

Parasol Laboratory, Dept. of Computer Science, Texas A&M University
Email: {marcom,ltapia,rap2317,sor8786,amato}@cs.tamu.edu

Abstract. Although there are many motion planning techniques, there is no method that outperforms all others for all problem instances. Rather, each technique has different strengths and weaknesses which makes it best-suited for certain types of problems. Moreover, since an environment can contain vastly different regions, there may not be a single planner that will perform well in all its regions. Ideally, one would use a suite of planners in concert and would solve the problem by applying the best-suited planner in each region.

In this paper, we propose an automated framework for feature-sensitive motion planning. We use a machine learning approach to characterize and partition C-space into regions that are well suited to one of the methods in our library of roadmap-based motion planners. After the best-suited method is applied in each region, the resulting region roadmaps are combined to form a roadmap of the entire planning space. Over a range of problems, we demonstrate that our simple prototype system reliably outperforms any of the planners on their own.

1 Introduction

The *motion planning (MP)* problem is to find a sequence of valid states to move a movable object (often called a robot) from an initial configuration to a goal configuration. In robotics path planning, valid states are typically collision-free robot configurations. Besides robotics, instances of the MP problem can be found in a range of applications from robotics and CAD to computational biology.

The MP problem is thought to be intractable except for robots with few *degrees of freedom (*DOF *)* as there is strong evidence that any complete planner will require exponential time in the number of DOFs of the robot [25,26,9]. The complexity of the MP problem has led researchers to investigate heuristic planners, initially based on cell decomposition [7] and potential fields [15]. After the introduction of the Randomized Path Planner (RPP) [2], research has focused on randomized methods. Notable examples include the roadmap-based probabilistic roadmap methods (PRMs) [14] and several tree-based methods that explore the planning space starting from one or two points [3,17,12]. These methods have achieved many remarkable results, including solving several previously unsolved MP problems.

M. Erdmann et al. (Eds.): Algorithmic Foundations of Robotics VI, STAR 17, pp. 361–376, 2005.

Although many good randomized techniques have been developed, no single method performs optimally for every MP problem. Rather, the performance of each technique depends on the problem instance. For example, some motion planning techniques are better suited for mostly free environments while others show their real strength in cluttered regions. Moreover, since an environment can contain vastly different regions, in many cases there may not even exist a single planner that will perform well in all its regions.

An ideal motion planner would be a meta-planner using a suite of more specialized planners to cooperatively solve an MP problem. It would automatically apply the best-suited planner for each distinct region in the planning space and would produce regional solutions that could be composed to solve the overall MP instance.

We propose an automated framework for feature-sensitive motion planning which uses a machine learning approach for *characterization and partitioning* of C-space. Partitioning is accomplished by recursively subdividing C-space until obtaining regions classified as homogeneous according to a set of features measured for each region. Next, each homogeneous region is matched with the best-suited planner in a library of roadmap-based planners and a regional roadmap is produced. Then, the regional roadmaps are combined to produce a roadmap of the entire planning space. We have implemented a simple prototype system that uses only basic techniques for partitioning C-space and combining regional roadmaps, and which contains only a few motion planning methods. Over a range of problems, we show that our simple prototype system reliably outperforms any of the planners on their own.

2 Preliminaries

A robot is a movable object that can be controlled through n parameters or *degrees of freedom (*DOF *)*, each corresponding to a unique component, e.g., object positions, object orientations, link angles, or link displacements. A robot's placement, or configuration, can be uniquely described by a point $(x_1, x_2, ..., x_n)$ in an n dimensional space (x_i being the ith DOF). This space, consisting of all robot configurations (feasible or not) is called *configuration space* (C-space) [18]. The subset of all feasible configurations is the *free C-space* (C-free), while the union of the unfeasible configurations is the *blocked C-space* (*C-obstacle*). Thus, the MP problem becomes that of finding a continuous trajectory for a point in C-space connecting the start and the goal configurations that completely lies in C-free. And, although it is intractable to compute C-space, we can often determine whether a configuration is feasible or not quite efficiently, e.g., by performing a collision detection test in the *workspace*, the robot's natural space.

Randomized Motion Planning Algorithms Randomized motion planners randomly explore C-space and produce a data structure containing feasible configurations and some information about the connectivity of C-free.

Roadmap-based PRMs [14,22,23] build a graph called a *roadmap* in two steps: *node generation* and *node connection.* In node generation, feasible configurations are generated. In node connection, simple *local planners* identify which pairs of selected nodes are connectable and should be noted at roadmap edges. Many PRM-variants exist. Basic PRM [14] uniformly samples nodes. Other sampling methods increase the proportion of feasible configurations in some regions [21,4,27,17]. For example, OBPRM [1] generates configurations on or near C-obstacle surfaces and Gaussian PRM [5] only retains free samples if they are close to C-obstacles. The performance of these methods varies in different kinds of environments. For example, Basic PRM performs well in open areas while OBPRM performs better in cluttered regions.

Roadmap components can be connected using (combinations of) roadmap-based and tree-based planners [20]. Tree-based planners grow a tree starting with a known feasible configuration and add new nodes to the tree according to some strategy that either makes the tree approach the goal or uniformly cover the space reachable from the nodes already in the tree. Among these techniques we note RPP [2], Ariadne's Clew Algorithm [3], RRTs [17], and a planner for expansive spaces [12].

C-space Subdivision Some of the early heuristic motion planners [7] performed uniform C-space subdivision by dividing each of the n DOFs into m equally-sized pieces. The result is a grid of m^n cells which is only feasible to compute for problems with few DOFs .

Deterministic sampling over C-space grids has been studied in [16]. A different approach using randomized methods, non-uniform C-space subdivision, has been explored in two stage planners. Decomposition-based Motion Planning [6] decomposes a high-dimensional problem into subproblems with reduced complexity for real-time planning. It searches for homotopic paths in the workspace, effectively subdividing the first three parameters of C-space while ignoring higher DOFs . Once a workspace path is found, it determines the motions of the joints (the higher DOFs).

We believe that C-space subdivision can be used to identify regions suitable for a particular motion planner. To be general and scalable for high DOF problems, the subdivision cannot be exhaustive or uniform. Instead, it should be feature-sensitive so that subdivision is performed only until the resulting regions have "uniform" features.

C-space Characterization Some basic studies of the performance of planners in different types of environments have been performed. A study of several PRM planners [11] in five distinct environments offered criteria for deciding which PRM variant to use. An iterative metaplanner based on region characterization is proposed in [10]. In a first iteration, four Basic PRM roadmap features help in determining whether the space is free or cluttered or if a surface is rough. Then, in a second iteration, more powerful methods could be applied. Finally, filters would remove redundant nodes to keep the remaining as seeds for tree-based planners. Environment subdivision is suggested, but no mechanism for performing it is proposed. As evidence that automatic environment classification is feasible, several features are measured and their values are shown for several different types of environments (e.g., free vs. cluttered, and rough vs. smooth).

Characterizing C-space regions requires constructing a representation for complex and high dimensional data. Machine learning methods have gained attention over the past few years because of their ability to interpret patterns directly from such data. For example, determining the three-dimensional structure of a protein given only knowledge of the sequence of a protein is a difficult problem addressed by many bioinformatics groups. Some progress has been made in this area using knowledge obtained by a learning method about the sequence to structure relationship [8].

Learning methods are also attractive when the scope of data analysis is too complex for a human (as in C-space characterization). For example, the FIRST dataset is a collection of images of radio sources collected by the Very Large Array. In six years this dataset grew to consist of 32,000 images of two million pixels each. For this difficult problem, the decision tree algorithm C4.5 [24] was trained to identify galaxies with the bent-double morphology. The decision tree was able to classify bent-doubles and find galaxies that had been overlooked during manual searches [13].

3 Feature-Sensitive MP Framework

We propose a framework for mapping the connectivity of the C-space for motion planning problems through the coordinated operation of a library of randomized planning methods. Easy regions are recognized and handled quickly while difficult regions are treated with more powerful planners. This framework, sketched in Fig. 1, subdivides the C-space of a problem into a set of overlapping homogeneous regions. It then produces a roadmap in each region using the best-suited planner as determined by the regional features. Finally, the regional roadmaps are integrated to form a roadmap for the general problem.

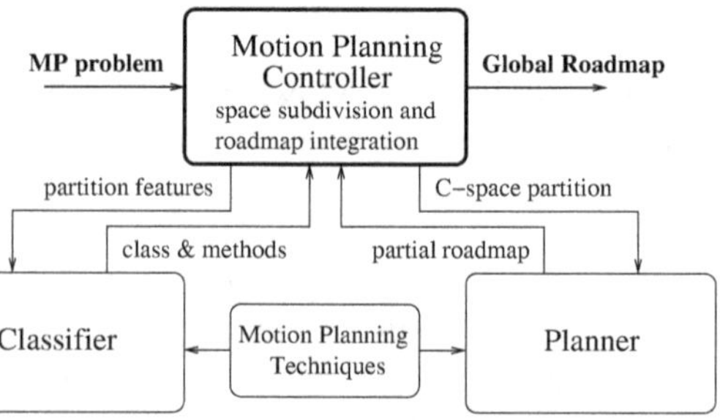

Fig. 1. Block diagram of the framework

At the core of this framework is a *motion planning controller* that recursively subdivides the environment and creates a tree representing the subdivision. Tree nodes represent C-space regions and the children of a node correspond to the regions that result from the subdivision of their parent node. Each node stores features extracted from the C-space of its associated region. These features are fed to the *classifier* that matches them to a class and a set of *motion planning techniques*. Based on this output, the size of the subdivision, and the granularity of the subdivisions already performed, the motion planning controller decides whether to further subdivide the current C-space region or mark it as a leaf node. Leaf node regions are sent to the *Planner* which produces a regional roadmap with the assigned motion planner. Then, regional roadmaps corresponding to the children of intermediate nodes are merged into a roadmap representing the parent node's region, so that larger and larger regions are integrated until the global roadmap is obtained in the root node.

Features should help to describe the complexity of the C-space region to be used in characterization and subdivision. They should not depend on the DOFs of the problem nor on the volume of the C-space of the partition. Also, feature extraction should use only a fraction of the resources needed to generate the roadmap.

Subdivisions are made to ease classification and to improve the chances of using the proper method efficiently, and they should be stopped once a good classification is made. Our approach differs from cell decomposition in that we do not perform uniform subdivision, instead we control the subdivision of C-space by the needs of the classification. Also, regions are allowed to overlap; indeed, some overlap may assist in "stitching" together regional roadmaps.

C-space subdivision can be obtained with different methods. Ideally, region boundaries should be decided so that resulting subregions are more homogeneous than the parent region. Information theory, such as that used by the machine learning algorithm C4.5 [24] for creating trees, gives interesting avenues for future exploration. Also, a component based partitioning method might obtain more homogeneous subdivisions.

We can characterize C-space to match it to some environment type as in [10], or to match it to the best-suited planner directly. The former requires studying the features of representative instances of the MP problem that match recognizable classes. The latter is the ideal case, it requires studying the performance of available planners on representative instances of the MP problem. Our prototype system presented here studies the former case, but we are also investigating the latter case.

Roadmap integration should use nodes, if available, in the overlapping portion of the respective regions as starting points for connecting graph components of different subregions with tree-based and roadmap-based planners as in [20].

4 Feature-Sensitive MP Framework Modules

In this section we describe the current implementation of the modules of our prototype system. Parts of the framework, such as the machine learning algorithm or the subdivision strategy, can be implemented in different ways. Our implementation, as described here, covers the basic components of the framework. We also discuss how we intend to further enhance the system.

4.1 Motion Planning Controller (Subdivision and Integration)

As described in Sect. 3, the motion planning controller creates a tree of C-space subdivisions of the MP instance, builds regional roadmaps in each subdivision, and then integrates them into a global map. First, Algorithm 1 gathers features from the C-space of the entire MP instance and the classifier uses these features to find a matching class (Line 1). Then, the environment and the matching class are sent to the recursive Algorithm 2 which creates a tree of classified C-space subdivisions (Line 2). It uses the selected planning method in each homogeneous region (as defined in Sect. 4.2) and it integrates regional roadmaps into a general roadmap covering the entire C-space. Finally, if Algorithm 2 fails to produce a roadmap, the planner makes one with a default roadmap-based method (Lines 3-4).

Algorithm 1 Motion Planning controller

Require: Environment $E : (robot, obstacles)$, subdivision limit L
Ensure: $Tree.roadmap$: a C-space Roadmap for E

```
1: E.class = Classifier.match(getFeatures(E))
2: Tree = Branch(E,class,L,0)
3: if Tree.roadmap == "noroadmap" then
4:    Tree.roadmap = Planner.makeRoadmap(E)
5: end if
```

Algorithm 2, Branch, has three main stages: C-space subdivision (Line 1), child generation (Lines 3-11), and regional roadmap generation and integration (Lines 12-21). This algorithm returns a regional roadmap for the region of C-space covered by the node; a roadmap will be generated in every subdivision found.

Subdivision of C-space Regions In the first stage, the C-space of the current node is subdivided into two or more overlapping *subdivisions* (Line 1). If the node is too small or the tree has reached its maximum height it is not subdivided and a leaf with no roadmap is returned (Lines 1-2 and 22-23).

In this implementation we have an elementary partitioning strategy that randomly chooses one of the parameters of the robot to partition. A random

Algorithm 2 Branch

```
Require: C-space region D, class C, subdivision limit L, node height H
Ensure: node = [region, class, roadmap]
 1: (subdivisions, error) = subdivide(D,L)
 2: if no error then
 3:   for all region_i in subdivisions do
 4:     class_i = Classifier.match(getFeatures(region_i))
 5:     if MeetLeafCriteria(region_i,class_i,C,L,H) then
 6:       child_i = [region_i, C, "noroadmap"]
 7:     else
 8:       child_i = Branch(region_i,class_i, L, H+1)
 9:     end if
10:     add child_i to children
11:   end for
12:   if MeetClassMergingCriteria(subdivisions,C) then
13:     node = [D, MergeClasses(subdivisions,C), "noroadmap"]
14:   else
15:     for all child_i in children do
16:       if child_i.roadmap == "noroadmap" then
17:         child_i.roadmap = Planner.makeRoadmap(child_i)
18:       end if
19:     end for
20:     node = [D, MergeClasses(children), Planner.mergeRoadmaps(children)]
21:   end if
22: else
23:   node = [D, class, "noroadmap"]
24: end if
25: return node
```

point within the range of the chosen dimension is selected, a small ε (around 10% of the range) is used to divide it into two overlapping ranges. Region overlap is important for roadmap integration as discussed in Sect. 4.3. In our prototype system, we only subdivided on the first positional parameters.

Children Generation During the second stage, in each C-space region, features are extracted and fed to the classifier that matches them to a type of environment (Line 4). The region is checked to determine whether it should represent a leaf node (Lines 5-6) or a branch (Lines 7-8). In any case, the result is a child for each subdivision of the current node (Line 10).

Features should capture characteristics of C-space that will give insight into the best planning method for the corresponding region. We create a uniform and quick Basic PRM roadmap from which we extract several features. In addition to giving some insight into C-space structure, features extracted this way will also depend on resolution and sampling strategy. These factors should be taken into account during classification. Our prototype classifier currently recognizes C-space regions as free, cluttered, narrow passage, and non-homogenous. This along with the features used for classification will be discussed in Sect. 4.2.

A region *meets leaf criteria* (Line 5) when it has reached its limit in size so that there is no room to vary any robot parameter, or when its class is the same as that of the parent whose classification is other than non-homogeneous. If these criteria are not met, the region is branched with a recursive call to Algorithm 2.

Regional Roadmap Generation and Integration In the last stage, subdivisions are evaluated to determine whether they should be merged back into their parent without making a roadmap (Lines 12-13) or roadmaps should be produced for leaf subregions through the planner (Lines 15-19) to integrate regional roadmaps into the parent node (Line 20).

Subdivisions *meet class merging criteria* (Line 12) when all of them are leaf nodes belonging to the same class and their parent is non-homogeneous. This means that the parent could not be classified succesfully, but the children were. Since all of them belong to the same class, their regions are merged back into their parent and assigned to its children's class. This prevents having to spend extra effort in merging regional roadmaps. However, in some types of C-space or for certain planners, it may be better to not merge those subregions; this is a topic for future study.

When the subdivisions do not meet class merging criteria the planner is called to make a roadmap for each leaf child (which does not have one yet) and to merge it's children's regional roadmaps into a single roadmap for the current node. Hence, the final roadmap obtained for the root of the tree is a global roadmap. The planner is described in Sect. 4.3.

4.2 Classifier

The classifier is given a set of features of a C-space partition, and it outputs a matching class and a confidence level. It is hard to derive analytical expressions defining classes of C-space or rankings for choosing good planners for several reasons: many features that only represent a fraction of C-space can be extracted; there are many MP strategies to map C-space; and the range of MP problems is too diverse in obstacle placement and robot dimensionality. Machine learning methods can be used to perform the classification. For our particular needs of understandable decisions, discrete classifications, and ease of use, decision tree learning methods are appropriate.

Several measurements are made in a small roadmap to describe the complexity of the partition. In our prototype system we used Basic PRM to create the feature roadmap; our future work will investigate other methods as well.

Roadmap Features Our objective is a general method that can classify a wide range of environment types and complexities. In our prototype system,

Table 1. Features for roadmap $G : (V, E)$, vertices $v \in V$, edges $e(v_j, v_k) \in E$, $CC \equiv \{components \in G\}$, $C_M(cc_k) = \sum v_i/\|cc_k\| \forall v_i \in cc_k$, $E_l = \sum \|e_j\| \forall e_j \in E$

General Features

Free node ratio: $V_r = \|V\|/V_{att}$	$V_{att} \equiv$ number of attempted nodes	(1)
Components: $C_n = \|CC\|$		(2)
Edge ratio: $E_r = \|E\|/E_{att}$	$E_{att} \equiv$ number of attempted edges	(3)

Component-Based Features for connected component cc_k, 4 features per definition[a]

Node distance: $d(v_i, v_j) = \|v_i - v_j\|$	$v_i, v_j \in cc_k, \quad i \neq j$	(4)
Edge length: $E_l = d(v_i, v_j)$	$(v_i, v_j) \in E, \quad v_i, v_j \in cc_k$	(5)
Distance to Center: $d_c = d(v_i, C_M(cc_k))$	$v_i \in E$	(6)
Component distance: $D(cc_i, cc_j) = d(C_M(cc_i), C_M(cc_j))$		(7)
Component length: $C_l = \sum \|e_i\|/E_l$	$\forall e_i \in cc_k$	(8)

[a] Average of the maximum, minimum, mean, and std. deviation over all $cc_k \in CC$
Notation: $|\cdot| \equiv$ cardinality of $\cdot$, $\|\cdot\| \equiv$ length of $\cdot$.

we extract a set of 23 features; ideally we want to capture the characteristics of each type of space while not being impacted by size and complexity of C-space (Table 1). Specifically, these features should be independent of the dimensionality and volume of the C-space region analyzed.

Three *General features* capture an overall picture of roadmap generation. A *Free node ratio* gives insight into the "freeness" of C-space. In partitions with no obstacles, this ratio will be one whereas in regions filled with obstacles it will be close to zero. *Components* gives an idea of the connectivity of C-space. A count of one would indicate easy connectability. In highly cluttered areas this count will approach the number of nodes sampled. An *Edge ratio* will be low when nodes are difficult to connect such as in cluttered regions, close to the surface of complex C-obstacles, and in narrow passages.

Twenty *Component-based features* are related to the complexity, extension, and coverage of components that map C-free, and the space between them. Each description corresponds to four features: its maximum, minimum, mean, and standard deviation in each component, averaged over the number of components.

Component-based features describe the connectivity of C-free and point to interesting areas of the problem. For example, *Node distance* gives an idea of the distance between nodes in a component. Because packed components will have smaller values than components covering elongated areas, node distance might be important in distinguishing between a cluttered space and a narrow passageway. While the features of *Edge length* and *Distance to center* give an idea of the span and coverage within a single component, *Component distance* contributes a measure of the distance between components that is related to the amount of infeasible space. The last feature, *Component length*, gives an idea of the volume covered by components. This particular feature is normalized by dividing it by the length of the entire roadmap.

Features are normalized to make them independent of C-space dimensionality and size. With the exception of *Component length*, that is already normalized, each feature is divided by a constant

$$k_{\mathrm{n}} = \max |dimension_i| * |dof|, \quad 1 \le i \le |dof| \tag{9}$$

where $|dimension_i|$ is the length of dimension i, so to relate k_{n} to the volume of C-space covered in the sampling. It may be needed to relate the number of sampled nodes for feature collection to k_{n} to keep it small, but large enough to capture features that properly classify the corresponding C-space region.

C-space Classes While MP environments are often non-homogeneous, if partitioned properly, they can be viewed as composed of homogeneous subcomponents. We currently use four classes of C-space that we plan to extend in the future:

- *Free:* very few obstacles in the region.
- *Cluttered:* obstacles clutter the region.
- *Narrow passage:* regions with passages between other types of region.
- *Non-homogeneous:* regions that do not fit in any other class.

Machine Learning Techniques There are many learning algorithms that work well for discrete-valued classification. From our learning method we need an algorithm that provides us with understandable decisions for verification, relatively quick learning time, and ease of use. For these needs, we have selected the decision tree method, C4.5 [24].

C4.5 creates a series of decisions based on feature values that leads to a classification of an example. The decision nodes or branches in the trees are selected by choosing the feature that "best" splits the space. Every path along the tree leads to a leaf node or classification based on the set of decisions that lead to that leaf. After a tree is constructed, C4.5 removes paths that do not apply to a general portion of the data through a process called *pruning*.

Classifier training. The classifier is trained with examples of free, cluttered, narrow passage, and non-homogeneous environments. Features are extracted, a decision tree is trained, ordered rules are composed from this tree, and the performance evaluated. Because of our small training set, evaluation is performed k times, each time using a different subset of the data for training and evaluation. This process of k-fold cross-validation [19], gives a better idea of how the method is working on average. Although the current set of classification rules do not take into account all of the features we examined, these features might be utilized for other environment types such as isolated or blocked spaces.

4.3 Planner and Pool of Motion Planning Methods

The planner works as a regular motion planner with a collection of roadmap-based and tree-based planners. It can be called to either make a roadmap for a C-space region or to merge a set of regional roadmaps.

When making a roadmap for a region, it receives the environment, region boundaries, region type as defined by the classifier, and the roadmap used for extracting features in that region. It starts with the feature roadmap to which it applies planners well-suited for that roadmap class. The decision of which planners to use in each class calls for a deeper study as in [11]. In this work, we assign planners to C-space type based on recommendations in [10] and on the strengths of some methods as described by their authors. In future work we will add more planners to our library and will investigate automated methods to match planners to regions. In free regions we do not do additional planning. In narrow passages we apply OBPRM (normals) with about ten times as many nodes as in the feature roadmap. In clutter regions we apply OBPRM (triangles) with about five times as many feature roadmap nodes. In non-homogeneous regions we apply Basic PRM followed of OBPRM (triangles), each method generating as many nodes as in the feature roadmap. In all regions we use a straight-line local planner and a k-closest connecting strategy with $k = 5$.

When merging regional roadmaps into their parent region the planner "stitches" them by attempting to connect subregion components focusing on their overlapping sections. This can be done with roadmap-based and tree-based methods as in [20] starting from the nodes in the overlap. But for simplicity, we applied a component connection and a k-closest connection. By construction, Algorithm 2 merges only neighboring regions. Our subdivision ensures that regions overlap by small ε. Again, we add the resulting roadmap to the feature roadmap of the parent region.

More investigation is needed to make the classifier learn also about the motion planning methods so that both methods and parameters can be automatically selected and assigned to each partition.

5 Experiments

We tested our implementation on an Intel Pentium 4 2.8 GHz processor with 512 MB of RAM and 512 kB of cache, running Linux 2.4.20-9 OS. We used our group's C++ motion planning library compiled with gcc 3.2.2 and Perl to control the partitioning. MySQL was used for storing features for training.

Training the Classifier We designed a set of 200 environments: 50 free, 50 cluttered, 50 narrow passages, and 50 non-homogeneous. Features were extracted from each environment and saved in a database from which we obtained data to train decision trees. Ten different sets of trees were created using 180 examples for training and a distinct set of 20 examples for testing.

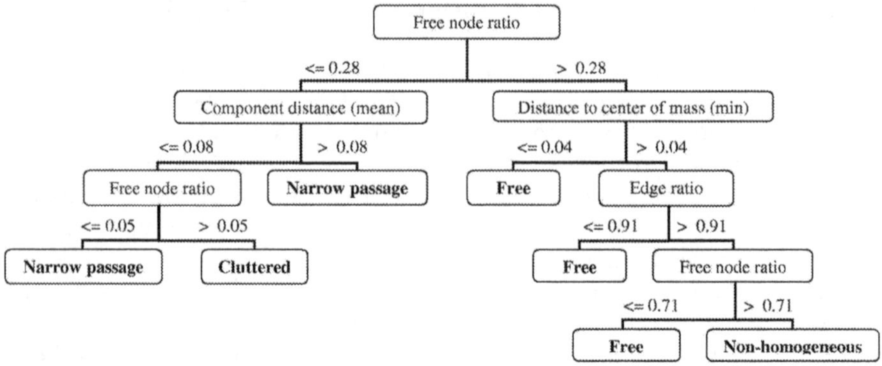

Fig. 2. Decision tree selected for classification

Accuracy rates produced by these cross-validation tests ranged from 100% to 85% with an average testing accuracy of 95%. We selected the tree that was able to classify 98.9% of our training and 100% of our testing examples to use for our classification (Fig. 2).

Test Environments For testing, we chose four environments with different kinds of obstacles and robots. First, a house environment (Fig. 3) that has three rooms with some furniture. The robot is a table to be moved from outside the house to the smallest room. It has different kinds of regions: free outside the house, cluttered close to the furniture, and a narrow passage must be traversed to reach the last room. Second, the walls-rigid environment (Fig. 4) has 5 chambers connected by varying-size openings. Three of the chambers are filled with small obstacles and the other two are empty. The robot is a stick that must traverse all the chambers. Third, the walls-serial environment (Fig. 5) is similar to the walls-rigid, but all the chambers are free of obstacles. The robot, however, is articulated with two long links, making it more difficult to maneuver. Last, the narrow-clutter environment (Fig. 6) is an open area with a big obstacle that can be traversed through a narrow passage followed by some open space and a clutter of small obstacles. The robot is articulated with 4 small links. It must traverse the entire environment including a narrow passage and a highly cluttered area.

Experimental Design We applied Basic PRM, OBPRM, and our framework (C&P, for classify and partition) to the environments to find the minimum number of nodes needed to solve the problem. The minimum number of nodes was found by a binary search, and it was repeated 4 times for each environment and method pair.

Basic PRM employs uniform sampling and k-closest connection. OBPRM generates nodes with normal alignment and k-unconnected-closest connection (a variant of the standard k-closest which only attempts connections to

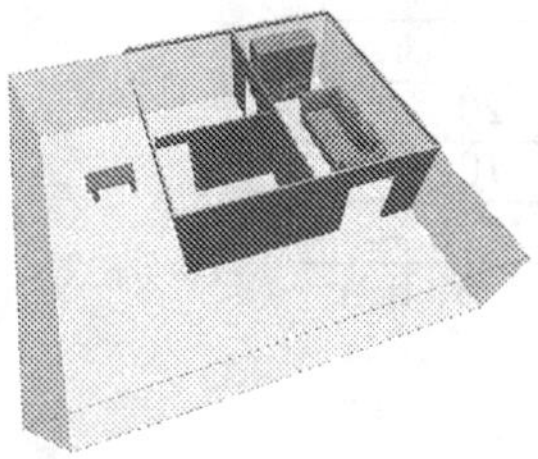

Method	Nodes$_{act/att}$	#CDs	#CCs	solved?
Basic PRM	5798/5798	1,385,004	9	yes
OBPRM	489/491	3,562,679	20	yes
C&P:total	226/375	724,861	3	yes
partitioning	102/225	90,917	*29	n/a
roadmap gen.	124/150	326,076	30	n/a
roadmap int.	0/0	307868	3	yes

* 1 narrow, 4 free, many non-homogeneous

Fig. 3. House environment (6 DOF) and performance metrics

Method	Nodes$_{act/att}$	#CDs	#CCs	solved?
Basic PRM	3000/3000	2,422,749	11	yes
OBPRM	1200/1179	11,107,232	48	yes
C&P:total	257/420	7,635,344	10	yes
partitioning	143/285	164,302	*35	n/a
roadmap gen.	114/135	72,979	30	n/a
roadmap int.	0/0	7,398,063	10	yes

* 1 narrow, 3 free, many non-homogeneous

Fig. 4. Rigid Body Walled environment (6 DOF) and performance metrics

different components). C&P's planner is used as described in Sect. 4.3. All experiments use the straight-line and rotate-at-s 0.5 local planners.

Figures 3, 4, 5, and 6 present results of representative runs. We show the number of nodes generated (act), attempted (att), the number of collision detection calls (#CDs), the number of connected components (#CCs), and we indicate whether the query was solved.

We can see the benefit of our framework in Fig. 3. The number of collision detection calls to solve the house environment using C&P is far smaller than when using Basic PRM and OBPRM. Also, fewer nodes are needed to solve the problem with the C&P framework than with individual methods.

The house environment is classified as a whole as a non-homogeneous environment. When partitioned and classified, useful partitions are found. When the narrow partition was identified, our feature-sensitive framework was able to focus many more OBPRM nodes in that difficult area. Similarly, when the free partitions were found, simple planners were used.

In the Walls Rigid environment (Fig. 4) OBPRM solves the query with far fewer nodes than Basic PRM, but requiring many more collision detection calls. C&P also needs many more collision detection calls than Basic PRM, but it solves the query with only 257 nodes, considerably less than the nodes needed by OBPRM and Basic PRM. During characterization, cluttered and narrow areas of the environment are also discovered as well as free areas that can be considered as already well-mapped.

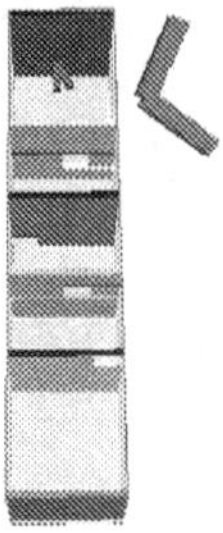

Method	Nodes$_{act/att}$	#CDs	#CCs	solved?
Basic PRM	17801/17801	3,183,940	20	no
OBPRM	816/821	2,830,528	10	yes
C&P:total	590/795	1,460,845	2	yes
partitioning	168/345	67,695	*35	n/a
roadmap gen.	422/450	81,263	145	n/a
roadmap int.	0/0	1,311,887	2	yes

* 2 narrow, 5 free, many non-homogeneous

Fig. 5. Serial Walls (2-link) environment (7 DOF) and performance metrics

Method	Nodes$_{act/att}$	#CDs	#CCs	solved?
Basic PRM	19461/19461	40,056,265	110	no
OBPRM	350/350	23,142,065	49	yes
C&P:total	271/435	5,966,310	60	yes
partitioning	68/225	1,017,267	*29	n/a
roadmap gen.	203/210	2,129,146	143	n/a
roadmap int.	0/0	2,819,897	60	yes

* 2 narrow, 1 cluttered, many non-homogeneous

Fig. 6. Narrow-Clutter environment (9 DOF) and performance metrics

As can be seen in Fig. 5, the Serial Walls environment is quickly solved by C&P. Significant narrow regions are identified and C&P thus focus on those difficult areas. It is important to note that the robot in this environment was articulated and narrow passages are still identified.

C&P shows its strength in the Narrow-Clutter environment (Fig. 6). After the entire environment is classified as non-homogeneous, the planner focuses on the second half of the environment around the narrow passage. The concentrated effort on the narrow passage allows C&P to use considerably fewer collision detection calls than OBPRM to map the narrow region. Basic PRM was never able to find a solution for the narrow region in the alloted number of nodes (20,000).

As mentioned before, C&P joins neighboring subdivisions that belong to the same class. This allows it to focus on entire regions of the same type and prevents oversampling. In general, C&P is able to solve the environments with fewer collision detection calls and nodes than Basic PRM and OBPRM. This is due in large part to its ability to identify difficult regions and focus on these areas. Also, free areas can be discovered and mapped without over-sampling.

6 Conclusions

We have proposed an automated framework for feature-sensitive motion planning. A prototype implementation of our machine learning classification and partitioning approach was shown to achieve results superior to those obtainable by any of the individual planners on their own.

Throughout this paper, we have discussed possible improvements and future directions for our framework. For example, we plan to extend the current framework to recognize and handle more C-space classes such as blocked and isolated regions. Another area of focus is the exploration of partitioning techniques. We suggested two ideas based on information theory and component based partitioning. Throughout the improvement process, we plan on extending our library of motion planning methods and continuing to develop methods of selecting an appropriate planner and its parameters based on previous experience.

Acknowledgement

This research supported in part by NSF Grants ACI-9872126, EIA-9975018, EIA-0103742, EIA-9805823, ACR-0113971, CCR-0113974, EIA-9810937, EIA-0079874, and by the Texas Higher Education Coordinating Board grant ATP-000512-0261-2001. Morales supported in part by a Fulbright/Garcia Robles (CONACYT) Fellowship. Rodriguez supported in part by a LSAMP Bridge to Doctorate Fellowship.

References

1. N. M. Amato, O. B. Bayazit, L. K. Dale, C. V. Jones, and D. Vallejo. OBPRM: An obstacle-based PRM for 3D workspaces. In *Robotics: The Algorithmic Perspective*, pages 155–168, Natick, MA, 1998. A.K. Peters. Proceedings of the Third Workshop on the Algorithmic Foundations of Robotics (WAFR), Houston, TX, 1998.
2. J. Barraquand and J. C. Latombe. Robot motion planning: A distributed representation approach. *Int. J. Robot. Res.*, 10(6):628–649, 1991.
3. P. Bessiere, J. M. Ahuactzin, E. G. Talbi, and E. Mazer. The Ariadne's clew algorithm: Global planning with local methods. In *Proc. IEEE Int. Conf. Intel. Rob. Syst. (IROS)*, volume 2, pages 1373–1380, 1993.
4. R. Bohlin and L. E. Kavraki. Path planning using Lazy PRM. In *Proc. IEEE Int. Conf. Robot. Autom. (ICRA)*, pages 521–528, 2000.
5. V. Boor, M. H. Overmars, and A. F. van der Stappen. The Gaussian sampling strategy for probabilistic roadmap planners. In *Proc. IEEE Int. Conf. Robot. Autom. (ICRA)*, volume 2, pages 1018–1023, 1999.
6. O. Brock and L.E. Kavraki. Decomposition-based motion planning: A framework for real-time motion planning in high-dimensional configuration places. In *Proc. IEEE Int. Conf. Robot. Autom. (ICRA)*, pages 1469–1475, 2001.
7. R. A. Brooks and T. Lozano-Pérez. A subdivision algorithm in configuration space for findpath with rotation. In *Proc. Int. Conf. Artif. Intel.*, pages 799–806, 1983.
8. C. Bystroff and D. Baker. Prediction of local structure in proteins using a library of sequence-structure motifs. In *J. Mol. Biol.*, volume 281, pages 565–577, 1998.

9. J. F. Canny. *The Complexity of Robot Motion Planning.* MIT Press, Cambridge, MA, 1988.
10. L. K. Dale and N. M. Amato. Probabilistic roadmaps – putting it all together. In *Proc. IEEE Int. Conf. Robot. Autom. (ICRA)*, pages 1940–1947, 2001.
11. R. Geraerts and M. H. Overmars. A comparative study of probabilistic roadmap planners. In *Proc. Int. Workshop on Algorithmic Foundations of Robotics (WAFR)*, December 2002.
12. D. Hsu, R. Kindel, J. C. Latombe, and S. Rock. Randomized kinodynamic motion planning with moving obstacles. In *Proc. Int. Workshop on Algorithmic Foundations of Robotics (WAFR)*, pages SA1–SA18, 2000.
13. C. Kamath, E. Cantu-Paz, I. Fodor, and N. Tang. Classifying bent-double galaxies. In *IEEE Comput. in Sci. & Eng.*, volume 4, pages 52–60, 2002.
14. L. E. Kavraki, P. Svestka, J. C. Latombe, and M. H. Overmars. Probabilistic roadmaps for path planning in high-dimensional configuration spaces. *IEEE Trans. Robot. Automat.*, 12(4):566–580, August 1996.
15. O. Khatib. Real–time obstacle avoidance for manipulators and mobile robots. *Int. J. Robot. Res.*, 5(1):90–98, 1986.
16. S. M. LaValle and M. S. Branicky. On the relationship between classical grid search and probabilistic roadmaps. In *Proc. Int. Workshop on Algorithmic Foundations of Robotics (WAFR)*, December 2002.
17. S. M. LaValle and J. J. Kuffner. Randomized kinodynamic planning. In *Proc. IEEE Int. Conf. Robot. Autom. (ICRA)*, pages 473–479, 1999.
18. T. Lozano-Pérez and M. A. Wesley. An algorithm for planning collision-free paths among polyhedral obstacles. *Communications of the ACM*, 22(10):560–570, October 1979.
19. T. Mitchell. *Machine Learning.* McGraw-Hill, 1997.
20. M. Morales, S. Rodriguez, and N. M. Amato. Improving the connectivity of PRM roadmaps. In *Proc. IEEE Int. Conf. Robot. Autom. (ICRA)*, volume 3, pages 4427–4432, September 2003.
21. C. L. Nielsen and L. E. Kavraki. A two level fuzzy PRM for manipulation planning. Technical Report TR2000-365, Computer Science, Rice University, Houston, TX, 2000.
22. M. Overmars. A random approach to path planning. Technical Report RUU-CS-92-32, Computer Science, Utrecht University, The Netherlands, 1992.
23. M. Overmars and P. Svestka. A probabilistic learning approach to motion planning. In *Proc. Int. Workshop on Algorithmic Foundations of Robotics (WAFR)*, pages 19–37, 1994.
24. J.R. Quinlan. *C4.5: Programs for Machine Learning.* Morgan Kaufmann, 1993.
25. J. H. Reif. Complexity of the mover's problem and generalizations. In *Proc. IEEE Symp. Foundations of Computer Science (FOCS)*, pages 421–427, San Juan, Puerto Rico, October 1979.
26. J. T. Schwartz and M. Sharir. On the "piano movers" problem II: General techniques for computing topological properties of real algebraic manifolds. *Adv. Appl. Math.*, 4:298–351, 1983.
27. S. A. Wilmarth, N. M. Amato, and P. F. Stiller. MAPRM: A probabilistic roadmap planner with sampling on the medial axis of the free space. In *Proc. IEEE Int. Conf. Robot. Autom. (ICRA)*, volume 2, pages 1024–1031, 1999.

Probik: Protein Backbone Motion by Inverse Kinematics

Kimberly Noonan[1], David O'Brien[1], and Jack Snoeyink[1]

University of North Carolina
Chapel Hill NC 27599, USA

Summary. To investigate the parameters of the protein design problem that we are exploring in collaboration with biochemists, we developed a tool that uses inverse kinematics to support moving small fragments of protein backbone, while respecting biochemists' desires to "remain in favorable regions of the Ramachandran plot" and "preserve ideal geometry." By presenting estimates of derivatives in response to motion, we are able to refine these qualitative desires as we work with our collaborators. We then explore low dimensional bases to parameterize the space of backbone motions.

1 Introduction

Proteins are fascinating polymers. Their various structures at the molecular level create the different macroscopic properties like that in hair and skin or shell and bone. Protein structure is a critical determinant of their function as recognizers, cleavers, splicers, motors, pumps, and many more. Many of the models of proteins (hard sphere, kinematic chain with rotatable bonds, HP-lattice; see [9]) are very geometric, so it is no surprise that computer scientists and roboticists have been tackling problems such as protein folding [18, 22], docking [19], rigidity and flexibility [20], and computing volume and surface area [5] and normal modes [1, 4, 8, 20].

Because the biological sciences have a history of being descriptive, there is a gap to be bridged when trying to apply mathematical and engineering techniques to their models. Biologists tell stories about their favorite proteins, reactions, or mechanisms. One consequence is that their models are often overly simplified—they are chosen as good verbal descriptors or analogies, and not for their mathematical completeness. When working with experienced biologists, there are always unstated qualifiers or exceptions: backbone bonds are rigid, except that all the atoms are in motion; peptide bonds are planar, plus or minus 5 degrees; atom interpenetration doesn't happen, except when forced by strain or improper modeling in some other area of the protein. The

M. Erdmann et al. (Eds.): Algorithmic Foundations of Robotics VI, STAR 17, pp. 377–392, 2005.

only way to bridge this gap is by close collaboration that makes the biologist part of the solution.

In this paper, we bridge these gaps to apply inverse kinematics to the problem of manipulating protein backbones in the context of protein design. After a brief introduction to protein geometry and protein design (Sec. 2), we review exact inverse kinematics solutions for 6 dof rotational manipulators (Sec. 3). Our contribution (Sec. 4) is to apply exact inverse kinematics to backbone manipulation because it allows us to communicate not just a single solution, but families and spaces of solutions to our biochemistry collaborators. We can then learn why they would consider one family of solutions preferable to another, and refine our models and solution criteria. The final validation will have to come from new biological hypotheses under test, which we hope will be published in the biological literature (even though the mathematics and computation will be barely sketched in the "Methods" section – another gap). Since we believe that our methods that fill gaps have merit, we use this paper to describe the computational steps in our tool, probik. Finally, (Sec. 5) we explore how few parameters are sufficient to cover the space of motions for backbone design tasks.

2 Background: Protein Geometry and Protein Design

In this section, we provide basic definitions and terminology of protein geometry and discuss the ideal versus observed geometric parameters. We also provide an overview of the protein design process discussing protein function, the existence of binding pockets, and the computational tool Dezymer, from the Hellinga lab [10].

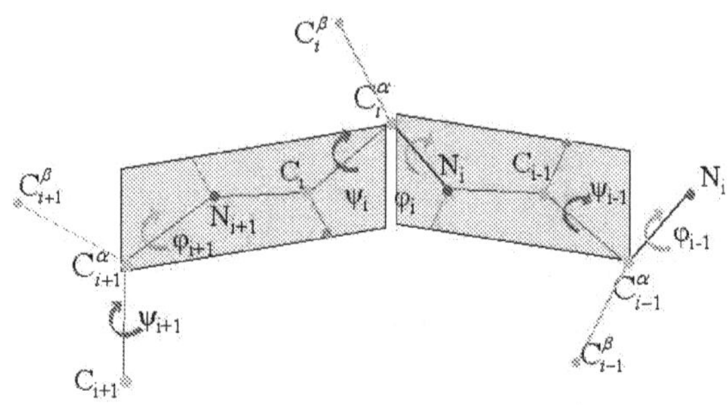

Fig. 1. Polypeptide chain

2.1 Protein Geometry

Proteins are built from amino acids that are chemically linked by peptide bonds to form a polypeptide chain. Each amino acid contributes three atoms to the *backbone*, creating a chain of repeating -N-C^α-C- units (nitrogen-carbon-carbon). The carbonyl carbon C is double bonded to an oxygen atom and the nitrogen atom N is bonded to a hydrogen atom. The side chain attached to the C^α is one of 20 different organic compounds that determines the amino acid type. In this paper we focus on the protein backbone, so we treat each residue as an alanine, whose side chain consists of one carbon atom bonded to the C^α atom, denoted by C^β and decorated with its hydrogen atoms.

The backbone can also be partitioned into peptide units, C_i^α-C-N-C_{i+1}^α (Figure 1). The bonds of the peptide units are generally considered to be planar and of fixed length. The rotations of the peptide planes about the bonds formed at the C^α atom, however, exhibit large variation within proteins, allowing the peptide chain to assume a variety of secondary structures like alpha helices and beta sheets [2]. By convention, the rotation around N-C^α and C^α-C bonds are denoted by ϕ and ψ (Figure 1). Many combinations of ϕ and ψ are infeasible due to inter-atomic collisions, called *sterics* in chemistry. The ϕ, ψ distribution of data points for the non-glycine and non-proline residues of the "top500" known protein structures are plotted against one another in a Ramachandran plot, Figure 2. The "top500" proteins were selected from the Protein Data Bank (PDB) by the Richardson Lab based on the resolution and accuracy of the structures [16, 17].

Although the peptide units are often considered to have a fixed geometry, variation of the bond lengths and bond angles is observed in known protein structures; Table 1 includes statistics obtained from 331 of the "top500" proteins. (169 were not considered due to sequence gaps or other inconsistencies.) For each of the 331 structures, we calculated the average value and standard deviation of the backbone bond angles and bond lengths. Displayed in Table 1 are the corresponding average values and average standard deviations of these parameters across the 331 structures. The average standard deviation represents the expected variation of the parameter value for a given protein structure and a given amino acid in the protein. Table 1 also includes the bond parameter variations allowed in probik and the standard ideal bond geometries from Engh and Huber [6].

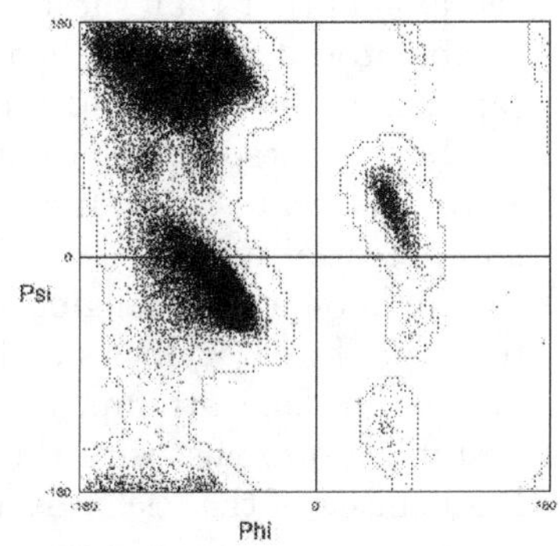

Fig. 2. Ramachandran plot of allowed ϕ, ψ

Table 1. Bond angles and lengths: ideal values, observed values and variations, and allowed variations in probik.

	Bond angle (°)			Bond length (Å)			dihedral (°)
atoms	$NC_\alpha C$	$C_\alpha CN$	CNC_α	NC	$C_\alpha C$	CN	$C_\alpha NCC_\alpha$
ideal value	111.2	116.2	121.7	1.47	1.53	1.33	180.0
avg value	111.700	117.326	121.931	1.447	1.511	1.317	179.533
avg stdev	3.066	1.581	1.815	0.011	0.012	0.010	8.172
allowed dev	4.0	2.0	2.0	0.02	0.02	0.02	8.0

2.2 Protein Design

Protein design has enormous potential application for the development of drugs and biosensors. Homme Hellinga's computational tool, Dezymer, for redesigning the binding site for a protein ligand complex to create a protein with new function. Among other results, his lab has successfully redesigned ribose-binding protein to bind TNT (trinitrotoluene) [10].

The general principles for the formation of specific protein/ligand complexes are often depicted as a lock and key fit, determined primarily by short-range interactions (steric contacts and hydrogen bonds). For a protein to bind a new ligand, the chemistry and geometry of the binding pocket must be changed without destabilizing the protein (i.e. preventing its fold). This change can be effected only indirectly, by changing the sequence of amino acids. The Dezymer software identifies a set of amino acids to change under the assumption that the backbone will remain fixed, then performs a large combinatorial search of ligand translations and rotations inside the pocket (approximately 10^8) and mutations of amino acid residues (typically 12-18 residues, corresponding to 10^{45} to 10^{68} mutant structures representing 10^{15} to 10^{25} sequences) [10]. It ranks feasible pocket designs, using a series of energy and potential functions.

We know that changing the sequence will move the backbone of the protein, too, but Dezymer's analysis does not currently include backbone modification. In fact, attempt to modify protein backbones have always tended to destablize proteins, with the exceptions of some recent manipulations by the biochemists of the Richardson lab [17], which are designed to correct incorrect crystallographic misfittings in the PDB. To preseve the stability of the protein, manipulation of the wild-type backbone must be conservative (< 1Å), and satisfy geometric constraints. We need an accurate geometric model of local backbone motions that can be realized by rotation of ϕ, ψ angles and possibly small perturbations of other bond parameters. Probik, our computational tool, provides the biochemist with the range of motions possible for a fragment of the protein backbone as a result of variations in bond parameters.

3 Review: Exact Inverse Kinematics for Proteins

A protein uses the relatively free rotation of the ϕ, ψ dihedral angles to fold into a stable and functional configuration, while it tends toward ideal bond lengths, bond angles and peptide dihedrals. For a small segment of the backbone, we want to model the free rotation of the dihedrals while maintaining fixed bond lengths, bond angles, and peptide dihedrals. Mathematically, if we fix coordinate frames for an anchor and a target atom, assume fixed bond angles, bond lengths, and peptide dihedral angles, and allow the ϕ and ψ dihedral angles to rotate freely, what are all the possible positions of the backbone? What happens if we additionally allow one bond parameter to vary within some tolerance range? This is the well-known n-atom loop closure problem in inverse kinematics; in this section, review exact analytic solutions and implementations.

3.1 Loop Closure Problem

Loop closure problems have been central in both computational protein modeling and robotics. The n-atom loop closure problem is, given a chain of n atoms linked by rigid bonds with fixed bond and peptide torsion angles, and given the (fixed) position and orientation of the first and last atoms, determine all possible positions of the $n-2$ intervening atoms that satisfy the bond length and angle constraints. Given the position and orientation of the first atom, 6 degrees of freedom (DOF) are required to specify the position and orientation of the last atom. The inverse kinematics problem is over-constrained for < 6 DOF and under-constrained for > 6 DOF.

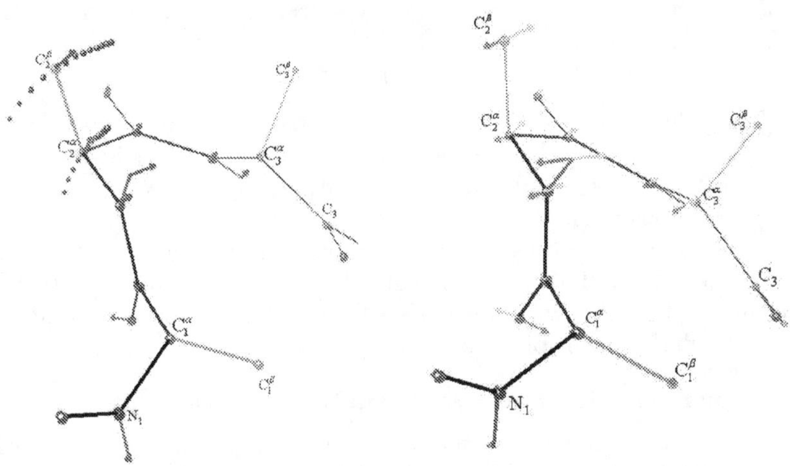

Fig. 3. A 3-residue loop and corresponding derivative vectors for wiggled C_2^α bond angle (left) and wiggled C_2^α and C_2 bond angles (right).

The degrees of freedom in the n-atom loop closure problem correspond to the dihedral angles, ϕ and ψ, and therefore depend on the number of atoms in the loop and the type of the first atom. In this paper, we will consider loops of a protein with 6 major degrees of freedom, corresponding to 6 consecutive ϕ or ψ dihedral angles. We will "wiggle" a 7th degree of freedom to generate continuous families of solutions. Figure 3 shows 3-residue loops.

3.2 Denavit-Hartenberg Local Frames

The Denavit-Hartenberg (DH) local frames are widely used for modeling molecular configurations and serial manipulators. By defining a local coordinate system for each atom in terms of the orientation of its adjacent bonds, this method provides a systematic way of describing the rotational and translational relationship between adjacent atoms in a chain.

Consider a chain of atoms, $u_1, u_2, \ldots, u_n$, linked by the bonds $b_1, b_2, \ldots, b_{n-1}$. We define the *DH local frame* associated with an atom, u_i, to be $F_i = \{x_i, y_i, z_i, u_i\}$, where z_i is in the direction of the previous bond, b_{i-1}, x_i is perpendicular to bonds b_{i-1} and b_i, and y_i is perpendicular to x_i and z_i (Figure 4). We call u_i the origin of F_i.

The DH local frames, F_i and F_{i+1}, of the adjacent atoms, u_i and u_{i+1}, are related by a series of rotations and translations with the following geometric *DH parameters*.

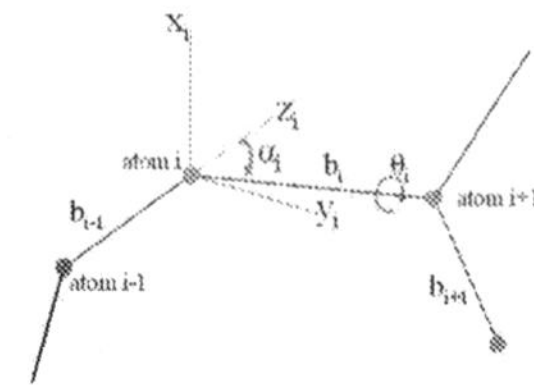

Fig. 4. Denavit-Hartenberg Local Frame

- α_i = bond angle = angle from z_{i-1} to z_i.
- d_i = bond length = distance from u_i to u_{i+1}.
- a_i = offset = distance from the intersection of z_i axis with the x_{i+1} axis to the origin F_i
- θ_i = dihedral angle = measured from x_i to x_{i+1} about z_i.

Let $R_x(\omega)$ denote the 4×4 homogeeous matrix for rotation about the x-axis by angle ω and $T_z(d)$ denote the 4×4 homogeneous matrix for translation along the z-axis by amount d. The DH frames F_i and F_{i+1} are related by $F_{i-1} = A_i F_i$ where $A_i = T_z(d) \cdot R_z(\theta) \cdot R_x(\alpha) \cdot T_x(a)$.

Let A_1 and A_{hand} denote the DH frames for the anchor atom and target atom in the chain. Then we can write the known coordinate system, A_{hand}, in terms of the unknown coordinate systems of the intermediate atoms with a series of matrix multiplications. $A_1 A_2 \cdots A_n = A_{hand}$.

3.3 Implementation and Exact Analytic Solutions

Raghavan and Roth consider the 6 revolute manipulator problem, analogous to the 6 DOF atom loop problem [15]. The chain consists of 6 revolute joints connected by 6 links, where each link contributes one free torsion angle. The previous equation becomes $A_1 A_2 A_3 A_4 A_5 A_6 = A_{hand}$. The parameters α_i, d_i, and a_i $(1 \leq i \leq 6)$ are fixed and the torsion angles $\theta_1, \ldots, \theta_6$ are free. Thus, the matrix equation gives 12 multivariable equations, only six of which are independent because the coordinate frame vectors are orthogonal.

To reduce degree and complexity of the system of six equations, they rewrite as $A_3 A_4 A_5 = A_2^{-1} A_1^{-1} A_{hand} A_3^{-1}$. The variables appearing in the matrix entries are

$$\begin{pmatrix} (\theta_3,\theta_4,\theta_5) & (\theta_3,\theta_4,\theta_5) & (\theta_3,\theta_4,\theta_5) & (\theta_3,\theta_4,\theta_5) \\ (\theta_3,\theta_4,\theta_5) & (\theta_3,\theta_4,\theta_5) & (\theta_3,\theta_4,\theta_5) & (\theta_3,\theta_4,\theta_5) \\ (\theta_4,\theta_5) & (\theta_4,\theta_5) & (\theta_4,\theta_5) & (\theta_4,\theta_5) \\ 0 & 0 & 0 & 1 \end{pmatrix} = \begin{pmatrix} (\theta_1,\theta_2,\theta_6) & (\theta_1,\theta_2,\theta_6) & (\theta_1,\theta_2,\theta_6) & (\theta_1,\theta_2) \\ (\theta_1,\theta_2,\theta_6) & (\theta_1,\theta_2,\theta_6) & (\theta_1,\theta_2,\theta_6) & (\theta_1,\theta_2) \\ (\theta_1,\theta_2,\theta_6) & (\theta_1,\theta_2,\theta_6) & (\theta_1,\theta_2,\theta_6) & (\theta_1,\theta_2) \\ 0 & 0 & 0 & 1 \end{pmatrix} \tag{1}$$

Raghavan and Roth proceed with the 6 independent equations from the third and fourth columns, sequentially eliminating four of the five variables by dialytic elimination. After symbolically expanding the determinant of the resulting 12×12 matrix, they obtain a sixteen degree polynomial in $\tan(\theta_3)$.

Mathematically, this method provides exact analytic solutions for the general 6 revolute joint manipulator problem, however, computationally it is subject to significant numerical errors as discussed in [13].

Manocha and Canny replace the root-finding of Raghavan and Roth with more robust eigenvalue computions in a 24×24 matrix [13]. Eight of the 24 eigenvalues are constants and the remaining 16 eigenvalues correspond to the undetermined dihedrals angles. There can be at most 16 real solutions and empirically < 8 real solutions are observed.

As input, their IK solver requires (1) the DH frame for the target, A_{hand}, and (2) the known DH parameters, $\{\alpha_i,\ d_i,\ a_i\ :\ 1 \leq i \leq 6\}$, for each of the 6 links in the chain. As output, the IK solver returns a finite number of solution sets for the free DH parameters, $\theta_1, \ldots, \theta_6$. We call each feasible set of dihedrals, $\theta_1, \ldots, \theta_6$, a *solution pose*.

Their solver uses symbolic preprocessing, rank computations, and dialytic elimination to obtain a 12×12 matrix whose entries are quadratic polynomials in terms of $\tan(\theta)$. This provides us an efficient and numerically stable solution to the 6-revolute manipulator problem.

Wedemeyer and Scheraga, in a 1999 J. Computational Chemistry paper, present exact, elegant solutions to several loop closure problems including the tripeptide loop [21]. They used spherical geometry and dialytic elimination to derive a 16 degree polynomial in the tangent of the half angle of one of the dihedral angles. Their method assumes that peptides are planar, however.

Coutsias, Seok, Jacobson, and Dill just published (2004) an independent extension of the Wedemeyer and Scheraga analysis that does handle the "oblique rotators" required by non-planar peptides [3]. They also incorporate Canny and Manocha's ideas to improve robustness. They apply this in Monte Carlo minimization for longer loops, including demonstrating improved efficiency on an 8-residue loop. An appendix shows some graphs of how perturbations change the solutions of the polynomial, but they do not exploit this as a tool for understanding manipulations as we have done in our work.

4 Probik: backbone manipulation algorithm

Probik is the computational tool we have developed for modeling local protein backbone motion. The PDB file for a given protein contains a list of atoms with residue numbers, atom numbers, types, and $\mathbf{R}^3$ coordinates. For a region of the backbone with 6 specified degrees of freedom (usually 3- or 4-residue loop) probik computes derivatives of each atom position with repect variation in the bond parameters in the region. In this way, probik provides a numeric and geometric indication of the range of local backbone motions available by varying geometric parameters. In this section, we describe the four steps of the algorithm: calculate wiggled loops, obtain exact solutions, classify poses, and estimate derivatives. We begin by defining notation.

4.1 Protein backbone representations

Given the PDB file for a protein and a specified region of the backbone, probik can iterate through all 6 DOF loops in the region. In this paper we consider one such loop, which we call the *6DOF loop*, and we fix one bond parameter to vary, which we call the *fuzz parameter.*

Probik extracts, from the PDB file, the coordinates and types of all backbone atoms (N-H, C^α-C^β, C-O) in the 6DOF chain. We call the list of main chain coordinates (N, C^α, C, ...) the *AA chain* for the 6DOF loop. In order to use the IK solver discussed in section 3.3, we compute another representation of the 6DOF loop with only 6 links, one for each free dihedral angle ϕ and ψ. For each full peptide, ${C^\alpha}_1$-C-N-${C^\alpha}_2$, in the AA chain, let l_1 and l_2 be the lines through the atoms C_1^α-C and N-C_2^α respectively. Let P_1 be the point on l_1 closest to line l_2. Define P_2 to be the corresponding point on l_2. We can represent the peptide by the three joints, C_1^α, P_2, and C_2^α, where the center joint P_2 is offset from the joint C_1^α by the distance between P_1 and P_2. We call the resulting list of coordinates, C^α, P_2, C^α, P_2, ..., the *IK chain.*

For both AA and IK chain representations of the 6DOF loop, we calculate corresponding sets of DH parameters $\{\theta_i, \alpha_i, d_i, a_i\}$, and denote them *AA parameters* and *IK parameters*, repectively. (Note that in the AA parameters offset $a_i = 0$ for all i, where in the IK parameters a_i is the distance between P_1 and P_2 for the C_1^α-P link and $a_i = 0$ for the P-C_2^α link.) Probik uses both representations: modifying AA parameters to wiggle or fuzz an angle parameter, sending IK chain parameters to the IK solver, and applying returned θ_i parameters to the AA chain to compute new amino acid coordinates.

4.2 Calculate wiggled loops

The first step in the algorithm is to compute the geometric representations and parameters of the 6DOF loop and to generate an array of values for the fuzz parameter. From the PDB file, we obtain the AA chain coordinates for the 6DOF loop and calculate the corresponding AA parameters, or *actual* DH parameters.

Using the actual value, the ideal value, and the standard deviation of the fuzz parameter, we generate an array of values for the parameter. We choose a range of values that includes at least one standard deviation around the actual value and (2/3) of a standard deviation about the ideal value. This ensures that we sample fuzz parameter values all around the actual and ideal values. The actual AA parameters, combined with this array of values, yields our array of wiggled AA parameters.

The corresponding array of wiggled IK parameter set is calculated as follows: for each set of wiggled AA parameters, we rebuild the AA chain, convert to its IK chain representation, and compute the corresponding IK parameters. Note that a change in a bond angle in the AA chain corresponds to a change in all the DH parameters (θ, α, d, a) of the corresponding adjacent links in the IK chain representation. The result is an array of *wiggled IK parameters.*

4.3 Get exact solutions - using Manocha and Canny

The second step in the algorithm is to generate all possible configurations that close the 6DOF loop for small variations in the fuzz parameter. We use the IK Solver (section 3.3,[13]) to obtain the solution poses for each set of wiggled IK parameters. Recall that there are at most 16 feasible configurations to close the 6DOF loop for a set of known DH parameters, $\{\alpha_i, d_i, a_i\}_{1 \leq i \leq 6}$. We call the loop closure configurations corresponding to the original IK parameters the *original solution poses*. Of these, the *actual pose* is the solution pose that corresponds to the initial AA coordinates.

Typically, the IK solver returns between 2 and 6 sets of DH dihedral angles. We denote these solutions by Θ_1, $\Theta_2, \ldots,$ Θ_s. For each solution, $\Theta_i = \{\theta_1, \ldots, \theta_6\}$, we apply the corresponding DH transformation matrices to rebuild the IK chain. We discard any solutions whose final DH frame, A_{end}, does not hit the target frame due to within a specified tolerance ($\|A_{target} - A_{end}\|_2 > 10^{-4}$) due to numerical errors in the solver. For each solution that closes the loop, we convert the IK chain to the AA chain and order the resulting poses in increasing distance from the actual pose.

4.4 Classify families of poses

The next step is to organize the solution poses obtained from the IK solver to determine the range of motion allowed by varying the fuzz parameter. Recall that each solution pose corresponds to a zero of the underlying 16 degree polynomial for the actual IK parameter set. A small variation in the fuzz parameter corresponds to a small perturbation of the polynomial, which generally yields a small change in the roots by properties of transversality. As we vary the fuzz parameter and obtain the associated solution poses, we are sampling along the zero level set of a manifold in R^3. The solution poses over the range of fuzz values correspond to branches of the solution space. These branches move at different speeds and may merge into or appear from a complex solution over the range of fuzz values, which can result in a decrease or increase in the number of solution poses from one fuzz value to the next.

To classify solution branches, we consider solution poses ordered by increasing fuzz value. For each pose, we compute the distance using the vector norm to each of the poses for the next fuzz value and store the minimum distance in the corresponding entry of an adjacency matrix. We align the solution poses for the next fuzz value with the original poses according to the adjacency matrix. We continue in this way, aligning the solution poses for the increasing fuzz values, and then repeat the process for decreasing fuzz values.

Ambiguity arises when the number of solutions increases or decreases from one fuzz value to the next. We use the adjacency matrix and the second closest solution pose to classify the branches as continuing, appearing, or disappearing. After processing all the solutions for increasing and then decreasing fuzz values, we obtain the feasible range of motion for each of the original branches of solution.

4.5 Estimate the derivative

The final step in the algorithm is to estimate the derivative of motion for each atom in the 6DOF chain motion with respect to the fuzz parameter. The geometric constraints within the loop force the atoms to move along smooth curves. For each branch, we model the motion of each atom in the loop with a second order polynomial. The coefficients of the quadratic polynomial are determined by computing the least squares fit on the coordinate points for each atom in the loop. The approximate magnitude and direction of motion is obtained for each atom (N-H, C^{α}-C^{β}, C-O) in the 6DOF loop, for each of the original solution poses.

4.6 Preliminary Results

Figure 3 shows example results of probik in mage, the protein viewer from the Richardson lab [17]. Probik can be used to study a specific motion in detail, with the derivatives and atom position dots corresponding to one wiggle parameter (left subfigure), or to offer choices of motion, by displaying derivatives corresponding to two (or more) bond parameter variations (right subfigure). Probik allows us to compare the motion resulting from varying different parameters, whether this be to choose one automatically, or to present the options to a biochemist and to see and understand why she makes the choice that she does.

In collaboration with biochemists, we have used probik to reduce clashes in crystallographic refinement and to increase the space of rotamers allowed for residue changes in Dezymer, These have suggested new sequence designs to increase the stability for the TNT-binding protein. These sequences are being created and tested in the wet lab.

5 Covering the Space of Protein Backbone Motions

The algorithm described above allows us to move a protein backbone loop by varying one or more of the individual bond angle, bond length, or peptide dihedral parameters. In this section, we use simple, linear principal component analysis to find combinations of parameters that can act as *control handles* to move the backbone. A small number of control handles can move the backbone throughout the same range of motion as that created by perturbing all individual parameters. This reduction in the number of controls reduces the complexity of protein backbone manipulation and design. Our results show that four control handles are sufficient to move a loop to within 0.1 Å cRMSD of any wiggled conformation.

5.1 Principal Component Analysis on Backbone parameters

Principal component analysis (PCA) reduces the dimensions of a set of interrelated variables, producing a set of derived variables known as the principal components [7]. We use the most common PCA technique, which is to use singular value decomposition (SVD) to successively find and remove the best linear approximation to the residuals. Although our protein model is not linear, our data suggests that it is a sufficient method to test our hypothesis that the flexibility of protein backbones can be represented to a high degree of accuracy with just four degrees of freedom. We can reconstruct the data from the projections on all principal components, but we can also use the projections on the first few principal components as a lower-dimensional representation of the data, since these components are chosen to greedily preserve as much of the original data variation as possible.

For this work, we apply PCA to a randomly generated *cloud set* consisting of a large number of wiggled loops centered around a three-peptide loop extracted from a pdb file. We collect eight bond angle parameters, two peptide dihedral parameters, and six ϕ, ψ parameters, from the cloud set and store their deviations from the mean values in a matrix $\mathbf{M}$. The SVD procedure calculates matrices $\mathbf{U}$, $\mathbf{S}$, and $\mathbf{V}$ such that $\mathbf{M} = \mathbf{USV^T}$. The columns of $\mathbf{V}$ are the principal components of $\mathbf{M}$ and represent a new set of orthogonal basis vectors for the loop parameters. The diagonal entries of $\mathbf{S}$ represent the relative weights of the principal components. Matrix $\mathbf{U}$ contains the original loop parameter deviations transformed into the new basis.

Principal component analysis gives us a lower dimensional approximation of the sixteen backbone parameters for each loop in the cloud set. To measure accuracy, we could easily compare the approximation to the original loop parameters. In design applications, however, accuracy is better measured by how well atom positions can be reconstructed by the IK solver (respecting the closure constraints). For this second measure, we calculate the coordinate root-mean-squared deviation error (cRMSD, a common measure in biochemistry) of the reconstructed loop from the original. We consider two loops with a cRMSD of less then 0.1 Å to be similar.

When reducing the dimension of the space of loops, it is more important that the space remain accessible with fewer handles than that the parameterization remain the same. Therefore, we calculate a third measure of how well the first few principal components approximate the full range of motion. We build a *probe set* of loops by varying the test loop parameters along the first few principal components and then measure how well the probe set corresponds to the cloud set. We find that four principal components are sufficient to construct a probe loop similar to each loop in the cloud set.

5.2 Experimental Setup

To perform principal component analysis on a given loop, we first generate a large number of perturbed loops. We wiggle each of the starting loop's bond angles and peptide dihedral angles in a normal distribution around their starting values. Normal sampling insures that few of the chains have the majority of their parameters far from ideal values, thus we obtain a cloud that covers the space of motions close to the starting loop. Because our starting loops come from real pdb files, many contain parameters that are beyond the allowable ranges. To compensate, we extend the ranges by 1.5 degrees in each direction. After the inverse kinematics solver generates a set of solutions from the new parameters, we keep all branches that are within the extended ranges and whose ϕ, ψ angles meet the additional requirements described below. This is repeated until we have 5,000 perturbed loops that form a cloud around the starting loop.

We must calculate the mean value of each of the parameters in the cloud set to perform principal component analysis. Calculating the mean of the peptide unit dihedrals and the bond angles is straightforward, because they are confined to small ranges around their ideal values. The ϕ, ψ dihedral angles do not have a specified range and can lie anywhere on the unit circle. Because it is not clear how to average angles that are distributed across the whole unit circle, we filter out all loops that have any ϕ, ψ angles that differ by more 90 degrees from the starting values. This allows us to calculate a well-defined mean value. This restriction should not be significant, because in tightly packed molecules, inter-atom collisions should restrict the movement of ϕ, ψ dihedrals to within a similar range.

We perform the above calculations on 150 test loops from the following pdb files, 1tre.pdb, 1r69.pdb, 1byi.pdb, 2ci2.pdb, and 2mhu.pdb. We use the tripeptide loop and wiggle the eight bond angles and the two peptide units. The inverse kinematics solver calculates the remaining six ϕ, ψ angles, for a total of sixteen parameters. These are used to create the 5,000×16 matrix for the PCA calculation described in the previous section.

5.3 Direct Reconstruction of the Cloud Set

A first test of how well a PCA-derived lower-dimensional approximation can reconstruct the cloud set is to measure how well each individual loop can be reconstructed from the approximation of its parameters. For each loop in the cloud set, we approximate the parameters using one to sixteen principal components and then use them to rebuild atom positions. Because the loop end positions and orientations will have changed, we use the inverse kinematics solver to close the reconstructions, then calculate the cRMSD of each reconstruction against the original atom positions. Figure 5 plots for 150 test loops the average cRMSD error of the 5,000 cloud set loops, as a function of the number of principal components used to approximate the individual loop parameters.

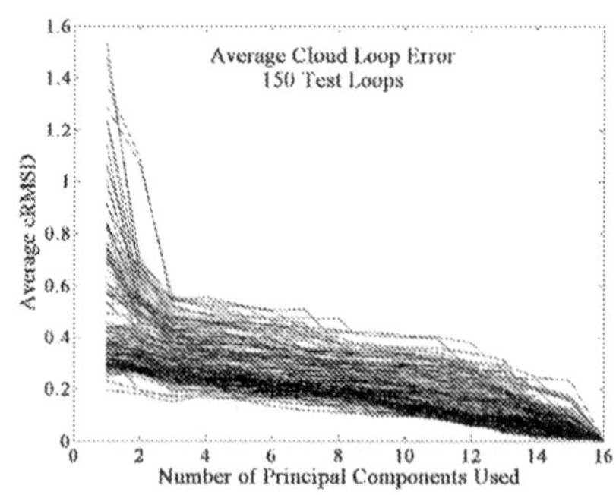

Fig. 5. Average error for 150 test loops

The plot shows that many more than five principal components are necessary for accurate reconstruction. The average cRMSD for most of the loops does not reach 0.2 Å until eight principal components are used, which is too many for interactive control of backbone motion. The error is high because closing the loop with the IK solver changes phi and psi angles and coordinate positions, making many of the reconstructed loops miss their original targets. Because it is not necessary to keep the same parameterization of the cloud, the next section describes a better measure.

5.4 Exploring the Range of Motion of Control Handles

A control handle adjusts a loop's angle parameters in the direction of its associated principal component. To test how well we can cover the space of the original loop cloud, we randomly adjust the control handles to create a *probe set* of 10,000 loops: Starting from the average angle parameters, we successively add the product of a principal component with a random scalar chosen so that no individual angle parameter moves beyond the extended ranges. For this experiment, we add 2, 3, 4, and 5 control handle movements, and run the inverse kinematics solver to enforce closure constraints.

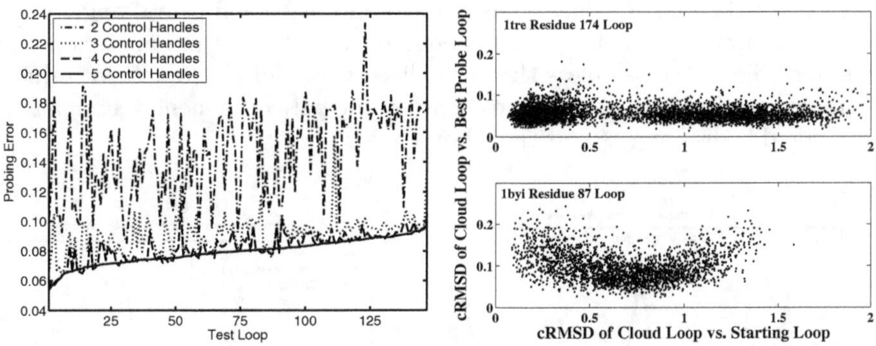

Fig. 6. (left) Average probing error for 150 test loops; (right) Relationship between a cloud loop's deviations from the original and its closest probe loop.

We then calculate, for each loop in the cloud set, the lowest cRMSD to our set of 10,000 probe loops. The average value across the 5,000 cloud loops is our estimate of *probing error.* Probing error decreases to a limit as the number of probes increases; In our experiments, we found that the probing error decreased only a few hundredths of an angstrom when the number of probes increased from 10,000 to 20,000. Figure 6 (left) plots the probing error for each of the 150 test loops. Each line on the plot shows the results for a different number of control handles. The y-axis shows the probing error for each of 150 test loops along the x-axis. The loops on the x-axis are sorted by their probing error when using 5 control handles. For most loops, three control handles produce average cRMSD error below 0.1 Å. Overall, the low errors demonstrate that PCA-based control handles can reproduce the space of allowed motions.

A final check shows that probe loops land close to cloud loops that are both near to and far from the starting position. Figure 6 (right) shows scatter plots for two examples: the upper, 1tre.pdb residue 174, has one of the smallest probing errors and the lower, 1byi residue 87, has one of the greatest. For each cloud loop we plot a point whose x coordinate is the cRMSD to the original and y-coordinate is the cRMSD to the closest probe loop constructed using four control handles. The lower, 1byi, shows a bit of a curve, probably because its initial poose is strained so that perturbations take it toward a more relaxed position. Since the range of x is 10 times the range of y, both are essentially flat—we cover all areas of the space.

5.5 Control Handles for Loops in α-Helices

The previous section demonstrated that we can construct control handles for most loops that provide a good range of motion. Ideally we should be able to precalculate sets of control handles that will work well for common loop conformations and store these in a library. We examined the principal components of the 71 loops from the previous section that were from α-helix structures. These loops have similar bond and peptide dihedral angle parameters. We found that the principal components of these parameters are also similar. Figure 7 (left) plots the normalized values of the first three principal components for the α-helix loops. Each principal component vector has two peptide dihedral angles and eight bond angles.

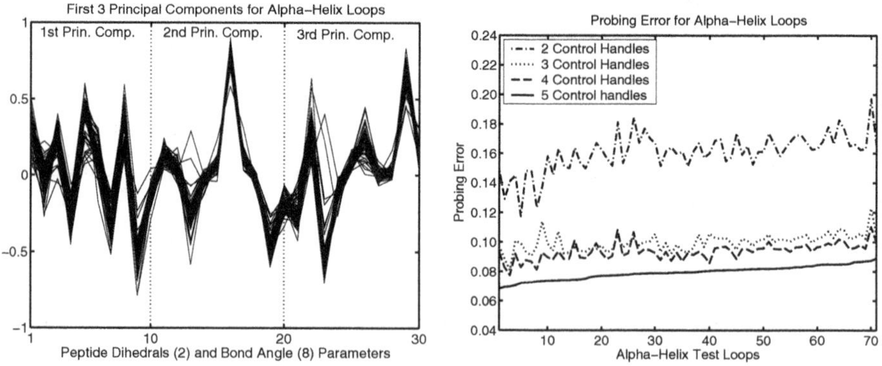

Fig. 7. First three PC for helix loops.

We ran the probing experiment from the previous section on these 71 loops using the average of their principal components as control handles. Figure 7 (right) shows the results, which were comparable to the original probing experiment in which each loop used its own individual control handles. The first four average control handles produce mean cRMSD errors of less than 0.1 Å for almost all loops. The error is slightly higher using average control handles, but still within our goal of reaching all cloud set loops to within 0.1 Å using four or fewer control handles. Furthermore, we no longer need to construct a large cloud set or perform principal component analysis to find good control handles for α-helix loops.

6 Conclusion

Geometric modeling of protein structure is fundamental to the task of computational protein design, crystalographic structure refinement, and computational protein folding. Solutions to inverse kinematic problems help to provide better geometric models for molecular chains. In general, the solution to an inverse kinematic problem is complex and computationally expensive. Probik, our computational tool, models local backbone motion by combining the exact kinematic solutions for a segment of a protein backbone with the observed variation of bond angles, lengths, and dihedrals.

To obtain the exact solutions to the 6 DOF loop closure problem, we use the Manocha and Canny IK solver[13]. This computational tool is an efficient and numerically robust 6 DOF IK solver. Our technique avoids many geometric assumptions, like planar peptides, that are often made to reduce the complexity of the inverse kinematics problem. We combine the actual geometric measures of the given backbone segment with controlled variation of a bond parameter, to model the natural motion of a protein backbone.

By estimating the derivative of motion with respect to variable bond parameters, we provide a concise numeric and geometric notion of the range of feasible motions. Our algorithm generates a kinemage file, roughly the size of the original PDB file, containing the 3-dimensional representation of the protein and the derivative vectors, which allows interactive computer display. The computation of and file generation for the motions resulting from one wiggle parameter in a given protein structure takes approximately 3-5 seconds. The geometric model produced by probik allows biochemists to investigate, understand, and predict the motion of a protein backbone.

Acknowledgements

We are grateful to our collaborators, Prof. Homme Hellinga and students Loren Looger and Yaji Tian, and Profs. David and Jane Richardson, and student Ian Davis, for discussions and ideas that have lead the development of this work. This research has been partially supported by grants from NSF ITR (0076984) and DARPA.

References

1. Amadei, A., et al. (1996). An efficient method for sampling the essential subspace of proteins. *J Biomol Struct Dyn*, 13(4): 615–625.
2. Branden C., Tooze J. (1999), *Introduction to Protein Structure*, 2nd ed., Garland Publishing.
3. Coutsais E. A., Seok C., Jacobson M. P., Dill K. A. (2004) A kinematic view of loop closure. *J Comput Chem*, 25: 510–528.

4. de Groot, B.L., et al. (1996). Toward an exhaustive sampling of the configurational spaces of the two forms of the peptide hormone guanylin. *J Biomol Struct Dyn*, 13(5): 741–751.
5. Edelsbrunner H., Koehl P. (2003) The weighted volume derivative of a space-filling diagram. *Proc. Natl. Acad. Sci.* 100: 2203–2208
6. Engh R. A., Huber, R. (1991) Accurate bond and angle parameters for X-ray protein structure refinement. *Acta Cryst*, A47, 392–400.
7. Jolliffe, I. T. (2002) Principal Component Analysis, New York: Springer-Verlag.
8. Kobayashi N., Yamato T., Go N. (1997) Mechanical property of a TIM-barrel protein *Proteins* 28(1): 109–116.
9. Leach A. R. (1996) *Molecular Modelling: Principles and Applications.* Addison-Wesley Longman, Reading, Massachusetts.
10. Looger L., Dwyer M., Smith J., Hellinga H. (2003) Computational design of receptor and sensor proteins with novel functions. *Nature*, May 8; 423(6936):185–90.
11. Lovell, S., et al. (2003) Structure Validation by Ca Geometry: ϕ, ψ and CB Deviation, PROTEINS: Structure, Function, and Genetics, 50: 437–450. http://kinemage.biochem.duke.edu/validation/model.html#rama.
12. Manocha, D., Canny, J. (1992) Real Time Inverse Kinematics for General 6R Manipulators. *Proceedings of IEEE International Conference on Robotics and Automation*, 383-389.
13. Manocha, D., Canny, J. (1994) Efficient Inverse Kinematics of General 6R Manipulators. *IEEE Transactions on Robotics and Automation*, 10(5): 648–657.
14. Manocha, D., Zhu, Y. (1994) A Fast Algorithm and System for Inverse Kinematics of General Serial Manipulators. *Proceedings of IEEE Conference on Robotics and Automation*, San Diego, CA. 3348–3354.
15. Raghavan M., Roth B. (1989) Kinematic analysis of the 6R manipulator of general geometry. *International Symposium on Robotics Reseach*, Tokyo, 314–320.
16. The Research Collaboratory for Structural Bioinformatics (RCSB) Protein Data Bank. http://www.rcsb.org/pdb/
17. Richardson Laboratory, Duke Univ. http://kinemage.biochem.duke.edu/
18. Song G., Amato N. M. (2001) A Motion Planning Approach to Folding: From Paper Craft to Protein Folding *ICRA'01*, 948–953.
19. Teodoro M. L., Phillips Jr. G.N., Kavraki L. E. (2001) Molecular Docking: A Problem with Thousands of Degrees of Freedom. *ICRA 2001*, 960–965.
20. Teodoro M. L., Phillips Jr. G.N., Kavraki L. E. (2003) Understanding Protein Flexibility Through Dimensionality Reduction. *Journal of Computational Biology*, 10(3–4), 617–634.
21. Wedemeyer, W., Scheraga, H. (1999) Exact Analytical Loop Closure in Proteins Using Polynomial Equations. *J. Comput. Chem.*, 20, 819–844.
22. Zhang, M., Kavraki, L. E. (2002) Solving Molecular Inverse Kinematics Problems for Protein Folding and Drug Design *Currents in Computational Molecular Biology, RECOMB '02*, 214–215.

Randomized Algorithms for Minimum Distance Localization

Malvika Rao, Gregory Dudek, and Sue Whitesides

McGill University, Montréal, Canada H3A 2A7 {rao,dudek,sue}@cs.mcgill.ca

Summary. *We address the problem of minimum distance localization in environments that may contain self-similarities. A mobile robot is placed at an unknown location inside a 2D self-similar polygonal environment P. The robot has a map of P and can compute visibility data through sensing. However, the self-similarities in the environment mean that the same visibility data may correspond to several different locations. The goal, therefore, is to determine the robot's true initial location while minimizing the distance traveled by the robot. We present two randomized approximation algorithms that solve minimum distance localization. The performance of our algorithms is evaluated empirically* [1].

1 Introduction

This paper addresses the problem of minimum distance localization. That is, the problem of estimating a robot's position in an environment without a prior position estimate and of selecting a minimum length path along which to acquire sensor data to accomplish this localization as quickly as possible. Global localization refers to estimating the robot's position where the space of possibilities is the entire environment [2]. In general, global localization must contend with the fact that the observations used to estimate position can be ambiguous (even with a perfect map and a perfect sensor, this ambiguity can be serious [6]). Thus, global localization intrinsically entails combining measurements from multiple vantage points along some path. The problem of selecting an optimal (minimum length) trajectory for estimating position is NP-hard [3]. Simple heuristic strategies have been used in practice but these can be shown to have strikingly poor performance. This paper presents algorithms for determining a localizing path that are efficient to compute and yet also efficient in their performance.

Existing deterministic solutions for calculating even nearly optimal localizing paths are computationally costly in complex environments. Further,

[1]Research supported in part by NSERC and FCAR.

M. Erdmann et al. (Eds.): Algorithmic Foundations of Robotics VI, STAR 17, pp. 393–408, 2005.

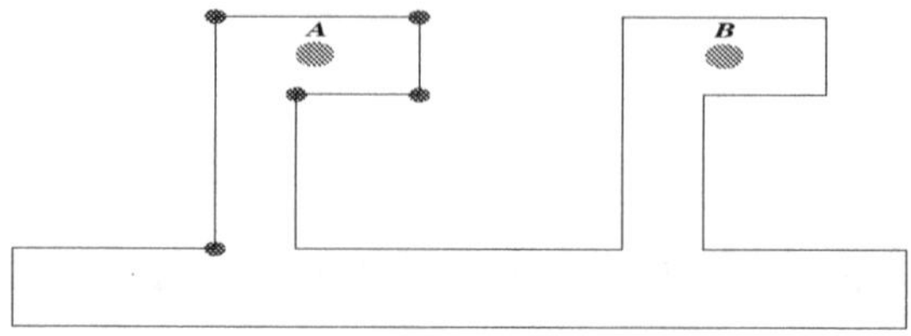

Fig. 1. Visible vertices from location A. The visibility data at locations A and B are the same.

previous work in minimum distance localization [3] may require the robot to make observations that are arbitrarily difficult to achieve in practice. That is, the robot is directed to visit a series of visibility cells - places where specific combinations of landmarks in the environment can be seen - even though these cells may be arbitrarily small. For any real robot with any odometry error whatsoever, maneuvering into such cells may not be feasible.

The algorithms we propose are based on visiting a series of randomly selected sample points in the environment from which distinguishing landmarks may be observed. In the context of our environment model, and with full generality, we consider landmarks to be combinations of vertices and edges that define the perimeter of the environment. Out of a set of randomly selected points in the environment, the robot visits the location that promises the most information gain, where it collects sensor data in order to disambiguate its initial position. We argue that by virtue of random sampling, our strategy would direct the robot to visit a particular visibility cell with a probability directly proportional to the area of that cell. Hence vanishingly small visibility cells have vanishingly small probabilities of being visited.

The performance of our randomized localization algorithms is validated and explored via experiments on a range of simulated environments. Our results show that path length improves rapidly as the number of random sample points is increased, and eventually settles at the near-optimum value. Moreover, for a given environment, computing time appears to be linear with respect to the number of sample points.

Although our work deals with the idealized case of a perfect polygonal map and an ideal range sensor, our algorithms are flexible and can generalize easily to more realistic and challenging environments. These results are significant for many problems in which pose estimation must take place in the context of ambiguity, including certain cases of SLAM (simultaneous localization and mapping) where pose errors may become large.

The rest of this paper is organized as follows: The next section discusses related work. Section 3 provides a formal description of the problem and states the assumptions underlying our approach. Our randomized localization algorithms are presented in Section 4. Experimental results are described in Section 5. Finally, we close with a discussion of our work and directions for future research.

2 Related Work

Dudek, Romanik, and Whitesides [3] present an approximation algorithm in which the localization problem is treated in two phases: hypothesis generation and hypothesis elimination. For the hypothesis generation phase, the solution proposed by Guibas, Motwani, and Raghavan [6] is used. Guibas *et al.* show how to preprocess a map polygon P, by computing its visibility cell decomposition, so that given the robot's visibility polygon V, the set of points in P whose visibility polygon is congruent to V and oriented similarly can be returned. This technique preprocesses P in $O(n^5 \log n)$ time and $O(n^5)$ space, where n is the number of vertices of P. For hypothesis elimination, an overlay arrangement is computed yielding $O(n^6)$ cells. The robot then greedily travels to the nearest distinguishing visibility cell from its initial location in order to rule out hypotheses. By measuring distance with respect to the robot's initial location, rather than its current location, the paper is able to show that the robot travels a path of at most $(k-1)d$, where k is the number of initial hypothetical locations generated and d is the length of an optimal verification tour. The authors prove that this competitive ratio is the best possible. However, with space and time complexity of $O(n^6)$, their algorithm is impractical to use. The authors also demonstrate that simple heuristic localization strategies can sometimes exhibit strikingly poor performance.

Schuierer [8] proposes a technique that uses geometric overlay trees to speed up the implementation of the greedy localization strategy put forth by [3]. While his approach reduces the time complexity to $O(kn^2)$ and space complexity to $O(kn)$, no implementation results are given and it is unclear how to extend the technique to address more practical issues. Kleinberg [7] approaches robot localization by modeling the environment as a bounded geometric tree. Brown and Donald [1] describe algorithms for robot localization which allow for uncertainty in the data returned by the range sensor. Fox, Burgard, and Thrun [4] use Markov localization to determine the position of a mobile robot from sensor data. In that work, global localization is achieved, but the length of the localizing trajectory relative to the optimum is not considered.

Gonzalez-Banos and Latombe [5] present randomized algorithms that use a $2D$ map to estimate the locations where sensing will be most effective through random sampling of the workspace. Their work does not address the problem of robot localization.

3 Problem Specification

In this section we formally define the localization problem and state our assumptions about the robot and its environment. We are given a random environment modeled by an n-vertex simple polygon P without holes positioned somewhere in the $2D$ plane. A mobile robot is placed at an unknown initial location within P, of which it has a map. First, the robot must determine if its initial location is unique by sensing its surroundings and matching the resulting visibility data W to the map of the environment. Given P and W, the robot must generate the set H of all hypothetical locations $p_i \in P$ such that the visibility at p_i is congruent under translation to W. Next, the robot must determine its true initial location by sensing and traveling in order to eliminate all hypothetical locations but one from H, while minimizing the distance traveled.

The robot's sensor behaves as a "perfect" sensor in that it can detect distances to those points on the boundary of the environment for which the robot has an unobstructed line of sight. The visibility data W sensed by the robot is composed of the counter-clockwise ordering of vertices and edges seen by the robot (see Figure 1). Geometric relationships amongst the data such as vertex angles, distances, adjacencies, and the robot's relative position with respect to the data sensed are available. W can also be regarded as a visibility polygon.

The robot is assumed to be a point robot moving in this static $2D$, obstacle-free environment. The robot is able to make error-free motions between arbitrary locations in the environment. The movement of the robot is restricted to the inside and along the boundary of the environment. As well, it is able to determine its orientation. Otherwise it would be impossible for the robot to uniquely determine its exact location in an environment with non-trivial symmetry such as a square [2].

4 Approach

Our localization algorithms (like that of [3]) comprise two phases: hypothesis generation and hypothesis elimination. The hypothesis generation phase computes the set of hypothetical locations that match the observations sensed by the robot at its initial location. The hypothesis elimination phase rules out incorrect hypotheses thereby determining the true initial location of the robot. However, unlike the hypothesis generation phase of [3], ours does not involve any preprocessing of the map polygon P, nor is the visibility cell decomposition of P computed. Instead we generate hypotheses using an online method. Our hypothesis elimination phase also differs from that of [3]. In contrast to the deterministic evaluation visibility cells as a potential probe locations, carried out in [3], we choose potential probe locations by randomly sampling points in certain regions of P. We then check if the sampled location provides any new information. This avoids the computational complexity of calculating the visibility cell decomposition together with the overlay arrangement as performed in [3]. Moreover, rather than pursue a greedy choice as in [3], our decision strategy directs the robot to make a weighted choice with respect to its initial location. The utility or information gain of each potential probe

[2]Note: the robot could still solve the problem up to rotational symmetry.

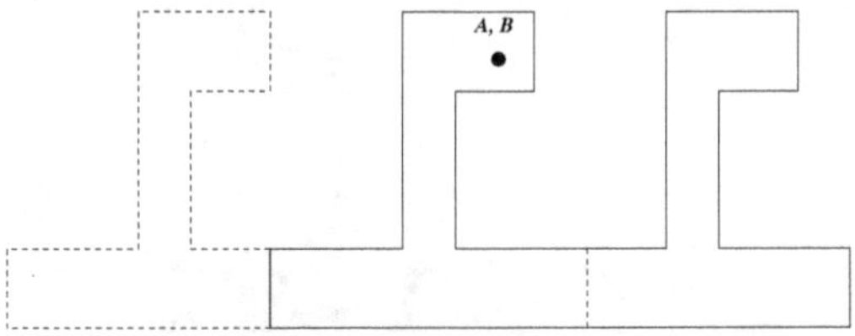

Fig. 2. Overlay arrangement for the situation shown in Figure 1.

destination is weighted by its distance from the robot's initial location. Before we continue with a description of the localization algorithms, we describe some key definitions and procedures.

4.1 Hypothesis Generation

Given an environment P and visibility data W, we want to determine the set H of all hypothetical locations $p_1, p_2, ..., p_k \subset P$ such that the visibility data computed at p_i, $Vis(p_i)$, matches W for $i = 1, ..., k$ (see Figure 1). This is done by tracing the pattern of vertices and edges listed in W in the set of vertices and edges of P. We search for as many sets of vertices and edges in P that match W as possible. Note that k refers to the number of *initial* hypothetical locations generated.

4.2 Overlay Intersection

The hypothesis generation phase generates a set $H = p_1, p_2, ..., p_k \subset P$ of hypothetical locations in P at which the robot might be located initially. Without loss of generality, we select an arbitrary hypothetical location p_i from H to serve as a reference point or origin. Next, for each hypothetical location $p_j, 1 \leq j \leq k$, a translation vector $t_j = p_i - p_j$ is defined that translates location p_j to p_i. As a result, we compute a set of copies $P_1, P_2, ..., P_k$ of the environment polygon P, corresponding to the set of hypothetical locations H, such that P_j is congruent to P translated by the vector t_j. Copy P_i is translated by the zero vector. The point in each translated polygon P_j corresponding to the hypothetical location p_j is now located at the origin p_i. We can now define the overlay arrangement as follows:

Definition 1 (Overlay Arrangement [3]).
The overlay arrangement *for the environment polygon P corresponding to the set of hypothetical locations H is the structure obtained by taking the union of the edges of each translated polygon $P_j, 1 \leq j \leq k$.*

Figure 2 illustrates an overlay arrangement for the situation shown in Figure 1. We consider only the *connected component of the intersection region of the overlay arrangement that contains the origin* since it is the area known to

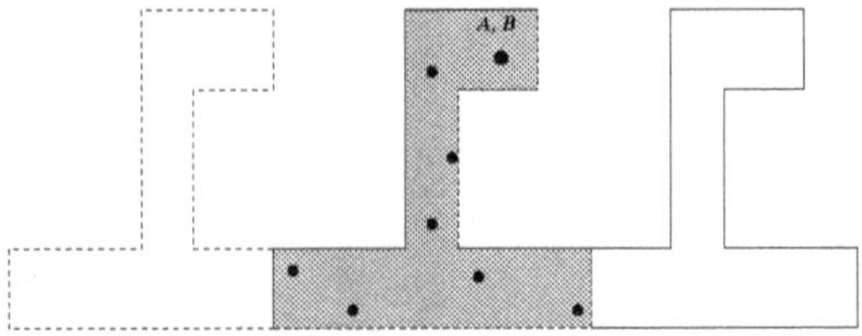

Fig. 3. Shaded area represents the overlay intersection region OI. Random points are chosen within this region.

exist in all hypotheses (see Figure 3). We will refer to this connected overlay intersection component containing the origin as OI. In this respect, we differ from [3] in recognizing that clearly, we do not need to examine visibility cells lying outside this overlay intersection region, common to all hypothetical locations, for new information. We compute OI by performing a counter-clockwise traversal of the edges of the polygons containing the origin. The algorithm begins with an arbitrary edge e which is either partly or fully visible to the origin. Every time an intersection point is encountered, the edge that makes the sharpest counter-clockwise turn with respect to the origin is selected. The algorithm terminates when it returns to the starting point on e.

4.3 Hypothesis Elimination

A set R of points is picked randomly from the connected intersecting region of the overlay arrangement surrounding the origin (see Figure 3). R is then evaluated to see if any of the points contained in R yield new information that could help to disambiguate the robot's initial location. Our decision strategy weights the utility of a potential destination point with its distance from the robot's initial location. Before we proceed we will define some terms:

Definition 2 (Random Point).
A random point *refers to a location in the connected component OI of the overlay intersection region containing the origin for a given P, H, and origin, chosen by randomly sampling the interior of OI according to a uniform distribution.*

Definition 3 (Useful Point).
A random point q is termed useful *if sensing at this location is guaranteed to yield new information that distinguishes amongst the different hypothetical locations. In other words, the $Vis(q)$ with respect to the different hypotheses are not all congruent under translation.*

For each random point picked, $r \in R$, a value function $Value(r) = info/distance_{OI}$ is computed, where $info$ is the expected number of hypotheses that could be eliminated at r, assuming all the hypothesized initial locations are equally likely, and $distance_{OI}$ is the shortest path trajectory, constrained to lie within OI, from the robot's initial location at the origin of the overlay to r. We calculate $info$ for a point r as follows:

We assume that all hypotheses are equally likely. We say two hypotheses h_i and h_j are equivalent at r if $Vis(h_i, r)$ is congruent to $Vis(h_j, r)$ and has the same orientation. $Vis(h_i, r)$ refers to the visibility data computed at a point z such that the relative position of z with respect to the hypothetical location p_i is equivalent to the relative position of r with respect to the overlay origin. If there exist b equivalence classes of hypotheses at r of sizes $s_1, s_2, ..., s_b$ respectively, where the total number of hypotheses $k = s_1 + s_2 + ... + s_b$, then

$info(r) = (s_1/k)(k - s_1) + (s_2/k)(k - s_2) + ... + (s_b/k)(k - s_b)$.

The robot is moved to the random point r' in OI with the highest non-zero value of $Value(r')$. Those hypotheses h_i where $Vis(h_i, r')$ does not match the visibility data sensed by the robot at its new location are ruled out.

4.4 Common Overlay Localization Algorithm

We can now present the *common overlay localization (COL)* algorithm. Given an input polygonal environment P and a robot placed at an unknown initial location in P, the COL algorithm proceeds as follows:

1. Sense visibility data W from the robot's current unknown initial location.
2. Generate the set of all hypothetical locations H in the environment P that match the visibility data sensed W.
3. Choose an arbitrary hypothetical location in H as the origin.
4. Construct an overlay arrangement centered on the origin.
5. Compute the connected overlay intersection component containing the origin, OI.
6. Randomly choose a predetermined number of points within OI.
7. For each random point picked, r, compute the value function $Value(r) = info/distance_{OI}$.
8. Observe that at each overlay intersection, there is latent information to be gained that is guaranteed to eliminate some hypotheses. Therefore, if none of the random points yield non-zero $info$, then the number of random points required is increased and chosen all over again within the current overlay intersection area OI. Steps 6, 7 and 8 are repeated for a predetermined number of trials [3].
9. The robot moves to the random point r' in the overlay with the highest non-zero value of $Value(r')$.
10. Now, eliminate hypotheses by comparing visibility data sensed by the robot at r' with the visibility data computed at all the equivalent random points corresponding to all the active hypotheses.

[3] In our implementation, we terminate the algorithm if no useful points are obtained after this predetermined number of trials. Hence we proceed to the next step only if useful points exist.

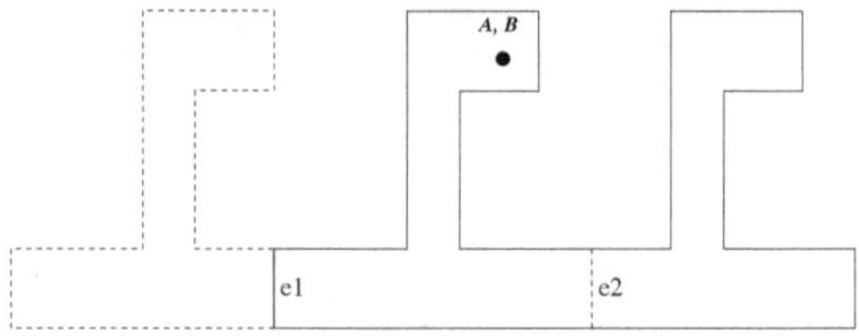

Fig. 4. e1 and e2 are internal edges.

11. Let us call the set of eliminated hypotheses E. We repeat the overlay arrangement with the reduced set of hypotheses $H - E$. Steps 3-10 are repeated until only 1 hypothesis, corresponding to the true initial location of the robot, is left in $H - E$.

4.5 Useful Region Localization Algorithm

The *useful region localization (URL)* algorithm differs from the COL algorithm with respect to the region where random points are chosen. Recall that the COL algorithm randomly chooses points within the overlay intersection area OI. The COL algorithm then selects only those random points that are expected to yield new information. It turns out that we can do even better. In fact, we can determine precisely the portions of OI where any random point chosen is guaranteed to provide new information.

Let us first observe that in large sections of OI, the visibility cells provide zero information gain for unambiguous localization. We must look to the boundaries of OI which is where the area common to all hypotheses ends and the"geography" changes. But the robot can spot this change from a distance as it approaches the boundaries. We refer to these boundaries as *internal edges*, which we define as follows:

Definition 4 (Internal Edge).
An internal edge *of an overlay intersection area OI is defined as an edge (one of many) that separates the inside of OI from other parts of the overlay arrangement, as opposed to those edges of OI that pertain to the outer silhouette of the overlay arrangement which separates the inside of OI from the rest of the 2D plane (see Figure 4).*

Before we proceed to the description of the URL algorithm, we will examine the notion of *weak* visibility from an edge. A set of points Q is said to be *weakly visible* from an edge e if for each point $q \in Q$ there exists a point $z \in e$ such that q and z are visible (z may depend on q) [9].

Definition 5 (Weak Visibility Polygon).
Given a polygon P the weak visibility polygon $W(e)$ *of an edge $e \in P$ is defined as the set of all points $y \in P$ that are visible from some point on e (see Figure 5).*

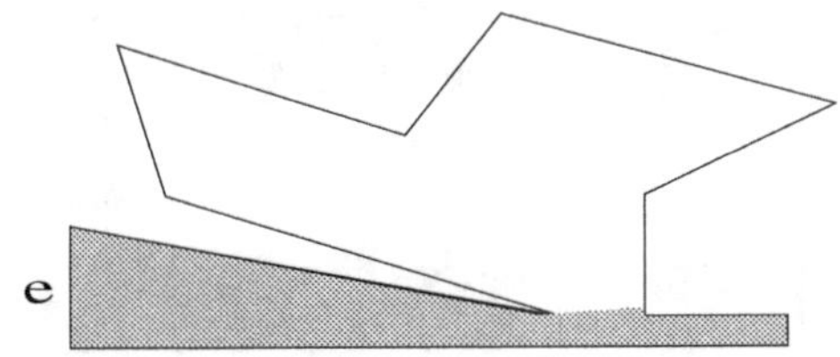

Fig. 5. Shaded region represents the weak visibility polygon of edge e.

Once we have determined the set of internal edges of *OI*, the useful portions *U* of *OI* can be computed using the following procedure:

- Let us call the set of internal edges of *OI*, *B*. For each edge in *B*, compute its weak visibility polygon within *OI*. The union of all such weak visibility polygons should give us a region or set of disjoint regions *U* where *U* is a subset of *OI*. We claim that any random point chosen in *U* provides non-zero information whereas any random point chosen in *OI*-*U* provides zero information.

Figure 6 depicts the useful region of polygon P. Steps 1-5 of the URL algorithm remain the same as in the COL algorithm. Following these steps, the URL algorithm proceeds as described below:

6. Compute the useful region U of OI.
7. Randomly choose a predetermined number of points within U (Figure 6).
8. For each random point picked, r, compute the value function $Value(r) = info/distance_{OI}$.
9. The robot moves to the random point r' in the overlay with the highest non-zero value of $Value(r')$. Note that we are guaranteed that all the random points chosen provide non-zero information for hypothesis elimination. As a result, we do not need to choose more random points repeatedly as is done in the COL algorithm.
10. Now, eliminate hypotheses by comparing visibility data sensed by the robot at r' with the visibility data computed at all the equivalent random points corresponding to all the active hypotheses.
11. Let us call the set of eliminated hypotheses E. We repeat the overlay arrangement with the reduced set of hypotheses $H - E$. Steps 3-10 are repeated until only 1 hypothesis, corresponding to the true initial location of the robot, is left in $H - E$.

It must be noted that once we have computed the useful region *U*, we need choose only one random point lying within *U* at each stage in order to guarantee that we eventually localize successfully (we could even just move the robot to the closest vertex or edge of *U*). A larger number of random points only serves to improve the performance of the algorithm by eliminating several hypotheses in one shot or by reducing the distance traveled by the robot.

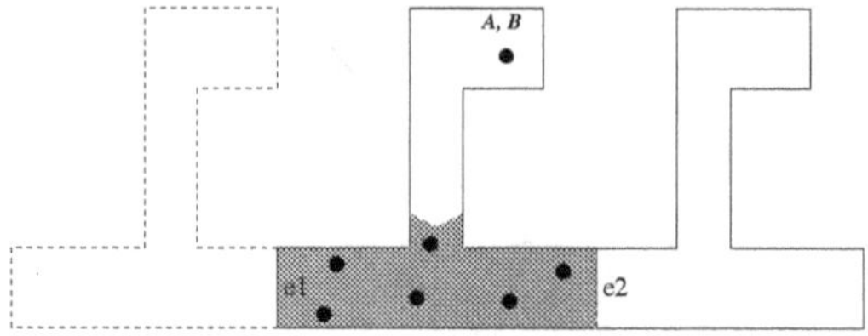

Fig. 6. Shaded area represents the useful region. Random points are chosen within this region.

Unlike the COL algorithm, the URL algorithm has computable and finite time bound. Consider a situation where the useful region comprises a small fraction of the entire overlay intersection area OI (this need not mean that the useful region is in itself a small area where a robot can not navigate - simply that it is relatively small compared to the entire overlay intersection area). In such cases, choosing points randomly from the entire overlay intersection region may yield a useful point only with a large number of random points and after several trials. The advantage of the URL algorithm is that we can instantly access the useful portions of OI and hence the useful points [4].

4.6 Complexity Analysis

In this section, we provide an estimate of the overall time complexity for the two localization algorithms. The number of initial hypothetical locations k is bounded above by the number of reflex vertices in the polygonal environment [5]. Let f be the number of active hypotheses remaining ($f \leq k$). Let n denote the number of vertices in the map polygon P.

Hypothesis generation takes time $O(mn)$ where m is the number of vertices in the visibility data W. Calculating the overlay intersection area takes time $O(fn \log n)$. Observe that as hypotheses are ruled out the value of f will decrease. If we rule out only one hypothesis per intersection then the overlay intersection is computed a maximum of k times. It takes $O(fn)$ time per random point in order to calculate the information gain at any random point. Computing the shortest path distance of a random point from the initial location amounts to $O(n \log n)$ time.

Common Overlay Localization Algorithm

At each intersection let us say that we select X random points on average. Let only Y be the number of useful points where $Y \leq X$. Assuming that we only eliminate one hypothesis at each overlay intersection, we will have to calculate the overlay intersection area, compute and compare visibility data of all the

[4]A proof of correctness for the URL algorithm is presented in [10].

[5]For a proof see [6].

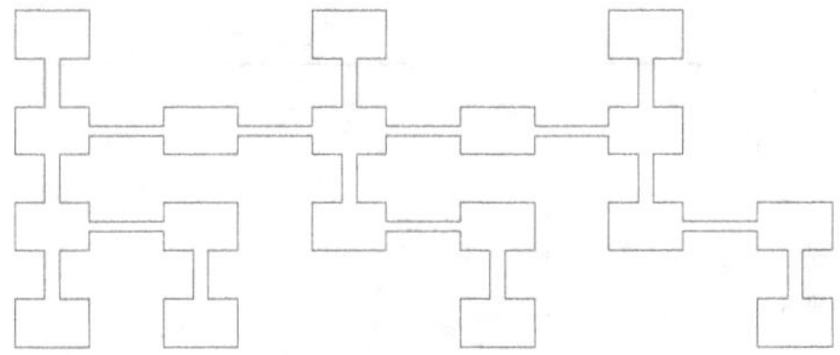

Fig. 7. A simulated office environment.

random points, and calculate the shortest path distance of the useful random points, a total of k times. Therefore we estimate the overall time complexity of the COL algorithm to be $O(mn)+k(O(fn\log n)+O(Xfn)+O(Yn\log n))$. Since $k=O(n)$ and $f=O(k)$, a looser but simpler bound for the overall time complexity of the COL algorithm is obtained: $O(n^3\log n+Xn^2)$.

Useful Region Localization Algorithm

The maximum number of internal edges possible at any overlay intersection is $O(fn)$. Let Y be the number of useful random points. Computing the weak visibility of an edge takes $O(n\log n)$ time [9]. As a result, the total cost of computing the useful region is $O(fn^2\log n)$. Therefore we estimate the overall time complexity of the URL algorithm to be $O(mn)+k(O(fn\log n)+O(fn^2\log n)+O(Yfn)+O(Yn\log n))$ which reduces to $O(n^4\log n+Yn^3)$.

5 Experimental Results

Experiments were carried out on random environments in order to obtain an empirical measure of the average path length for different numbers of random points and with respect to the expected optimum path length. We also compare the performance of our strategy with competing strategies. In order to test the localization algorithms, we used random simulated "office" environments that were generated algorithmically (see Figure 7).

We generated 73 simulated office environments with an average of 400 vertices each. For each environment an initial robot location was randomly selected. We then ran the COL algorithm with these environments and their respective initial locations for a series of different quantities of random points. The objective was to measure the average distance traveled by the robot as the number of random points was increased and to compare these average path lengths to the estimated optimum result. The number of random points was varied from 20 to 1000 and the average path length obtained for 1000 points was used as the estimated optimum result. Figure 8 shows the error margin of the average path length with respect to the estimated optimum path length plotted against the number of random points. Our results indicate that the average path length gets significantly shorter initially as the number of

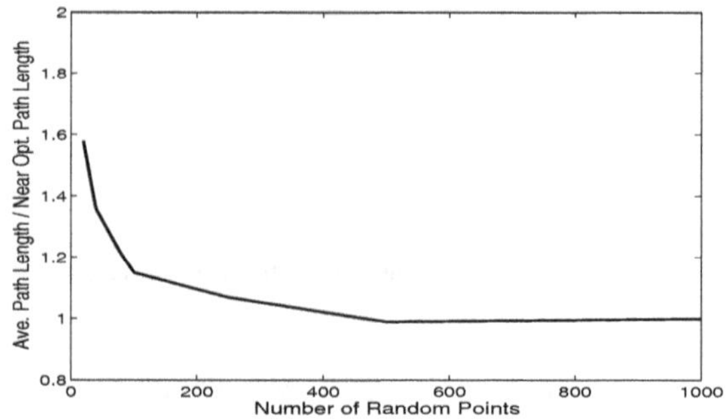

Fig. 8. Ratio of path length obtained with COL algorithm with respect to the expected optimum path length based on 73 trials. Near-optimal path length is achieved with a relatively small number of points.

random points is increased. Eventually, the incremental reduction in path length decreases and the path length settles at the near-optimum value [6].

We ran experiments to compare the performance of the *weighted decision strategy* used in the COL algorithm (which we will refer to as the weighted strategy) versus its greedy version, where the robot is directed to move to the nearest location to its initial location that provides any non-zero information. This greedy alternative *approximates* the greedy technique of [3] but is not equivalent to it since we require an appropriate minimum number of points in order to adequately cover all the visibility cells in the overlay intersection region. We generated 20 simulated office environments with an average of 400 vertices each. For each of the 20 environments, 3 initial locations were randomly selected. We then ran both weighted and greedy strategies with each of the 3 initial locations for each environment for a series of different quantities of random points, repeating each quantity twice to balance out any abnormal distributions. The total number of experimental trials was, therefore, $20 \times 3 \times 2 = 120$ trials [7]. Figure 9 shows the average path length for weighted and greedy strategies plotted against the number of random points. Even with very large numbers of random points (see 1000 points case), where we may reasonably assume an adequate sampling of most of the visibility cells in the overlay intersection region, on average the weighted strategy still outperforms its greedy counterpart.

Theoretically, the greedy localization strategy has been shown to have the best possible worst case bound on the distance traveled by the robot [3]. In our weighted strategy as well as its greedy variant, it is possible that a tiny visibility cell which might hold the key to an optimally short path was never sampled. This is a theoretical disadvantage but a practical advantage as the

[6] Recall that computing the optimum path is NP-hard.

[7] This set of trials will be used repeatedly for the experiments described in the remainder of this paper.

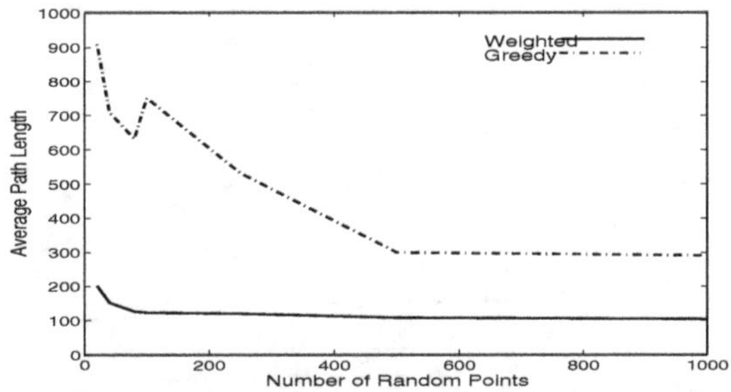

Fig. 9. Performance of weighted decision strategy of COL algorithm vs. greedy strategy based on 120 trials. Weighted strategy gives shorter path lengths on average.

probability that the robot is directed to visit an arbitrarily tiny, and hence inaccessible, visibility cell is small.

A set of experiments, involving 120 trials, was performed where the weighted decision strategy of the COL algorithm was compared with a traditional heuristic used in many localization papers. The heuristic directs the robot to simply visit the closest point from its *current* location which provides any non-zero information. Figure 10 shows the average path length for our weighted strategy and the traditional heuristic plotted against the number of random points. The weighted strategy produces shorter path lengths than the heuristic.

Experiments were carried out to evaluate the performance of the URL algorithm with respect to different numbers of useful random points. We performed the experiments over 120 trials as described above. Figure 11 depicts the average path length obtained for number of useful random points ranging from 1 to 500. Since we are choosing points directly from the useful region, the algorithm is able to effectively localize the robot with just 1 random point, although the path length is understandably high in that case. We observe that path length values for the URL algorithm are somewhat higher than those obtained for the same number of points when running the COL algorithm. This is because the COL algorithm chooses a more dense sampling of random points each time none of the existing points prove useful. In particular, the overlay intersection areas get larger as hypotheses are eliminated thereby requiring a more dense sampling. On the other hand, the URL algorithm adheres to exactly the same number of points initially specified, regardless of the size of the subsequent overlay intersection regions.

Figures 12 and 13 depict the localization path taken by the robot for a staircase-like environment and a simulated office environment respectively, using the COL algorithm. The shaded area in these figures represents the overlay intersection region OI. The square black points labeled $H0, H1, ..., Hk$ indicate the different hypothetical locations. The small round dots scattered across the shaded region represent the random points.

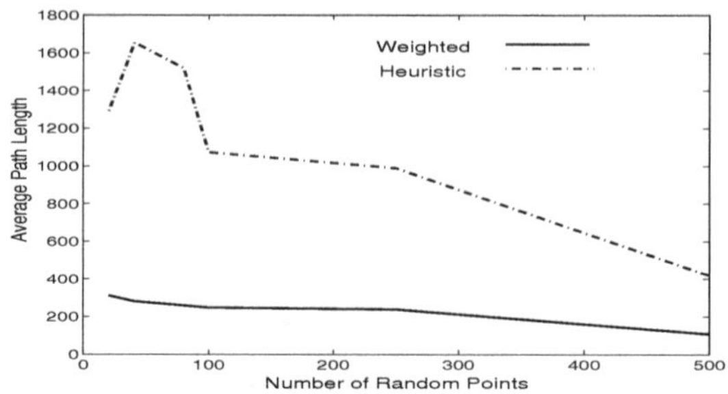

Fig. 10. Performance of weighted decision strategy of COL algorithm vs. traditional heuristic based on 120 trials. Weighted strategy gives shorter path lengths on average.

6 Discussion and Conclusions

In this paper we have presented a new algorithmic formulation of minimum distance localization based on randomized sampling and determined its complexity. We demonstrate that although minimum distance localization is a hard problem, a good approximation can be achieved in most cases with limited computation.

The experimental results show that for the ensemble of environments we have evaluated, our algorithms are effective. The performance of our algorithms improves rapidly with the number of random sample points used. However, the incremental improvement decreases as the number of samples increases, so that a fairly limited number of samples typically is sufficient to obtain a near-optimal length localization trajectory.

In the case of the URL algorithm, we require as little as 1 random point in order to unambiguously localize the robot. In situations where the useful region comprises a small fraction of the entire overlay intersection area, the URL algorithm seems to be the better choice (than the COL algorithm). Choosing points randomly from the entire overlay intersection region may yield a useful point only with a relatively large number of points. The advantage of the URL algorithm is that we can instantly access the useful portions of the overlay intersection and hence the useful points.

In addition, experimental results indicate that on average, our weighted decision strategy is clearly superior to alternative greedy strategies which simply visit the nearest points to the robot's current, or even starting, location, that offer any information at all. The strategy of picking points stochastically leads to a visibility cell having a probability of being visited that is proportional to its area. This implies that large cells, that are accessible to the robot, will be chosen for visits more frequently than small ones. This in itself makes our algorithm more feasible for real implementation than its predecessors.

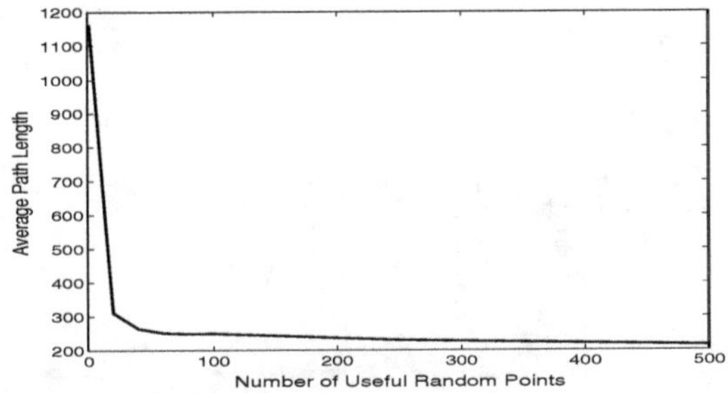

Fig. 11. Performance of URL algorithm based on 120 trials. More dense sampling of useful random points produces rapid reduction in path length.

While the validity of these results is formally limited to the class of environments presented here, there is no reason why these conclusions should not generalize to broader classes of environments. Our algorithms appear to be appropriate to use in contexts where the sensor is more limited, given suitable modifications to the definition of a landmark. This suggests a natural extension to an even more realistic context.

References

1. R.G. Brown and B.R. Donald, *Mobile robot self-localization without explicit landmarks*, Algorithmica **26** (2000), no. 3/4, 515–559.
2. G. Dudek and M. Jenkin, *Computational principles of mobile robotics*, Cambridge University Press, 2000.
3. G. Dudek, K. Romanik, and S. Whitesides, *Localizing a robot with minimum travel*, SIAM J. Computing **27** (1998), no. 2, 583–604.
4. D. Fox, W. Burgard, and S. Thrun, *Active markov localization for mobile robots*, Robotics and Autonomous Systems **25** (1998), 195–207.
5. H.H. Gonzalez-Banos and J.C. Latombe, *Planning robot motions for range-image acquisition and automatic 3d model construction*, In Proc. AAAI Fall Symposium Series, 1998.
6. L. Guibas, R. Motwani, and P. Raghavan, *The robot localization problem*, SIAM J. Computing **26** (1997), no. 4, 1120–1138.
7. J. Kleinberg, *The localization problem for mobile robots*, In Proc. 35th IEEE Conference on Foundations of Computer Science (Santa Fe, NM), IEEE Computer Society Press, 1994, pp. 521–533.
8. S. Schuierer, *Sensing, modelling and planning*, Intelligent Robots, ch. Efficient robot self-Localization in simple polygons, pp. 129–146, World Scientific Publ., 1996.
9. G.T. Toussaint, *A linear-time algorithm for solving the strong hidden-line problem in a simple polygon*, Pattern Recognition Letters **4** (1986), 449–451.
10. M. Rao, *A Randomized Approach to Minimum Distance Localization*, MSc. thesis, Dept. of Computer Science, McGill University, Montreal, Canada, 2004.

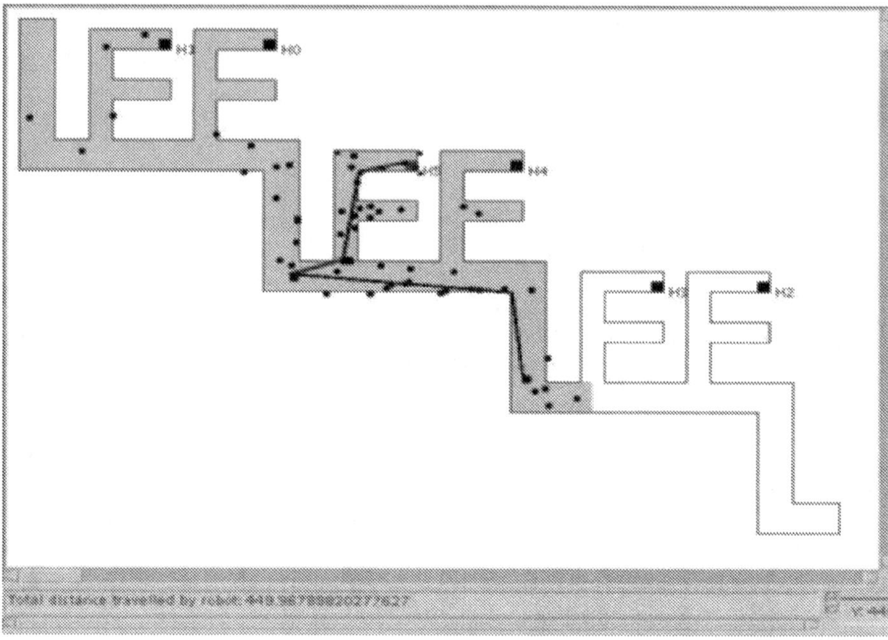

Fig. 12. Localization trajectory in staircase environment.

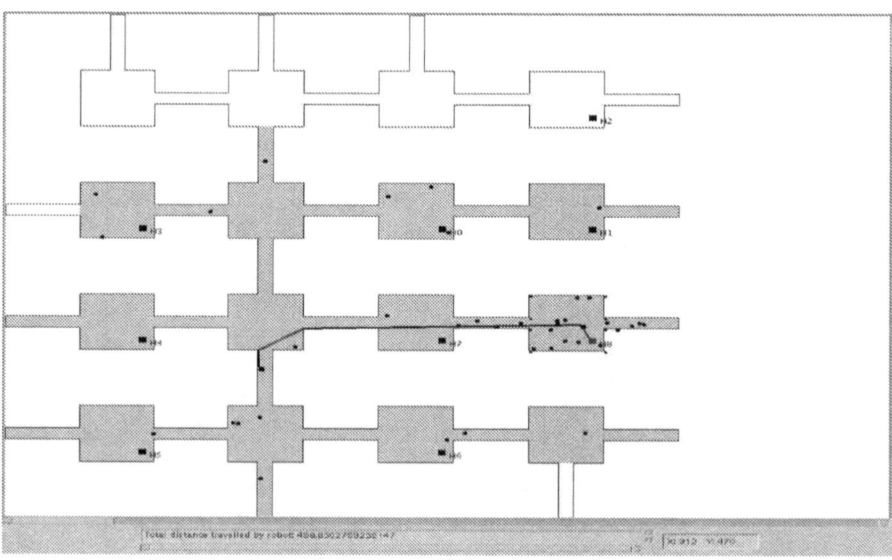

Fig. 13. Localization trajectory in simulated office environment.

Multi-Point Contact Models for Dynamic Self-Righting of a Hexapod*

Uluç Saranli[1], Alfred A. Rizzi[1], and Daniel E. Koditschek[3]

[1] Robotics Institute, Carnegie Mellon University, Pittsburgh, PA 15213, USA
[2] Dept. of EECS, The University of Michigan, Ann Arbor, MI 48109-2110, USA

Summary. In this paper, we report on the design of a model-based controller that can achieve dynamical self-righting of a hexapod robot. Extending on our earlier work in this domain, we introduce a tractable multi-point contact model with Coulomb friction. We contrast the singularities inherent to the new model with other available methods and show that for our specific application, it yields dynamics which are well-defined. We then present a feedback controller that achieves "maximal" performance under morphological and actuation constraints, while ensuring the validity of the model by staying away from singularities. Finally, through systematic experiments, we demonstrate that our controller is capable of robust flipping behavior.

Keywords: legged robot, model based control, contact modeling, flipping, RHex

1 Introduction

RHex is an autonomous hexapod robot that negotiates badly irregular terrain at speeds better than one body length per second [12]. In this paper, we report on efforts to extend RHex's present capabilities with a self-righting controller. Based on our earlier work in this domain [13, 14], we introduce a multi-point contact model based on Coulomb friction so as to derive the maximum benefit of our robot's limited power budget. In designing a model-based feedback controller, we highlight fundamental singularity problems in handling multiple intermittent contacts under external actuation and illustrate a particular solution for our application domain. In particular, we present experimental results that demonstrate the practical implementation of a controller based on the model which yields a reliable self-righting maneuver.

The primary motivation for this problem is to achieve autonomous operation of robotic platforms in unstructured environments. In addition to physical

*Portions of the material in this paper, in combination with material excerpted from [14] have been submitted for publication to the International Journal of Robotics Research.

M. Erdmann et al. (Eds.): Algorithmic Foundations of Robotics VI, STAR 17, pp. 409–424, 2005.

necessities such as onboard power and computation, the survivability of such robots in the real world requires at least basic manueuverability capabilities to remain functional in the absence (or inattention) of a human operator [1]. The problem of recovering correct body orientation is one of the simplest among such tasks and constitutes the focus of the present paper.

RHex's morphology is roughly symmetric with respect to the horizontal plane, and allows nearly identical upside-down or right-side up operation, a solution adopted by other mobile platforms [7]. However, many application scenarios such as teleoperation and vision based navigation entail a nominal orientation arising from the accompanying instrumentation and algorithms. In order to secure self-righting capabilities in such settings, most robotic platforms manipulate their body orientation through special kinematic structures (for example, long extension arms or reconfigurable wheels [3, 4, 18]). In contrast, the imperatives of dynamical operation that underly RHex's design and confer its unusual mobility performance [12] preclude such structural appendages. RHex must rely on its existing morphology and dynamic maneuvers to achieve the necessary self-righting ability.

In this light, the central contributions of this paper include: i) introducing a frictional multiple point collision/contact model that characterizes RHex's behavior during the flipping maneuver; and ii) the description of a new torque control strategy that uses the model to maximize the energy injected into the system in the face of these constraints (i.e., consistent with maintaining a set of postural invariants integral to the task at hand). We contrast analytical properties of the newly introduced model to a number of other approaches proposed in our earlier work and highlight fundamental problems in their application to the design of feedback controllers. We also present experimental evidence to establish the validity of the present model.

2 Flipping with RHex

RHex is a relatively simple hexapod robot with only six actuators, directly coupled to individual passively compliant legs. The rotation axes for the legs are all parallel and aligned with RHex's transverse horizontal body axis (see Fig. 1). Consequently, the most natural backflip strategy for RHex pivots the body around one of its endpoints. Pitching the body in this manner, while keeping one of the body endpoints in contact with the ground, maximizes contact of the legs with the ground for the largest range of pitch angles and thus promises to yield the best utilization of available actuation. In contrast, flipping by producing a sideways rolling motion suffers from early liftoff of three legs on one side as well as the longer protrusion of the middle motor shafts.

For surfaces with sufficiently low lateral inclination, RHex's lateral symmetry and the line of body contact restrict this flipping motion to the saggital plane, admitting the use of planar models for its analysis.

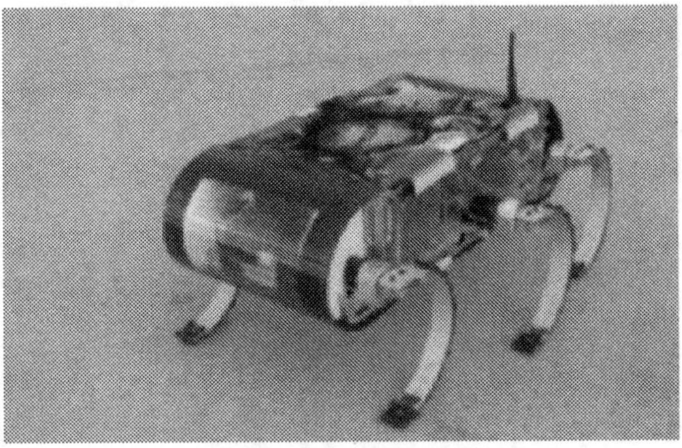

Fig. 1. RHex 1.5

2.1 Basic Controller Structure

The model-based controller we present in this paper has a finite state machine structure, illustrated in Fig. 2. Starting from a stationary position on the ground, the robot quickly thrusts itself upward while maintaining contact between the ground and the endpoint of its body (poses I and II in Figure 2) as the front and middle legs successively leave the ground. This thrust results in some initial kinetic energy of the body that, depending on the frictional properties of the leg/ground contact, may be sufficient to allow "escape" from the gravitational potential well of the initial configuration, allowing the robot to fall into the desired configuration. In cases where a single thrust is not sufficient to flip the body over, the robot reaches some maximum pitch, and falls back toward its initial state. Our controller then brings the legs back to Pose I of Figure 2 and waits for the impact of the front legs with the ground, avoiding negative work – a waste of battery energy given the familiar power-torque properties of RHex's conventional DC motors. The

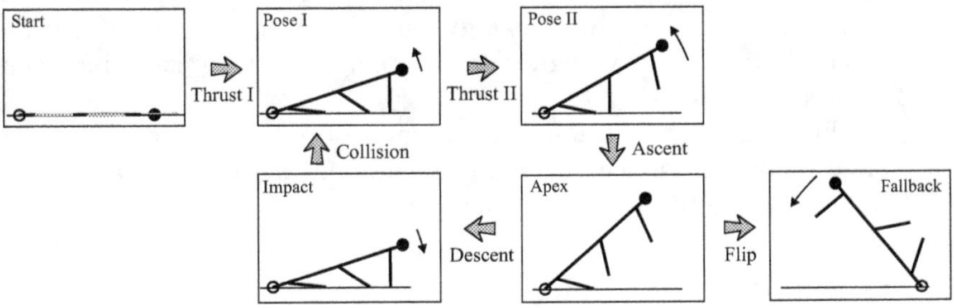

Fig. 2. Sequence of states for the flipping controller

impact of the compliant front legs with the ground in their kinematically singular configuration recovers some of the body's kinetic energy, followed by additional thrust from the middle and back legs, during the period of decompression and flight of the front leg – i.e., during a phase interval when it is possible for the legs in contact to perform positive work on the robot's mass center. The maximum pitch attained by the body increases with each bounce up until the point where the robot flips or the energy that can be be imparted by the thrust phase balances collision losses at which point it must follow that flipping is not possible.

3 Planar Flipping Models

3.1 Assumptions and Constraints

Two assumptions constitute the basis for our modeling and analysis of the flipping behavior.

First, we assume that the flipping behavior is primarily planar. The controller structure described in Sec. 2.1 operates contralateral pairs of legs in synchrony. On flat terrain, the robot's response lies almost entirely in the saggital plane and departures are rare enough to be negligible. Our models and analysis are thus constrained to the saggital plane.

Second, we assume that the tail of the body should maintain contact with the ground throughout the flipping action. This is motivated by a number of observations gathered during our empirical flipping experiments. During the initial thrust phases, the front and middle legs provide most of the torque. Configurations where the tail endpoint of the body is in contact with the ground yield the longest duration of contact for these legs, harvesting greatest possible benefit from the associated actuators. Collisions of the body with the ground, which introduce significant losses due to the high damping in the body structure designed to absorb environmental shocks, can also be avoided by preserving contact with the ground throughout the flipping action. Finally, the body ground contact is essential for maintaining the planar nature of the behavior and eliminating body roll. This is especially important for repeated thrust attempts of the hybrid energy pumping scheme, which rely on the robot body being properly aligned with as much of the impact kinetic energy recovered as possible.

In light of these assumptions, the design of thrust controllers has to satisfy two major constraints: keeping the tail endpoint of the body on the ground and respecting the torque limitations of the actuators.

3.2 Generic Planar Flipping Model

In this section, we introduce a generic planar model that constitutes the basis for our presentation for a number of more constrained variations.

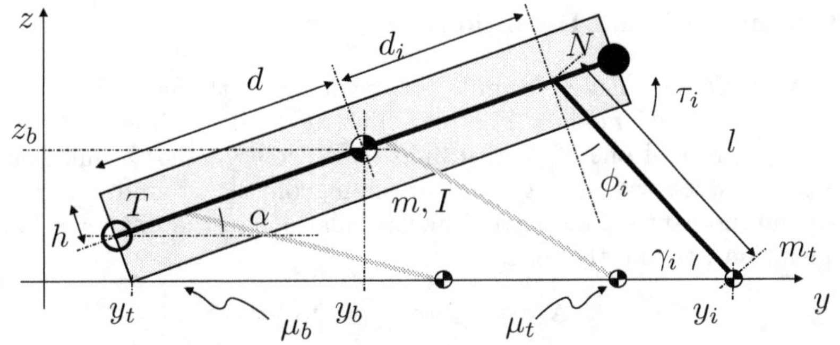

Fig. 3. Generic 3DOF planar flipping model

Figure 3 illustrates the model. Three rigid legs with small toe masses m_t – each representing a pair of RHex's legs – are attached to a rectangular rigid body with mass m and inertia I. The attachment points of the legs are fixed at d_i, along the midline of the rectangular body. This line also defines the orientation of the body, α, with respect to the horizontal. The center of mass is midway between the points N and T, defined to be the "nose" and the "tail". The body length and height are $2d$ and $2h$. Finally, the body-ground and toe-ground contacts experience Coulomb friction with coefficients μ_b and μ_t, respectively. Table 1 summarizes the notation used throughout the paper.

Table 1. Notation used throughout the paper

	States and dependent variables
y_b, z_b, α	Body center of mass coordinates and pitch
$\mathbf{c} \in \mathcal{X}$	System configuration vector
$\mathbf{q} := [\,\mathbf{c},\ \dot{\mathbf{c}}\,]^T$	System state vector
ϕ_i, γ_i	Hip and toe angles for i^{th} leg
$y_i, \dot{y}_i$	position and velocity of the ith toe
	Forces and model parameters
$\tau \in \mathbb{R}^3$	Hip torque control vector
$\mathcal{T}(\mathbf{q}) \subseteq \mathbb{R}^3$	Set of allowable torque vectors
F_i^y, F_i, F_c^y, F_c^z	GRF components on i^{th} toe and the tail
d, h, m, I	Body length, height, mass and inertia
d_i, l, m_t	Leg attachment offset, leg length and leg mass
μ_t, μ_b	Coulomb coeff. for toes and body

Our model requires both endpoints of the body as well as the toes to be above the ground, and that a leg must reach the ground before it can apply any torque to the body. We assume that the inertial effects of the toe masses on the body during their flight phase is negligible.

3.3 Framework and Definitions

We use a Newton-Euler formulation in deriving the equations of motion for all constrained models. Figure 4 illustrates the free-body diagrams for the body link and one of the leg links. Based on whether a link is in flight, in fixed contact with the ground or sliding on the ground, the associated force and moment balances yield linear equations in the unknown forces and accelerations, taking the form

$$\mathbf{A}(\mathbf{c})\mathbf{v} = \mathbf{b}(\mathbf{c}, \dot{\mathbf{c}}) + \mathbf{D}(\mathbf{c})\, \boldsymbol{\tau} \, . \tag{1}$$

The definitions of the configuration vector $\mathbf{c}$ and the vector of unknown forces $\mathbf{v}$, as well as the matrices $\mathbf{A}(\mathbf{c})$, $\mathbf{b}(\mathbf{c}, \dot{\mathbf{c}})$ and $\mathbf{D}(\mathbf{c})$ are dependent on the particular contact state and will be made explicit in subsequent sections.

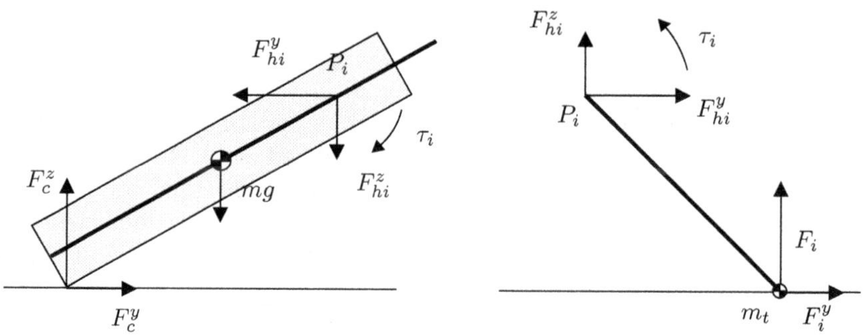

Fig. 4. Free body diagrams for the body and one of the legs

3.4 Simple Model with No Friction and No Toe Mass

As described in Sec. 2.1, flipping structurally relies on the slipping of the toes and the body to achieve thrust. The simplest model that comes to mind is a rigid body with massless and frictionless legs (i.e. $m_t = 0$, $\mu_t = 0$ and $\mu_b = 0$). This yields a one degree of freedom system where the ground reaction forces and thus the motion of the center of mass is strictly vertical. The configuration and unknown force vectors are defined as $\mathbf{c} := \alpha$ and $\mathbf{v} := [\ F_1,\ F_2,\ F_3,\ \ddot{\alpha},\ F^z_c\ ]^T$, respectively.

Unfortunately, in trying to derive the dynamics of such a system through a Newton-Euler analysis, one quickly realizes that configurations in which any one of the legs is vertical are singular and yield infinite ground reaction forces. Intuitively, this singularity arises from the faulty assumption that we can apply nonzero torque against a massless leg where the only external force is the vertical ground reaction force. Clearly, this model is physically inaccurate and is hence inappropriate for the analysis of the flipping behavior.

3.5 Extended Model with No Toe Friction and Small Toe Masses

Based on the observations of the previous section, we attempt to add small masses at the toes to counteract the action of the hip torque Assuming sufficient friction the body-ground contact and zero friction at the toes, this results in a one degree of freedom system. Defining the configuration and force vectors as $\mathbf{c} := \alpha$ and $\mathbf{v} := [\, F_1,\ F_2,\ F_3,\ \ddot{\alpha},\ F_c^z,\ F_c^y \,]^T$, the moment balance for leg i takes the form

$$(l \cos \gamma_i + l \bar{\mu}_i \sin \gamma_i) F_i + (l m_t b_i^y \sin \gamma_i) \ddot{\alpha} = l m_t a_i^y \sin \gamma_i - \tau_i \, ,$$

where $\bar{\mu}_i := -\mu_t \operatorname{sign}(\dot{y}_i)$ and $\ddot{y}_i = a_i^y - b_i^y \ddot{\alpha}$ is obtained by differentiating the kinematic equations for the toe position y_i twice. Note that the contact states of the toes no longer depend only on the configuration, but also on the velocities in the system due to the coriolis terms in a_i^y as well as the coupling with the pitch acceleration through b_i^y. This primarily results from the fact that the hip torque necessary to produce the horizontal toe acceleration required to maintain contact goes to infinity as the leg approaches its vertical configuration. As a consequence, the leg must "lift off" before it reaches its vertical configuration, at an angle which is a function of both the pitch angle and the pitch velocity. Detailed discussions of the algorithms necessary to uniquely identify contact states for this model and the derivation of its dynamics can be found in [13].

For this model, the ability to uniquely determine the contact state allows us to use the framework of Sec. 3.3 for computing the unknown forces. Surprisingly, there are still configurations for which the matrix $\mathbf{A}(\mathbf{c})$ may become singular. These singularities are much less intuitive and we have not been able to identify a physical explanation for their characterization. Even though for restricted ranges of operation on a particular morphology (such as flipping with RHex) it is possible to show that the system stays away from these singularities, they seem to be "inherent" to the family of models we have investigated and persist even when we reintroduce toe friction. A unifying explanation for this family of singularities remains an open question.

In any case, both the nonlinear kinematics of toe positions and the algorithmic complexities in determining contact state significantly impair the practicality of this model for an embedded implementation. Even though we previously showed simulation evidence for the performance of the model [14], its practical utility as a basis for implementing model-based flipping on RHex is limited.

Fortunately, it turns out that reintroducing Coulomb friction at the toes but maintaining the assumption of massless legs still yields a model wherein the typical range of operation for RHex's flipping is found to be away from singularities. The models we describe in the following two sections constitute the basis for our experimental flipping controllers.

3.6 Frictional Dynamics with Sliding Body, Sliding Toe Contacts

In this section, we present a model with massless legs and Coulomb friction, wherein both the leg and body contacts are sliding on the ground. Defining the configuration and unknown force vectors as

$$\begin{aligned} \mathbf{v} &:= [\, F_1,\ F_2,\ F_3,\ \ddot{\alpha},\ F_c^z,\ \ddot{y}_t \,]^T \\ \mathbf{c} &:= [\, \alpha,\ y_b \,]^T \ , \end{aligned} \tag{2}$$

yields a system with two degrees of freedom – the body pitch and the horizontal position of the tail. For each leg, we can write the moment balance equations as

$$(l \cos\gamma_i + l\bar{\mu}_i \sin\gamma_i)F_i = -\tau_i \ , \tag{3}$$

where $\bar{\mu}_i := -\mu_t \operatorname{sign}(\dot{y}_i)$ is the effective Coulomb friction coefficient and γ_i corresponds to the toe angle as shown in Fig. 3. In the operational range of the flipping controller, these equations are solvable. However, there are interesting "jamming" singularities in the remaining parts of the state space, which we investigate in Sec. 3.8.

Similarly, force and moment balances for the body link yield

$$\begin{aligned} \bar{\mu}_1 F_1 + \bar{\mu}_2 F_2 + \bar{\mu}_3 F_3 - \bar{\mu}_b\, F_c^z - m\ddot{y}_b &= 0 \\ F_1 + F_2 + F_3 + F_c^z - m\ddot{z}_b &= mg \\ \sum_{i=1}^{3} (d_i \cos\alpha - d_i \bar{\mu}_i \sin\alpha)F_i + ((h + \bar{\mu}_b d)\sin\alpha & \\ +(\bar{\mu}_b h - d)\cos\alpha)F_c^z - I\ddot{\alpha} &= \sum_{i=1}^{3} \tau_i \ , \end{aligned} \tag{4}$$

where, once again, system kinematics yields the body accelerations $\ddot{y}_b$ and $\ddot{z}_b$ as functions of $\ddot{\alpha}$ and $\ddot{y}_t$. As before, the combination of (3) and (4) yields the matrices $\mathbf{A}(\mathbf{c})$, $\mathbf{b}(\mathbf{c}, \dot{\mathbf{c}})$ and $\mathbf{D}(\mathbf{c})$.

3.7 Frictional Dynamics with Sliding Body, Fixed Rear Toe

The last model we consider still has massless legs and Coulomb friction, and corresponds to cases where the rear toe is stationary under the influence of stiction. This model is primarily motivated by the observed behavior of RHex's flipping, where the rear toe stops sliding following the liftoff of the front and middle pairs of legs. We incorporate this model into our feedback controller to be activated when the measured (or estimated) system state indicates that the rear toe is indeed stationary. Defining

$$\begin{aligned} \mathbf{v} &:= \left[F_1\ F_2\ F_3\ \ddot{\alpha}\ F_c^z\ F_1^y \right] \\ \mathbf{c} &:= \alpha \ , \end{aligned} \tag{5}$$

leaves us with a system with one degree of freedom. The moment balance for the rear leg includes the unknown horizontal ground reaction force, yielding

$$l \cos \gamma_1 F_1 + l \sin \gamma_1 F_1^y = -\tau_1 \,, \tag{6}$$

while the moment balance equations for the middle and front legs remain the same as (3). Finally, the balance equations for the body link take the form

$$\begin{aligned}
F_1^y + \bar{\mu}_2 F_2 + \bar{\mu}_3 F_3 - \bar{\mu}_b \, F_c^z - m\ddot{y}_b &= 0 \\
F_1 + F_2 + F_3 + F_c^z - m\ddot{z}_b &= mg \\
\sum_{i=1}^{3} d_i \cos\alpha F_i - \sum_{i=2}^{3} d_i \bar{\mu}_i \sin\alpha F_i - d_1 \sin\alpha F_1^y & \\
+((h + \bar{\mu}_b d)\sin\alpha + (\bar{\mu}_b h - d)\cos\alpha) F_c^z - I\ddot{\alpha} &= \sum_{i=1}^{3} \tau_i
\end{aligned} \tag{7}$$

Similar to the previous two models, system kinematics yields the body accelerations $\ddot{y}_b$ and $\ddot{z}_b$ as functions of $\ddot{\alpha}$ we use (6) and (7) to compute the matrices $\mathbf{A}(\mathbf{c})$, $\mathbf{b}(\mathbf{c}, \dot{\mathbf{c}})$ and $\mathbf{D}(\mathbf{c})$.

3.8 Existence of Solutions and Leg Jamming

It is not immediately obvious that the equations presented in Sects. 3.6 and 3.7 can be solved. In fact, a major singularity arises in computing the ground reaction forces on sliding legs using the moment balance equation (3). To illustrate the inconsistency, suppose that leg i is sliding forward with $\dot{y}_i > 0$ and the leg is within the friction cone with $\cot\gamma_i < \mu_t$. When $\tau_i < 0$, the massless legs in our model require a positive vertical component for the ground reaction force, $F_i > 0$. However, solution of the leg moment balance equation yields $F_i = -\tau_i/(l\cos\gamma_i - l\mu_t \sin\gamma_i) < 0$, resulting in an inconsistency. Consequently, when the leg is sliding forward and is inside the friction cone, there are no consistent solutions for the unknown forces and accelerations.

It turns out that this problem is a special case of the well known Painlevé's problem of a rigid rod sliding on a frictional surface [6, 10]. For certain parameter and state combinations, it is impossible to find any consistent set of finite forces and accelerations and one needs to seek impulsive solutions for the unknown quantities. This problem and its variations stimulated a large body of work in frictional collisions [2, 16, 17, 19], which hypothesize that the rigid rod would "jam" in such cases and start pivoting around its toe.

Fortunately, empirical evidence accumulated over months of physical experiments with the robot reveals that in the absence of dramatic external disturbances, RHex operates in regions of its state space away from these singularities. Starting from a stationary position, the front four legs always slide backward which guarantees a solution for the associated ground reaction forces. Furthermore, even though the rear legs usually slide forward, RHex's kinematics ensure that the orientation of the rear two legs is always outside the friction cone, yielding a consistent solution for the associated reaction forces. Finally, the body link always slides forward and admits a consistent solution once the toe reaction forces are identified.

4 Model Based Control of Flipping

Our goal in designing a model-based controller for flipping is to maximize the acceleration of the body pitch, while maintaining contact of the body endpoint with the ground and respecting actuator torque constraints. Depending on the measured (or estimated) state of the rear toe, the appropriate model is chosen among those presented in Sects. 3.6 and 3.7 in formulating the maximization problem. The resulting feedback controller implicitly defines a switching law based on the physical state of the rear toe, with no explicit discrete internal states. On RHex, direct measurement of toe stiction is not possible and we instead use an empirically designed estimator [15].

For both planar models, when the system is far from singular regions described in Sec. 3.8, the unknown forces and accelerations can be computed by directly solving (1), yielding

$$\mathbf{v} = \mathbf{A}^{-1}(\mathbf{c})\mathbf{b}(\mathbf{c}, \dot{\mathbf{c}}) + \mathbf{A}^{-1}(\mathbf{c})\mathbf{D}(\mathbf{c})\,\boldsymbol{\tau}\,. \tag{8}$$

Both constrained systems are underactuated and direct inversion of these dynamics to obtain torque solutions is generally not possible. Furthermore, our task is not specified in terms of particular choices of ground reaction forces and accelerations. Rather, we are interested in the (in)stability properties of particular degrees of freedom in the system, particularly the body pitch, as well as various constraints arising from our assumptions in Sec. 3.1. As a consequence, our controller is based on a constrained optimization formulation informed by the underlying dynamics.

4.1 Constraints on Control Inputs

The first set of constraints we impose in solving (8) are motivated by physical limitations of RHex's actuators. Torque limitations for the simplest, resistive model of a geared DC motor arise from the interaction between the back EMF voltage, the maximum available supply voltage and the armature resistance. Our model based controller is designed to respect constraints based on this simple model for each motor, yielding decoupled torque limits for each leg.

We introduce a second constraint to explicitly enforce body-ground contact throughout the progression of the remaining degrees of freedom. Fortunately, this requirement is easily captured through the constraint $F_c^z > 0$, an inequality that is linear in the input torques.

Definition 1. *For a particular state* $\mathbf{q} \in \mathcal{Q}$*, we define the corresponding set of allowable torques,* $\mathcal{T}(\mathbf{q})$ *as the set of all torque input vectors* $\boldsymbol{\tau} \in \mathbb{R}^3$ *such that*

$$\begin{aligned} F_c^z(\mathbf{q}, \boldsymbol{\tau}) &\geq 0 \\ \forall i,\ F_i(\mathbf{q}, \boldsymbol{\tau}) &\geq 0 \\ \forall i,\ \tau_i^{min}(\mathbf{q}) \leq \tau_i &\leq \tau_i^{max}(\mathbf{q})\ . \end{aligned} \tag{9}$$

4.2 Maximal Thrust Control

For both models of Sec. 3.6 and 3.7, the solutions for $\ddot{\alpha}$ and $\ddot{y}_t$ are continuous functions of the input torques. For any given state, this functional relationship is defined through our hybrid toe contact model and the solutions for the ground reaction forces. As a consequence, the problem of choosing hip controls to maximize thrust becomes a constrained optimization problem over the allowable input torque space.

Definition 2. *Given the current state* $\mathbf{q} \in \mathcal{Q}$*, we define the maximal torque input* $\boldsymbol{\tau}^*$ *as the torque vector that yields the maximum pitch thrust:*

$$\boldsymbol{\tau}^*(\mathbf{q}) := \underset{\boldsymbol{\tau} \in \mathcal{T}(\mathbf{q})}{\operatorname{argmax}} \left(\ddot{\alpha}(\mathbf{q}, \boldsymbol{\tau})\right)\ .$$

Fortunately, the solutions of (8) depend linearly on the input torques. Consequently, the constraints in Defn. 1 as well as the objective function, $\ddot{\alpha}(\mathbf{q}, \boldsymbol{\tau})$, are linear in the input torques as well. As a result, standard linear programming techniques can be employed to identify efficiently the maximal torque solution $\boldsymbol{\tau}^*$. In particular, we use a simple geometric solution that exploits the low dimension and the largely decoupled structure of the constraints [11].

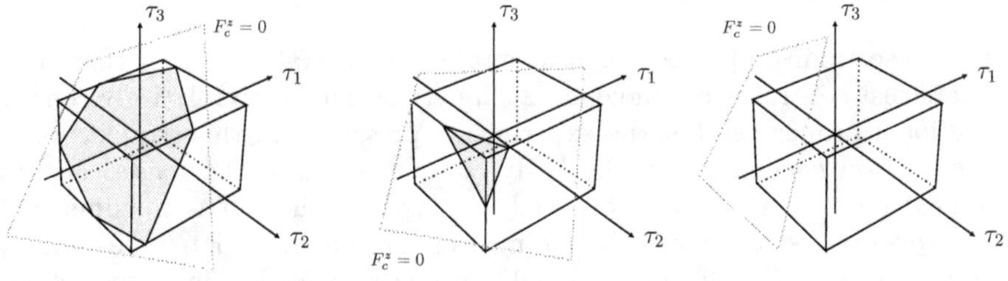

Fig. 5. All possible arrangements of volumes for motor torque limits and the body liftoff constraint. Simple geometric analysis yields the solution that yields the maximum thrust, while satisfying all the constraints.

As illustrated in Fig. 5, motor torque limits can be expressed as an axis-aligned constraint volume in the hip torque space, which can then be intersected with the half-space resulting from the inequality constraint of (9). The optimal solution can be obtained by simply evaluating the objective function on the vertices of the resulting volume. There are only three possible arrangements of these volumes corresponding to different geometric arrangements of the cube relative to the body constraint half space. The only case where there are no solutions that satisfy all the constraints is when these volumes do not intersect at all, in which case we prioritize the motor torque limit constraint and choose a solution that yields the smallest vertical force on the tail.

5 Experimental Results

5.1 Experimental Platform

The most recent version of the robot, RHex 1.5, adopted for the present experiments, has a rigid body that measures 50x20x15 cm, and houses all the computational and motor control hardware. The total weight of the robot is roughly 8.5 kg. Each of RHex's six leg is directly actuated by a Maxon RE118751 20 W brushed DC motor combined with a Maxon 114473 two-stage 33:1 planetary gear, delivering an intermittent stall torque of 6 Nm at 24 V. We achieve approximate software torque control at 1 KHZ by adopting an approximate feedforward model of the motor/amplifier stacks[3] – a variant of the model described in [8]. RHex 1.5 also incorporates a three-axis gyro for inertial sensing of the body orientation in addition to the motor encoders.

The legs on the current RHex are monolithic pieces of compliant fiberglass, attached to motor shafts through aluminum hip fixtures. Each of the legs in the set used for the experiments in this paper is roughly 16.5 cm long, weighs 80 g and has a radial compliance[4] of 1900 N/m [9].

5.2 Thrust Phase Model Performance

In this section, we present experimental data to establish the performance of our thrust controller on linoleum, a slippery surface with relatively consistent frictional properties. For the experiments presented in this section, we fixed the friction coefficient for the body contact as $\mu_b = 0.4$, based on ranges indicated in [5] for plastic on linoleum type surfaces. The maximal thrust controller of Sec. 4.2 explicitly minimizes the ground reaction force on the tail. Hence, the frictional force on the tail is also very small, significantly decreasing the importance of our particular choice for the associated coefficient of friction on the overall performance of the controller.

In contrast, in order to estimate the much more important toe friction coefficient and assess the corresponding model performance, we ran a number

[3]See [15] for details.

[4]Even though compliance is critical in RHex's dynamic locomotion performance, it is not nearly as dominant for the flipping behavior.

of experiments using approximate measurements of RHex's kinematic and dynamic parameters for different settings of the toe friction coefficient μ_t. Four runs were recorded for 15 different settings in the range $\mu_t \in [0.1, 0.6]$. Figure 6 displays a sequence of snapshots for one of these experiments, extracted from the high speed video footage.

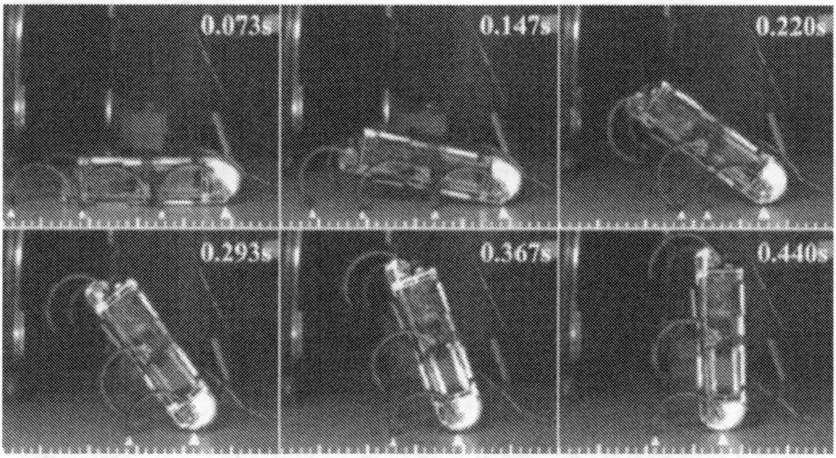

Fig. 6. Sequence of snapshots during flipping on linoleum with $\mu_t = 0.39$, for the period where the maximal thrust controller is active. Subsequent frames in which the robot falls back and recovers are not included. White arrows indicate contact points for the toes and the body

The best model performance was obtained for $\mu_t = 0.39$ (see Fig. 7) with a mean RMS error of $0.4 rad/s$ between model predicted pitch velocity and experimental measurements; surprisingly small considering various levels of approximations used in our model, including the inaccuracies in the kinematic and dynamic parameters, unmodeled leg compliance and the approximate torque control.

These results suggest that our model provides an accurate representation of the thrust phase. Nevertheless, there are a number of inaccuracies in its prediction, visible in the pitch acceleration plots. Most significantly, our model fails to predict the large overestimation of the initial acceleration and the subsequent, relatively large oscillations in the measured acceleration.

We believe that the origin of both discrepancies is the compliance in RHex's legs. The initially uncompressed legs introduce some delay in responding to the torque commands, resulting in a delayed acceleration of the pitch. A similar effect is visible subsequent to the liftoff of the middle legs which also causes oscillations due to the sudden loading of the rear legs.

Nevertheless, most of these differences do not significantly influence the average performance of the model. In addition to the accurate prediction of

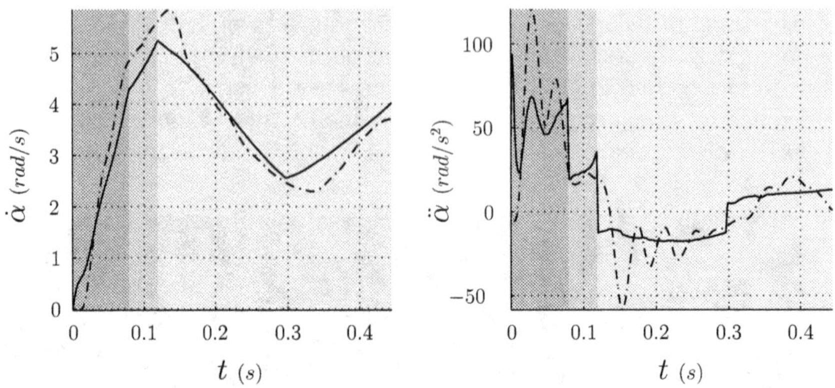

Fig. 7. Model predictions for the pitch velocity(left) and acceleration(right) compared to the experiment with best model performance ($\mu_t = 0.39$). Solid lines indicate model prediction whereas dashed lines show the actual measurements. Shaded regions indicate different number of legs in contact with the ground: dark(3), middle(2) and light(1)

the pitch velocity, the robot successfully keeps its tail on the ground and consistently performs flips on linoleum in a single thrust (see Fig. 6).

6 Conclusion

In robotic locomotion research, autonomy is likely to impose some of the most demanding constraints on design and limitations on behavior. It is very difficult, often impossible to achieve in systems otherwise designed for non-autonomous operation. RHex, our hexapedal platform, demonstrated that autonomy as a design goal can achieve significant advances in real world performance and robustness.

In this paper, we present a new controller to implement self-righting behavior on RHex, which is perhaps the simplest instance of maneuverability other than locomotion itself. To this end, we investigate different options for modeling the flipping behavior, each with different practical as well as analytic properties. In doing so, we review a number of models we had introduced in earlier work, and highlight their analytical properties and singularities with respect to their potential implementation on RHex. A new, simpler model still offers the advantages of some of these earlier approaches in its singularity properties while being much more feasible for practical deployment. The main focus of the paper is to identify common fundamental problems that arise in models of dynamical systems with multiple contacts and external actuation, with an emphasis on the implications of these problems in the design of feedback controllers.

Our modeling efforts reveal two distinct types of singularities in the models proposed for the self-righting behavior. The simplest approach with nonfrictional contacts and massless legs results in singular configurations which are fairly straightforward to characterize and yield physically inaccurate model predictions. The same problem persists even when we recover physical accuracy by introducing leg masses at the expense of model simplicity due to nonlinear kinematics and algorithmic complexities in determining contact state.

In contrast, a new class of singularities appear when we consider contacts with Coulomb friction, resulting in ranges of inconsistent states wherein one cannot find solutions for the dynamics. It turns out that this class of singularities is an instance of the relatively well-known Painlevé's problem of a rigid rod sliding on a frictional surface. Our flipping controller overcomes the problematic consequences of this singularity by ensuring that the progression of the robot state throughout the behavior stays away from singular regions. The feasibility of this approach is demonstrated by systematic experiments on RHex, which both demonstrate the performance of the controller and show that model predictions closely match with experimental measurements during flipping.

In parallel with most of the literature in modeling multiple contacts, our approach is to present domain specific characterizations of various models and equally specialized solutions for their usage in designing feedback controllers. However, extensions to the flipping behavior such as uninterrupted rolling or handstands are likely to require a much better analytical understanding of the various proposed models. In particular, a more unified characterization of singularities for generic classes of models with multiple contacts is critical for the ability to design robust controllers that explicitly avoid or exploit singular configurations. We believe that understanding and resolving fundamental dynamical properties of robotic systems arising from their morphology in conjunction with multiple intermittent contacts will be instrumental in the design and deployment of robust and practical controllers for such systems.

Acknowledgments

We thank Prof. Matt Mason for his insight on friction models and the connection to Painlevé's problem. This work was supported in part by DARPA/ONR Grant N00014-98-1-0747.

References

1. R. Altendorfer, N. Moore, H. Komsuoglu, M. Buehler, H. B. Brown Jr., D. McMordie, U. Saranli, R. J. Full, and D. E. Koditschek. RHex: A Biologically Inspired Hexapod Runner. *Autonomous Robots*, 11:207–213, 2001.
2. D. Baraff. Coping with Friction for Non-penetrating Rigid Body Simulation. *Computer Graphics*, 25(4):31–40, 1991.

3. P. Fiorini and J. Burdick. The Development of Hopping Capabilities for Small Robots. *Autonomous Robots*, 14(2-3):239–254, May 2003.
4. E. Hale, N. Schara, J. W. Burdick, and P. Fiorini. A Minimally Actuated Hopping Rover for Exploration of Celestial Bodies. In *Proceedings of the IEEE International Conference On Robotics and Automation*, pages 420–7, San Francisco, CA, April 2000.
5. L. S. Marks. *Marks' standard handbook for mechanical engineers.* McGraw-Hill, New York, 1996.
6. M. T. Mason. *Mechanics of Robotic Manipulation.* MIT Press, 2001.
7. L. Matthies, Y. Xiong, R. Hogg, D. Zhu, Rankin A., B. Kennedy, M. Hebert, R. Maclachlan, C. Won, T. Frost, G. Sukhatme, M. McHenry, and S. Goldberg. A Portable, Autonomous, Urban Reconnaissance Robot. In *Proceedings of The 6th International Conference on Intelligent Autonomous Systems*, Venice, Italy, July 2000.
8. D. McMordie, C. Prahacs, and M. Buehler. Towards a Dynamic Actuator Model for a Hexapod Robot. In *Proceedings of the IEEE International Conference On Robotics and Automation*, 2003. accepted for publication.
9. E. Z. Moore, D. Campbell, F. Grimminger, and M. Buehler. Reliable Stair Climbing in the Simple Hexapod RHex. In *Proceedings of the IEEE International Conference On Robotics and Automation*, volume 3, pages 2222–2227, Washington, DC., May 2002.
10. P. Painlevé. Sur le lois de frottement de glissement. *C. R. Acad. Sci. Paris*, 121:112–115, 1895.
11. U. Saranli. *Dynamic Locomotion with a Hexapod Robot.* PhD thesis, Computer Science and Engineering. The University of Michigan, 2002.
12. U. Saranli, M. Buehler, and D. E. Koditschek. RHex: A Simple and Highly Mobile Robot. *International Journal of Robotics Research*, 20(7):616–631, July 2001.
13. U. Saranli and D. E. Koditschek. Design and Analysis of a Flipping Controller for RHex. Technical Report CSE-TR-452-02, UM, Ann Arbor, MI, 2001.
14. U. Saranli and D. E. Koditschek. Back Flips with a Hexapedal Robot. In *Proceedings of the IEEE International Conference On Robotics and Automation*, volume 3, pages 128–134, Washington, DC., May 2002.
15. U. Saranli, A. Rizzi, and D. E. Koditschek. Model-Based Dynamic Self-Righting Maneuvers for a Hexapedal Robot. *International Journal of Robotics Research*, 2004. Accepted for publication.
16. D. E. Stewart. Convergence of a Time-Stepping Scheme for Rigid-Body Dynamics and Resolution of Painleve's Problem. *Archive for Rational Mechanics and Analysis*, 145(3):215–260, 1998.
17. D. E. Stewart and J. C Trinkle. Dynamics, Friction, and Complementarity Problems. In M. C. Ferris and J. S. Pang, editors, *Complementarity and Variational Problems*, pages 425–439. SIAM, 1997.
18. E. Tunstel. Evolution of Autonomous Self-Righting Behaviors for Articulated Nanorovers. In *Proceedings of the 5th International Symposium on Artificial Intelligence, Robotics and Automation in Space*, pages 341–6, Noordwijk, The Netherlands, June 1999.
19. Y. Wang and M. Mason. Two Dimensional Rigid Body Collisions with Friction. *Trans ASME; J Applied Mechanics*, 59(3):635–642, September 1992.

Gap Navigation Trees: Minimal Representation for Visibility-based Tasks

Benjamín Tovar, Luis Guilamo, and Steven M. LaValle

Dept. of Computer Science
University of Illinois
Urbana, IL 61801, USA
{btovar,lguilamo,lavalle}@uiuc.edu

Abstract. In this paper we present our advances in a data structure, the Gap Navigation Tree (GNT), useful for solving different visibility-based robotic tasks in unknown planar environments. We present its use for optimal robot navigation in simply-connected environments, locally optimal navigation in multiply-connected environments, pursuit-evasion, and robot localization. The guiding philosophy of this work is to avoid traditional problems such as complete map building and exact localization by constructing a minimal representation based entirely on critical events in online sensor measurements made by the robot. The data structure is introduced from an information space perspective, in which the information used among the different visibility-based tasks is essentially the same, and it is up to the robot strategy to use it accordingly for the completion of the particular task. This is done through a simple sensor abstraction that reports the discontinuities in depth information of the environment from the robot's perspective (gaps), and without any kind of geometric measurements. The GNT framework was successfully implemented on a real robot platform.

1 Introduction

This paper centers on the development of mobile robotics systems that perform sophisticated visibility-based tasks with minimal sensing requirements. The goal is to provide autonomous mobile robots for applications such as surveillance, search-and-rescue, firefighting, law enforcement, and remote visual presence. Classical approaches often lack reliability when applied in practice due to problems such as mapping uncertainty, registration, segmentation, localization errors, and unpredictable control errors. A primary cause is that previous algorithmic efforts have often assumed the availability of perfect geometric models. The guiding philosophy in this work is to avoid most of these difficulties by developing algorithms and mobile robots that minimize the information requirements. By constructing an algorithm and control law that use the information directly from the robot sensors, it may be possible to complete a task while eliminating the need to make potentially flawed measurements. This can provide low-cost solutions to challenging problems, while achieving greater reliability in the face of uncertainties.

M. Erdmann et al. (Eds.): Algorithmic Foundations of Robotics VI, STAR 17, pp. 425–440, 2005.

In this work we present our developments of a dynamic data structure called the *Gap Navigation Tree* (GNT). The GNT is constructed entirely from online sensor measurements made by the robot. Once constructed, it encodes paths from the current position of the robot to any place in the environment. As the robot moves, the GNT is updated as to always reflect the path information from the current position of the robot. These paths are globally optimal in distance if the environment is simply-connected, even though geometric information, such as lengths, angle measurements and robot orientation, is not available. The only restriction is that the components of the environment's boundary must be piecewise smooth closed curves, with only a finite number of nonsmooth points. The approach is based on modeling an abstract sensor that reports the order of depth discontinuities with respect to the heading of the robot. These discontinuities are called *gaps*. Encoding the changes of the gaps, the information state of the robot evolves, until a goal state is reached. Such goal state is task dependent, and it may be, for example, to explore the whole environment, to find all evaders in pursuit tasks, etc. The gap changes, called gap critical events, correspond to combinatorial changes in the visibility region of the robot. Given that no exact geometric information is needed for its construction or use, the GNT provides a robust framework for solving robotic tasks in the plane. Moreover, it provides a simple sensor-feedback strategy if the path that the robot transverses is encoded in the GNT.

The paper is organized as follows. The remainder of this section is a brief review of the previous work related to our approach. Section 2 presents the sensing model, followed by a description of the GNT in an information space framework in Section 3. Section 4 sketches the use of the GNT in robot localization, navigation and pursuit-evasion tasks.

1.1 Sensorless and minimal sensing settings

In this paper we are concerned with visibility-based tasks in the plane. More specifically, those robotics tasks for which sensor information can be modeled as a visibility region. The visibility region is the set of points that can be joined with a line segment to the robot's position, without intersecting the environment's boundary. As the robot moves, its visibility region changes, modifying the information that the robot knows about the environment or the progress towards a goal. We are interested in determining the minimal sensing capabilities a robot should have in order to complete a visibility-based task. One approach that considers minimal sensing settings is known as *bug algorithms* [16,20]. Combining global knowledge with local information, a robot is able to navigate among obstacles and reach a given goal. The robot navigation capabilities are simple (movement towards obstacles and wall-following), no representation of the environment is kept, and the global information may consist only of the position of the goal to reach. These characteristics allow the use of bug algorithms in robots with very limited sensing capabilities and unreliable motion control.

In general, the initial state of the robot is unknown, and as it moves in the environment, it might reach a goal state without determining its exact current state. In robotics, the problem of driving a system from an unknown state to a goal state was first considered in the context of manipulation [10]. For example, up to convex hull symmetry it is possible to manipulate polygonal parts to a final configuration without any sensor information [11]. Of course, not all robotic tasks can be solved without sensors, but it is very interesting, and scientifically relevant, to determine the minimum information necessary to complete a given task [2,6]. Moreover, one may go a step further and design a sensor that exactly suits the robotic task. One can think of an *abstract sensor* that gives the "ideal" minimal information to the robotic system to work correctly, and its physical implementation in a "subideal" sensor [9]. This philosophy can be used for solving the classical path-planning problem, as presented in [28]. Using a critical-point detector and a passage-point detector as abstract sensors, collision avoidance is achieved with sensor-based repulsive potentials.

1.2 Visibility-based tasks

The changes in the visibility region have been extensively studied, from the art-gallery problem [23], to decompositions of the environment into regions of *similar* visibility. In [26], a cell complex decomposition is presented, the *visibility complex*, in which points inside a cell *see* the same set of objects in the environment. The space can be decomposed also into equivalence classes of similar visibility of an object. Elements inside a class have a similar qualitative view of the object, that is, see the same *aspect*. An aspect here may be defined as the set of views of an object that share the same combinatorial structure. This leads to the study of the *aspect graph* [4]. In [13], a planar environment is decomposed into cells that see the same aspect of the environment's boundary. The combinatorial structure of visibility regions in that work is defined in terms of *spurious* and *non-spurious* edges. A spurious edge is an edge of the visibility region that does not exist in the environment's boundary, but it is produced by occlusions (i.e., by a corner). The supporting line of a spurious edge is always collinear with the robot's position. The sequence of spurious and nonspurious edges define the *visibility skeletons*, and the environment is decomposed into cells of points that share the same visibility skeleton. Such decomposition is called the *visibility cell decomposition.*

In the decompositions mentioned before, if the robot moves inside a cell there is not significant change in information. The robot receives about the same information from the sensors. Such movements are called *conservative* in the sense that they preserve the current robot's information. In contrast, when the robot crosses one of the cells' boundary edges, the structure of the visibility region suffers a drastic change, and the robot's information may be modified. Such sudden changes are called *visual events*[8].

In the plane, if the components of the environment's boundary are represented as a regular piecewise smooth closed curves, there are two kinds of visual events. One kind is triggered when the robot crosses an environment's boundary generalized inflection ray, and the other when it crosses the complement of bitangent line segments of the boundary[1]. An *inflection* is a change in the sign of the curvature of the environment's boundary. We use the term "generalized," as in [19], to include polygonal boundaries. Given a generalized inflection, an *inflection ray* is found by extending a ray from the inflection until it hits another point of the environment's boundary. A bitangent line segment is a segment completely contained in the environment representation, whose supporting line is tangent to two points of the boundary, and whose endpoints are these points of tangency. A common general position assumption is that no line is tangent to more than two points of the boundary (thus the term *bitangent*). For each bitangent, its *complement* is found by extending outward from each point of tangency until the environment's boundary is hit again (see Figure 2).

Since the inflection and bitangents are so relevant for visibility-based problems, one may wonder if the robotic tasks may be described *solely* in terms of inflections and bitangents, forgetting all the rest of geometric information of the visibility region. Visibility-based tasks may have then the form of states of information that are modified each time a visual event is detected. This relation between the visual events and the robotic tasks has been exploded before for different visibility-based tasks. Our aim here in the rest of this Section is to give an unified view of visibility-based tasks. Particularly, we will try to convince the reader that the tasks of optimal navigation, robot localization, and pursuit-evasion share the same information structure. It is up to the robot's strategy to choose the way of manipulating this structure to solve the particular task.

The first task we describe is optimal navigation in the plane. The classical algorithm is to perform a search in the visibility graph, or in the bitangent graph [22]. The bitangent graph is obtained from the visibility graph by deleting all the edges, that if grew by a small epsilon, at least one of the endpoints lies outside the boundary, or inside an object[2]. The bitangent graph is entirely defined in terms of inflections and bitangents. Taking advantage of this, in [1], shortest path trees are used to efficiently update the visibility region of a moving point. On the other hand, each time a visual event occurs, a portion of the shortest-path structure is revealed to the robot. It then becomes

[1] In polygons there is a third kind of visual event, in which a non-reflex vertex comes into view. Although this is relevant for defining, for example, the visibility skeletons, it is particular for polygons, making generalization to curved environments difficult.

[2] Thus, the bitangent graph is called the *reduced visibility graph* in [18].

possible to retrieve the shortest-path trees of an unknown environment, as it is presented in [31,32]. Note, however, that the task of getting the shortest-path information may use longer paths itself. This is because, in general, for unknown environments, the competitive ratio between the length of the path generated by any navigation strategy and the length of the optimal path is unbounded [24].

In the robot localization problem, the robot must infer its position based on local sensing [3,5,13,17]. One approach is to generate position hypotheses by comparing the visibility skeleton of the robot's visibility region with the skeletons of the visibility cell decomposition[13]. If there is more than one position hypotheses, the robot should move around to determine its position with certainty. In [7], a strategy that computes an overlay arrangement of the visibility cells is used to determine the optimal movement strategy to reduce the set of position hypotheses. In that approach, a decision tree is simulated, in which each decision node is a visibility skeleton comparison that discards some position hypotheses. From the robot perspective, the localization problem can be stated as series of visual events that (must) occur to uniquely identify the robot's position. For example, if the robot detects two bitangents and one inflection, in that order, it can discard all the possible positions where the same movements lead to three bitangents in a row. This sequence of visual events is available from the shortest-path tree[1]. Using only visual events (i.e., no compass) will localize the robot up to a rigid rotation transformation of the environment. In fact, the assumption of a compass has a big impact in robotic tasks [2,6].

The last visibility-based task we describe here is the pursuit-evasion problem. In this task, one or several *pursuers* should find all possible *evaders* that may be hiding in the environment[30]. In this task, a region of the environment is said to be *contaminated* if it might contain an evader, otherwise it is referred as *cleared.* A cleared region that is contaminated again is called *recontaminated.* The pursuit-evasion problem is solved when all the regions of environment are cleared. Cleared and contaminated regions are separated from the robot's perspective by the current visibility region. Cleared regions may only get recontaminated if they *merge* with a contaminated one, which can only be the result of a combinatorial change in the visibility region. These combinatorial changes are the base of the search algorithm presented in [12]. In terms of visual events, an strategy should dictate the sequence of inflection rays and bitangent complements to cross in order to clear the whole environment. In [29] a strategy is presented that explicitly avoids some visual events of being triggered to clear an unknown environment. The order of how the visual events should occur also dictates the structure of the "T", "S" and "Z" patterns in [15], in which a 6-state automaton restricted to move along the environment boundary serves as a pursuer. In [14] an algorithm is presented in which clearing the environment is equivalent to clearing the shortest-path structure from the current robot position in an unknown environment.

2 Sensing model

In this Section we formalize the model of the robot and the model of the information used from the sensors. The robot is modeled as a moving point in a connected open set R in the plane. Let $O = \{o_1, o_2, ..., o_n\}$ be the possibly empty set of pairwise disjoint obstacles, in which $o_i \subset R$ for $i = [1, 2, \ldots, n]$ is a closed set. Let ∂o_i be the boundary of $o_i \in O$. We assume that ∂o_i has a single connected component. Let $F = R \setminus \bigcup_{i=1}^{n} o_i$, with $o_i \in O$, be the free space. Let ∂F be the boundary of F. The components of ∂F are piecewise smooth closed curves with only a finite number of nonsmooth points. The robot is only able to move in F. In Section 5 we will briefly discuss the case when the robot is not a point.

The problem here is to design a sensor that detects the combinatorial changes in the visibility region. As we will detail later, the changes of the visibility region can be defined as a function of the spurious edges. When a spurious edge either appears, disappears, splits or merges with another, a combinatorial change in the visibility region occurs. From the robot perspective, the spurious edges are the discontinuities in depth information in the environment. Note that geometric information of the spurious edges is not relevant for the visual event detection. The events will be the same in spite of the exact length and angular position of the spurious edges. Their order is relevant, though, since we are interested in which discontinuity disappeared, or merged with another, for example. Although the precise distances to the walls may be unknown, the robot only needs a type of edge detector that can detect each of the discontinuities, and return their direction relative to the robot's heading. From now on, each of this discontinuities will be referred to as a *gap*, and the sensor as a *gap sensor* [27]. Besides the gap sensor, no other sensing ability is assumed, i.e., it has neither a compass nor a reliable odometer. The ideal gap sensor can be easily realized with through a range sensor (i.e., laser or sonar) or using computer vision techniques.

Each gap *hides* a connected region of the environment that is occluded to the robot from its current position. A label of "L" or "R" is assigned to a gap to indicate the direction of the part of R that is hidden behind the gap. This corresponds to transitions of the gap sensor from "far to near" (left) or "near to far" (right), if the gaps are detected by a counterclockwise scan with respect to the robot's heading (see Figure 1.(a)).

2.1 Gap Tracking

When the robot moves in the environment, the gaps, as reported by the gap sensor, may change. We refer to *gap tracking* to the process by which the gap sensor monitors the current visible gaps and their respective changes. It is assumed that the robot can track the gaps at all times and record any topological change. There are four possible ways in which gaps change:

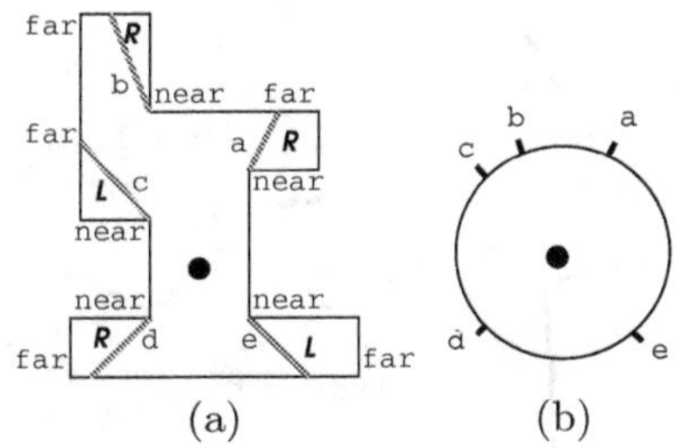

Fig. 1. The robot's view of the environment. The position of the robot is shown with a black disk. (a) The environment and the respective labeling of the gaps detected. (b) Angular position of the gaps detected in the visibility region.

Gap appearance: A gap, not detected before, is now tracked by the gap sensor. The gap is said to be visible.

Gap disappearance: A gap is no longer detected by the gap sensor. The given gap is not visible for the gap sensor.

Gaps merge: Several gaps merge into a single one.

Gap split: One gap splits into several gaps.

If a gap appears, the region behind it was just visible to the robot, but now is "hidden" by the gap. Similarly, when a gap disappears, the region of the environment behind the gap is now visible to the robot. With bitangents, exactly two gaps may merge into one, and one gap splits exactly into two gaps. These four gap topological changes are called the *gap critical events.* We say 'critical events' instead of 'visual events' to make clear that the complete information of the visibility region is not available.

Appearances and disappearances of gaps are related to generalized inflections of ∂F. As illustrated in Figure 2.(a), appearances and disappearances of gaps occur when the robot crosses inflection rays. Merges and splits of gaps, are related to the bitangents of ∂F, and they occur when the robot crosses bitangent complements. (Figure 2.(b)).

With bitangents, exactly two gaps may merge into one, and one gap splits exactly into two gaps. With the minimal capabilities assumed now for the gap sensor, the robot cannot discriminate between a gap splitting in three, or a gap splitting into two, and an appearance close together. The general position assumption may be relaxed if a more sophisticated gap sensor is available (i.e., measuring exact distances or angles).

Given the gaps that the robot detects at a given time, it is possible to command the robot to move toward a given gap. This sensor-feedback movement is defined as *chasing a gap.* Note that with only the gap sensor information the only movements that are guaranteed to be collision free are gap chasing movements.

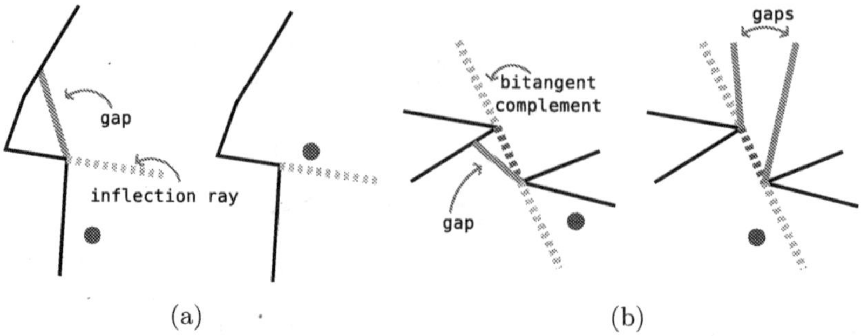

Fig. 2. Inflections and bitangents of ∂F. (a) Appearance and disappearance of gaps occur when the robot crosses inflection rays. (b) Splits and merge occur by crossing bitangent complements.

3 The GNT from an information state perspective

Before introducing the GNT data structure, we define what we mean by an information space and an information state. As a state space captures the dynamics of a system, allowing the design of control inputs, or the study of the system trajectories, an *information space* captures the evolution of a system in terms of previous sensor readings and control inputs. Formally, let $X \subset \Re^n$ be the state space, $U \subset \Re^m$ the input space, and $u : [0, \infty) \rightarrow U$ the input history. Also, $Y \subset \Re^k$ is defined as the sensor space, which models the set of all of the possible instantaneous readings from the sensors. Let $y :\rightarrow [0, \infty) \rightarrow Y$ be the sensor history, in which $y(t) = h(x(t))$ is the instantaneous sensor reading for the state $x(t) \in X$ for some mapping $h : X \rightarrow Y$. An *information state* at time t is given by $\eta_t = (u_t, y_t)$, that is, the input and sensor history up to time t. The *information space* I is defined as the set of all possible information states. Given an information state η_t and an initial set of states $X_0 \subset X$, a *derived information state* $F(\eta_t, X_0)$ is the set of all the possible $x(t) \in X$ given η_t and X_0.

In our sensing model, the sensing space Y is defined by the set of all the of ordered circular sequences of possible readings of gaps. Thus, $\{L, L, R\} \in Y$ correspond to a sensor reading where two "left" gaps and a "right" gap are detected. Note this sensor reading is indistinguishable from $\{R, L, L\}$ and $\{L, R, L\}$, since a compass is not available. This means that the mapping h, from states to sensor readings is not injective. Even more, with only gaps readings, the exact position of the robot cannot be determined, and different neighborhoods of points will generate the same sensor reading across the whole environment. The input space is determined by the gap chasing movements, that is, the commands to the robot to move towards a gap, or a sequence of gaps.

3.1 Encoding information states

Remember that the robot can track the gaps all of the time and record any of their topological changes. Thus, it can detect that from the transition $\{L_1, R_1, R_2, L_2\}$ to $\{L_1, R_2, L_2\}$, the gap R_2 disappeared. The gap sensor only will report to the robot that a gap, detected before in this order, disappeared, for example. This identification of gaps is implicit at the sensor level, and it is possible if we assume coherency between the robot's motion and gap changes (i.e., small position changes of the robot will produce small angular position changes in the gaps).

The gaps and their topological changes are encoded into a tree, hereafter referred to as T. The tree T is the *Gap Navigation Tree* of the environment. The root of T moves along with the robot. Each child of the root represents a gap that is currently visible, and they are maintained in the circular order of the gaps they represent. In T, we will use the terms gaps and nodes interchangeably because each node encodes a gap.

As the robot moves, critical events are triggered. As events occur, T is updated as follows: if a gap disappears, the corresponding node is removed from T. If a gap appears, it is added as a child of the root of T in a location that preserves the circular ordering of gaps. Any node that is added in this way is designated as a *primitive node*, which indicates that a portion of the environment that was once visible is now occluded. If a gap splits, then the corresponding child of the root will be replaced with two children. If two gaps merge, the two corresponding children of the root become the children of a new node, d, and d becomes a child of the root (see Figure 3.a). Merging can only occur between a pair of gaps that are adjacent in the circular ordering produced by the gap sensor. Each node also keeps the label associated with the gap.

A sequence of nodes from the root of T to a leaf define a sequence of gaps, that if chased, follows a path in the bitangent graph. The proof of this result is presented in Section 4.2. The relation of T with an information state is immediate. In fact, T is nothing more than the sensor history of the current information state. For example, assume that at time t_1 the state of T is T_1. At t_1 the robot is commanded to chase the sequence of gaps α, which brings T into the state T_2. Comparing T_1 and T_2 we can readily obtain the sequence α followed (input history), with the respective changes as reported by the gap sensor (sensor history). Note, however, that we are assuming that α is the shortest sequence of gaps to take T_1 into T_2. In this sense, all sequences of gaps that take T from one state to another are *equivalent* to the shortest one, since at the end, they modify T in the same way.

3.2 Building the GNT of an unknown simply-connected environment

We say that the robot is *exploring* if it is building the GNT of an environment. When the robot is placed in an unknown environment, all of the leaves of

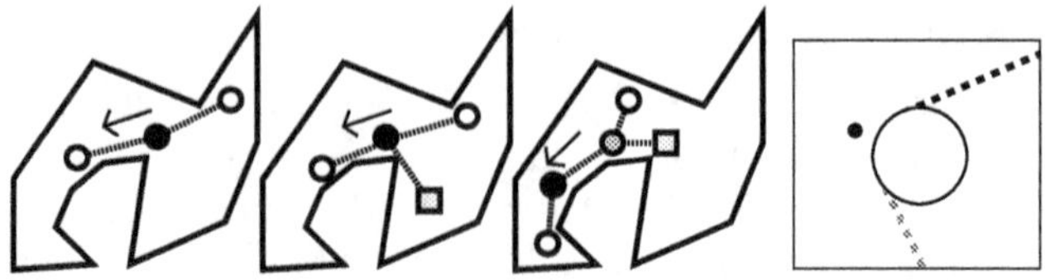

(a) (b)

Fig. 3. (a) Encoding of critical events into a Gap Navigation Tree. The black disk denotes the root of T and the current position of the robot. As the robot chases the gap on the left, a gap appearance, and a gap merge are triggered. The structure of T is updated accordingly. (b) Absence of gap critical events in multiply connected environments. If the robot chases any of the to gaps, it will go around the obstacle forever, because no gap chasing movement modifies the information state of the robot.

T are marked as nonprimitive. This is because the robot has not yet seen what is behind the corresponding gaps. The robot can arbitrarily choose to chase any one of the gaps, and one of the following will occur when it crosses the critical event related to the gap: (1) the gap will disappear, or (2) the gap will split. If the gap disappears, this means that the robot has seen the entire portion of the environment that was originally behind the gap. In this case, the robot must choose another nonprimitive gap to chase. If the second condition occurred, and the gap splits, then the robot can choose to chase either one of the new gaps. This process can be applied repeatedly until the first condition is met and a gap finally disappears. There is one further complication. In some cases, there may be a nonprimitive leaf node, d, that is not a child of the root. To handle this situation, the robot must first chase the child of the root that is an ancestor of d in T. Once a split occurs, it must then follow the next ancestor. This process repeats until d finally disappears or splits. Note that during motion, critical events may occur in other gaps, particularly appearances and merges, and T must be updated accordingly. It is worth nothing that only the angular order, but not the precise angular position of the gaps, is taken into account to build the tree. The construction of T ends when all of the leaves are marked as primitive. This condition indicates that the robot has seen the whole environment.

3.3 The GNT for multiply-connected environments

We briefly mention the complications that multiply-connected environments present to the gap sensor model. For a more extensive description, the reader is referred to [31]. The most interesting complication is that for some environments there may be no critical events at all. Consider Figure 3.b, where the environment's boundary does not have either inflections or bitangents. If the robot tries to follow either one of the gaps, expecting it to disappear, it will go around the obstacle forever. In terms of information states, there is no gap chasing movement that modifies the current information state of the robot. Even if there were critical events, no gap topological change gives the information to the robot that it surrounded the obstacle once already. To solve this, an *artificial* critical event must be introduced, that indicates to the robot that the obstacle has been surrounded already. This serves as a *pebble*, or marker, that the robot drops and picks as needed.

Another complication is present. In general, the obstacles cannot be discriminated one from the other with only gap sensing information. If the obstacles are assumed to be uniquely identifiable (i.e., they have different *color*), a second sensor can indicate to the robot which gaps have been already explored. For example, a gap may be associated with the obstacle where it begins (the 'near' of the transition). A simple strategy to construct T may command the robot to 'wall follow' each of the obstacles of the environment. In this case, the environment is completely explored when each of the obstacles has been surrounded. This strategy assumes that the robot has the capability of navigating by wall following, in addition to navigating by chasing gaps.

4 Using the GNT: Solving some visibility-based tasks

4.1 Robot Localization

For brevity, we only sketch a straightforward algorithm for robot localization. Given a map of the environment, the cells formed by the inflection rays and bitangent complements can be computed. For simplicity, let the state space X be the set of all of these cells. For each cell, a shortest path tree can be obtained using a breath-first search in the bitangent graph. To each of the nodes of the shortest-path tree a label of “L” or “R” is assigned. The label will indicate the side to which the environment's boundary lies, with respect to an observer standing at the node, and looking away from the root. A localization strategy here will try to match a partial construction of the T with a shortest-path tree of one of the cells. Note that the GNT can be obtained from a shortest-path tree by following the methods to find visibilty polygons from shortest-path trees presented in [1].

At the beginning, the possible position of the robot belongs to $X_0 = X$. After the first sensor reading, a partially constructed T, T_0, is used to compute $X_1 = F(X_0, T_0)$. The set X_1 consist of all the cells to which the sequence of 'L' and 'R' of the gap sensor reading correspond with the first level of nodes of the respective shortest-path trees. As the robot moves, a set of possible positions X_i is computed from the partial constructed T_i, until a set X_n, where no robot movement can decrease the position uncertainty. If $|X_n| = 1$, the robot will be localized, up to a visibility cell, and up to a rotation of the environment.

To reduce the position uncertainty, the localization movements can be computed at least in two ways. The first one is using a modified version of the spiral search in the Kleinberg's algorithm [17], using the gaps as the incident edges of a tree representation of the environment. The movements will be similar if the strategy presented by Dudek et. al.[7] is used. Instead of using the arrangement of visibility cells, the shortest-path trees are used for the comparisons in the movement decisions. Note that the localization strategy sketched here works with only the information provided by the edges not used in the visibility skeleton, which is the opposite case from previous approaches.

4.2 Optimal and locally optimal navigation

We state the optimality of the paths encoded in T by chasing sequences of gaps with the following two propositions:

Proposition 1 *Let $q \in F$, and assume that F is simply-connected. If $V = (v_1, v_2, ..., v_n)$ is the sequence of tangency points of ∂F of the shortest path between q and v_n in the bitangent graph, and $G = (\alpha_1, \alpha_2, ..., \alpha_n)$ is the sequence of gaps where α_i is the gap produced by the tangency point v_i when it is visible, then the path generated by chasing the gap sequence G is the shortest path between q and v_n.*

Proof. It is enough to prove that the path between v_i and v_{i+1} is optimal, since the sequence V is optimal by definition. The shortest path between two points in the Euclidean plane is unique and is a straight line. Following α_{i+1} the trajectory is a straight line that ends in v_{i+1}. We conclude that the path between v_i and v_{i+1} is optimal. □

Proposition 2 *The paths encoded in T between the root of T and any point $q \in F$, if F is simply-connected, are optimal.*

Proof. Let $p \in F$. Let $T(p)$ refer to the configuration of T when the robot is at p. From $T(p)$ we generate the sequence of gaps $G = \{\alpha_1, \alpha_2, ..., \alpha_n, \alpha_q\}$ that must be chased to reach q. By following G, the tangency points $V = \{v_1, v_2, ..., v_n, q\}$ of ∂F are visited. Let G_o be the sequence of gaps that generates the shortest path between p and q. Let $\alpha_d \in G$ (of point $v_d \in V$) be the first gap in which G and G_o differ. Since critical events are recorded in T as they become visible, this means that the point v_d is visible before the rest of the path in G_o, which therefore, *hides* behind α_d. The shortest path between the current position of the robot and the rest of the path encoded in G_o is the one following α_d (by Proposition 1). Therefore, G_o contains α_d, and we conclude that $G = G_o$. □

By Proposition 2, the paths encoded in T are optimal, and in this sense, a shortest-path tree and the GNT encode the same path information. Optimal navigation occurs by executing motions that follow the path in T from the root to the desired node d. During these motions, the robot records all changes to T that occur from critical gap events. At every time during the navigation, the robot chases the gap that corresponds to the child of the root that is an ancestor of d. At several points, the gap may split, and the robot must chase the new child of the root that is an ancestor of d. This results in a sequence of gap chasing commands, which ends when d is not detected anymore. As it is presented in [32], *interesting* objects in the environment can be associated with gaps, and the robot can manipulate (i.e., move) them through optimal paths.

When T is constructed for a multiply-connected environment, global optimality cannot be guaranteed. This is because T is a tree, which encodes only one path between the robot's position and any place in the environment. Since no distance information was taken into account in T, the path encoded may not be the optimal one. When multiple paths are found to the same place, a heuristic is used and only the paths with less gaps to chase are kept. This guarantees that the robot will travel through less cluttered areas, but tells nothing about optimality. Still, by a similar argument used in Proposition 1 they are still locally optimal. For more details, the reader is referred to [31].

4.3 Pursuit-Evasion

We next consider augmenting the GNTs to handle pursuit-evasion. A gap may be labeled as *cleared*, *contaminated* or *recontaminated* as a function of the region that hides behind it. Initially, all of the gaps in T are labeled as contaminated. If a gap appears, it is labeled as cleared, since if an evader is behind the gap, the pursuer would already have detected it. A cleared gap is recontaminated if it merges with a contaminated or a recontaminated gap. If T has at least one node that is not cleared, it is labeled as contaminated; otherwise, it is said that T is cleared. Since each node in T encodes a connected region of the environment, solving the pursuit-evasion problem is equivalent to clearing each node of T.

In [14] we present an online strategy that reads "cleaning" schedules from T. Each schedule is an ordered sequence of gaps that should be cleared. In other words, it is a sequence of changes in the visibility region that guarantees that some region of the environment is cleared. The strategy works by growing a set of adjacent children of the root of T. When a gap recontamination occurs, this is naturally reflected in T by the merge critical event. Schedules are then defined recursively, where for a node encoding a merge, its children should be cleared first. Circular dependencies of contamination merges can be detected, as in [25], to conclude that the environment cannot be successfully cleared with a single pursuer. The pursuer can place a static *sentry*, or guard, to prevent a critical event from happening, breaking the circular dependency. As proved in [14], by using a GNT the number of sentries needed is asymptotically optimal, $O(\log n)$, where n is the number of bitangents in the environment.

Although the GNT keeps all the contamination information to solve a pursuit-evasion problem, it may not encode the paths needed to clear some regions. The GNT only provides shortest-paths, but maybe longer paths are needed to avoid some critical event (i.e., a merge) from happening. Figure 4 shows an environment that can be cleared by a single pursuer, but not by following paths in the GNT.

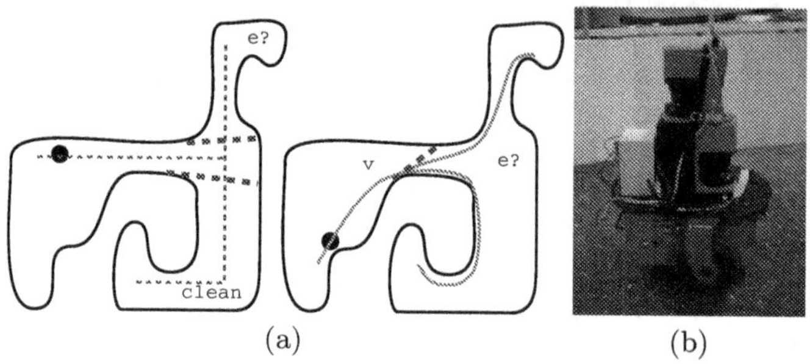

Fig. 4. (a) Not every environment searchable with one robot is searchable with one robot following sequences of gaps. A robot following the dashed path can find all of the evaders (left) . The thick paths, generated by chasing gaps, cause unavoidable recontaminations (right). (b) Robot platform used in the real experiments. The gap sensor was realized using two laser range sensors.

5 Implementation and conclusions

The GNT construction for a simply-connected environment was successfully implemented using the robot platform shown in Figure 4.b.. In the GNT model, the robot was considered as a point that can get arbitrarily close to the boundary before a critical event is triggered. Since this is not possible in the real robot, wall-following capabilities were provided such that the robot still chase the detected gaps, while taking into account the environment's boundary.

Although from a planning perspective the GNT may unify some of the visibility tasks, the navigation paths it offers may not be adequate for all robotic settings. For example, one may prefer the safety of maximum clearance paths [5], instead of the shortest-paths. Still, the GNT may be used to direct the motions of a robust navigation system, such as the one presented in [21]. An algorithm using the GNT can determine which discontinuity to follow, and the navigation system, at a lower level, will determine the safer motions for the robot.

Acknowledgements: The authors want to thank Boris Simov, Stjepan Rajko, Shai Sachs and Rafael Murrieta for their helpful discussions on the gap sensing model, to Javier Minguez for his comments on navigation on real robots, and to Anna Yershova and Robert Ghrist for their interesting discussions on information spaces. The robot equipment was provided by the ITESM Campus Ciudad de México and by the ITESM Campus Morelos, México. This work is supported in part by ONR Grant N000014-02-1-0488 and NSF-CONACyT Grant 0296126.

References

1. B. Aronov, L. Guibas, M. Teichmann, and L. Zhang. Visibility queries in simple polygons and applications. *Algorithms and Computation, 9th International Symposium, ISAAC '98*, 1533, 1998.
2. M. Blum and D. Kozen. On the power of the compass (or, why mazes are easier to search than graphs). In *Proc. Annual Symposium on Foundations of Computer Science*, pages 132–142, 1978.
3. J. Borenstein, B. Everett, and L. Feng. Navigating mobile robots: Systems and techniques. A.K. Peters, Wellesley, MA, 1996.
4. K. W. Bowyer and C. R. Dyer. Aspect graphs: An introduction and survey of recent results. *Int. J. of Imaging Systems and Technology*, 2:315–328, 1990.
5. H. Choset and K. Nagatani. Topological simultaneous localization and mapping (SLAM): toward exact localization without explicit localization. *IEEE Int. Conf. Robot. & Autom.*, 17(2):125–137, April 2001.
6. B. R. Donald. On information invariants in robotics. *Artif. Intell.*, 72:217–304, 1995.
7. G. Dudek, K. Romanik, and S. Whitesides. Localizing a robot with minimum travel. In *ACM-SIAM Symposium on Discrete Algorithms*, pages 437–446, 1995.
8. F. Durand. *3D Visibility: Analytical study and applications.* PhD thesis, Université Grenoble I – Joseph Fourier Sciences et Géographe, July 1999.
9. M. Erdmann. Understanding action and sensing by designing action-based sensors. *Int. J. Robot. Res.*, 14(5):483–509, 1995.
10. M. A. Erdmann and M. T. Mason. An exploration of sensorless manipulation. *IEEE Trans. Robot. & Autom.*, 4(4):369–379, August 1988.
11. K. Y. Goldberg. Orienting polygonal parts without sensors. *Algorithmica*, 10:201–225, 1993.
12. L. J. Guibas, J.-C. Latombe, S. M. LaValle, D. Lin, and R. Motwani. Visibility-based pursuit-evasion in a polygonal environment. *International Journal of Computational Geometry and Applications*, 9(5):471–494, 1999.
13. L. J. Guibas, R. Motwani, and P. Raghavan. The robot localization problem. In K. Goldberg, D. Halperin, J.-C. Latombe, and R. Wilson, editors, *Proc. 1st Workshop on Algorithmic Foundations of Robotics*, pages 269–282. 1995.

14. L. Guilamo, B. Tovar, and S. M. LaValle. Pursuit-evasion in an unknown environment using gap navigation graphs. In *IEEE/RSJ Int. Conf. on Intelligent Robots & Systems*, 2004. Under review.
15. T. Kameda, M. Yamashita, and I. Suzuki. On-line polygon search by a six-state boundary 1-searcher. Technical Report CMPT-TR 2003-07, School of Computing Science, SFU, 2003.
16. I. Kamon and E. Rivlin. Sensory-based motion planning with global proofs. *IEEE Trans. Robot. & Autom.*, 13(6):814–822, December 1997.
17. J. M. Kleinberg. The localization problem for mobile robots. In *IEEE Symposium on Foundations of Computer Science*, pages 521–531, 1994.
18. J.-C. Latombe. *Robot Motion Planning.* Boston, MA, Kluwer Academic Publishers, 1991.
19. S. M. LaValle and J. Hinrichsen. Visibility-based pursuit-evasion: The case of curved environments. *IEEE Transactions on Robotics and Automation*, 17(2):196–201, April 2001.
20. V. J. Lumelsky and A. A. Stepanov. Path planning strategies for a point mobile automaton moving amidst unknown obstacles of arbitrary shape. *Algorithmica*, 2:403–430, 1987.
21. J. Minguez and L. Montano. Nearness diagram navigation (ND): Collision avoidance in troublesome scenarios. *IEEE Transactions on Robotics and Automation*, February 2004.
22. N. J. Nilsson. A mobile automaton: An application of artificial intelligence techniques. In *1st International Joint Conference on Artificial Intelligence*, pages 509–520, 1969.
23. J. O'Rourke. *Art Gallery Theorems and Algorithms.* Oxford University Press, New York, NY, 1987.
24. C. Papadimitriou and M. Yannakakis. Shortest paths without a map. *Theoretical Computer Science*, 84:127–150, 1991.
25. S.-M. Park, J.-H. Lee, and K.-Y. Chwa. Visibility-based pursuit-evasion in a polygonal region by a searcher. Technical Report CS/TR-2001-161, KAIST, Dept. of Computer Science, Korea, January 2001.
26. M. Pocchiola and G. Vegter. The visibility complex. *Int. J. Comput. Geom. & Appl.*, 6(3):279–308, 1996.
27. S. Rajko and S. M. LaValle. A pursuit-evasion bug algorithm. In *Proc. IEEE Int'l Conf. on Robotics and Automation*, pages 1954–1960, 2001.
28. E. Rimon. Construction of C-Space roadmaps from local sensory data. What should the sensors look for? *Algorithmica*, 17(4):357–379, 1997.
29. S. Sachs, S. Rajko, and S. M. LaValle. Visibility-based pursuit-evasion in an unknown planar environment. Submitted to *International Journal of Robotics Research*, 2003.
30. I. Suzuki and M. Yamashita. Searching for a mobile intruder in a polygonal region. *SIAM J. Computing*, 21(5):863–888, October 1992.
31. B. Tovar, S. M. LaValle, and R. Murrieta. Locally-optimal navigation in multiply-connected environments without geometric maps. In *IEEE/RSJ Int'l Conf. on Intelligent Robots and Systems*, pages 3491–3497, 2003.
32. B. Tovar, S. M. LaValle, and R. Murrieta. Optimal navigation and object finding without geometric maps or localization. In *Proc. IEEE International Conference on Robotics and Automation*, pages 14–19, 2003.

A Simple Algorithm for Complete Motion Planning of Translating Polyhedral Robots

Gokul Varadhan[1], Shankar Krishnan[2], T.V.N. Sriram[3], and Dinesh Manocha[4]

[1] University of North Carolina at Chapel Hill `varadhan@cs.unc.edu`
[2] AT&T Labs - Research `krishnas@research.att.com`
[3] University of North Carolina at Chapel Hill `tvn@cs.unc.edu`
[4] University of North Carolina at Chapel Hill `dm@cs.unc.edu`
http://gamma.cs.unc.edu/motion

Summary. We present an algorithm for complete path planning for translating polyhedral robots in 3D. Instead of exactly computing an explicit representation of the free space, we compute a roadmap that captures its connectivity. This representation encodes the complete connectivity of free space and allows us to perform exact path planning. We construct the roadmap by computing deterministic samples in free space that lie on an adaptive volumetric grid. Our algorithm is simple to implement and uses two tests: a *complex cell test* and a *star-shaped test*. These tests can be efficiently performed on polyhedral objects using max-norm distance computation and linear programming. The complexity of our algorithm varies as a function of the size of narrow passages in the configuration space. We demonstrate the performance of our algorithm on environments with very small narrow passages or no collision-free paths.

1 Introduction

Path planning is an important problem in algorithmic robotics. The basic problem is to find a collision-free path for a robot among rigid objects and it has been well-studied for over three decades. Some of the earlier interest was in developing algorithms for *complete* path planning. An algorithm is complete, if it is guaranteed to find a solution when one exists and to return failure otherwise. It is well known that any complete planner will run in exponential time in the number of degrees-of-freedom (dofs) of the robot [13]. Most practical algorithms for complete path planning are restricted to 2D polygonal objects or 3D convex polytopes or special objects e.g. ladders, discs or spheres.

Given the complexity of a complete path planner, most of the effort in the last two decades has been on development of approximate approaches including those based on cell decomposition and potential field [13]. These approaches can be *resolution complete* if the resolution parameters are selected properly, but not exact or complete. Other algorithms are based on probabilistic roadmaps [11], which have been successfully applied to many high-dof robots. However, they may not terminate in a deterministic manner when there is no collision-free path.

M. Erdmann et al. (Eds.): Algorithmic Foundations of Robotics VI, STAR 17, pp. 441–456, 2005.

In this paper, we restrict ourselves to complete path planning for a 3D polyhedral robot undergoing translation motion among 3D polyhedral obstacles. The configuration space of the robot can be computed based on Minkowski sum of the robot and the obstacles. The Minkowski sum of two convex polytopes (with n features) can have $O(n^2)$ combinatorial complexity and is relatively simple to compute. On the other hand, the Minkowski sum of non-convex polyhedra can have complexity as high as $O(n^6)$. A commonly used approach to compute Minkowski sums decomposes the two non-convex polyhedra into convex pieces, computes their pairwise Minkowski sums and finally the union of the pairwise Minkowski sums. The main bottleneck in implementing such an algorithm is computing the union of pairwise Minkowski sums. Given m polyhedra, their union can have combinatorial complexity $O(m^3)$ and m can be high in the context of Minkowksi sum computation. Furthermore, robust computation of the boundary of the union and handling all degeneracies remains a major open issue. As a result, no good algorithms are known for robust computation of exact Minkowski sum of 3D polyhedral models and complete path planning.

Main Contributions: We present a novel algorithm for complete path planning for translating polyhedral robots in 3D. We perform convex decomposition and compute pairwise Minkowski sums of resulting convex polytopes. Instead of exactly computing the union of these polytopes, we compute a connectivity roadmap that captures the connectivity of the free space. We generate the connectivity roadmap by taking deterministic samples on an adaptive volumetric grid. We employ two main tests during sample generation: *complex cell test* and *star-shaped test.* These tests can be efficiently performed for polyhedral objects using max-norm computation and linear programming. The complexity of our algorithm varies as a function of the size of narrow passages in the configuration space.

Our algorithm is simple to implement in practice. We highlight its performance on two environments with few hundred polygons with either very small narrow passages or no collision-free paths. In these configurations, our algorithm takes a few seconds to either compute a collision-free path or guarantees that no path exists.

Organization: The rest of the paper is organized in the following manner. We give a background on motion planning in Section 2 and briefly survey related work. Section 3 gives an overview of our approach. We describe connectivity roadmaps in Section 4 and present an algorithm to compute the roadmap in Section 5. We highlights its performance on some environments in Section 6. In Section 7, we give an analysis of our approach and discuss few of its limitations. We conclude in Section 8.

2 Background and Prior Work

In this section, we define the general motion planning problem. Let $\mathcal{R}$ be a robot consisting of a collection of rigid subparts moving in a Euclidean space W, called *workspace*, represented as $\mathbb{R}^d$, with $d = 2$ or 3. Let $\mathcal{O}_1, \ldots, \mathcal{O}_q$

be fixed rigid objects, hereafter referred as *obstacles* embedded in W. Assume that the geometry of $\mathcal{R}, \mathcal{O}_1, \ldots, \mathcal{O}_q$ is accurately known, and that there are no kinematic constraints to limit the motion of $\mathcal{R}$. The position and orientation of the subparts define the *placement* of $\mathcal{R}$ (also referred to as a *configuration* of $\mathcal{R}$). The set of all placements of $\mathcal{R}$ defines a configuration space $\mathcal{C}$. The motion planning problem is defined as: given an initial and goal placement of $\mathcal{R}$, generate a *path* ρ specifying a continuous sequence of placements of $\mathcal{R}$ avoiding contact with the $\mathcal{O}_i$'s, starting at initial placement and terminating at the goal placement. Report failure if no such path exist. Every obstacle $\mathcal{O}_i, i = 1, \ldots, q$, in W maps to the region

$$\mathcal{CO}_i = \{\mathcal{Z} \in \mathcal{C} : \mathcal{R}(\mathcal{Z}) \cap \mathcal{O}_i \neq \phi\},$$

in $\mathcal{C}$, where $\mathcal{R}(\mathcal{Z})$ is the subset of W occupied by $\mathcal{R}$ at the placement $\mathcal{Z}$. The union of all $\mathcal{CO}_i$, $\bigcup_{i=1}^q \mathcal{CO}_i$ is called *C-obstacle region* or *forbidden region.* The *free configuration space* is defined as the set

$$\mathcal{F} = \mathcal{C} \setminus \bigcup_{i=1}^{q} \mathcal{CO}_i.$$

A *free path* between two free configurations $\mathcal{Z}_{init}$ and $\mathcal{Z}_{goal}$ is a continuous map $\rho : [0,1] \rightarrow \mathcal{F}$, $\rho(0) = \mathcal{Z}_{init}$ and $\rho(1) = \mathcal{Z}_{goal}$.

2.1 Previous Work

Motion planning has been extensively studied in the literature for more than three decades. In this section, we limit our discussion to algorithms for exact motion planning or to polyhedral objects undergoing translation motion in 3D. At a broad level they can be classified into the following techniques: *roadmaps*, *cell decomposition*, and specialized algorithms for 2D and 3D objects. A comprehensive survey of motion planning results is presented in [13].

Roadmaps

The idea underlying this approach is to convert the path planning problem in k-dimensional configuration space to path planning in network of 1-dimensional curves maintaining the connectivity in the robot's free space. The various types of roadmaps proposed to achieve this task are *visibility graph*, *Voronoi diagram or retraction approach*, and *silhouettes.*

Visibility graph method [13] reduces the problem of motion planning to a graph search. Using this approach, Lozano-Perez and Wesley [15] proposed an $O(n^3)$ algorithm which was improved to $O(n^2 \log n)$ by Lee [14] and to $O(n^2)$ by Guibas and Hershberger [9]. An output sensitive algorithm of $O(k+n \log n)$, where k is output size, was proposed by Ghosh and Mount [7]. In practice, this technique is mostly used for motion planning in two dimensional configuration spaces.

The retraction approach uses the concept of retraction in topology by defining a continuous mapping of the robot's free space $\mathcal{F}$ onto one-dimensional network of curves lying in $\mathcal{F}$. When $\mathcal{C} = R^2$ and the robot and obstacles are polygonal, the Voronoi diagram is a roadmap obtained by retraction [21]. This approach provides the additional property that the obtained paths maximize the clearance between the robot and the obstacles.

Silhouette method was proposed by Canny [4]. Unlike approaches surveyed earlier, this method does not make any assumptions about the configuration space and is a complete path planning algorithm that runs in single exponential time in the configuration space dimension.

Cell Decomposition

These methods are most extensively applied for motion planning problem [13]. The crux of this approach is to partition the robot's free space into a collection of non-overlapping cells and to construct a connectivity graph representing the cell adjacency. One of the first exact cell decomposition methods for solving the general motion planning problem was by Schwartz and Sharir [17].

Motion Planning Algorithms for 2D and 3D Robots

Different exact algorithms have been proposed for complete motion planning of 2D and 3D robots. Many of them are based on computing the Minkowski difference and the exact representation of C-obstacle region [15]. Kedem and Sharir [12] presented an efficient algorithm for a polygonal robot among polygonal obstacles with 3-dof configuration space. A similar algorithm for polygonal objects was also developed by Avnaim and Boissonant [3]. Sacks [16] presented a practical configuration space computation algorithm for pairs of curved planar parts. Halperin [10] presented efficient and robust algorithms to compute the Minkowski sum of 2D polygonal objects and used them for exact motion of planning of 2D objects undergoing translation motion. Aronov and Sharir [1] present a randomized algorithm to plan the motion of a convex polyhedron translating in 3-space amidst convex polyhedral obstacles. Vleugels and Overmars [21] presented a spatial subdivision algorithm to approximate the Voronoi diagram of an environment with convex primitives and used it for motion planning using retraction.

3 Overview

In this section, we formulate the problem of computing a collision free path for a polyhedral object undergoing translation motion in 3D using Minkowski sums, and provide an overview of our approach.

3.1 Motion Planning of Translating Robot

In this paper, our focus is on complete motion planning of translating polyhedral robots in the presence of polyhedral obstacles. It is well known that for a translating robot $\mathcal{R}$ and an obstacle $\mathcal{O}$, the C-obstacle $\mathcal{CO} = \mathcal{O} \oplus (-\mathcal{R})$, where $\oplus$ is the Minkowski sum and $-\mathcal{R}$ is $\mathcal{R}$ reflected about the origin. This formulation shrinks the robot to a point object, and the obstacles $\mathcal{O}_i$s are transformed to the respective Minkowski sums $\mathcal{CO}_i$.

3.2 Minkowski Sum

The Minkowski sum of two subsets P and Q of an affine space is defined as $P \oplus Q = \{\mathbf{x} | \ \mathbf{x} = \mathbf{p} + \mathbf{q}, \mathbf{p} \in P, \mathbf{q} \in Q\}$. It is relatively easier to compute Minkowski sums of convex polytopes as compared to general polyhedral models. For convex polytopes in 3D, the Minkowski sum can be computed in $O(n \log n + k)$ time, where k is the total number of features of the Minkowski sums [8]. In the worst case, $k = O(n^2)$. However, for non-convex polyhedra in 3D, the Minkowski sum can have $O(n^6)$ worst-case complexity [5].

One common approach for computing Minkowski sum of general polyhedra is based on *convex decomposition*. It uses the following property of Minkowski sum. If $P = P_1 \cup P_2$, then $P \oplus Q = (P_1 \oplus Q) \cup (P_2 \oplus Q)$. The resulting algorithm combines this property with convex decomposition for general polyhedral models:

1. Compute a convex decomposition for each polyhedron
2. Compute the pairwise convex Minkowski sums between all possible pairs of convex pieces in each polyhedron.
3. Compute the union of pairwise Minkowski sums.

After the second step, there can be $O(n^2)$ pairwise Minkowski sums. The pairwise convex Minkowski sums are convex. Their union can have $O(n^6)$ complexity [2].

3.3 Our Approach

Our algorithm for computing the Minkowski sum is based on the decomposition property. We have a set of convex primitives consisting of the pairwise convex Minkowski sums whose union is the Minkowski sum. Although the above approach provides a simple algorithmic framework, the union computation is non-trivial. This is because the combinatorial complexity can be very large. Exact computation of the boundary of the union is prone to robustness problems and degeneracies. It is very difficult to compute the exact union in 3D. As a result, it is difficult to compute an exact representation for the free space $\mathcal{F}$.

Our goal is to obtain a representation for $\mathcal{F}$ that captures its connectivity and allows us to perform exact path planning. The *connectivity roadmap* is one such representation. The roadmap consists of a graph that encodes the complete connectivity of $\mathcal{F}$. We construct this graph by computing deterministic samples in $\mathcal{F}$ that lie on an adaptive volumetric grid. The ability to perform exact path planning using this approach relies critically on the sampling of the volumetric grid. We provide a simple sampling criterion for adaptive grid generation based on linear programming and max-norm distance computation. We construct a weighted graph using the vertices and edges of the grid that lie in $\mathcal{F}$. The weight of each edge is the Euclidean distance between the corresponding vertices. Then the path planning problem reduces to computing shortest paths in this graph.

4 Connectivity Roadmaps

In this section, we define a connectivity roadmap that captures the connectivity of the free space $\mathcal{F}$. We begin by introducing some notation.

Given a set S, two points $\mathbf{p}, \mathbf{q} \in S$ are connected if there exists a path between $\mathbf{p}$ and $\mathbf{q}$ that lies in S. We use the shorthand notation $\mathbf{p} \overset{S}{\longleftrightarrow} \mathbf{q}$ to mean that $\mathbf{p}$ and $\mathbf{q}$ are connected in S. The connectivity relation is symmetric. Given an undirected graph $G = (V, E)$ and two vertices $\mathbf{v}, \mathbf{w} \in V$, $\mathbf{v} \overset{G}{\longleftrightarrow} \mathbf{w}$ means that $\mathbf{v}$ and $\mathbf{w}$ are connected in G, i.e., there exists a path between $\mathbf{v}$

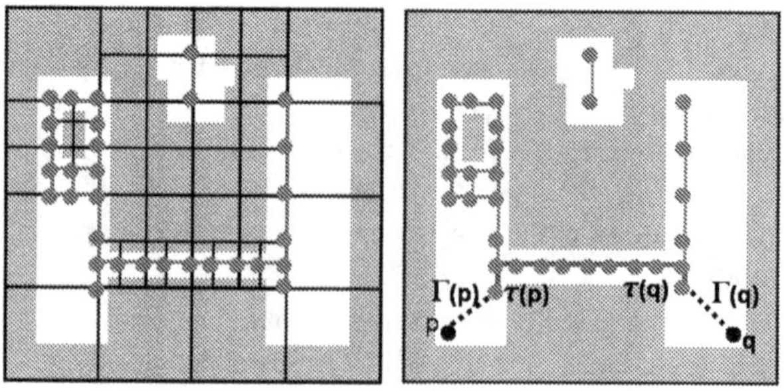

Fig. 1. Connectivity Roadmap: *We construct a connectivity roadmap by generating an adaptive voxel grid. The roadmap consists of a connectivity graph which is obtained by considering the set of grid points and grid edges that lie in free space (shown in green in the left figure). Connectivity roadmap also consists of a transfer function τ that maps points in free space to a vertex in the connectivity graph. As shown in the right figure, the source $\mathbf{p}$ and destination $\mathbf{q}$ get mapped to vertices $\tau(\mathbf{p})$ and $\tau(\mathbf{q})$ respectively. The problem of path planning between $\mathbf{p}$ and $\mathbf{q}$ reduces to simple graph search between $\tau(\mathbf{p})$ and $\tau(\mathbf{q})$ in the connectivity graph.*

and $\mathbf{w}$ consisting of a sequence of edges in E. As before, $\mathcal{F}$ refers to the free space. $\partial\mathcal{F}$ denotes the boundary of $\mathcal{F}$. The sign of a point is positive if it lies in $\mathcal{F}$, negative otherwise.

A connectivity roadmap consists of a connectivity graph G.

Definition 1. *The connectivity graph $G = (V, E, w)$ is a weighted undirected graph defined by a set V of points in free space, a set E of line segments in free space that connect pairs of points in V, and a weight function $w : E \rightarrow \mathbb{R}$. V and E correspond to the set of vertices and edges of G, respectively.*

Definition 2. *A transfer function $\tau : \mathcal{F} \rightarrow V$ is a mapping such that*

$$\forall \mathbf{p} \in \mathcal{F},\ \mathbf{p} \overset{\mathcal{F}}{\longleftrightarrow} \tau(\mathbf{p})$$

The connectivity graph satisfies the following two properties:

- **Property 1.** It encapsulates the connectivity of the free space. Every point $\mathbf{p} \in \mathcal{F}$ is connected to atleast one vertex in the connectivity graph. This implies that there exists a *transfer function* τ that maps a point in free space to a vertex in the connectivity graph. There also exists a *transfer path function* Γ such that $\Gamma(\mathbf{p})$ returns the path between $\mathbf{p}$ and $\tau(\mathbf{p})$ in $\mathcal{F}$.
- **Property 2.** It captures the connectivity of free space. Two points $\mathbf{p}, \mathbf{q} \in \mathcal{F}$ are connected in $\mathcal{F}$ if and only if the two vertices $\tau(\mathbf{p}), \tau(\mathbf{q}) \in V$ are connected in G, i.e.,

$$\mathbf{p} \overset{\mathcal{F}}{\longleftrightarrow} \mathbf{q} \iff \tau(\mathbf{p}) \overset{G}{\longleftrightarrow} \tau(\mathbf{q})$$

Definition 3. *The connectivity roadmap is defined as the tuple (G, τ, Γ)*

Fig. 1 shows an example of a connectivity roadmap. By combining the above two properties, we see that the connectivity roadmap satisfies the following property:

$$\mathbf{p} \overset{\mathcal{F}}{\longleftrightarrow} \mathbf{q} \iff \mathbf{p} \overset{\mathcal{F}}{\longleftrightarrow} \tau(\mathbf{p}) \quad \tau(\mathbf{p}) \overset{G}{\longleftrightarrow} \tau(\mathbf{q}) \quad \tau(\mathbf{q}) \overset{\mathcal{F}}{\longleftrightarrow} \mathbf{q}$$

This property enables us to perform complete path planning. As long as the source and destination are connected, we can find a path between them using the connectivity roadmap. Suppose we wish to find a path between two points $\mathbf{p}, \mathbf{q} \in \mathcal{F}$. Assume they are connected. We compute a mapping on the connectivity graph using the transfer function τ. The transfer path function Γ gives us the path between $\mathbf{p}$ and $\tau(\mathbf{p})$, and $\mathbf{q}$ and $\tau(\mathbf{q})$. Moreover, there exists a path between $\tau(\mathbf{p})$ and $\tau(\mathbf{q})$ in the connectivity graph. This path can be found easily by performing a simple graph search. Call this path γ. Thus we obtain the path $\Gamma(\mathbf{p}) :: \gamma :: \Gamma(\mathbf{q})$ between $\mathbf{p}$ and $\mathbf{q}$, where :: denotes a path concatenation. In this manner, the connectivity roadmap reduces the problem of path planning to computing a graph shortest path between $\tau(\mathbf{p})$ and $\tau(\mathbf{q})$.

The above property also provides us with a test for non-existence of a path. If no path exists between $\mathbf{p}$ and $\mathbf{q}$ in $\mathcal{F}$, we can detect this by testing if $\tau(\mathbf{p})$ and $\tau(\mathbf{q})$ are disconnected in G (by Property 2).

5 Connectivity Roadmap Construction

In this section, we describe our algorithm to compute the connectivity roadmap.

5.1 Sampling

We construct a roadmap by performing a sampling of the free space. Unlike previous approaches such as probabilistic roadmaps (PRMs) that generate samples randomly [11], we construct a roadmap in a deterministic fashion. Our goal is to sample the free space sufficiently to capture its connectivity. If we do not sample the free space adequately, we may not detect valid paths that pass through the narrow passages in the configuration space.

In our prior work [20], we proposed a sampling algorithm to generate an octree grid for the purpose of topology preserving surface extraction. We use this sampling algorithm to capture the connectivity of free space. We provide a brief description of the octree generation algorithm. We refer the reader to [20] for a detailed description. The algorithm starts with a single grid cell that is large enough to capture relevant features of $\mathcal{F}$. It performs two tests, *complex cell test* and *star-shaped test*, to decide whether to subdivide a grid cell.

Complex Cell Test

A cell is *complex* if it has a *complex voxel*, *face*, *edge*, or an *ambiguous sign configuration.* We define a voxel (face) of a grid cell to be *complex* if it intersects $\partial\mathcal{F}$ and the grid vertices belonging to the voxel (face) do not exhibit a sign change (see Figs. 2(a) & 2(b)). The sign of a vertex is positive if it lies within $\mathcal{F}$, negative otherwise. An edge of the grid cell is said to be *complex* if $\partial\mathcal{F}$ intersects the edge more than once (see Fig. 2(c)).

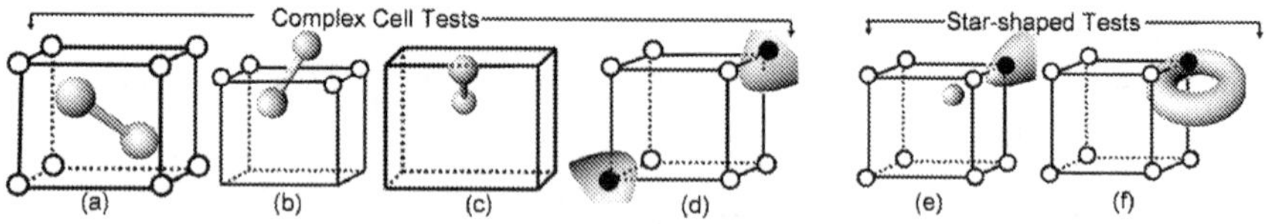

Fig. 2. *This figure shows the different cases corresponding to the complex cell and star-shaped test. Figs (a), (b), (c) and (d) show cases of complex voxel, complex face, complex edge, and ambiguous sign configuration. The white and black circles denote positive and negative grid points respectively. Fig. (e) shows the case where the surface is not star-shaped w.r.t a voxel. In Fig (f), the restriction of the surface to the right face of the cell is not star-shaped.*

There are two types of sign ambiguities — *face ambiguity* and *voxel ambiguity* [22]. When the signs at the vertices of a single face alternate during counterclockwise (or clockwise) traversal, the resulting configuration is a face ambiguity. A voxel ambiguity results when any pair of diagonally opposite vertices have one sign while the other vertices have a different sign (see Fig. 2(d)). Either of these cases is defined as an ambiguous sign configuration. We classify grid cells with such sign configurations as complex.

Intuitively, the complex cell criterion ensures that the surface intersects the grid cell in a simple manner in most cases. We use max-norm distance computation to perform the complex cell test [20]. Max-norm distance is used to determine whether $\partial\mathcal{F}$ intersects a voxel/face/edge of the cell. It can be computed efficiently for polyhedral primitives. If a grid cell is complex, it is subdivided and the algorithm is recursively applied to each of its children.

Star-shaped Test

Let S be a nonempty subset of $\mathbb{R}^n$. The set Kernel(S) consists of all $\mathbf{s} \in S$ such that for any $\mathbf{x} \in S$, we have $\mathbf{s} + \lambda(\mathbf{x} - \mathbf{s}) \in S, \forall\lambda \in [0, 1]$. S is *star-shaped* if Kernel(S) $\neq \emptyset$. We refer to a point belonging to Kernel(S) as an origin of S. A star-shaped primitive has at least one representative point (origin) such that all the points in the primitive are visible from it.

$\mathcal{F}$ is defined to be *star-shaped with respect to (w.r.t) a voxel* v if $\mathcal{F}_v = \mathcal{F} \cap v$ is star-shaped. Similarly, $\mathcal{F}$ is defined to be *star-shaped w.r.t a face* f if the two-dimensional set, $\mathcal{F}_f = \mathcal{F} \cap f$, is star-shaped. $\mathcal{F}$ is said to be *star-shaped w.r.t a cell* if it is star-shaped w.r.t the cell's voxel, and each of its six faces. If $\mathcal{F}$ is not star-shaped w.r.t the cell (see Figs. 2(e) & 2(f)), then the cell is subdivided and the algorithm is recursively applied to the children cells.

We use linear programming to perform the star-shaped test [20]. By definition, a polyhedron is star-shaped if it has a non-empty kernel. If $\mathbf{p}$ is a point belonging to the kernel, then each face of the polyhedron with centroid $\mathbf{c}$ and outward normal $\mathbf{n}$ defines the linear constraint $\mathbf{n} \cdot (\mathbf{c} - \mathbf{p}) > 0$ on $\mathbf{p}$. As a result, the kernel is non-empty if the set of constraints admits a feasible solution for $\mathbf{p}$.

The star-shaped and complex cell tests can be performed very efficiently. Moreover, performing these tests does not require an explicit representation of $\mathcal{F}$. It works even when the surface is represented as a Boolean combination

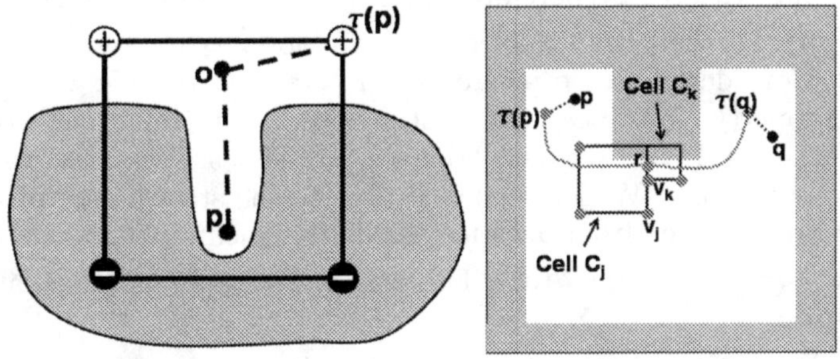

(a) Transfer Function (b) Connectivity

Fig. 3. *The left figure shows how to define the transfer function τ. We wish to define a path from* $\mathbf{p}$ *to one of the positive vertices of the grid cell that lies completely in $\mathcal{F}$. Due to the star-shaped property, there exists a straight line path from* $\mathbf{p}$ *to the origin* $\mathbf{o}$ *and from* $\mathbf{o}$ *to a grid vertex. This defines the transfer function τ and the transfer path function Γ for point* $\mathbf{p}$*. The right figure highlights the connectivity graph. If there exists a path $\mathcal{P}$ between $\tau(\mathbf{p})$ and $\tau(\mathbf{q})$ in free space $\mathcal{F}$, then there exists a path between them in the connectivity graph G. This is because the set of connectivity graph vertices belonging to the cells along path $\mathcal{P}$ is connected in G*

of other primitives. This fits with our representation of $\mathcal{F}$ as the complement of the union of individual Minkowski sums.

5.2 Roadmap Computation

The recursive subdivision algorithm discussed above generates an adaptive voxel grid. In this section, we show how to use the grid to construct a connectivity roadmap. In particular, we use the grid to define a connectivity graph $G = (V, E, w)$ and a transfer function τ.

Connectivity Graph

We extract a graph from the adaptive grid as follows. Let V be the set of grid vertices that lie in $\mathcal{F}$ and E be the set of grid edges that lie in $\mathcal{F}$. The weight function $w : E \to \mathbb{R}$ is defined as the Euclidean distance between the edge vertices. The connectivity graph is defined as the undirected graph $G = (V, E, w)$.

Transfer Function

We define a transfer function by defining a mapping from any arbitrary point in free space to a grid vertex. We note that a naive function that maps the point by snapping to the closest grid vertex need not guarantee a collision free path.

Consider a point $\mathbf{p} \in \mathcal{F}$ belonging to a grid cell C. We consider two cases. **Case 1**: Suppose C is not intersected by $\partial\mathcal{F}$. Since C contains point $\mathbf{p} \in \mathcal{F}$, we have $C \subseteq \mathcal{F}$. In particular, all the grid vertices of C lie in $\mathcal{F}$. We pick any one of the grid vertices of C, say $\mathbf{v}$. By definition of the connectivity graph $G = (V, E, w)$, we have $\mathbf{v} \in V$. We let $\tau(\mathbf{p}) = \mathbf{v}$. Since C is convex, the straight line segment between $\mathbf{p}$ and $\mathbf{v}$, $\mathbf{pv}$, is contained within C, and therefore lies within $\mathcal{F}$. Therefore, the transfer function satisfies the property $\mathbf{p} \overset{\mathcal{F}}{\longleftrightarrow} \tau(\mathbf{p})$. Further, the transfer path function $\Gamma(\mathbf{p}) = \mathbf{pv}$.

Case 2: Consider the case where C is intersected by $\partial\mathcal{F}$. Due to the star-shaped property, $\mathcal{F}_C = \mathcal{F} \cap C$ is star-shaped. Let $\mathbf{o}$ be the origin of $\mathcal{F}_C$. Because the cell is not complex, there exists at least one grid vertex in $\mathcal{F}_C$. Let this vertex be $\mathbf{v}$. We let $\tau(\mathbf{p}) = \mathbf{v}$. Due to the star-shaped property, both $\mathbf{v}$ and $\mathbf{p}$ are "visible" from $\mathbf{o}$. Since the line segments $\mathbf{po}$ and $\mathbf{ov}$ lie in $\mathcal{F}$, we have $\mathbf{p} \overset{\mathcal{F}}{\longleftrightarrow} \tau(\mathbf{p})$ (see Fig. 3(a)). This also gives us the transfer path function, $\Gamma(\mathbf{p}) = \mathbf{po} :: \mathbf{ov}$.

5.3 Connectivity Guarantee

In this section, we show that the connectivity roadmap defined above captures the connectivity of $\mathcal{F}$. This is formally expressed in Theorem 1. We begin by presenting a lemma. Given a cell C, the connectivity graph restricted to C is given by $G_C = (V_C, E_C)$ where V_C and E_C denote the set of grid points and grid edges respectively of cell C that lie in $\mathcal{F}$. We have dropped the weight function from the graph notation for convenience. It is assumed to be the Euclidean distance function.

Lemma 1. *Given a cell C, the graph $G_C = (V_C, E_C)$ is connected.*

Proof: Consider any two grid vertices $\mathbf{v}, \mathbf{w} \in V_C$. We prove that $\mathbf{v} \overset{G_C}{\longleftrightarrow} \mathbf{w}$. It suffices to prove that there exists a sequence of grid cell edges connecting $\mathbf{v}$ and $\mathbf{w}$ that do not intersect $\mathcal{F}$. We consider three cases:

- **Case 1**: The grid points $\mathbf{v}$ and $\mathbf{w}$ are the endpoints of an edge of the cell. Since both $\mathbf{v}$ and $\mathbf{w}$ have the same (positive) sign and the edge is not complex, this edge cannot intersect $\partial\mathcal{F}$. Therefore the edge belongs to E_C and we have $\mathbf{v} \overset{G_C}{\longleftrightarrow} \mathbf{w}$.
- **Case 2**: The grid vertices $\mathbf{v}$ and $\mathbf{w}$ lie diagonally opposite on a cell face. The case where the other two grid vertices on the face are negative corresponds to a case of face ambiguity. Therefore, at least one of the other two grid vertices (say $\mathbf{u}$) on the face has a positive sign. $\mathbf{v}$ and $\mathbf{u}$ are two positive grid vertices on the endpoints of an edge of the cell. Therefore by Case 1, we have $\mathbf{v} \overset{G_C}{\longleftrightarrow} \mathbf{u}$. Similarly, we have $\mathbf{u} \overset{G_C}{\longleftrightarrow} \mathbf{w}$. This implies $\mathbf{v} \overset{G_C}{\longleftrightarrow} \mathbf{w}$.
- **Case 3**: The grid points $\mathbf{v}$ and $\mathbf{w}$ are diagonally opposite vertices of the cell. The case where all the other grid vertices are negative corresponds to a case of voxel ambiguity. Therefore, at least one other grid vertex $\mathbf{u}$ has a positive sign. Depending on $\mathbf{u}$'s position, the vertices $\mathbf{v}$ and $\mathbf{u}$ either reduce to Case 1 or 2 ($\mathbf{u}$ and $\mathbf{w}$ will belong to the other case). Therefore, $\mathbf{v} \overset{G_C}{\longleftrightarrow} \mathbf{u}$ and $\mathbf{u} \overset{G_C}{\longleftrightarrow} \mathbf{w}$ which implies $\mathbf{v} \overset{G_C}{\longleftrightarrow} \mathbf{w}$.

□

Since G_C is a subgraph induced by G, any two vertices $\mathbf{v}$, $\mathbf{w} \in V_C$ satisfy $\mathbf{v} \overset{G}{\longleftrightarrow} \mathbf{w}$.

Theorem 1.

$$\mathbf{p} \overset{\mathcal{F}}{\longleftrightarrow} \mathbf{q} \iff \tau(\mathbf{p}) \overset{G}{\longleftrightarrow} \tau(\mathbf{q})$$

Proof: We first prove that if $\tau(\mathbf{p}) \overset{G}{\longleftrightarrow} \tau(\mathbf{q})$, then we have $\mathbf{p} \overset{\mathcal{F}}{\longleftrightarrow} \mathbf{q}$. By construction of the transfer function, we have $\mathbf{p} \overset{\mathcal{F}}{\longleftrightarrow} \tau(\mathbf{p})$ and $\mathbf{q} \overset{\mathcal{F}}{\longleftrightarrow} \tau(\mathbf{q})$. Moreover, because graph G consists of vertices and edges in $\mathcal{F}$, we have

$$\tau(\mathbf{p}) \overset{G}{\longleftrightarrow} \tau(\mathbf{q}) \implies \tau(\mathbf{p}) \overset{\mathcal{F}}{\longleftrightarrow} \tau(\mathbf{q})$$

Therefore, we have $\mathbf{p} \overset{\mathcal{F}}{\longleftrightarrow} \mathbf{q}$.

We now prove the converse. Let $\mathbf{p} \overset{\mathcal{F}}{\longleftrightarrow} \mathbf{q}$. We have

$$\tau(\mathbf{p}) \overset{\mathcal{F}}{\longleftrightarrow} \mathbf{p}, \quad \mathbf{p} \overset{\mathcal{F}}{\longleftrightarrow} \mathbf{q}, \quad \mathbf{q} \overset{\mathcal{F}}{\longleftrightarrow} \tau(\mathbf{q})$$

Therefore, we have $\tau(\mathbf{p}) \overset{\mathcal{F}}{\longleftrightarrow} \tau(\mathbf{q})$. If $\tau(\mathbf{p})$ and $\tau(\mathbf{q})$ belong to the same grid cell, then Lemma 1 ensures that $\tau(\mathbf{p}) \overset{G}{\longleftrightarrow} \tau(\mathbf{q})$.

Consider the case where $\tau(\mathbf{p})$ and $\tau(\mathbf{q})$ belong to different cells, $C_{\mathbf{p}}$ and $C_{\mathbf{q}}$ respectively. There exists a path between $\tau(\mathbf{p})$ and $\tau(\mathbf{q})$ in $\mathcal{F}$. Let $C_i, i = 0, \ldots n$, be the set of cells that are intersected by this path such that $C_0 = C_{\mathbf{p}}$ and $C_n = C_{\mathbf{q}}$. Suppose the path passes from a cell C_j into an adjacent cell C_k. Let the corresponding connectivity graphs restricted to C_j and C_k be $G_{C_j} = (V_j, E_j)$ and $G_{C_k} = (V_k, E_k)$ respectively. According to Lemma 1, both V_j and V_k are connected in graph G. The path passes from cell C_j to C_k through a face of the cell. Let f_j be the face of C_j that is incident on C_k and f_k be the face of C_k that is incident on C_j. Since grid cells C_j and C_k can be at different resolutions, f_j and f_k need not be identical. The path penetrates faces f_j and f_k at a common point $\mathbf{r}$ that lies in $\mathcal{F}$ (see Fig. 3(b)). Since both faces f_j and f_k are not complex and have at least one point in $\mathcal{F}$, they contain positive vertices $\mathbf{v_j} \in V_j$ and $\mathbf{v_k} \in V_k$ respectively that lie in $\mathcal{F}$. We will show below that $\mathbf{v_j} \overset{G}{\longleftrightarrow} \mathbf{v_k}$. As a result, $V_j \cup V_k$ is connected in graph G. This is true of all the cells along the path. Therefore, $\tau(\mathbf{p})$ and $\tau(\mathbf{q})$ are connected in graph G and we have $\tau(\mathbf{p}) \overset{G}{\longleftrightarrow} \tau(\mathbf{q})$.

We now return to the proof of the result $\mathbf{v_j} \overset{G}{\longleftrightarrow} \mathbf{v_k}$. The octree subdivision ensures that one of the faces f_j and f_k is a subset of the other. Let f denote the larger of the two faces. Vertices $\mathbf{v_j}$ and $\mathbf{v_k}$ lie on f. Using the facts that f is not complex and that $\mathcal{F}$ is star-shaped w.r.t f, it can be shown that the set $\mathcal{F}_f = \mathcal{F} \cap f$ is connected [20]. This implies that vertices $\mathbf{v_j}$ and $\mathbf{v_k}$ are connected in $\mathcal{F}_f$, i.e., $\mathbf{v_j} \overset{\mathcal{F}_f}{\longleftrightarrow} \mathbf{v_k}$. We wish to prove that $\mathbf{v_j} \overset{G}{\longleftrightarrow} \mathbf{v_k}$.

This is a two-dimensional version of the above theorem. Since our constraints of complex cell and star-shapedness extend to all the faces and edges of the cell, we can apply our argument recursively. The base case of our recursion is the one-dimensional case where we need to show that two points on an edge connected in $\mathcal{F}$ are also connected through G. This is readily shown by observing that the edges of the grid cells are not complex.

□

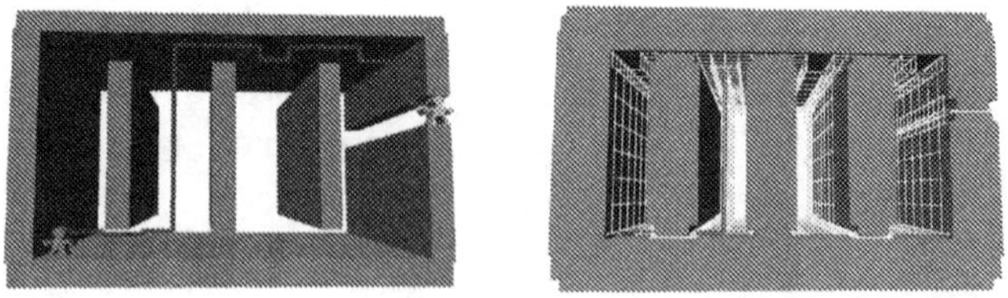

Fig. 4. Maze problem: *The left image shows a robot navigating within a maze model from a source (shown in red) to a destination (shown in green). The right image show the configuration space obstacle along with the connectivity graph. A path exists and passes through a number of narrow passages. Our algorithm generated a connectivity roadmap with* $18K$ *vertices in* 6 *secs and was able to find a path (shown in blue) in* 0.07 *secs.*

6 Implementation and Results

In this section, we describe the implementation of our algorithm and demonstrate its performance on path planning examples with narrow passages that test our algorithm. We used C++ programming language with the GNU g++ compiler under Linux operating system. Table 1 highlights the performance of our algorithm on these models. All timings are on a 2 GHz Pentium IV PC with a GeForce 4 graphics card and 1 GB RAM.

We compute a convex decomposition of the two polyhedra and compute pairwise Minkowski sums between the convex pieces. We used a modification of the convex decomposition scheme available in a public collision detection library, SWIFT++ [6]. We used a *convex hull* algorithm to compute the pairwise Minkowski sums. This algorithm adds the vectors of each vertex of one polyhedron with that of every vertex of the other polyhedron to get a point cloud. It computes a convex hull of the point cloud to obtain the pairwise Minkowski sum. Its time complexity is $O(f^2)$ where f is the number of features in the two convex polyhedra. In practice, this step doesn't take too much time (see Table 1). We used Dijkstra's single source shortest path algorithm to perform the graph search on the connectivity graph. We used the routine provided by a public domain library, Boost Graph Library [18]. Its running time is $O(|V| + |E|) \log |V|$ where $|V|$ and $|E|$ are the number of vertices and edges in the graph.

Table 1 provides a breakup of the total time. It shows that most of the time is spent on roadmap construction and a very small fraction of the total time is spent on graph search and pairwise Minkowski sum computation. Given an environment with static obstacles and a robot, we need to construct the connectivity roadmap just once. We can perform planning between a new pair of initial and final positions without having to recompute the roadmap.

Fig. 4 shows a robot navigating within a maze model. A path exists and passes through a number of narrow passages. Our algorithm successfully found a path through the narrow passages. It generated a connectivity roadmap with $18K$ vertices in 6 secs and was able to find a path in 0.07 secs. We also considered a scenario where the maze was modified such that there was no collision-free path. Our algorithm took 7 secs to generate a connectivity roadmap with $25K$ vertices and found out in 0.07 secs that no path exists.

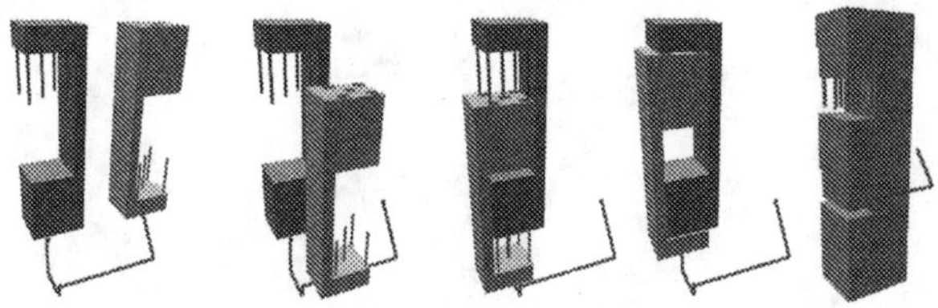

Fig. 5. Assembly problem: *This benchmark shows application of our algorithm to assembly planning. The four images on the left shows a path that the robot can take so that the two parts could be assembled. Our algorithm took* 12 *secs to generate a connectivity roadmap with* $28K$ *vertices and was able to find a path (shown in blue) in* 0.11 *secs. The rightmost image shows the configuration space obstacle and the path of the robot in configuration space. This is a challenging example because the goal configuration is lodged within a narrow passage in the configuration space.*

Fig. 5 shows application of our algorithm to assembly planning. It consists of two parts each with pegs and holes. The goal is to assemble the two parts so that the pegs of one part fit into the holes of the other. This problem can be reduced to a motion planning problem by treating one of the parts as a robot and the other as the obstacle. Our algorithm took 12 secs to generate a connectivity roadmap with $28K$ vertices and was able to find a path (shown in blue) in 0.11 secs.

Table 1. *This table highlights the performance of our algorithm on different models. The model complexity is indicated in terms of the triangle count and number of convex pieces for the obstacles and robot. The table shows the time taken to compute pairwise convex Minkowski sum, construct the connectivity roadmap and to perform the graph search. The rightmost column shows the size of the connectivity roadmap, which is equal to the number of vertices in the connectivity graph.*

Model	Combinatorial Complexity				Performance			Roadmap size
	Num Tris		Num Convex		Convex Mink	Roadmap	Graph	
	Obstacle	Robot	Obstacle	Robot	(s)	(s)	(s)	
Maze (Path exists)	96	136	8	10	0.08	6	0.07	18,453
Maze (No Path)	100	136	8	10	0.08	7	0.07	25,920
Assembly	224	224	16	16	0.1	12	0.11	28,815

Fig. 6 shows the performance of our roadmap construction algorithm on the maze model (Fig 4) as a function of the robot size. As we vary the robot size, the size and the number of narrow passages in the configuration space change, resulting in varying performance.

7 Discussion

7.1 Performance

The overall performance is dominated by the roadmap construction step of our algorithm. Its time complexity is output-sensitive — it is $O(N)$ where N is the size of the roadmap. The roadmap size is primarily dependent on the narrow passges in the free space. Our algorithm ensures that it captures the connectivity of free space by performing additional sampling in the vicinity of narrow passages. The amount of sampling is related to the size and number of narrow passages (see Fig. 6). For example, suppose two configuration space

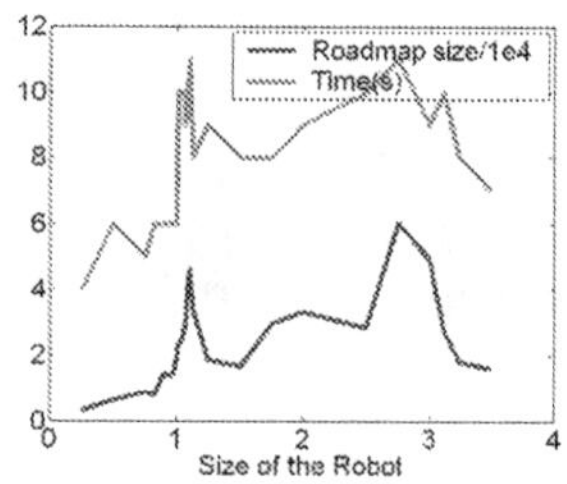

Fig. 6. *This plot shows the performance of our roadmap construction algorithm on the maze model as a function of the robot size. We have shown two measures of performance: roadmap construction time and roadmap size. The x-axis shows the scaling factor applied to the robot shown in Fig. 4.*

obstacles are at a distance ϵ apart such they form a narrow passage of width ϵ. In order to obtain samples within the narrow passage, the grid cells in the vicinity may need to be subdivided until they become smaller than ϵ.

Moreover, the volumetric grid generated by our sampling algorithm is coordinate system dependent. A change of coordinate system can result in a volumetric grid with different levels of subdivision. Consequently, the size of the roadmap depends on the coordinate system.

7.2 Translational and Rotational Motion Planning

Our basic approach is also applicable to a 2D robot undergoing rotation and translation motion in a plane. We can reformulate the problem of computing the connectivity of its configuration space by reducing rotation to a swept volume computation and the translational part remains a Minkowski sum computation. Let the robot $\mathcal{R}$ and obstacles $\mathcal{O}$ be in a $2D$ workspace. We can think of this $2D$ workspace to be embedded in a $3D$ workspace where the z-axis represents the rotation (θ-axis). We map both the robot and obstacles to an alternate space, the swept volume space, as follows: we extrude the robot along the θ-axis while rotating it about θ-axis simultaneously. In other words, we are computing the swept volume of the robot under screw motion. We extrude the obstacles $\mathcal{O}$ along the θ-axis, which is the swept volume of the obstacles under translation. As a result, we obtain representations $\mathcal{R}_{SV}$ and $\mathcal{O}_{SV}$ for the robot and obstacles, respectively, in the swept volume space. Then the configuration space obstacle corresponds to performing a *sliced Minkowski sum* operation on $\mathcal{R}_{SV}$ and $\mathcal{O}_{SV}$. This is related to the concept of computing critical slices in the configuration space [12, 16]. By using the above formulation, we can use our path planning algorithm to compute a collision-free path.

7.3 Tangential Contact

The sampling algorithm presented earlier uses an octree to perform the adaptive subdivision. This presents a problem when the configuration space obstacles are in tangential contact. In such a case, if we use axis-aligned hyperplanes to subdivide the cell, the subdivision process may not terminate. Dealing with such degenerate configurations is a difficult problem. One possible approach to handle the termination issue is to augment our original subdivision strategy by including supporting planes that arise from the individual Minkowski primitives themselves. This is akin to binary space partitioning (BSP) technique. Cells generated by this approach are no longer cubical, but general convex polyhedra. It is easy to prove that such a technique will always terminate [19].

7.4 Limitations

Our sampling algorithm uses two criteria: complex cell test and star-shaped test to guide the subdivision. These criteria are conservative. Consequently, our algorithm may perform unnecessary subdivision. This can reduce the performance of our algorithm and result in a larger roadmap than necessary. The convex decomposition method can result in a large number of convex pieces. Given two polyhedra each with n convex pieces, we obtain n^2 pairwise convex Minkowski sums. Since this set of pairwise convex Minkowski sums is an input to our algorithm, its large size affects the performance of the overall algorithm.

8 Conclusion and Future Work

We have presented an algorithm for complete path planning for translating polyhedral robots in 3D. Our algorithm is based on constructing a connectivity roadmap that captures the connectivity of the free space. It is guaranteed to find a collision-free path if one exists. Otherwise it detects non-existence of any collision-free path.

Our algorithm is simple to implement in practice. It uses two tests: a complex cell test and a star-shaped test. These tests can be efficiently performed for polyhedral objects using max-norm distance computation and linear programming. The complexity of our algorithm varies as a function of the size of the narrow passages in configuration space. We highlight the performance of our algorithm on two environments with very small narrow passages or no collision-free paths.

There are many avenues for future work. For some applications, a robot is allowed to be in contact with the obstacles. We would like to extend our algorithm to accommodate this. We are interested in application of our algorithm to the problem of motion planning for a robot with translational and rotational degrees of freedom. Finally, we would like to improve the roadmap generation algorithm by making our approach less conservative and improve the overall performance.

9 Acknowledgments

This research is supported in part by ARO Contracts DAAD19-02-1-0390 and W911NF-04-1-0088, NSF awards ACI 9876914 and ACR-0118743, ONR Contracts N00014-01-1-0067 and N00014-01-1-0496, DARPA Contract N61339-04-C-0043 and Intel. We thank Danny Halperin for many fruitful discussions on the complexity and practical implementations of Minkowski sums in 3D.

References

1. B. Aronov and M. Sharir. On translational motion planning in 3-space. In *ACM Symposium on Computational Geometry*, pages 21–30, 1994.

2. B. Aronov, M. Sharir, and Boaz Tagansky. The union of convex polyhedra in three dimensions. *SIAM Journal on Computing*, 26:1670–1688, 1997.
3. Francis Avnaim and J.-D. Boissonnat. Practical exact motion planning of a class of robots with three degrees of freedom. In *Proc. of Canadian Conference on Computational Geometry*, page 19, 1989.
4. J.F. Canny. *The Complexity of Robot Motion Planning.* ACM Doctoral Dissertation Award. MIT Press, 1988.
5. D. Dobkin, J. Hershberger, D. Kirkpatrick, and S. Suri. Computing the intersection-depth of polyhedra. *Algorithmica*, 9:518–533, 1993.
6. S. Ehmann and M. C. Lin. Accurate and fast proximity queries between polyhedra using convex surface decomposition. *Computer Graphics Forum (Proc. of Eurographics'2001)*, 20(3):500–510, 2001.
7. S. K. Ghosh and D. M. Mount. An output sensitive algorithm for computing visibility graphs. In *Proc. 28th Annu. IEEE Sympos. Found. Comput. Sci.*, pages 11–19, 1987.
8. L. Guibas and R. Seidel. Computing convolutions by reciprocal search. *Discrete Comput. Geom*, 2:175–193, 1987.
9. L. J. Guibas and J. Hershberger. Computing the visibility graph of n line segments in $O(n^2)$ time. *Bull. EATCS*, 26:13–20, 1985.
10. D. Halperin. Robust geometric computing in motion. *International Journal of Robotics Research*, 21(3):219–232, 2002.
11. L. Kavraki, P. Svestka, J. C. Latombe, and M. Overmars. Probabilistic roadmaps for path planning in high-dimensional configuration spaces. *IEEE Trans. Robot. Automat.*, pages 12(4):566–580, 1996.
12. K. Kedem and M. Sharir. An efficient motion planning algorithm for a convex rigid polygonal object in 2-dimensional polygonal space. *Discrete Comput. Geom.*, 5:43–75, 1990.
13. J.C. Latombe. *Robot Motion Planning.* Kluwer Academic Publishers, 1991.
14. D. T. Lee. Proximity and reachability in the plane. Report R-831, Dept. Elect. Engrg., Univ. Illinois, Urbana, IL, 1978.
15. T. Lozano-Pérez and M. Wesley. An algorithm for planning collision-free paths among polyhedral obstacles. *Comm. ACM*, 22(10):560–570, 1979.
16. E. Sacks. Practical sliced configuration space for curved planar pairs. *International Journal of Robotics Research*, 18(1), 1999.
17. J. T. Schwartz and M. Sharir. On the piano movers probelem ii, general techniques for computing topological properties of real algebraic manifolds. *Advances of Applied Maths*, 4:298–351, 1983.
18. J. G. Siek, L. Lee, and A. Lumsdaine. *The Boost Graph Library: User Guide and Reference Manual.* Addison-Wesley, Boston, 2002.
19. TVN Sriram, Gokul Varadhan, and Dinesh Manocha. Handling degeneracies during adaptive subdivisions. In *UNC Technical Report TR04-14*, 2004.
20. G. Varadhan, S. Krishnan, T. V. N. Sriram, and D. Manocha. Topology preserving surface extraction using adaptive subdivision. Technical report, Department of Computer Science, University of North Carolina, 2004.
21. J. Vleugels and M. Overmars. Approximating Voronoi diagrams of convex sites in any dimension. *International Journal of Computational Geometry and Applications*, 8:201–222, 1997.
22. Jane Wilhelms and Allen Van Gelder. Topological considerations in isosurface generation extended abstract. *Computer Graphics*, 24(5):79–86, 1990.